Advances in Isotope Geochemistry

Series Editor
Jochen Hoefs, Geowissenschaftliches Zentrum, Universität Göttingen, Göttingen, Germany

In the last few decades, isotope geochemistry has become an essential part of geochemistry and has contributed significantly to the solution of a wide variety of geoscientific problems, which span the whole field of earth sciences. Continued improvements in mass spectrometry and the invention of new mass-spectrometer systems, such as multicollector-ICP mass spectrometers (MC-ICP-MS), has enabled investigations of isotope variations of a wide range of transition and heavy elements that could not previously be measured with adequate precision. This has allowed many of the stable and radioactive isotopic systems to be investigated and applied to a huge variety of inorganic and organic samples.

Advances in Isotope Geochemistry, seeks to provide in-depth reviews of isotopic systems, methods and applications to a degree which is not possible within journal articles. Methods are described in detail, from sample collection and preparation to the fine tuning and subtleties of the mass spectrometric methods, data reduction and interpretation. The AIG series is the first stop when establishing new methods and an excellent reference for every isotope laboratory, serving as textbooks in university courses as well as a source of information for professionals.

Hans Eggenkamp

The Geochemistry of Stable Chlorine and Bromine Isotopes

Second Edition

Springer

Hans Eggenkamp
Onderzoek en Beleving
Laren (NH), The Netherlands

ISSN 2364-5105 ISSN 2364-5113 (electronic)
Advances in Isotope Geochemistry
ISBN 978-3-031-75632-0 ISBN 978-3-031-75633-7 (eBook)
https://doi.org/10.1007/978-3-031-75633-000

This Springer imprint is published by the registered company Springer Nature Switzerland AG
The registered company address is: Gewerbestrasse 11, 6330 Cham, Switzerland

Preface

In 2024 it is 40 years ago that the PhD thesis written by Ronald Kaufmann "Chlorine in ground water: Stable isotope distribution" was published. This thesis definitely is the landmark publication of all articles, book chapters, conference proceedings, and other papers on halogen stable isotopes because it, for the very first time, showed that natural chlorine stable isotope variations were clearly, and remarkably easily, measurable. Earlier papers were published on chlorine (and even bromine) isotopes, but these were published at very low frequencies of mostly less than one paper per year on average. These papers, that were mostly aimed at the development of analytical methods and to measure fractionation that was obtained through laboratory experiments, suggested that, unfortunately, natural variations were too small to be measured. Ronald Kaufmann was the first to develop a method by which it was shown that not only natural variations existed, but that they also showed very clear systematics.

Once it was proven that small natural chlorine isotope variations could be measured, it was expected that the number of published studies would rise significantly and quickly. However, it took 10 years until the average number of chlorine isotope publications per year rose above 5. By the time of publication of the first edition of this book, in 2014, the total number of papers on chlorine and, since the year 2000, bromine isotopes was about 300. In the last 10 years this number has approximately doubled to a total bibliography of about 600 peer-reviewed papers in which chlorine and bromine isotopes are the main subject of the research.

This results in the possibility, probably for the last time, that the full knowledge on stable halogen geochemistry can be summarised in a single, relatively small, book volume in the series "Advances in Isotope Geochemistry". In this book I present the development of our knowledge on chlorine and bromine stable isotope geochemistry. The book starts with an introduction on chlorine and bromine, their history, chemistry, and isotope behaviour. A large part of the book is filled with the numerous methods that have been developed to analyse chlorine and bromine stable isotope ratios. Strikingly with this regard is that several methods have been developed within laboratories that presented very little (or even none) scientific chlorine or bromine isotope work thereafter. It could be questioned what would be the reason for this striking observation.

The book then describes the various processes that fractionate chlorine and bromine isotopes. It is important to realise that chlorine and bromine are predominantly present on Earth in the -I oxidation state as chloride and bromide, and as a result oxidation-reduction processes that impose large isotope fractionation on other light stable isotopes are mostly absent. The range of measured isotope ratios rarely goes beyond 10‰, and about 80% of all published chlorine and bromine isotope measurements are between −1‰ and +1‰ of the internationally accepted standard (ocean chloride and ocean bromide). Finally, I describe in this book the chlorine and bromine isotope variations that have been observed during the measurement of geological samples. This includes not only measurements on chlorine and bromine samples from Earth, but also extra-terrestrial material such as meteorites, rocks sampled on the moon, and in-situ measurements on extra-terrestrial bodies like Mars. Also samples with different oxidation states such as perchlorates and organic molecules are presented here. New in this edition is the discussion on chlorine isotope ratio measurements made of stars and interstellar space.

Of course a book this size could not be easily written by a single author. I was lucky enough to have been able to work on the forefront of halogen isotope geochemical developments for a long time, from the start of my PhD appointment in May 1988 until the end of my final academic contract in December 2022. This was not possible without the enthusiasm of my teachers and senior collaborators, the most important of which were Rob Kreulen, Guus Koster van Groos, Max Coleman, Pierre Agrinier, José Manuel Marques, and Michael Marks. Thanks to them I was able to conduct my research at various renowned institutions in different countries in Europe. Since I am no longer working at an academic institution since 2023 I am very thankful to Utrecht University, especially to Prof. Jack Middelburg (coincidentally co-author on my first published paper from 1994), for allowing me to have digital access to the library of the University, so that I was able to actually access the non-open access scientific papers that are still being published. I also am very thankful to Pierre Agrinier for being the first to read the complete, finished manuscript and suggest many improvements to the manuscript. Finally, I am very grateful to my wife, Nina, who always supported me in my international jobs, being away from home most of the years past and for her continued support in writing this second edition. Although being home again since 2023, most of my free time I was working on writing this second edition, and for that reason it was not always easy to live a family life together. Nina, heel veel dank voor je geduld en je steun en je liefde al deze tijd, vooral in de moeilijke laatste paar jaren.

The number of laboratories that have been involved in chlorine and bromine isotope studies has never been very large. I believe the number has hardly been more than ten to fifteen worldwide at any given time. It is hoped that the data provided in this book will further increase the interest in the study of chlorine and bromine isotopes and that more laboratories are going to realise the opportunities that investing time and resources in the study of chlorine and bromine isotopes can give a healthy reward. The equipment is present in virtually every isotope laboratory and with only small adoptions and a small investment every isotope geochemist who wishes can become part of the halogen isotope community.

Onderzoek en Beleving, Laren (NH) Hans Eggenkamp
The Netherlands

Contents

Part IV Natural Chlorine and Bromine Isotope Variations on Earth

About the Author

Hans Eggenkamp obtained his PhD from Utrecht University (The Netherlands). His research interest and scientific expertise focuses on the geochemistry of stable chlorine and bromine isotopes. He studied both isotope systems during his PhD, and during (postdoctoral) stays at the University of Reading (UK), the Instituto Superior Técnico (Lisbon, Portugal), the Institut de Physique du Globe de Paris (France) and the Eberhard-Karls Universität in Tübingen (Germany). He applied his knowledge on stable halogen isotope geochemistry mostly on studies of groundwater and salt systems and more recently also on magmatic systems. Further, he has worked on geothermal energy solutions for industrial and agricultural heating projects as well as isotopic and chromatographic analysis solutions for the oil and gas exploration and, more recently, the food industry. At this moment he is working as Senior Stable Isotope Scientist at Isolab BV, Geffen, The Netherlands.

Part I

Introduction

1 The Halogen Elements

Group 17 of the periodic table of the elements (see Table 1.1) consists of the so called halogen elements fluorine (F), chlorine (Cl), bromine (Br), iodine (I), astatine (At) and tennessine (Ts). Halogen elements are characterised by the fact that their outer electron shell just misses one electron to form a stable noble gas configuration. As a result most halogen elements form stable anions in the −I oxidation stat, so called halide ions. However, in the halogen group the stability of the halide ion decreases from fluorine to astatine. It is not known yet if tennessine, discovered only in 2010 (Organessian et al. 2010), behaves like a typical group 17 element, which can be the result of possible relativistic effects (see e.g. Pyykkö 2011a, 2011b). Most probably tennessine behaves much more like a metal, with the +1 oxidation state being expected the most stable oxidation state and the −1 oxidation state, which is the most stable oxidation state in the other halogens, being unstable (Norseev 2021).

H. Eggenkamp, *The Geochemistry of Stable Chlorine and Bromine Isotopes*, Advances in Isotope Geochemistry, https://doi.org/10.1007/978-3-031-75633-7_1

Table 1.1 Periodic table of the elements. Chlorine and bromine are shown in bold. Within the cells the number at the top is the atomic number, in the middle the symbol and the number at the bottom is the rounded (two decimals) abridged standard atomic weight after Prohaska et al. (2022). For elements without a stable isotope the mass of the longest living isotope according to Prohaska et al. (2022) is taken

	Group 1	Group 2	Group 3	Group 4	Group 5	Group 6	Group 7	Group 8	Group 9	Group 10	Group 11	Group 12	Group 13	Group 14	Group 15	Group 16	Group 17	Group 18
Period 1	1 H 1.01																	2 He 4.00
Period 2	3 Li 6.94	4 Be 9.01											5 B 10.81	6 C 12.01	7 N 14.01	8 O 16.00	9 F 19.00	10 Ne 20.18
Period 3	11 Na 22.99	12 Mg 24.31											13 Al 26.98	14 Si 28.09	15 P 30.97	16 S 32.06	**17** **Cl** **35.45**	18 Ar 39.95
Period 4	19 K 39.10	20 Ca 40.08	21 Sc 44.96	22 Ti 47.87	23 V 50.94	24 Cr 52.00	25 Mn 54.94	26 Fe 55.85	27 Co 58.93	28 Ni 58.69	29 Cu 63.55	30 Zn 65.38	31 Ga 69.72	32 Ge 72.63	33 As 74.92	34 Se 78.97	**35** **Br** **79.90**	36 Kr 83.80
Period 5	37 Rb 85.47	38 Sr 87.62	39 Y 88.91	40 Zr 91.22	41 Nb 92.91	42 Mo 95.95	43 Tc (97)	44 Ru 101.07	45 Rh 102.91	46 Pd 106.42	47 Ag 107.87	48 Cd 112.41	49 In 114.82	50 Sn 118.71	51 Sb 121.76	52 Te 127.60	53 I 126.90	54 Xe 131.29
Period 6	55 Cs 132.91	56 Ba 137.33	La-Lu	72 Hf 178.49	73 Ta 180.95	74 W 183.84	75 Re 186.21	76 Os 190.23	77 Ir 192.22	78 Pt 195.08	79 Au 196.97	80 Hg 200.59	81 Tl 204.38	82 Pb 207.20	83 Bi 208.98	84 Po (209)	85 At (210)	86 Rn (222)
Period 7	87 Fr (223)	88 Ra (226)	Ac-Lr	104 Rf (267)	105 Db (268)	106 Sg (269)	107 Bh (270)	108 Hs (269)	109 Mt. (277)	110 Ds (281)	111 Rg (282)	112 Cn (285)	113 Nh (286)	114 Fl (290)	115 Mc (290)	116 Lv (293)	117 Ts (294)	118 Og (294)

Period 6	57 La 138.91	58 Ce 140.12	59 Pr 140.91	60 Nd 144.24	61 Pm (145)	62 Sm 150.36	63 Eu 151.96	64 Gd 157.25	65 Tb 158.93	66 Dy 162.50	67 Ho 164.93	68 Er 167.26	69 Tm 168.93	70 Yb 173.05	71 Lu 174.97
Period 7	89 Ac (227)	90 Th 232.04	91 Pa 231.04	92 U 238.03	93 Np (237)	94 Pu (244)	95 Am (243)	96 Cm (247)	97 Bk (247)	98 Cf (251)	99 Es (252)	100 Fm (257)	101 Md (258)	102 No (259)	103 Lr (262)

1.1 General Chemical Properties of the Halogen Elements

Group 17 elements are characterised by an outer electron shell which contains seven electrons, so that only one electron needs to be added to give it a full noble gas configuration. This property results in a dominant oxidation state, especially for the lighter elements of this group, of −I and these elements predominantly form ionogenic compounds. In the Earth's surface reservoir chlorine and bromine are most commonly found in aqueous solutions in the oceans, because most of their salts are readily soluble in water, and in evaporite deposits. On the contrary, fluorine is most common in a number of fluoride rich minerals such as fluorite (CaF_2), fluorapatite ($Ca_5(PO_4)_3F$), and cryolite (Na_3AlF_6) because fluoride salts of the alkaline earth metals (such as CaF_2 and MgF_2) have low solubilities and precipitate out of waters with significant alkaline earth metals concentrations. This results in the decoupling of fluorine from chlorine and bromine in most surface reservoirs. Iodine has a relatively low concentration in seawater because it is efficiently removed from seawater by certain brown algae which are able to accumulate iodine heavily (Küpper and Carrano 2019). Also, iodine is relatively easy to oxidise to iodate (IO_3^-). As a result iodate is, in seawater, generally the dominant iodine species (Chance et al. 2020). It has however been suggested that also hypoiodous acid (HOI) may be an important species (Shaw and Cooper 1957) in seawater. Iodate minerals also are the only minerals containing oxidised halogen ions and these rare minerals are found primarily in the Atacama desert in Northern Chile. This geochemical behaviour also results in the decoupling of iodine from chlorine and bromine in surface reservoirs on Earth. Chemical characteristics of chlorine and bromine are similar enough to prevent the occurrence of separate bromine minerals in nature (except bromargyrite) and bromide is always found, in varying amounts, in chloride minerals. Both chlorine and bromine can be oxidised under natural circumstances (to chlorate and perchlorate or bromate), but the concentrations of these oxidised species are not sufficiently high enough to form minerals of their own. Due to their toxicity however, their concentrations in the environment can be high enough to be of environmental concern.

In their elemental form group 17 elements form diatomic molecules with a single covalent bond between the atoms. At higher temperatures these molecules dissociate, and from fluorine to iodine to an increasing extent. Hydrogen compounds from group 17 elements are strong acids when dissolved in water. When combined with metals they easily form salts that are called halides (fluoride, chloride, bromide, iodide, astatide), in which the halide ion is monovalent. Except for fluorine they form oxides and oxoacids in which their valency state is 1, 3, 5, or 7 (Holleman and Buchner 1949). At standard temperature and pressure fluorine and chlorine are gaseous elements, bromine is a liquid and iodine and astatine are solids. No chemical and physical properties are known yet from tennessine, although these have been predicted in the past (Seaborg 1968; Fricke et al. 1971; Fricke and Waber 1971). These authors even predicted some properties for the next group 17 element (element 171) which may be so unstable that it will probably never be discovered. The electronegativity of the group 17 elements decreases from fluorine to astatine. Fluorine is the most electronegative element known. As a result fluorine gas is one of the strongest oxidising agents known. This also is the reason that no oxidised species of fluorine are known, except perhaps for hypofluorous acid (HOF; Poll et al. 1988), although in this compound the formal oxidation state of fluorine is −1 and of oxygen it is 0. In the same direction ($F_2 \rightarrow Cl_2 \rightarrow Br_2 \rightarrow I_2 \rightarrow At_2$) the elemental molecules become weaker oxidising agents. Some characteristic chemical and physical properties for the group 17 elements can be found in Table 1.2.

Table 1.2 Characteristics of the group 17 elements. Most data (for F, Cl, Br and I) in this table are taken from Holleman and Buchner (1949) with additional data from Wikipedia. Data for Ts and element 171 are mostly taken from Seaborg (1968), Fricke (1971) and Fricke et al. (1971). The redox potential from astatine is taken from Champion et al. (2010), the atomic weights of the known elements are taken from Prohaska et al. (2022) and their representative isotopic abundance from Meija et al. (2016)

	F	Cl	Br	I	At	Ts	E171
Colour	Light yellow	Yellow green	Brown	Black, gas is violet			
Atomic number	9	17	35	53	85	117	171
Atomic weight	18.998403162(5)	[35.446;35.457]	[79.901; 79.907]	126.90447(3)	[210][a]	[294][b], [313][c]	[500][c]
Stable isotopes	19 (100%)	35 (75.5–76.1%), 37 (23.9–24.5%)	79 (50.5–50.8%), 81 (49.2–49.5%)	127 (100%)			
Melting point (°C)	−223	−101.5	−7.3	113.7	302	350–550	
Boiling point (°C)	−187	−34	58.8	184.3	337	610	
Critical temp. (°C)	−129	144	311	553			
Density (liquid)	1.11	1.33	3.12	4.93 (solid)	8.9	7.1–7.3	
Electronegativity (Pauling)	3.98	3.16	2.96	2.66	2.2		
Dissociation @ 1200 °C	–	Little	11%	40%			
Dissociation enthalpy (kJ. mol^{-1})	–	239	193	147			
Oxidation states	−I	−I,0,+I,+III,+IV, +V, (+VI), +VII	−I,0,+I,+III,+V, (+VII)	−I, 0, +I, +III, +V, +VII	−I, 0, +I, +III, +V, (+VII)	+I, +III	+I, +III, +V
Oxidation potential ($2X^- \leftrightarrow X_2 + 2e^-$, V)	+3.05	+1.36	+1.07	+0.54	+0.36	+0.25 – +0.5	
Ionic radius (X^-, pm)	119	167	182	206			
Atomic radius (pm)	73	99	114	136	127	138	
Year of discovery	1810, 1886	1774	1826	1811	1939	2010	Not discovered

[a]Isotope with longest half life
[b]Isotope with longest known half life
[c]Undiscovered isotope with the longest expected half life

1.2 Discovery of the Halogens

As the group 17 elements are reactive elements they are not normally found as "native" elements in nature. As a consequence they were discovered at a relatively late date. The discovery history of the group 17 with stable isotopes is described in detail by Weeks (1942).

The first group 17 element that was discovered was chlorine. It was discovered by Scheele (1774) who, while studying manganese, in the reaction between hydrochloric acid and pyrolusite (MnO_2) noticed that a gas with the odour of warm aqua regia escaped. He was able to detect several of the properties of this gas, such as its green colour, its solubility in water which produces a weakly acid taste of the water, and its bleaching capabilities. However, as back in those days all acids were supposed to contain oxygen, he considered this gas as the oxide obtained from hydrochloric acid, which was then called muriatic acid, and as such failed to establish it as a new element. In spite of this he still is credited with the discovery of the element. Gay-Lussac and Thénard (1809) tried to remove the oxygen from this gas, but failed to do so. Although they indicated that, because of this observation oxygenated muriatic acid might be a chemical element they did not yet consider it as such as they believed that their observations on it could still be explained more appropriate as it being a compound substance. Finally Davy (1811) after failing again to decompose "oxygenated muriatic acid" realised that it could only be a "simple body" (element) and he proposed the name of (eu)chlorine for it after its bright yellow green colour.

The second halogen coming to surface was iodine. It was discovered in 1811 by Bernard Courtois who was a producer of saltpetre (KNO_3). Considering that France was a country at war in those days, and saltpetre was used as one of the main gunpowder ingredients, saltpetre was in great demand. Production of saltpetre required Na_2CO_3 which was made from ash prepared from seaweed. Seaweed was burned to isolate the Na_2CO_3, and the ash was washed with water. To destroy the remaining waste sulphuric acid was added. When Courtois added an excess of sulphuric acid accidentally a cloud of a purple vapour rose. It was observed that this vapour condensed on cold surfaces thereby forming dark crystals. He suspected that these crystals might be an at that time unknown element but he was unable to continue his research because of lack of funding. Courtois supplied the new substance to his friends Charles Bernard Desormes and Nicolas Clément for further analyses after which they presented the discovery to the scientific world (Courtois 1813). This new substance was further investigated by Joseph Louis Gay-Lassac and Humphry Davy who independently concluded that it was a newly discovered element and was subsequently called "iode" after the purple colour of iodine vapour by Gay-Lassac. Davy also discovered the chemical similarities between this newly discovered element and chlorine which he studied a few years previously (Gay-Lussac 1813a, 1813b, 1814; Davy 1813, 1814a).

Bromine has been discovered independently by Antoine Jérôme Balard and Karl Löwich in 1825/1826. Because Balard was the first to publish his observations (Balard 1826) he is credited with the discovery of this element although Löwich might have prepared it a little earlier (Löwich 1827, 1828) than Balard. Balard discovered bromine when he extracted iodine from seaweed. During this process he distilled a substance from which he discovered that it has properties that were intermediate between chlorine and iodine. He tried to prove that this substance was iodinechloride (ICl) but failed in this process. He concluded that the substance was an element that he called muride. However, this was renamed bromide as the name proposed by Balard did not get approval by the French Academy of Sciences and it was renamed bromide after its "bad odour". Löwich discovered bromine when he oxidized salty springwater with elementary chlorine. He found a red liquid that he showed to his professor (Leopold Gmelin). Gmelin asked Löwich to produce some more of the substance so that it could be studied in more detail. While he was working on this task Balard's discovery was made public, indicating the same properties as the red liquid discovered by Löwich.

Due to its extremely reactive behaviour the discovery of fluorine was delayed for a very long time. Hydrogen fluoride was already described for the first time in 1768 (Marggraf 1768) and the results of these experiments were confirmed by Scheele 3 years later (Scheele 1771). As in that period it was believed that all acids contain oxygen it was concluded that hydrofluoric acid should also contain oxygen. However Davy (1814b) showed that this was not the case. It was Ampere (1816) who realised that hydrofluoric acid must be a compound between hydrogen and an, at that moment, unknown element. Searching for the pure element was a very delicate task. Considering the fact that fluorine is the most electronegative element known and it is extremely reactive, it prefers to form chemical compounds with virtually all other elements making it very difficult to produce it in its pure form. It was believed that pure fluorine could be obtained by electrolysis of hydrogen fluoride. Electrolysis of solutions of hydrogen fluoride in water did however not proceed to produce fluorine gas due to the extreme corrosiveness of both hydrogen fluoride and fluorine gas. Attempts by Edmond Frémy to use pure hydrogen fluoride to obtain fluorine by electrolysis failed due to the fact that pure hydrogen fluoride does not conduct electricity (Frémy 1850). The experiments were continued by Henri Moissan who experimented with mixtures of potassium bifluoride and hydrogen fluoride, a mixture that conducts electricity and which, finally, made it possible to produce free fluorine by electrolysis (Moissan 1887).

Fluorine was the last group 17 element containing at least one stable isotope that was isolated. Element 85, the later astatine consists solely of short living radioactive isotopes. The longest living isotope is ^{210}At with a half-life of only 8.1 h. Although present (in minute quantities) on Earth it was first discovered after synthesis of the element at the University of California, Berkeley by Corson et al. (1940). They found the element after bombarding ^{209}Bi with alpha particles. Three years later the element was finally discovered as occurring naturally in the uranium decay chains (Karlik and Bernert 1943a, 1943b).

The last group 17 element discovered is tennessine. This element was discovered in 2010 by an international team of scientists from Russia and the USA (Organessian et al. 2010). The element was prepared after bombarding a ^{249}Bk target with ^{48}Ca particles during which process the isotopes ^{293}Ts (5 events) and ^{294}Ts (1 event) where detected. The existence of element 117 was confirmed in 2014 (Khuyagbaatar et al. 2014) using the same fusion reaction at GSI in Darmstadt where also the isotope ^{294}Ts was observed. Its discovery was officially announced by Karol et al. (2016) and its name by Öhrström and Reedijk (2016).

It is not expected that any more group 17 elements will ever be discovered. The next group 17 element is expected to be element 171. It is predicted however that elements above number 137 may not exist as such because according to the Bohr model electrons in the 1 s shell would get a speed faster than that of light in these elements. Also according to the semi-relativistic Dirac equation problems would arise above element number 137. This effect was first noticed by Feynman (1948) for which reason element 137 is sometimes called Feynmanium (Elert 1998–2012). Khazan (2007) made calculations on where the end of the periodic table could be found and concluded that the upper limit of the periodic table would be at element 157 which he called after himself Khazanium. On the other hand, calculations that have been made to define the electron configurations of elements up to 164 (Nefedov et al. 2006) and 172 (Pyykkö 2011b) indicate that at least in theory element 171 might exist. Early predictions on chemical and physical characteristics of this element were published by Fricke and Waber (1971) and Fricke et al. (1971) and these can be found in Table 1.2.

1.3 Nucleosynthesis of the Stable Halogen Nuclides (^{19}F, ^{35}Cl, ^{37}Cl, ^{79}Br, ^{81}Br, ^{127}I)

After the Big Bang mostly the lightest chemical elements (H, He) were formed as a primordial gas. When the first stars formed heavier elements were produced due to fusion of these light elements in a series of nucleosynthetic processes. The different elements are formed in a series of processes that continue to this day, and while the lighter elements are formed during fusion in active stars, the heavier elements mostly form during the processes that take place when the hydrogen present in a star is exhausted and stars are dying. The process of the production of the different nuclides, which is called nucleosynthesis, starting from the Big Bang is summarised in a very understandable review by Prantzos (2019). The halogens all contain an odd number of protons and are characterised with lower cosmical abundances than their even neighbours. In the solar system the abundance of chlorine and bromine are largely in line with their odd neighbours (P and K vs. As and Rb), while the abundance of iodine is clearly higher than its odd neighbours antimony and caesium. On the contrary the solar abundance of fluorine is significantly below that of its neighbours. These differences are the result of the processes that are responsible for the formation of the different nuclides.

Fluorine has only one stable isotope (^{19}F) and this nuclide is fragile and easily destroyed by proton, neutron and alpha particle reactions. As a result, any process that is able to form ^{19}F, must also enable it to escape from the system where it is formed, as otherwise it is quickly destroyed again. A possible environment to facilitate this may be an exploding supernova (Woosley and Haxton 1988). This also is the environment where enormous amounts of neutrinos originate. Of the various proposed production scenarios (Woosley et al. 1990; Forestini et al. 1992; Meynet and Arnould 2000; Longland et al. 2011) only production of fluorine in AGB (asymptotic giant branch) stars has been proven (Jorissen et al. 1992; Werner et al. 2005; Otsuka et al. 2008; Abia et al. 2010, 2015; Lucatello et al. 2011). The production of ^{19}F can then be proposed through excitation of ^{20}Ne by neutrinos [$^{20}Ne(\nu,\nu')^{20}Ne^*$] followed by proton or neutron decay of the excited neon, whereby in the case of neutron decay the ^{19}Ne that is formed decays to ^{19}F through β-decay [$^{20}Ne^* \rightarrow {}^{19}F + p$ or $^{20}Ne^* \rightarrow {}^{19}Ne + n$, $^{19}Ne \rightarrow {}^{19}F$] (Woosley and Haxton 1988).

Chlorine has two stable isotopes (^{35}Cl and ^{37}Cl) and these are formed in different reaction pathways from different parent nuclides. As a result the ratio of ^{35}Cl and ^{37}Cl may vary based on the occurrence of the parent nuclides in the stars where these isotopes are formed. ^{35}Cl may be formed through proton capture of ^{34}S during explosive oxygen burning ($^{34}S(p,\gamma)$). The free protons may originate from a reaction of two ^{16}O nuclides, or from photodisintegration (Woosley et al. (1973). Thielemann and Arnett (1985) suggested the forming of ^{35}Cl during oxygen burning following the pathway: $^{16}O(^{16}O,\alpha)^{28}Si(n,\gamma)^{29}Si(\alpha,n)^{32}S(\alpha,p)^{35}Cl$. ^{37}Cl, on the other hand may have formed from the decay of ^{37}Ar that was produced via neutron capture of ^{36}Ar during oxygen burning in core collapsing supernovae (CCSN; Woosley et al. 1973; Thielemann and Arnett 1985; Woosley and Weaver 1995). Maas and Pilachovski (2018) summarised the variability of the $^{35}Cl/^{37}Cl$ nucleosynthesis ratio based on the mass and metallicity of supernovae and noticed that the $^{35}Cl/^{37}Cl$ ratio can range from 1.20 to 4.23, compared to 3.13 for the solar system $^{35}Cl/^{37}Cl$. In massive stars the production of ^{37}Cl can be much higher and increases with He-core mass and neutron excess (Prantzos et al. 1990). In principle, Pignatari et al. (2010) showed that in a star of 25 solar masses ^{37}Cl can be overabundant with a factor of 47 compared to the solar system, theoretically decreasing the $^{35}Cl/^{37}Cl$ from 3.13 to less than 0.1, something that has never been observed yet. In general the $^{35}Cl/^{37}Cl$ ratio seems to be dependent on the mass of the star in which they are formed, with higher ^{37}Cl abundances in heavier stars.

Bromine is a rare element and very difficult to detect in extraterrestrial systems. The solar abundance values were actually not determined through direct measurement but established indirectly from meteoric data. Bromine nuclei are heavier than those from iron and they cannot be formed during charged particle reactions because at the temperatures required photodisintegration will break down all heavy nuclides that are formed that way. Nuclides heavier than Fe are produced via neutron capture on already existing nuclides, in processes summarised by Arnould and Takahashi (1999) and Prantzos (2019). Two different types of neutron capture processes do exist. The first is the s-process (s is for slow), that produces nuclei near the valley of nuclear stability and the r-process (r is for rapid) to produce the most neutron rich nuclei. In the s-process stable isotopes are bombarded by neutrons at low densities (~10^8 cm^{-3}) for longer timescales (>1 yr) where stable nuclides capture a neutron, and subsequently decay via β-decay to a stable daughter nucleus. In the case of ^{79}Br the process is $^{78}Se + n \rightarrow {}^{79}Se \rightarrow {}^{79}Br + \beta^-$ and ^{81}Br is produced via $^{79}Br(n,\gamma)^{80}Br(n,\gamma)^{81}Br$ and via $^{80}Se(n,\gamma)^{81}Se(\beta^-\nu)^{81}Br$ (Pignatari et al. 2010). In the r-process very large neutron densities (>1020 cm^{-3}) are bombarding the nuclei for very short timescales. This environment corresponds to explosive sites such as that exist in supernovae. With this process very highly neutron rich nuclei are formed that through a series of β-decays decay to stable nuclei. This process is especially important for neutron rich nuclei. This may be the pathway for the formation of the final stable halogen nuclide, ^{127}I. This nuclide may be the result from fission of very heavy nuclides formed during the r-process. Nuclides in the range of ^{127}Rh, ^{127}Pd, ^{127}Ag and ^{127}Cd may be fission products that ultimately decay to ^{127}I (Eichler et al. 2015; Thielemann et al. 2017). Iodine is however very rare in interstellar space and has (for as far as I know) not yet been observed in interstellar space.

1.4 Abundance of the Halogens in the Solar System

The processes described in Sect. 1.3 ultimately result in the different abundances of the halogen elements that are observed in the universe. Within the solar system, the halogens are relatively minor elements. Being elements with an odd proton number they are all less abundant than their neighbouring elements with even numbers of protons. 20th century studies (e.g. Goldschmidt 1937; Suess and Urey 1956; Cameron 1970; Anders and Ebihara 1982), agreed very much on the order of magnitude of the solar abundance of these elements although slightly different abundances were found as a result of increasing knowledge during the twentieth century. The most recent compilation, collected from all available data by Lodders et al. (2009) recommends that, relative to 10^6 silicon atoms, there are 1270 ± 275 fluorine atoms, 5290 ± 810 chlorine atoms, 12.3 ± 2.9 bromine atoms and 1.59 ± 0.64 iodine atoms in the solar system.

1.5 Abundance of the Halogens on Earth

On earth the distribution of the halogens is different from that in the solar system. As the Earth is mostly a rocky planet the most common elements are the rock forming elements, especially silicon, aluminium, iron, nickel, magnesium, calcium, sulphur and oxygen. Due to mass differentiation during the early age of the earth the heavy elements iron, the most common element on Earth by mass, and nickel, have been concentrated in the core. The halogen elements are relatively rare when considering the whole earth. The most common of them is chlorine with a content of 10 ± 5 ppm for the whole Earth, the concentrations of the other elements being: fluorine at 5.12 ± 0.5 ppm, bromine at 400 ± 150 ppb and iodine at 40.5 ± 15 ppb (Allegre et al. 2001). Data presented in Sect. 3.2.1 of this book suggest however that the chlorine content of the whole Earth might be significantly higher than the value presented by Allegre et al. (2001), and should be more in line with a value of 76 ppm as suggested by McDonough (2001). As the halogen elements are among the most volatile of all elements they are not very compatible with most rock types at low temperatures. Especially chlorine and bromine are so called excess volatiles and prefer to concentrate in fluid phases. On Earth this means that they are very heavily concentrated in the oceans. Except in oceans they are only found in high concentrations in evaporites, which are formed from the evaporation of ocean or lake waters. It has to be realised however, that, while chlorine and bromine at low temperatures do not replace hydroxyl groups (OH^-) in minerals such as amphiboles, at high temperatures they regularly do. As a result in high temperature minerals especially chloride is more common than in low temperature minerals (P Agrinier, pers. comm). Iodine is also concentrated in the oceans, but the iodide ion is too large to fit into common salt minerals (Boeke 1908) and stays either in the fluid phase, or forms minerals of its own. As iodine is more easily to oxidise than chlorine or bromine iodine minerals regularly are iodates, although iodides of transition minerals are also found. Furthermore, iodine is very much concentrated in certain brown algae such as *Laminaria* and *Fucus*, which can contain up to 0.4% iodine by dry weight (Lyday 2005). Bromine does not form minerals of its own (except for bromargyrite; AgBr) and is always found as solid solution in chloride minerals, especially in evaporites. Fluorine has a different behaviour than the other halogens as its alkaline earth salts (such as calcium and magnesium fluoride) are very little soluble in water. As a consequence fluoride concentrations in seawater and thus in evaporites is very low. Due to the small size of the fluoride ion fluorine is not found commonly in silicates, but forms minerals of its own, of which fluorite [CaF_2], fluorapatite [$Ca_5(PO_4)_3F$] and cryolite [Na_3AlF_6] are the most common, even to such an extent that they can be used as commodities and mined as natural fluorine resources. Summarising the concentrations of the halogens in the crust are: fluorine 300 ppm, chlorine 314 ppm, bromine 1.6 ppm and iodine 0.3 ppm (Mason 1952). The concentrations of the halogens in ocean salt is on average: fluorine 37 ppm, chlorine 54.3%, bromine 0.19% and iodine 1.8 ppm (recalculated from Turekian 1971). These data show that the halogens are mostly concentrated in the oceans and the crust, so that, as a consequence, the concentrations of these elements in the mantle must be very low.

Astatine, although found naturally on Earth is particularly rare in nature, and it is calculated that at any one moment less than 30 grams of this element is present on Earth. Astatine is produced naturally in the decay chains of ^{238}U (isotope ^{218}At) and ^{235}U (isotopes ^{215}At and ^{219}At) in minute quantities as it is formed solely on very low probability decay branches. An other isotope (^{217}At) theoretically exists in nature in even smaller amounts as decay product of the neptunium chain, however the parent isotope of this chain has virtually decayed completely on Earth due to its short half-life.

References

Abia C, Cunha K, Cristallo S, De Laverny P, Domínguez I, Eriksson K, Gialanella L, Hinkle K, Imbrianani G, Recio-Blanco A (2010) Fluorine abundances in galactic asymptotic giant branch stars. Astrophys J Lett 712:L94

Abia C, Cunha K, Cristallo S, De Laverny P (2015) The origin of fluorine: abundances in AGB carbon stars revisited. A&A 581:A88

Allegre C, Manhès G, Lewin É (2001) Chemical composition of the earth and the volatility control on planetary genetics. Earth Planet Sci Lett 185:49–69

Ampère AM (1816) D'une classification naturelle pour les corps simples. Ann Chim Phys 2:6–21

Anders E, Ebihara M (1982) Solar-system abundances of the elements. Geochim Cosmochim Acta 46:2363–2380

Arnould M, Takahashi K (1999) Nuclear astrophysics. Rep Prog Phys 62:395–464

Balard AJ (1826) Sur une substance particulière contenue dans l'eau de la mer. Ann Chim Phys 2me series, vol 32, pp 337–381

Boeke HE (1908) Über das Krystallisationsschema der Chloride, Bromide, Jodide von Natrium, Kalium und Magnesium, sowie über das Vorkommen des Broms und das Fehlen von Jod in den Kalisalzlagerstätten. Z Kristallogr 45:346–391

Cameron AGW (1970) Abundances of the elements in the solar system. Space Sci Rev 15:121–146

Champion J, Alliot C, Renault E, Mokili BM, Chérel M, Galland N, Montavon G (2010) Astatine standard redox potentials and speciation in acidic medium. J Phys Chem A 114:576–582

Chance R, Tinel L, Sarkar A, Sinha AK, Mahajan AS, Chacko R, Sabu P, Roy R, Jickells TD, Stevens DP, Wadley M, Carpenter LJ (2020) Surface inorganic iodine speciation in the Indian and Southern oceans from 12°N to 70°S. Front Mar Sci 7:621

Corson DR, MacKenzie KR, Segrè E (1940) Artificially radioactive element 85. Phys Rev 58:672–678

Courtois B (1813) Découverte d'une substance nouvelle dans le Vareck. Ann Chim 88:304–310

Davy H (1811) On a combination of oxymuriatic gas and oxygene gas. Philos Trans R Soc Lond 101:155–162

Davy H (1813) Sur la nouvelle substance découverte par M. Courtois, dans le sel de Vareck. Ann Chemie 88:322–329

Davy H (1814a) Some experiments and observations on a new substance which becomes a violet coloured gas by heat. Philos Trans R Soc Lond 104:74–93

Davy H (1814b) Mémoire sur la nature de l'acide fluorique, lu á la Societé Royale de Londres, le 8 julliet 1814. Ann Chim 88:271

Eichler M, Arcones A, Kelic A, Korobkin O, Langanke K, Marketin T, Martinez-Pinedo G, Panov I, Rauscher T, Rosswog S, Winteler C, Zinner NT, Thielemann FK (2015) The role of fission in neutron star mergers and its impact on the *r*-process peaks. Astrophys J 808:30

Elert G (1998–2012) The physics hypertextbook, atomic models. http://physics.info/atomic-models/. Accessed 4 Nov 2012

Feynman RP (1948) Relativistic cut-off for quantum electrodynamics. Phys Rev 74:1430–1438

Forestini M, Goriely S, Jorissen A, Arnould M (1992) Fluorine production in thermal pulses on the asymptotic giant branch. Astron Astrophys 261:157

Frémy E (1850) Recherches sur les fluorures. Ann Chim Phys 2me Serie, t, vol LXVII, p 5

Fricke B, Waber JT (1971) Theoretical predictions of the chemistry of superheavy elements. Actinides Rev 1:433–485

Fricke B, Greiner W, Waber JT (1971) The continuation of the periodic table up to Z = 172. The chemistry of superheavy elements. Theor Chim Acta (Berlin) 21:235–260

Gay-Lussac J (1813a) Sur un nouvel acide formé avec la substance décourverte par M. Courtois Ann Chim 88:311–318

Gay-Lussac J (1813b) Sur la combination de l'iode avec d'oxigène. Ann Chim 88:319–321

Gay-Lussac J (1814) Mémoire sur l'iode. Ann Chim 91:5–160

Gay-Lussac J, Thénard L (1809) De la nature et des propriétés de l'acide muriatique et de l'acide muriatique oxigéné. Mém Phys Chim Société d'Arcueil 2:339–358

Goldschmidt VM (1937) The principles of distribution of chemical elements in minerals and rocks. J Chem Soc 1937:655–673

Holleman AF, Buchner EH (1949) Leerboek der anorganische chemie. 15e druk. JB Wolters, Groningen – Batavia

Jorissen A, Smith VV, Lambert DL (1992) Fluorine in red giant stars: evidence for nucleosynthesis. Astron Astrophys 261:164–187

Karlik B, Bernert T (1943a) Eine neue natürliche α-Strahlung. Naturwissenschaften 31:298–299

Karlik B, Bernert T (1943b) Das Element 85 in den natürlichen Zerfallsreihen. Z Physik A 123:51–72

Karol PJ, Barber RC, Sherrill BM, Vardaci E, Yamazaki T (2016) Discovery of the elements with atomic numbers Z=113, 115 and 117 (IUPAC Technical Report). Pure Appl Chem 88:139–153

Khazan A (2007) Effect from hyperbolic law in periodic table of elements. Prog Phys 2:83–86

Khuyagbaatar J, Yakushev A, Düllmann CE, Ackermann D, Andersson LL, Asai M, Block M, Boll RA, Brand H, Cox DM, Dasgupta M, Derkx X, Di Nitto A, Eberhardt K, Even J, Evers M, Fahlander C, Forsberg U, Gates JM, Gharibyan N, Golubev P, Gregorich KE, Hamilton JH, Hartmann W, Herzberg RD, Heßberger FP, Hinde DJ, Hoffmann J, Hollinger R, Hübner A, Jäger E, Kindler B, Kratz JV, Krier J, Kurz N, Laatiaoui M, Lahiri S, Lang R, Lommel B, Maiti M, Miernik K, Minami S, Mistry A, Mokry C, Nitsche H, Omtvedt JP, Pang GK, Papadakis P, Renisch D, Roberto J, Rudolph D, Runke J, Rykaczewski KP, Sarmiento LG, Schädel M, Schausten B, Semchenkov A, Shaughnessy DA, Steinegger P, Steiner J, Tereshatov EE, Thörle-Pospiech P, Tinschert K, Torres De Heidenreich T, Trautmann N, Türler A, Uusitalo J, Ward DE, Wegrzecki M, Wiehl N, Van Cleve SM, Yakusheva V (2014) $^{48}Ca+^{249}Bk$ fusion reaction leading to element Z=117: long-lived α-decaying ^{270}Db and discovery of ^{266}Lr. Phys Rev Lett 112:172501

Küpper FC, Carrano CJ (2019) Key aspects of the iodine metabolism in brown algae: a brief critical review. Metallomics 11:756–764

Lodders K, Palme H, Gail HP (2009) Abundances of the elements in the solar system. In: Trümper JE (ed) Landolt-Börnstein, new series, vol VI/4B, chap 4.4. Springer, Berlin/Heidelberg/New York, pp 560–630

Longland R, Lorén-Aguilar P, José J, García-Berro E, Althaus LG, Isern J (2011) Nucleosynthesis during the merger of white dwarfs and the origin of r coronae borealis stars. Astrophys J Lett 737:L34

Löwich C (1827) Über Brombereitung und eine auffallende Zersetzung des Aethers durch Chlor. Mag Fur Pharm 21:31–36

Löwich C (1828) Über einige Bromverbindungen und über Bromdarstellung. Ann Phys Chem 14:485–499

Lucatello S, Masseron T, Johnson JA, Pignatari M, Herwig F (2011) Fluorine and sodium in C-rich low-metallicity stars. Astrophys J 729:40

Lyday PA (2005) Iodine and iodine compounds. In: Ullmann's encyclopedia of industrial chemistry, vol A14. Wiley-VCH, Weinheim, pp 382–390

Maas ZG, Pilachovski CA (2018) Chlorine isotope ratios in M giants. Aston J 156:2

Marggraf AS (1768) Observation concernant une volatilisation remarquable d'une partie de l'espèce de pierre, à laquelle on donne les noms de Flosse, Flusse, Flus-Spaht, et aussi celui d'hespéros; laquelle volatilisation a été effectuée au moyen des acides. Histoire de l'Académie Royale des Sciences et des Belles-Lettres de Berlin 1768:1–11

Mason B (1952) Principles of geochemistry. John Wiley and Sons, New York

McDonough WF (2001) The composition of the earth. In: Teisseyre R, Majewski E (eds) Earthquake thermodynamics and phase transformations in the earth's interior

Meija J, Coplen TB, Berglund M, Brand WA, De Bièvre P, Gröning M, Holden NE, Irrgeher J, Loss RD, Walczyk T, Prohaska T (2016) Isotopic compositions of the elements 2013 (IUPAC Technical Report). Pure Appl Chem 88:293–306

Meynet G, Arnould M (2000) Synthesis of ^{19}F in Wolf-Rayet stars. Astron Astrophys 355:176–180

Moissan H (1887) Recherges Sur l'isolement du fluor. Annal Chim Phys 6[me] serie, t. XII. Gauthiers-Villars, Paris, p 66
Nefedov VI, Trzhaskovskaya MB, Yarzhemskii VG (2006) Electronic configurations and the periodic table for superheavy elements. Dokl Phys Chem 408:149–151
Norseev YV (2021) Extrapolation estimates of the physicochemical parameters of Nichonium, Tennessin, and Oganesson. Radiochemistry 63:269–274
Öhrström L, Reedijk J (2016) Names and symbols of the elements with atomic numbers 113, 115, 117 and 118 (IUPAC recommendations 2016). Pure Appl Chem 88:1225–1229
Organessian YT, Abdullin FS, Bailey PD, Benker DE, Bennett ME, Dmitriev SN, Ezold JG, Hamilton JH, Henderson RA, Itkis MG, Lobanov YV, Mezentsev AN, Moody KJ, Nelson SL, Polyakov AN, Porter CE, Ramayya AV, Riley FD, Roberto JB, Ryabinin MA, Rykaczewski KP, Sagaidak RN, Shaughnessy DA, Shirokovsky IV, Stoyer MA, Subbotin VG, Sudowe R, Sukhov AM, Tsyganov YS, Utyonkov VK, Voinov AA, Vostokin GK, Wilk PA (2010) Synthesis of a new element with atomic number Z = 117. Phys Rev Lett 104:142502
Otsuka M, Izumiura H, Tajitsu A, Hyung S (2008) Detection of fluorine in the halo planetary nebula BoBn 1: evidence for a binary progenitor star. Astrophys J 682:L105
Pignatari M, Gallino R, Heil M, Wiescher M, Käppeler F, Herwig F, Bisterzo S (2010) The weak *s*-process in massive stars and its dependence on the neutron capture cross sections. Astrophys J 710:1557–1577
Poll W, Pawelke G, Mootz D, Appelman EH (1988) The crystal structure of Hypofluorous acid: chain formation by O-H˙O hydrogen bonds. Angew Chem Int Ed Engl 27:392–393
Prantzos N (2019) Nucleosynthesis: the origin of the chemical elements. In: Lique F, Faure A (eds) Gas-phase chemistry in space, Chapter 2. IOP Publishing Ltd, pp 1–44
Prantzos N, Hashimoto M, Nomoto K (1990) The *s*-process in massive stars – yields as a function of stellar mass and metallicity. Astron Astrophys 234:211–229
Prohaska T, Irrgeher J, Benefield J, Böhlke JK, Chesson LA, Coplen TB, Ding TP, Dunn PJH, Gröning M, Holden NE, Meijer HAJ, Moosen H, Possolo A, Takahashi Y, Vogl J, Walczyk T, Wang J, Wieser ME, Yoneda S, hu XK, Meija J (2022) Standard atomic weights of the elements 2021 (IUPAC Technical Report). Pure Appl Chem 94:573–600
Pyykkö P (2011a) Relativistic effects in chemistry: more common than you thought. Annu Rev Phys Chem 63:45–64
Pyykkö P (2011b) A suggested periodic table up to Z ≤ 172, based on Dirac–Fock calculations on atoms and ions. Phys Chem Chem Phys 13:161–168
Scheele CW (1771) Undersökning om fluss-spat och dess syra. Kongl Vetenskaps Academiens Handlingar 32:120–137
Scheele CW (1774) On Brun-sten eller Magnesia, och dess Egenskaper. Kongl Vetenskaps Academiens Handlingar 35:89–116
Seaborg GT (1968) Elements beyond 100, present status and future prospects. Annu Rev Nucl Sci 18:53–152
Shaw TI, Cooper LHN (1957) State of iodine in sea water. Nature 180:250
Suess H, Urey H (1956) Abundances of the elements. Rev Mod Phys 28:53–74
Thielemann FK, Arnett WD (1985) Hydrostatic nucleosynthesis – II – Core neon to silicon burning and presupernova abundance yields of massive stars. Astrophys J 295:604–619
Thielemann FK, Eichler M, Panov IV, Wehmeyer B (2017) Neutron star mergers and nucleosynthesis of heavy elements. Annu Rev Nucl Part Sci 67:253–274
Turekian KK (1971) McGraw-Hill Encyclopedia of Science and Technology, vol 4. McGraw-Hill, pp 627–630
Weeks ME (1942) The discovery of the elements. XVIII. The halogen family. J Chem Educ 9:1915–1939
Werner K, Rauch T, Kruk JW (2005) Fluorine in extremely hot post-AGB stars: evidence for nucleosynthesis. Astron Astrophys 433:641–646
Woosley SE, Haxton WC (1988) Supernova neutrinos, neutral currents and the origin of fluorine. Nature 334:45–47
Woosley SE, Weaver TA (1995) The evolution and explosion of massive stars. II explosive hydrodynamics and nucleosynthesis. Astrophys J Suppl 101:181
Woosley SE, Arnett WD, Clayton DD (1973) The explosive burning of oxygen and silicon. Astrophys J Suppl 26:231
Woosley SE, Hartmann DH, Hoffman RD, Haxton WC (1990) The ν-process. Astrophys J 356:272–301

2 The Chemistry of Chlorine and Bromine

2.1 Geochemical Separation of Chlorine and Bromine from Fluorine and Iodine

In Chap. 1 we discussed that, although the four geologically relevant halogen elements have comparative chemical characteristics, these characteristics are different enough to separate the elements into different geological environments. That results in fluorine being enriched in magmatic environments, while iodine is enriched in organic material such as in seaweed. Under natural circumstances it also is clear that iodate is more stable than iodide. The fact that fluorides of the alkaline earth metals are not very soluble in water results in low fluoride concentrations in seawater. Overall these effects result in iodine and fluorine that are decoupled from chlorine and bromine in most geological environments. Chlorine and bromine on the other hand appear to be largely coupled in geological environments. Although chlorine and bromine do appear in a large array of (mass/ionic) ratios in geological material, invariably they are always found together, a characteristic that has been studied regularly in the past (e.g. Boeke 1908, Behne 1953, Valyashko 1956, Yoshida et al. 1971, Alcala and Custodio 2008, Wang et al. 2020), because it supplies very important information on the provenance of the samples from which chlorine and bromine are studied.

2.2 The Chemistry of Chlorine

Chlorine is a very common element that is present in very large quantities in the oceans (at about 19,000 mg/L; 546 mmol/kg). Large amounts are also present in salt deposits where it contains up to 60% of halite (NaCl) and is also present in many other salt minerals such as sylvite (KCl), kainite ($K_4Mg_4Cl_4(SO_4)_4.11H_2O$), carnallite ($KMgCl_3.6H_2O$) and bischofite ($MgCl_2.6H_2O$) (Braitsch 1962). Chlorine does not exist in its elemental form on earth as it is too reactive and readily forms chlorides. Commercially it is prepared by the electrolysis of a sodium chloride solution, the so called chloralkali process. During this process an electric current is passed through the solution whereby chlorine gas is formed at the anode. At the cathode water is dissociated into hydroxide ions (OH^-) and hydrogen gas. The sodium hydroxide solution that forms at the cathode is actually the commercially most important product of this process. The chemical process can be summarised as:

$$2NaCl + 2H_2O \rightarrow Cl_2 + H_2 + 2NaOH \quad (2.1)$$

Due to the dissociation of water molecules this process originally needed to be stopped once the solution contains about 8% NaOH. The sodium hydroxide was then removed from the solution and replaced by a fresh NaCl solution (Holleman and Buchner 1949). Currently the

H. Eggenkamp, *The Geochemistry of Stable Chlorine and Bromine Isotopes*, Advances in Isotope Geochemistry, https://doi.org/10.1007/978-3-031-75633-7_2

process is applied in a diaphragm cell, where the anode and cathode areas are separated from each other by a permeable diaphragm. In this process the NaOH concentration can be increased to 50%, which is considerably more efficient than the method without the diaphragm (O'Brien et al. 2005; eurochlor.com). Small amounts of chlorine, for example for the use in laboratories, can also be prepared by the oxidation of hydrochloric acid by manganese dioxide:

$$4HCl + MnO_2 \rightarrow Cl_2 + MnCl_2 + 2H_2O \quad (2.2)$$

However, this process is currently little used due to the readily availability of chlorine gas in pressurised cylinders.

As with all halogen elements the most common oxidation state is -I and the most common form of chlorine in nature is the chloride ion (Cl^-). Negative ions are formed by the addition of electrons to the valence shell through a process called electron affinity. Of all the elements chlorine has the highest electron affinity (349 kJ. mol^{-1}), which is even higher than the electron affinity of fluorine. The reason for this is that the fluorine atom is very small and the high electron density in the valence shell compared to chlorine makes it slightly less favourable to add an electron to the shell as compared to chlorine.

Higher oxidation states for chlorine are very well known. They are all strong oxidisers and have different uses as such, mainly as bleaching agents in the case of the lower oxidation states and in pyrotechnics, for example as solid rocket fuel in the higher oxidation states. In the higher oxidation states chlorine is present in the oxyanions hypochlorite (ClO^-), chlorite (ClO_2^-), chlorate (ClO_3^-) and perchlorate (ClO_4^-) where chlorine is in the +I, +III, +V and + VII oxidation state respectively. Oxidation state +IV is also found in the gas chlorine dioxide, and oxidation state +VI in the compound Cl_2O_6 when in the gaseous state. Within the stability field of water the chloride ion is the only stable chlorine species (Brookins 1988). That means that in Pourbaix (Eh-pH) diagrams (Pourbaix et al. 1963; Garrels and Christ 1965), the other potentially stable species (mostly ClO_4^- (Takeno 2005) and possibly at very low pH also Cl_2 are only stable above the stability field of water (Schweitzer and Pesterfield 2010), and are thus able to oxidise and decompose water according to the equation:

$$2H_2O + 2Cl_2 \rightarrow 4H^+ + O_2 + 4Cl^- \quad (2.3)$$

As can be deducted from the Frost diagram (Frost 1951, Fig. 2.1) in acid solutions only chloride, chlorine and perchlorate are stable anions as the other forms tend to disproportionate, while in alkaline solutions chloride, hypochlorite, chlorite, chlorate and perchlorate can be stable while chlorine and chlorine dioxide tend to disproportionate.

The reduction potentials used in this figure were taken from www.webelements.com.

Well known is the hypochlorite ion which, in the form of sodium hypochlorite, is commonly used as a bleaching agent. It is the strongest oxidiser of all chlorine oxyanions and it is a very unstable ion that only exists in solutions. When water is removed from a sodium hypochlorite solution it converts to a mixture of sodium chloride and sodium chlorate. It is a very effective oxidizer used to kill germs at cleaning. It is also used, mainly in the form of calcium hypochlorite, as "chlorine" in swimming pools.

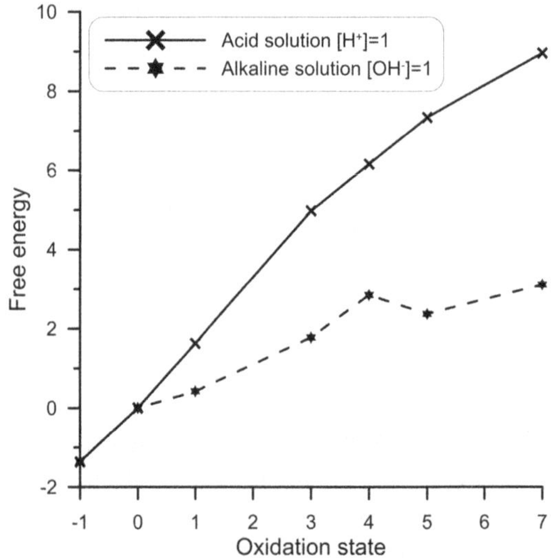

Fig. 2.1 Frost diagram of the chlorine species showing the relative stability of the species in acid and alkaline solutions

Chlorite is mainly used for the generation of chlorine dioxide which is also used as a bleaching agent. Sodium chlorite itself was used in the past as a chlorinating agent for the disinfection of drinking water, as it, contrary to chlorine did not produce trihalomethane organic contaminants. Since 2009 it is however phased out in the European Union.

Chlorates are mainly used as herbicides and oxidants in explosives and rock fuel. Little is known about its occurrence on Earth but it appears to be associated with perchlorate occurrences (Rao et al. 2010) which are known very well from the Atacama desert in Chile (Ericksen 1981) and also from more recent studies in the United States (Rajagopalan et al. 2006; Parker et al. 2008) and Antarctica (Kounaves et al. 2010). As these studies indicate that chlorate is almost always associated with perchlorate it is also likely that chlorate is present at the Phoenix landing site on Mars (Hanley et al. 2012) where perchlorate was detected. Chlorate salts have very strong freezing point depression capabilities and the eutectic point of magnesium chlorate dissolved in water is only 204 K (−66 °C). As a result it is theoretically possible that liquid water on Mars is present as concentrated chlorate solutions (Hanley et al. 2012). Chlorate is also common in municipal drinking water as a result of chlorination with chlorine dioxide. Concentration in drinking water can be as high as 3.2 to 7.0 mg/L (Michael et al. 1981) and the WHO (2005) has proposed a provisional guideline value of 0.7 mg/L. This guideline value is designated as provisional because use of chlorine dioxide as a disinfectant may result in the chlorate guideline value being exceeded, and difficulties in meeting the guideline value should never be a reason for compromising adequate disinfection according to the WHO (2005). Chlorine isotopes of chlorates have not yet been studied regularly. The first known chlorine isotope study of the chlorate ion was presented by Ader et al. (2001). They analysed the chlorine isotope composition of several man made chlorates and found that $\delta^{37}Cl$ values in commercially produced sodium and potassium chlorates are +0.2 to +0.3‰ vs. SMOC. More recently new methods were developed in order to be able to measure the much smaller amounts of chlorate present in natural environments (Jackson et al. 2021). Chlorine isotope ratios in natural chlorate showed significant variability, with values between −1.4 and + 1.3‰ in samples from the Atacama desert in Chile and Death Valley in the USA. Recently, also the isotope effects of the reaction pathways towards the formation of chlorate and perchlorate have been studied (Estrada et al. 2021).

Perchlorates do have comparable applications as chlorates. Unlike chlorates however there is considerable more knowledge on the occurrence and distribution of perchlorates on earth. It is at the present generally well known that it is much more widely distributed than was assumed in the past. Perchlorate is produced naturally by atmospheric oxidation of chlorine species with ozone (e.g. Kang et al. 2008) and due to the solubility of most perchlorate salts it is mostly concentrated in the driest regions on the earth. Perchlorates are also manufactured industrially by electrolysis of sodium chloride brines. The annual production is in the order of 10^7 kg.year^{-1} (Dasgupta et al. 2006). Its isotope behaviour is by now well established. Following preliminary studies by Ader et al. (2001) it is now possible to determine various sources of perchlorate by its chlorine and oxygen isotopes (e.g. Sturchio et al. 2011; Poghosyan et al. 2016; Cao et al. 2020), and it has also been possible to determine the isotope effects of microbial perchlorate reduction (Ader et al. 2008). In the study by Estrada et al. (2021) both the reaction pathways and corresponding isotope effects of chlorate and perchlorate were detected. This study indicated that the reaction pathways could be responsible for much of the observed isotope variations in both natural chlorate and natural perchlorate.

Chlorine is also present in organic compounds. Although minor amounts of organochlorine compounds are known to be from natural origins larger amounts are produced annually in industrial processes. Analyses of the chlorine isotopic composition of organochlorine compounds potentially found in geologic environments started with the development of several methods using dual inlet techniques in the 1990's (Van

Warmerdam et al. 1995, Holt et al. 1997; Jendrezejewski et al. 1997; Beneteau et al. 1999). During the early years of the twenty-first century techniques were developed using continuous flow techniques (Shouakar-Stash et al. 2003; Wassenaar and Koehler 2004) and with ICP-MS (Van Acker et al. 2006). More recently techniques were also developed where these isotopes are measured directly on whole molecules and fragments of them using isotope ratio mass spectrometry (e.g. Shouakar-Stash et al. 2006) and quadrupole mass spectrometry (Jin et al. 2011; Bernstein et al. 2011).

2.3 The Chemistry of Bromine

Bromine is significantly less common on Earth than chlorine. The most important reservoir for bromine is the ocean, where the average concentration is 65 mg/L or 0.81 mmol/L. The molar Br/Cl ratio is only $1.48*10^{-3}$. This indicates that for every bromide ion in the oceans 674 chloride ions are present. During precipitation of halite (NaCl) from evaporating brines bromide tends the concentrate in the brine as it does not crystallise easily with NaCl. This results in significantly increased bromide concentrations in the potassium and magnesium salts that crystallise after sodium in the brine has been exhausted. It is from these potassium and magnesium salts that bromine can be commercially exploited. Currently bromine is mostly extracted from salt lakes and brine wells in China, Israel, Jordan and the USA (USGS 2022). Br contents in the Dead Sea for example is extremely high and reaches to values from 5000 to over 6000 mg/L (Oren 2010; Al Bawab et al. 2018). Bromine is liberated from the most concentrated brine by exchange with chlorine. Chlorine gas oxidises the bromide to bromine according to the equation:

$$MBr + Cl_2 \rightarrow MCl + Br_2 \left(M = K,Mg\right) \qquad (2.4)$$

Bromine gas is subsequently condensed and purified trough distillation from the remaining chlorine.

As with all lighter halogen elements the most common oxidation state of bromine is -I and the most common form of bromine in nature is the bromide ion (Br^-). Negative ions are formed by the addition of electrons to the valence shell by a process called electron affinity, the value of which is 324 $kJ.mol^{-1}$ for bromine. According to Brookins (1988), the bromide ion is the only stable bromine species within the stability field of water. However, Pourbaix et al. (1963) showed that, especially at low bromine concentrations, Br_2 can also be stable within the stability field of water at low pH (<5) values. The other potentially stable species (BrO_3^- is only stable above the stability field of water (Pourbaix et al. 1963, Takeno 2005).

As for chlorine, higher oxidation states for bromine, except the oxidation state +VII, are well known. Oxidation states +I (hypobromite) and + V (bromate) are well established. Although salts of the +III oxidation state (bromites) are stable they are less well established and no data for their oxidation potential were published at www.WebElements.com. The relative stability of the bromine oxidation states is shown in a frost diagram (Frost 1951), see Fig. 2.2.

The reduction potentials used in this figure were taken from www.webelements.com. From this figure it can be deducted that in acid solu-

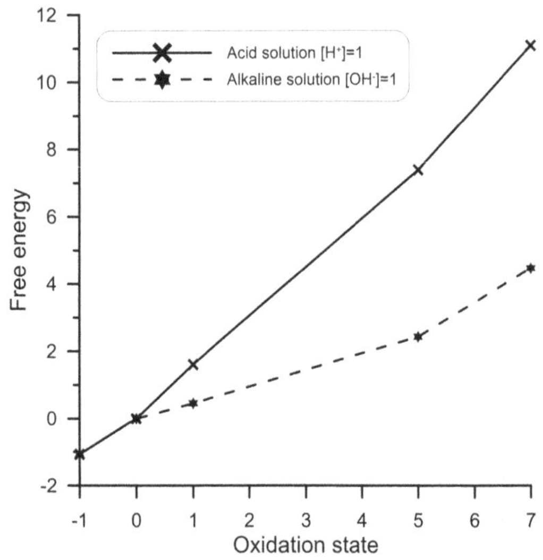

Fig. 2.2 Frost diagram of the bromine species showing the relative stability of the species in acid and alkaline solutions

tions bromide, bromine and bromate are stable, while hypobromite will disproportionate, in alkaline solutions bromine will disproportionate to bromide and hypobromite. Bromide and bromate are also stable. Contrary to the situation with chlorine the +VII oxidation state of bromine is very difficult to obtain and perbromates and perbromic acid were unknown prior to 1968. They were for the first time synthesised in 1968, originally through radioactive decay of ^{83}Se to ^{83}Br in $^{83}SeO_4^{2-}$ forming $^{83}BrO_4^-$ plus a β^- particle (Appelman 1968). A few years later Appelman (1973) managed to produce perbromate ions from bromate ions through the oxidation with fluorine and xenondifluoride gas. Due to the difficulty to oxidise bromate to perbromate, perbromate and perbromic acid are not found in nature. Hypobromite is formed by the disproportionation of bromine in alkaline solution. It has a very specific use in the defence system of higher animals as certain white blood cells prefer to oxidise bromide dissolved in the blood to hypobromite above the oxidation of chloride to attack certain parasites (Mayeno et al. 1989) even considering that chloride concentrations in blood are more than a thousand times higher than bromide concentrations.

The more stable oxyanions of bromine are sometimes found in drinking waters because bromide, which is always present in small concentrations in drinking water, can easily be oxidised by ozone or other oxidizing agents (such as chlorine or even the presence of sunlight). The formation of bromates is not desired in drinking waters as they are a suspected carcinogen (Kurokawa et al. 1990) and the EU maximum allowable concentration in drinking water is 0.01 $mg.L^{-1}$ (Diduch et al. 2011). Bromates are however often found in variable quantities in drinking water and considerable efforts are being taken to remove it from drinking water (e.g. Genuino and Espino 2012; Wisnieuwski et al. 2011).

The further existence of oxidised bromine species in nature is only recently established. Bobrowski et al. (2003) determined bromine oxide (BrO) in the plume of the volcano Monserrat. Not much is known about the occurrence of bromine oxyanions in other environments, but considering their occurrence in drinking water and their stability they must be present in other environments. No research is done yet on the bromine isotope composition of bromine oxyanions. Considering the low concentrations of bromine oxyanions in most natural environments and the current state of bromine isotope studies this does not come as a surprise. It is hoped however, that in the near future techniques will be developed that will make it possible to study the isotope behaviour of these environmentally important ions.

Bromine is also found in organic molecules. Compared to chlorine it has a much more important organic geochemistry and it is found in a considerable number of natural and artificial compounds. This last group of substances consists for a large part of fire retardants. From a relatively early date methods were developed to analyse stable isotope compositions of organically bound bromine (Sylva et al. 2007) and although the number of scientific papers on bromine isotopes is still modest a significant number of them is on the distribution of bromine isotopes in organic compounds (e.g. Carrizo et al. 2011; Hitzfeld et al. 2011; Holmstrand et al. 2010; Horst et al. 2019; Yankelson 2019, Tang et al. 2020; Hu et al. 2021).

References

Ader M, Coleman ML, Doyle SP, Stroud M, Wakelin D (2001) Methods for the stable isotopic analysis of chlorine in chlorate and perchlorate compounds. Anal Chem 73:4946–4950

Ader M, Chaudhuri S, Coates JD, Coleman M (2008) Microbial perchlorate reduction: a precise laboratory determination of the chlorine isotope fractionation and its possible biochemical basis. Earth Planet Sci Lett 269:604–612

Al Bawab A, Bozeya A, Abu-Mallouh S, Abu Irmaileh B, Daqour I, Abu-Zurayk RA (2018) The Dead Sea mud and salt: a review of its characterization, contaminants, and beneficial effects. Mat Sci Eng 305:012003

Alcala FJ, Custodio E (2008) Using the Cl/Br ratio as a tracer to identify the origin of salinity in aquifers in Spain and Portugal. J Hydrol 359:189–207

Appelman EH (1968) Synthesis of perbromates. J Am Chem Soc 90:1900–1901

Appelman EH (1973) Nonexistent compounds. Two case histories. Acc Chem Res 6:113–117

Behne W (1953) Untersuchungen zur Geochemie des Chlor und Brom. Geochim Cosmochim Acta 3:186–214

Beneteau KM, Aravena R, Frape SK (1999) Isotopic characterization of chlorinated solvents-laboratory and field results. Org Geochem 30:739–753

Bernstein A, Shouakar-Stash O, Ebert K, Laskov C, Hunkeler D, Jeannottat S, Sakaguchi-Söder K, Laaks J, Jochmann MA, Cretnik S, Jager J, Haderlein SB, Schmidt TC, Aravena R, Elsner M (2011) Compound-specific chlorine isotope analysis: a comparison of gas chromatography / isotope ratio mass spectrometry and gas chromatography / quadrupole mass spectrometry methods in an Interlaboratory study. Anal Chem 83:7624–7634

Bobrowski N, Hönninger G, Galle B, Platt U (2003) Detection of bromine monoxide in a volcanic plume. Nature 423:273–276

Boeke HE (1908) Über das Krystallisationsschema der Chloride, Bromide, Jodide von Natrium, Kalium und Magnesium, sowie über das Vorkommen des Broms und das Fehlen von Jod in den Kalisalzlagerstätten. Z Kristallogr 45:346–391

Braitsch O (1962) Entstehung und Stoffbestand der Salzlagerstätten. Springer-Verlag

Brookins DG (1988) Eh-pH diagrams for geochemistry. Springer-Verlag, Berlin Heidelberg

Cao FF, Struchio NC, Ollivier P, Deveau N, Heraty LJ, Jaunat J (2020) Sources and behavior of perchlorate in a shallow chalk aquifer under military (World War I) and agricultural influences. J Haz Mat 398:123072

Carrizo D, Unger M, Holmstrand H, Andersson P, Gustafsson Ö, Sylva SP, Reddy CM (2011) Compound-specific bromine isotope compositions of one natural and six industrially synthesised organobromine substances. Environ Chem 8:127–132

Dasgupta PK, Dyke JV, Kirk AB, Jackson WA (2006) Perchlorate in the United States. Analysis of relative source contributions to the food chain. Environ Sci Technol 40:6608–6614

Diduch M, Pokowska Ź, Namieśnik J (2011) Chemical quality of bottled waters: a review. J Food Sci 76:178–196

Ericksen GE (1981) Geology and origin of the Chilean nitrate deposits. USGS Prof Pap 1188:37 pp

Estrada NL, Anderson TA, Böhlke JK, Gu BH, Hatzinger PB, Mroczkowski SJ, Rao B, Sturchio NC, Jackson WA (2021) Origin of the isotopic composition of natural perchlorate: experimental results for the impact of reaction pathway and initial ClO_x reactant. Geochim Cosmochim Acta 311:292–315

Frost AA (1951) Oxidation potential – free energy diagrams. J Am Chem Soc 73:2680–2683

Garrels RM, Christ CL (1965) Solutions, minerals and equilibria. Harper & Row, New York

Genuino HC, Espino MPB (2012) Occurrence and sources of bromate in chlorinated tap drinking water in Metropolitan Manila, Philippines. Arch Environ Contam Toxicol 62:369–379

Hanley J, Chevrier JF, Berget DJ, Adams RD (2012) Chlorate salts and solutions on Mars. Geophys Res Lett 39:L08201. https://doi.org/10.1029/2012GL051239

Hitzfeld KL, Gehre M, Richnow HH (2011) A novel online approach to the determination of isotopic ratios for organically bound chlorine, bromine and Sulphur. Rapid Commun Mass Spectrom 25:3114–3122

Holleman AF, Buchner EH (1949) Leerboek der anorganische chemie. 15e druk. JB Wolters, Groningen – Batavia

Holmstrand H, Unger M, Carrizo D, Andersson P, Gustafsson Ö (2010) Compound-specific bromine isotope analysis of brominated diphenyl ethers using GC-ICP-multicollector-MS. Rapid Commun Mass Spectrom 24:2135–2142

Holt BD, Sturchio NC, Abrajano TA, Heraty LJ (1997) Conversion of chlorinated volatile organic compounds to carbon dioxide and methyl chloride for isotopic analysis of carbon and chlorine. Anal Chem 69:2727–2733

Horst A, Bonifacie M, Bardoux G, Richnow HH (2019) Isotopic characterization (^{2}H, ^{13}C, ^{37}Cl, ^{81}Br) of abiotic degradation of methyl bromide and methyl chloride in water and implications for future studies. Environ Sci Technol 53:8813–8822

Hu X, Hou XW, Liu Q, Liu JY, Jiang GB (2021) Techniques for stable chlorine/bromine isotope analysis and recent applications in environmental research. Environ Chem 40:331–342. (in Chinese with English abstract)

Jackson WA, Brundrett M, Böhlke JK, Hatzinger PB, Mroczkowski SJ, Sturchio NC (2021) Isotopic composition of natural and synthetic chlorate ($\delta^{18}O$, $\Delta^{17}O$, $\delta^{37}Cl$, $^{36}Cl/Cl$): methods and initial results. Chemosphere 274:129586

Jendrzejewski N, Eggenkamp HGM, Coleman ML (1997) Sequential determination of chlorine and carbon isotopic composition in single microliter samples of chlorinated solvent. Anal Chem 69:4259–4266

Jin B, Laskov C, Rolle M, Haderlein SB (2011) Chlorine isotope analysis of organic contaminants using GCqMS: method optimization and comparison of different evaluation schemes. Environ Sci Technol 45:5279–5286

Kang N, Jackson WA, Dasgupta PK, Anderson TA (2008) Perchlorate production by ozone oxidation of chloride in aqueous and dry systems. Sci Total Environ 405:301–309

Kounaves SP, Stroble ST, Anderson RM, Moore Q, Catling DC, Douglas S, McKay CP, Ming DW, Smith PH, Tamppari LK, Zent AP (2010) Discovery of natural perchlorate in the Antarctic Dry Valleys and its global implications. Environ Sci Technol 44:2360–2364

Kurokawa Y, Maekawa A, Takahashi M, Hayashi Y (1990) Toxicity and carcinogenicity of potassium bromate – a new renal carcinogen. Environ Health Perspect 87:309–335

Mayeno AN, Curran AJ, Roberts RL, Foote CS (1989) Eosinophils preferentially use bromide to generate halogenating agents. J Biol Chem 264:5660–5668

Michael GE, Miday RK, Bercz JP, Miller RG, Greathouse DG, Kraemer DF, Lucas JB (1981) Chlorine dioxide water disinfection: a prospective epidemiology study. Arch Environ Health 36:20–27

O'Brien TF, Bommaraju TV, Hine F (2005) History of the chlor-alkali industry. In: Handbook of Chlor-Alkali technology. Springer, Boston, MA, pp 17–36

Oren A (2010) Lakes and reservoirs. Res Manag 15:215–222

Parker DR, Seyfferth AL, Reese BK (2008) Perchlorate in groundwater: a synoptic survey of "pristine" sites in the conterminous Unites States. Environ Sci Technol 42:1465–1471

Poghosyan A, Morel-Espinosa M, Valentin-Blasini L, Blount BC, Ferreccio C, Steinmaus CM, Sturchio NC (2016) Chlorine isotopic composition of perchlorate in human urine as a means of distinguishing among exposure sources. J Expo Sci Environ Epidemiol 26:324–328

Pourbaix MJN, Van Muylder J, de Zhoubov N (1963) Atlas d'Équilibres Électrochimiques à 25°C. Gauthier-Villars, Paris

Rajagopalan S, Anderson TA, Fahlquist L, Rainwater KA, Ridley M, Jackson WA (2006) Widespread presence of naturally occurring perchlorate in high plains of Texas and New Mexico. Environ Sci Technol 40:3156–3162

Rao B, Hatzinger PB, Bohlke JK, Sturchio NC, Andraski BJ, Eckardt FD, Jackson WA (2010) Natural chlorate in the environment: application of a new IC-ESI/MS/MS method with a (ClO_3^-)-^{18}O internal standard. Environ Sci Tech 44:8429–8434

Schweitzer GK, Pesterfield LL (2010) The aqueous chemistry of the elements. Oxford University Press

Shouakar-Stash O, Drimmie RJ, Frape SK (2003) Stable hydrogen, carbon and chlorine isotope measurements of selected chlorinated organic solvents. J Contam Hydrol 60:211–228

Shouakar-Stash O, Drimmie RJ, Zhang M, Frape SK (2006) Compound-specific chlorine isotope ratios of TCE, PCE and DCE isomers by direct injection using CF-IRMS. Appl Geochem 21:766–781

Sturchio NC, Böhlke JK, Gu BH, Hatzinger PB, Jackson WA (2011) Isotopic tracing of perchlorate in the environment. In: Baskaran M (ed) Handbook of environmental isotope geochemistry, advances in isotope geochemistry. Springer-Verlag, Berlin Heidelberg

Sylva SP, Ball L, Nelson RK, Reddy CM (2007) Compound-specific $^{81}Br/^{79}Br$ analysis by capillary gas chromatography/multicollector inductively coupled plasma mass spectrometry. Rapid Commun Mass Spectrom 21:3301–3305

Takeno N (2005) Atlas of Eh-pH diagrams. Geol Surv Japan Open File Rep N° 419

Tang CM, Tan JH, Fan YJ, Zheng K, Ye ZQ, Peng XZ (2020) Theoretical evaluation of inter-ion and intra-ion isotope effects in fragmentation: insights into chlorine and bromine isotope effects of halogenated organic compounds occurring in electron ionization mass spectrometry. RSC Adv 10:13749

USGS (2022) Mineral commodity summaries 2022. U.S. Geological Survey, 202

Valyashko MG (1956) Geochemistry of bromide in the process of salt deposition and the use of the bromide content as a genetic and prospecting tool. Geochem Int 6:570–587

Van Acker MRMD, Shahar A, Young ED, Coleman ML (2006) GC/multiple collector-ICPMS method for chlorine stable isotope analysis of chlorinated aliphatic hydrocarbons. Anal Chem 78:4663–4667

Van Warmerdam EM, Frape SK, Aravena R, Drimmie RJ, Flatt H, Cherry JA (1995) Stable chlorine and carbon isotope measurements of selected chlorinated organic solvents. Appl Geochem 10:547–552

Wang LC, Hu WX, Wang XL, Cao J, Yao SP (2020) Halogens (Cl, Br, and I) geochemistry in middle Triassic carbonates: implications for salinity and diagenetic alteration of I/(Ca + Mg) ratios. Chem Geol 533:119444

Wassenaar LI, Koehler G (2004) On-line technique for the determination of the $\delta^{37}Cl$ of inorganic and total organic Cl in environmental samples. Anal Chem 76:6384–6388

WHO [World Health Organization] (2005) Chlorite and chlorate in drinking-water. Background document for development of WHO guidelines for drinking-water quality. http://www.who.int/entity/water_sanitation_health/dwq/chemicals/chlorateandchlorite0505.pdf. Accessed 4 Nov 2012

Wisniewski JA, Kabsch-Korbutowicz M, Lakomska S (2011) Donnan dialysis and electrodialysis as viable options for removing bromates from natural water. Desalination 281:257–262

Yankelson I (2019) Microbial and abiotic degradation assessment of 1-bromo-2-chloroethane using multi-elemental C-Br-Cl isotope analysis: laboratory and field study. MSc Thesis Ben-Gurion University of the Negev

Yoshida M, Takahashi K, Yonehara N (1971) The fluorine, chlorine, bromine, and iodine contents of volcanic rocks in Japan. Bull Chem Soc Jpn 44:1844–1850

3 Chlorine and Bromine Concentrations in Different Sample Types and their Distribution and Fluxes Between the Major Reservoirs on Earth

3.1 The Concentrations of Chlorine and Bromine and the Br/Cl Ratios in Geological Samples

3.1.1 Chlorine and Bromine Concentrations in Natural Waters and their Ion Ratio Variations

Within the Earth's crust most chlorine and bromine is present in solution in water. The single largest of these reservoirs are the interconnected oceans. This reservoir contains roughly one third of all chlorine and bromine that is present in the surface reservoirs and is very well mixed. As a result the chloride and bromide concentrations in the open oceans are constant within narrow limits at about 19,353 ppm Cl and 67 ppm Br, giving a Br/Cl ratio of $3.46*10^{-3}$. Only in more restricted parts of the oceans the concentrations significantly deviate from these numbers. In arid regions the concentrations are, due to evaporation of seawater higher, while in colder regions the concentrations may be, due to dilution with rain and river water lower. In the case of evaporation only water molecules are removed from the oceans so that the Br/Cl ratio remains constant during the progress of evaporation. Only when so much water has evaporated that the brine is saturated in NaCl, and halite starts to precipitate, the Br/Cl ratio will change. The reason for this is that bromide is incorporated in halite at a much lower ratio to chloride than present in seawater (Boeke 1908), resulting in a slowly increasing Br/Cl ratio in brine from which salt precipitates.

3.1.1.1 Distribution of Bromine Between Salt Minerals and Brine

The Br/Cl distribution between (seawater) brine and precipitating and dissolving salt minerals has been studied in great detail during the twentieth century, and many studies were published during the 1950s to 1970s (e.g. Valyashko 1956; Braitsch and Hermann 1963; Herrmann et al. 1973). In general however, the bromide distribution in salt minerals was already well understood during the 1960s when they were summarized in Braitsch (1962) and the English translation of the same book (Braitsch 1971). Rittenhouse (1967) summarised chloride and bromide concentrations in oil-field brines. He studied the origin of the systematics and divided the investigated samples into five groups based on their suggested origin. Valyashko (1956) already summarised the most important points that can be observed when bromide is incorporated in salt minerals during its precipitation from a saline brine. These observations were: (1) Bromide is incorporated almost exclusively in chloride salts. The remaining bromide concentrates (more and more) in the residual brine. (2) The amount of bromide that is incorporated in the solid salt is a function of the

H. Eggenkamp, *The Geochemistry of Stable Chlorine and Bromine Isotopes*, Advances in Isotope Geochemistry, https://doi.org/10.1007/978-3-031-75633-7_3

bromide concentration in the brine. (3) The distribution of bromide is characterised, in each different chloride salt, by its own distribution coefficient. (4) All distribution coefficients are less than unity, so stat irrespective of the salt mineral that precipitates, the bromide content in the brine increases. (5) If more than one mineral precipitates from a brine the distribution coefficient in each mineral is the same as if this mineral was the only mineral that crystallised from the brine.

The distribution coefficients of the different salt minerals differ significantly from one to another. Distribution coefficients were determined for saline solutions containing low (<1%) bromide contents by Valyashko and Mandrykina (1952) and later by Braitsch and Hermann (1963). Unfortunately Valyashko and Mandrykina (1952) is only available in Russian and the results from their experiments are cited in Valyashko (1956). In these studies the distribution coefficient is defined as:

$$b = \text{Weight}\%\text{Br}_{(\text{crystal})} / \text{Weight}\%\text{Br}_{(\text{solution})} \quad (3.1)$$

The results of the two studies unfortunately deviate significantly. According to Braitsch and Hermann (1963), Valyashko (1956) determined the bromide content in a brine on the dry residue, and it is not clear how much water has been evaporated exactly, because it is not known if there are any salt minerals with crystal water present in the dry residue. For that reason Braitsch and Hermann (1963) considered the determinations by Valyashko (1956) of limited meaning. Determinations of bromide distributions between salt minerals and brine solutions were at an earlier date also made by Boeke (1908). Boeke (1908) however did his experiments with significantly higher bromide contents (1 to 50%) in his test solutions. When bromide concentrations in a brine are above 1% distribution coefficients increase significantly (so that more bromide is incorporated relatively in the precipitated salt crystals than will be the case at low bromide concentrations) and this results in the observation that the determinations made by Boeke (1908), how important they are in order to understand $Na^+,K^+,Mg^{2+} \| Cl^-,Br^- + H_2O$ phase diagrams, are not meaningful for the evaluation of natural salt systems.

Table 3.1 summarises the distribution coefficients that were published by Braitsch and Hermann (1963), supplemented with a few determinations from Boeke (1908) and D'Ans and Kühn (1940). The distribution coefficients of Boeke (1908) and D'Ans and Kühn (1940) may be of lesser exactness, but were given by Braitsch and Hermann (1963) because no newer data were available. The table shows that the distribution coefficients are not constant irrespective of the chemical composition of the brine, as for example in the case NaCl the coefficient decreases with increasing Mg^{2+} content in the brine. It also

Table 3.1 Br distribution coefficients between salt minerals and brine for various salt minerals. Most data are defined by Braitsch and Hermann (1963). Values for tachhydrite are from Boeke (1908) and for kainite are interpolations of data from Boeke (1908) and D'Ans and Kühn (1940)

Mineral	Br distribution coefficients (salt-brine)		
	Braitsch and Hermann (1963)		
	25 °C	55 °C	83 °C
Halite (NaCl), first precipitate	0.14 ± 0.01		
Halite, at start Na, Mg sulphate precipitate	0.073 ± 0.004		
Halite, at start carnallite precipitate	0.073 ± 0.004	0.078 ± 0.004	0.079 ± 0.004
Sylvite (Na-rich solution)	0.81 ± 0.05		
Sylvite (Mg-rich solution	0.73 ± 0.04	0.77 ± 0.04	0.83 ± 0.04
Carnallite	0.52 ± 0.03	0.56 ± 0.03	0.59 ± 0.03
Kainite	0.25 (0.21–0.27) (Boeke 1908; D'Ans and Kühn 1940)		
Bischofite	0.66 ± 0.03		
Tachhydrite	0.29 (Boeke 1908)		

shows that distribution coefficients increase slightly with increasing temperatures. Most striking however are the large differences between the individual distribution coefficients. They range from 0.07 for halite (NaCl) precipitating from evolved (Mg-rich) brines to 0.8 for sylvite (KCl) precipitating from Na-rich brines.

3.1.1.2 Variations in Br/Cl Ratios in Different Ground Water Sample Types

The distribution coefficients of bromide between salt and brine are important in order to evaluate the Br/Cl ratios in ground waters. For example, because only little bromide is incorporated in NaCl this means that if ground water dissolves halite from a salt deposit the Br/Cl ratio in the resulting groundwater decreases. On the other hand, if late stage brine from which magnesium salts have been precipitated is diluted with fresh water a low concentration groundwater will be formed that still has a high Br/Cl ratio. Thus, the Br/Cl ratio in a ground water will give information on the evolution of the groundwater, from which the Br/Cl ratio is one of more important the determining factors.

Rittenhouse (1967) is one of the first studies to divide oil-filed waters into different groups based on their bromide and chloride (which, in this study, is represented by the total dissolved solids). Five groups were distinguished by Rittenhouse (1967). Group I, waters with Br/Cl ratios close to the ratio in seawater, that were formed by pure dilution or concentration of seawater. Group II, similar to I, but with increased bromide contents, that was supposedly added to the water during early diagenesis. Group III, these are samples with higher dissolved solids than seawater but with lower Br concentrations, presumably originated from the dissolution of halite. Group IV, like group III, but with lower dissolved solids than seawater and lower Br contents. These samples are either diluted group III waters or otherwise water samples that are diluted with water with low Br contents. Finally, group V, waters with high dissolved solids and high Br concentrations. The origin of this group was not always clear to Rittenhouse (1967), but appeared in cases to be altered bitterns.

In order to understand the Cl and Br variability a large number of samples were collected from the literature. Samples with high Cl and Br concentrations were selected based on the presence of Cl and Br isotope measurements too, and will be discussed later in this volume. Samples with lower Cl and Br concentrations were selected based on the study of the variability and due to low concentrations the isotope compositions are not known from these samples, but, from these samples it is very educational to understand the variability of the two halogens in (diluted) water samples.

Figure 3.1 summarises the chloride and bromide concentrations in a large number of water samples, to which also a significant number of mineral and rock samples have been added in order to compare differences in chloride and bromide characteristics between water and solid samples. The mineral samples will be discussed below in the next section. The water samples that are shown in Fig. 3.1 have been split in eight groups based on the origin of the samples. The eight groups are (1) NaCl dominated brines, data have been taken from Eggenkamp and Coleman (2000), Ziegler et al. (2001), Shouakar-Stash et al. (2007), Shouakar-Stash (2008), Stotler et al. (2010), Boschetti et al. (2011), Bagheri et al. (2014), Chen et al. (2014), Wang et al. (2014), Alexeeva et al. (2015), Eissa et al. (2016), Louvat et al. (2016), Giunta et al. (2017), Hanlon et al. (2017), Yu et al. (2019) and Stotler et al. (2021). From most of these samples also the chlorine and bromine isotope compositions were determined that will be discussed later in Chap. 11. (2) Potash (Mg,Ca-Cl dominated) brines. These data have been taken from Shouakar-Stash et al. (2007), Shouakar-Stash (2008), Stotler et al. (2010), Chen et al. (2014), Alexeeva et al. (2015) and Wang et al. (2014). From most of these samples also chlorine and bromine isotope data are known. (3) Dissolved halite brines. These samples were taken from Shouakar-Stash et al. (2007), Shouakar-Stash (2008), Stotler et al. (2010), Boschetti et al. (2011), Chen et al. (2014), Alexeeva et al. (2015), Wang et al. (2014) and Hanlon et al. (2017). Again, from most of these samples both the

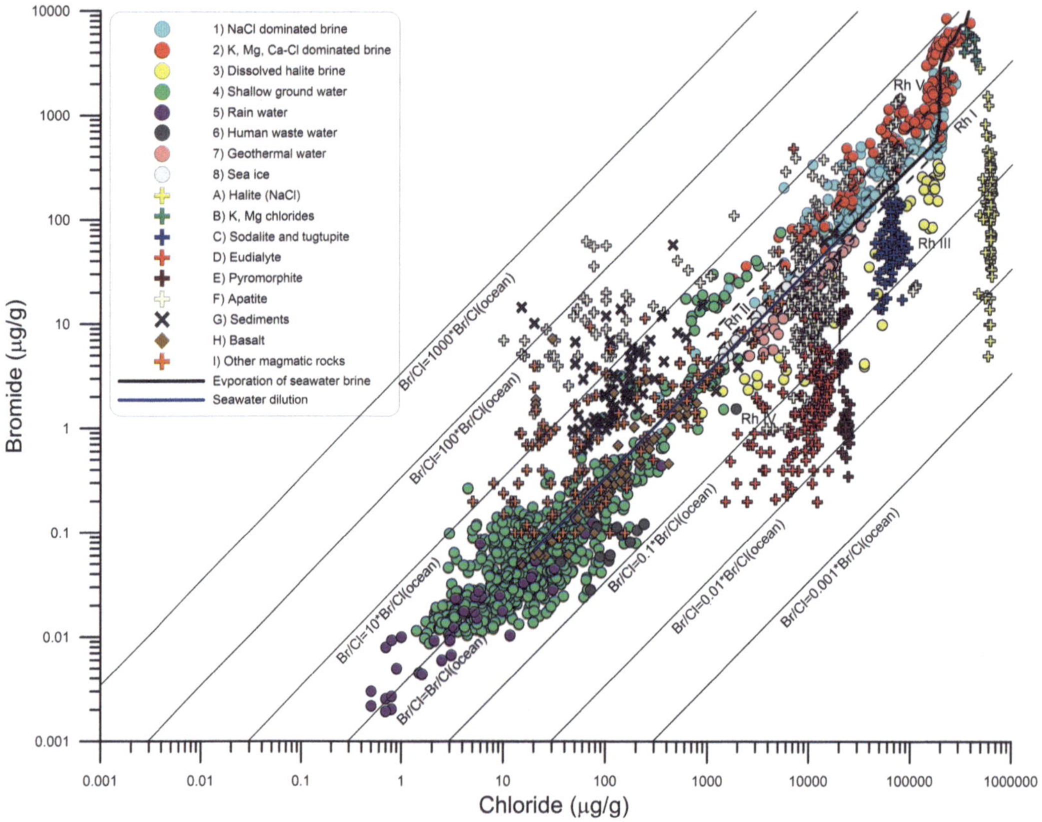

Fig. 3.1 Overview of chloride and bromide concentrations in water, mineral and rock samples

chlorine and bromine isotope compositions are known. (4) Shallow groundwaters, mostly wells. Samples are taken from Behne (1953), McArthur et al. (2012), Gwynne et al. (2013), Zhang et al. (2007), Eissa et al. (2016), Hanlon et al. (2017), Henkemans et al. (2018). Isotope data are only known from a limited subset of these samples. (5) Rain water, samples taken from Behne (1953) and Alcala and Custodio (2008). Samples from Alcala and Custodio (2008) are from Spain and no isotope data are known. (6) Human waste water (from India), these data are taken from McArthur et al. (2012) and represent human urine and septic tank-outlets. (7) Geothermal water, data from Mexico and taken from Pinti et al. (2019, 2020). From these samples both chlorine and bromine isotope data are known. And (8) Sea-ice, data from Vallelonga et al. (2021). Ice from Antarctica from which also isotope data are known.

In Fig. 3.1 it is visible that, with considerable overlap, the different sample groups cover different areas in chlorine-bromine space. In order to compare the collected samples with the groupings from Rittenhouse (1967), the boundaries between the different water types as defined by Rittenhouse (1967) are added to the figure too. The groups from Rittenhouse (1967) are indicated with the text Rh I, Rh II, Rh III, Rh IV and Rh V in the figure. Type 1 samples, NaCl brines do agree mostly with group I, but also group II, from Rittenhouse (1967). These samples, in our case defined as samples with Na > K + Mg + Ca, are mostly samples that are formed by evaporation of seawater, although the bromide content can be higher than was suggested by Rittenhouse (1967). In some cases the bromide content can even be so high that there is overlap with the field covered by Rittenhouse (1967) group V. When this is the case, the NaCl rich waters might have been mixed with potash brines, but only to such an extent that sodium is still the dominating cation. Type 2 samples, potash brines defined by Na < K + Mg + Ca agree for the most part, as expected, to Rittenhouse (1967) group V, and to a lesser extent overlap with the more bromide rich samples from group II. Type 3 samples, dissolved halite brines agree fully with group III and IV samples from Rittenhouse (1967). Type 4 samples have in general low bromide and chloride contents which are generally lower than the concentrations in oil-brines so that there is no overlap. The Br/Cl ratios of type 4 samples agree mostly with seawater, although with a relatively large spread (Br/Cl ratios can be from 10 times higher to 10 times lower than the ratio in seawater). This indicates that most groundwaters originate ultimately from seawater, but that significant mixing with water with different Br/Cl ratios has, in most circumstances, taken place. For example this can be type 6 water, human waste water that generally contains low bromide concentrations. These low bromide concentrations in human waste water can be the result of eating table salt that contains relatively little bromide as compared to seawater. This waste water might have leaked from the sewage puts into shallow groundwater. A few type 4 water samples have relatively high chlorine and bromine concentrations. These samples also have relatively high Br/Cl ratios and they are sampled from Egypt and taken from Eissa et al. (2016) as opposed to McArthur et al. (2012) who took their samples from India. The high chloride and bromide concentrations in these Egyptian samples are explained by deep saline upwelling caused by over-pumping. Where the origin of this deep water is strongly evaporated seawater brine (Eissa et al. 2016). Type 5 waters are rainwater samples which are characterised, in all cases, by low chloride and bromide concentrations. The Br/Cl ratios in rainwater samples vary considerably which may be the result from possible oxidation reactions in the atmosphere, that attack chlorine and bromine species in different amounts. Interestingly however in rain water samples, is that it is striking to observe that the total Br/Cl ratio variations are less than in groundwaters, indicating that in general a more pure seawater origin for rainwater than for groundwaters in general is suggested. The geothermal water from type 7 is taken from Pinti et al. (2019, 2020) and these samples are from Mexico. Based on the $^{3}He/^{4}He$ ratios the origin of these waters is the upper mantle. Br/Cl ratios are slightly below that of seawater, and are explained by the mixture of modified seawater with Br poor mantle water. Finally, type 8 is chloride and bromide measured in Antarctic sea ice (Vallelonga et al. 2021). The Br/Cl ratios are almost equal to that from standard ocean water and indicate that this ice has been formed from ocean water from which chloride and bromide have been frozen out in the same proportion as they were present in the original seawater. This also explains the low chloride and bromide concentrations that are present in sea ices as compared to ocean water.

3.1.2 Br/Cl Ratios Is Minerals and Rocks

Chloride and bromide concentrations have also been collected from a large set of chlorine rich minerals. These data can be compared to the chloride and bromide concentrations in water samples and give significant information on how chlorine and bromine are distributed in different minerals too. Most chlorine and bromine measurements are from minerals that contain chloride as a fixed component in their crystal lattices. For that reason their chloride concentrations plot only on a narrow range on the chloride (X-)axis in Fig. 3.1, and the total bromide concentration differences are in most cases much larger as can be seen on the bromide (Y-)axis in Fig. 3.1. Although chloride concentrations are regularly known from rock and mineral samples, bromide data are much scarcer. Data on chloride and bromide concentrations in minerals and rocks were collected from various literature sources and these are added to Fig. 3.1, so that in this figure Cl vs Br distributions in both water and solid (rock and mineral) samples can be compared. We collected data from the following types of minerals and rocks: (A) Halite (McArthur et al. 2012, Ercan et al. 2019, Eggenkamp et al. 2019a). In the last two studies chlorine and in part bromine isotopes data are also reported. Chloride is a fixed component in halite (NaCl) that nominally contains 60.66% Cl. This is visible in Fig. 3.1 where the range in chloride concentrations is really narrow, while the total range in bromide concentrations is several orders of magnitude (approximately from 2 to 2000 ppm). This range is so large because the bromide content is halite is low, especially in the earliest precipitated halite from a brine, due to its low distribution coefficient, but highly variable as the bromide content of a brine can vary substantially, and the incorporation of bromide in halite is dependent on a relatively constant distribution coefficient. (B) potassium and magnesium chlorides (sylvite [KCl], carnallite [$KMgCl_3.6H_2O$] and bischofite [$MgCl_2.6H_2O$]). These data were taken from Eggenkamp et al. (2019b) and from these samples also chlorine and bromine isotopes are known. These minerals also contain fixed (between 35 and 48%) amounts of chloride, which again is visible from the narrow range in Fig. 3.1, where they plot slightly to the left of the halite samples. Br contents of these salts are very high because they precipitate from brines with extremely high Br concentrations with high distribution coefficients. As a result they plot rather closely to the K,Mg,Ca-brines from which they precipitate. Considering the fact that large amounts of the water present in the brine actually will be incorporated in the salt minerals as crystal water the overall chemical composition of the salt deposits is comparable to the chemical composition of the brine. (C) Sodalite ($Na_4Si_3Al_3O_{12}Cl$) and tugtupite ($Na_4Si_4AlBeO_{12}Cl$), two related minerals from the sodalite group (Hassan and Grundy 1984). Data taken from Eggenkamp et al. (2020, in prep). These magmatic minerals both contain chloride in a fixed amount, at 7.3 and 7.6% respectively. As a result the samples also plot near each other on de chloride axis, while bromide contents again vary significantly more. The total bromide variability is however less than for halite with a total range of about 1 order of magnitude. Interestingly the Br/Cl ratios in sodalite and tugtupite is always less than the ratio in ocean water. (D) Eudialyte group minerals (ideally $Na_{15}Ca_6Fe_3Zr_3Si_{26}O_{72}(H_2O,OH,O)_4Cl_2$, with common replacement of Mn for Fe or Ca, F for Cl and Nb for 1 of the Si atoms). These data have been collected from Eggenkamp et al. (2020, in prep). The chloride content is variable, up to 1.5%, but due to common exchange with fluoride it is in most cases between 0.5 and 1%. This is visible in Fig. 3.1 where the chloride content tops at about 15,000 μg/g, but can be as low as 1000 μg/g in chloride poor samples. Bromide contents are normally low, with Br/Cl ratios significantly below those for sodalite. Br/Cl ratios are almost always less than 0.2 times the ratio in seawater and can be even less than 0.01 times the ratio in seawater. (E) Pyromorphite ($Pb_5(PO_4)_3Cl$), a lead-phosphate-chloride mineral from the apatite group. Data have been taken from Epp et al. (2019) and consist of pyromorphite that has been sampled from German soils. Pyromorphite contains nominally 2.6% Cl, which is visible in Fig. 3.1, where the pyromorphite data plot slightly, but clearly, to the right of the eudialyte samples, with only a small range in possible chloride concentrations. Bromide concentrations in pyromorphite are generally low, with Br/Cl ratios that are at most approximately 0.2 times the Br/Cl ratio in seawater, although they can be even lower than 0.01 times the Br/Cl ratio in seawater. In general in pyromorphite the range in Br/Cl ratios agrees rather well with the Br/Cl range in eudialyte. Pyromorphite is a mineral that forms due to supergene weathering of galena. The halogens (both chlorine and bromine) are incorporated in the pyromorphite structure based on their occurrence in the soil where it is formed, which is, unfortunately not well understood yet (Epp et al. 2019). However, as pyromorphite is formed from the reaction of (ground)water with galena the Br/Cl ratio of the water can be expected to be in the range of shallow groundwater, with Br/Cl between 0.1 and 10 times that of seawater. Considering that the Br/Cl ratio in pyromorphite is mostly less than 0.2 times that of seawater it appears that the distribution coefficient for the incorporation of bromide in pyromorphite is low and apparently comparable to the distribution coefficient of the incorporation of Br in halite, or about 0.1. (F) Apatite [$Ca_5(PO_4)_3(OH,F,Cl)$] has the same mineral structure as pyromorphite, but unlike in pyromorphite, the chloride location in the crystal structure can also be occupied with fluoride or a hydroxide ion [OH^-]. In fact, pure chlorapatite is relatively rare and in the more common apatites this location in occupied by fluoride or hydroxide. This indicates that the total chloride concentration variability in apatite is much larger than that in pyromorphite. Pure chlorapatite contains 6.8% Cl, but in pure fluorapatite or hydroxylapatite this can be reduced to almost zero. As a result the chloride range of apatite is much larger than for the earlier discussed minerals. Apatite data are taken from Behne (1953), Dong (2005) and Kusebauch et al. (2015). Added are also 57 chloride and bromide concentrations from Martian meteorites discussed by Belluci et al. (2017). Samples from Dong (2005) are mostly from China, while a few samples with very low chloride concentrations are from Oka (Canada). The samples from Kusebauch et al. (2015) are from Bamble and Ødegardens Verk in Norway. Most apatite samples are of magmatic origin and as their chlorine isotope composition is well measurable with the SIMS technique there is currently a lot of interest in the chlorine isotope geochemistry of this mineral. The bromide concentrations in apatite are in general relatively high. Br/Cl ratios are on average about 10 times higher than in pyromorphite and eudialyte and comparable to the ratio in sodalite. Br/Cl ratios in apatite can be, also in samples with high chloride contents, significantly above (up to 10 times) the Br/Cl ratio in ocean water. Interestingly, the bromide incorporation in samples with very little chloride can be very high, more than 100 times the Br/Cl ratio in seawater. This suggests that in low chloride apatites the bromide uptake is not governed by chloride, and that the uptake may even be higher. In general however, the high bromide uptake in all apatite samples suggests that the distribution coefficient for the uptake of bromide in apatite is about the highest of all magmatic minerals. Interestingly, the data from Martian meteorites (Belluci et al. 2017) agree well with the data from terrestrial apatite. (G) Chloride and bromide distribution in sedimentary rocks. Behne (1953) published many chloride and bromide concentrations in clay and limestone samples from Germany, Wang et al. (2020) published a significant number of chloride and bromide concentrations in limestone and dolomites from China. Interestingly the bromide content in most sediments is relatively high, with Br/Cl ratios significantly above the ratio in seawater. The Br/Cl ratio in sediments may originate from the Br/Cl ratio of the groundwater where the sedimentary rock formed. The Br/Cl ratio of this water is however expected to have the same characteristic as current shallow ground water. Since the bromide concentration is relatively increased relative to this value, it may be suggested that during compression of the sediment chloride is preferentially expelled from the remaining fluid. Experiments conducted by Strydom et al. (2020) expelling saline fluid from clay during compaction showed however that chloride and bromide are expelled from the clay without any change in ion ratio, so that a different explanation need to be found. No clear explanation is known at the moment, although Behne (1953) already suggested that bromide might have been enriched due to the presence of organic material. Wang et al. (2020) observed that chloride and bromide in limestones and dolomite increases linearly with increasing dolomitization of the rock, which is explained with a possible seawater evaporation effect. According to Wang et al. (2020), higher Br/Cl ratios as compared to seawater may be caused by the input of terrestrial water. However, Wang et al. (2020) do not discuss why this terrestrial water should have higher Br/Cl ratios than seawater. (H) Basalt. A small number of basalt samples with both chloride and bromide concentrations have been collected from Behne (1953), Yoshida et al. (1971), Schilling et al. (1978), Jambon et al. (1995) and Braunger et al. (2021). In most cases the Br/Cl ratio in basalt is close to the Br/Cl ratio in seawater. This suggests that both Cl and Br in basalt (most of them MORB samples) originate from seawater, as a result of seawater contamination. This may confirm the observation by Bonifacie et al. (2008) in their study on chlorine isotopes in MORB that indicate that uncontaminated MORB have very low chloride and bromide concentrations (and low chlorine isotope ratios), and that higher chloride and bromide concentrations (in samples that have chlorine isotope ratios intermediate between the values in low concentration MORB and seawater) are the result of seawater contamination. (I) Other magmatic rocks. Chloride and bromide concentrations were collected from a large range of relatively chlorine poor rock samples published in Behne (1953), Yoshida et al. (1971), Braunger et al. (2021), Han et al. (2023) and Eggenkamp et al. (in prep). Contrary to minerals that have high chloride concentrations, with the exception of apatite, the Br/Cl ratios in these low chloride magmatic rocks regularly show Br/Cl ratios higher than that ratio in seawater. This suggest that, while most minerals with high chloride contents, mostly minerals with a defined chloride location in their crystal lattice, have low Br/Cl distribution coefficients during formation of the mineral, minerals without such a defined chloride location in the crystal lattice incorporate bromide in a relatively larger amount, thus with a larger distribution coefficient. This may be due to the fact that there is no defined chloride location and that the space that can be occupied by both Cl and Br in these magmatic minerals has a higher tolerance for the bromide ion. Of course the distribution coefficient will be different from one to another mineral, but, since the Br/Cl ratios bias to values higher than the Br/Cl ratio in seawater in general the distribution coefficients must be higher than for most minerals with defined (fixed) chloride contents.

3.2 Distribution and Fluxes of Chlorine and Bromine Between Major Reservoirs on Earth

During the period Earth was formed from the solar nebula and was not condensed yet it was not differentiated in the main reservoirs. Once it was condensed it started to differentiate. Heavy elements (mostly iron) sank and Earth was at first differentiated into the core and the bulk silicate earth or primitive mantle. Later, this primitive mantle differentiated in the deeper primitive mantle, the depleted mantle and crust. Regarding the study of the halogens, the crustal reservoir can be split in the oceanic and continental lithospheres and the oceans, where the continental lithosphere is probably better described as being split into three very different reservoirs, being the silicate lithosphere, the (sedimentary) evaporites and the (saline) porewaters found in sediments and rocks. Compared to all other reservoirs, the atmosphere is only a very small reservoir.

3.2.1 Chlorine and Bromine in the Major Global Reservoirs

The study by Schilling et al. (1978) is probably still one of the most comprehensive studies on chlorine and bromine contents of the major reservoirs and fluxes between these reservoirs. In the following sections the chlorine and bromine contents and fluxes between the major reservoirs are described because chlorine and bromine are the main subject of this book. As chlorine and bromine are the only two halogen elements to have two stable isotopes they are the only halogens for which variations in the stable isotope ratios exist and thus for which the fluxes may cause part of the observed isotope variations. Schilling et al. (1978) attempted to establish the origin of chloride and bromide in the surface reservoir. They concluded that chlorine and bromine in the surface reservoirs (defined as the oceans, atmosphere and sediments, including the evaporites) reached their current concentrations due to continuous outgassing over the history of the Earth. Current knowledge more and more agrees on a relatively early and fast outgassing of the earth during its primordial state, so that already during Achaean times most chlorine (and bromine) was stored in the oceans, which would suggest that the chloride concentration in the oceans even decreased since these early times (Knauth 1998, 2005; Hay et al. 2006) due to removal of chloride and bromide through precipitation of evaporites that started approximately 2 Ga ago. These modern studies agree more and more on the fact that the total chlorine in the surface reservoirs has been more or less in steady state since early

Proterozoic times, suggesting also a steady state between the different reservoirs. Considering however the fact that the fluxes between the reservoirs are still not well understood in detail, estimations by Schilling et al. (1978) and Ito et al. (1983) can still be used as estimations for these fluxes.

With respect to the study of chlorine and bromine the following major reservoirs can be taken into account: the core, the primitive (lower) mantle, the depleted mantle, the oceanic lithosphere, the silicate continental lithosphere, the evaporites, saline porewater in sediments and rocks, the oceans and the atmosphere. Overall nine different reservoirs of vastly different sizes.

The core is one of the largest reservoirs, but not much chlorine and bromine exchange between the core and the shallower, silicate rich, reservoirs is expected. Schilling et al. (1978) considered the sedimentary and evaporite reservoirs as one reservoir, however I consider them as two different reservoirs for the purpose of chlorine and bromine isotope studies as within the evaporite reservoir chlorine and bromine are mostly present as solid salts, while in the sediment reservoir they are mostly present as dissolved species in saline porewaters. The known inventory of evaporites has increased considerably since the study of Schilling et al. (1978). The study by Hay et al. (2006), which meticulously prepared an inventory of all evaporites deposited since the Ediacaran period also attempted to estimate the amount of these evaporites eroded since their deposition. They conclude that the total mass of currently present evaporites (as halite) is $3.12*10^{19}$ kg. This amount is equivalent to $1.89*10^{19}$ kg solid chloride, which is even more than the $1.84*10^{19}$ kg chloride estimated by Schilling et al. (1978) for all chloride, both evaporites and dissolved chloride.

Chloride and bromide concentrations are not well known for all reservoirs. They are reasonably well known for the surface reservoirs (and for the oceans they are almost perfectly known), but they become less well known for the deeper reservoirs. Especially for the core concentrations are unknown and they are estimated based on average chlorine and bromine contents of the Earth and in part from concentrations in iron meteorites. Chlorine and bromine data from the various mantle reservoirs are also not well known, and mostly estimated from chlorine and bromine contents in minerals that derive from the mantle. The deeper the reservoir the more difficult it is to estimate its chlorine and bromine contents. For that reason it is also important to know the average chlorine and bromine concentrations of the whole Earth. Estimates for chlorine and bromine are not often given and unfortunately the few published estimations differ widely. According to Morgan and Anders (1980) for example, the global chlorine concentration is 19.9 and the global bromine concentration is 0.106 ppm, while according to McDonough (2001, 2014) the values are 76 and 0.3 respectively. These differences have significant influences on calculation of chlorine and bromine contents for the core and the mantle, which cannot be determined directly.

Table 3.2 shows the estimated amounts of chloride and bromide in the various reservoirs on Earth. Data from seawater are taken from Kendrick et al. (2017), from the atmosphere from Ganapathy and Anders (1974), from the oceanic crust from Gast (1972) and the continental crust from Rudnick and Gao (2003). The data for evaporites are recalculated from data taken from Hay et al. (2006) and Guo and Korenaga (2021). The amount of bromide in evaporites is estimated from average bromide contents in evaporites from Braitsch (1962). The amount of chloride in sediments (mostly present as chloride in formation brines) is estimated from data presented in Land (1995) and Knauth (1998) and is perhaps two to three times as high as the amount of chloride in evaporites. Bromide in these brines is calculated as if the bromide is the amount that is not precipitated in evaporites compared to seawater, so that the concentration of bromide in sediments (brines) is higher than in the oceans and evaporites. Guo and Korenaga (2021) also presented data on chloride and bromide in various sedimentary reservoirs. Their data suggest a much smaller (about 5 times) sedimentary reservoir than based on the data Land (1995) and Knauth (1998). This suggests that Guo and Korenaga (2021), who collected their data from Kendrick et al. (2011),

Table 3.2 Chlorine and bromine contents in the different Earth reservoirs based on different estimations for chlorine and bromine concentrations in the mantle and bulk Earth. The different estimations for the total amount of chlorine and bromine on Earth result in vastly different calculated chlorine and bromine concentration in the mantle and core. Overall however, it is suggested that the chlorine and bromine content in the mantle is low, so for the mantle the data by Kendrick et al. (2017) are followed here. This results in very high chlorine and bromine concentrations in the core that agree however well with the data presented by McDonough (2001). High concentrations in the core may be confirmed by high chloride concentrations observed in iron meteorites (see Chap. 19). Overall I consider the data from this table that are presented in bold as the most probable chlorine and bromine concentration in the different reservoirs. For further considerations I refer to the text in this section. References: a = Ganapathy and Anders (1974); b = Kendrick et al. (2017); c = Land (1995); d = Knauth (1998); e = Hay et al. (2006); f = Guo and Korenaga (2021); g = Rudnick and Gao (2003); h = Gast (1972); i = Morgan and Anders (1980); j = McDonough (2001)

	Reservoir	Mass (kg)	Cl (ppm)	Br (ppm)	Total Cl (kg)	Total Br (kg)	1000*(Br/Cl)	%Cl of total Earth	%Br of total Earth	Ref.
A	**Atmosphere**	**$5.10*10^{18}$**	**$1.6*10^{-3}$**	**$7.8*10^{-6}$**	**$8.16*10^{9}$**	**$3.98*10^{7}$**	**4.88**	**$1.8*10^{-11}$**	**$2.2*10^{-11}$**	a
B	**Oceans**	**$1.40*10^{21}$**	**$1.93*10^{4}$**	**66**	**$2.70*10^{19}$**	**$9.24*10^{16}$**	**3.46**	**6.0**	**5.2**	b
C	**Sediments**	**$1.97*10^{21}$**	**$1.92*10^{4}$**	**98**	**$3.78*10^{19}$**	**$1.92*10^{17}$**	**5.08**	**8.3**	**10.7**	c, d
D	**Evaporites**	**$3.12*10^{19}$**	**$5.50*10^{5}$**	**150**	**$1.71*10^{19}$**	**$4.67*19^{15}$**	**0.27**	**3.8**	**0.3**	e, f
E	**Continental crust**	**$2.09*10^{22}$**	**244**	**0.88**	**$5.10*10^{18}$**	**$1.84*10^{16}$**	**3.61**	**1.1**	**1.0**	g
F	**Oceanic crust**	**$4.80*10^{21}$**	**48**	**0.17**	**$2.30*10^{17}$**	**$8.16*10^{14}$**	**3.54**	**0.05**	**0.05**	h
G^1	**Mantle (1)**	**$4.02*10^{24}$**	**4.8**	**$1.8*10^{-3}$**	**$1.92*19^{19}$**	**$7.13*10^{15}$**	**0.37**	**4.2**	**0.4**	b
G^2	Mantle (2)	$4.02*10^{24}$	71	$5.0*10^{-2}$	$2.85*10^{20}$	$2.02*10^{17}$	0.71			f
G^2A	Depleted MORB mantle	$2.81*10^{24}$	26	$3.2*10^{-2}$	$7.31*10^{19}$	$8.99*10^{16}$	1.23			f
G^2B	Enriched mantle (deep)	$1.21*10^{24}$	175	$9.3*10^{-2}$	$2.12*10^{20}$	$1.13*10^{17}$	0.53			f
A to F + G^1 (BSE^1)	**Bulk silicate Earth (1)**	**$4.05*10^{24}$**	**26**	**$7.6*10^{-2}$**	**$1.05*10^{20}$**	**$3.08*10^{17}$**	**2.92**	**23.2**	**17.2**	**b**
A to F + G^2 (BSE^2)	Bulk silicate Earth (2)	$4.05*10^{24}$	92	0.13	$3.72*10^{20}$	$5.12*10^{17}$	1.38			f
T^1	Total Earth (1)	$5.97*10^{24}$	19.9	0.106	$1.19*10^{20}$	$6.33*10^{17}$	5.33			i
T^2	**Total Earth (2)**	**$5,97*10^{24}$**	**76**	**0.3**	**$4.54*10^{20}$**	**$1.79*10^{18}$**	**3.95**			**j**
T^1-BSE^1	Core (1)	$1.92*10^{24}$	7.1	0.17	$1.36*10^{19}$	$3.25*10^{17}$	23.97			
T^2-BSE^2	Core (2)	$1.92*10^{24}$	43	0.67	$8.17*10^{19}$	$1.28*10^{18}$	15.66			
T^2-BSE^1	**Core (3)**	**$1.92*10^{24}$**	**181**	**0.77**	**$3.49*10^{20}$**	**$1.48*10^{18}$**	**4.26**	**76.8**	**82.8**	

Worden (1996) and Muramatsu et al. (2007) did not take the full sedimentological record into account. Interestingly however is that they do suggest a significantly higher Br/Cl ratio in the sedimentary record as compared to most other reservoirs, indicating that they agree that Br is enriched relative to seawater in the sedimentary record. As they suggest such a small sedimentary reservoir their data on sedimentary reservoirs are not taken into account for the calculation of the total chloride and bromide contents in these reservoirs.

Mantle and core data are much more complex to establish. Recently, Guo and Korenaga (2021) published halogen data for the mantle that distinguished between the enriched mantle (EM) and the depleted MORB mantle (DMM). They established average chloride and bromide contents in EM of 175 and 0.093 respectively, and in DMM of 26 and 0.032 respectively. That results in average contents in the mantle of 71 ppm for chloride and 0.05 for bromide. Different chloride and bromide contents for the mantle can be estimated from Bulk Silicate Earth (BSE, that is the whole Earth with exception of the metallic core) data published by Kendrick et al. (2017) and McDonough (2001). Their estimations for the chloride and bromide concentrations in BSE are significantly lower than the estimates from Guo and Korenaga (2021). Also, some widely different estimates of the average chloride and bromide contents of the total Earth (TE) have been published (Morgan and Anders 1980, McDonough 2001). Data on the chloride and bromide contents in the different reservoirs on Earth are summarised in Table 3.2. Chloride and bromide contents in the core can then be estimated from the various BSE and TE estimates. These estimates, based on different BSE and different TE estimates are also shown in Table 3.2. Due to the large difference in the estimates the suggested chloride contents of the core could be anywhere from less than 0 to 200 and the bromide contents from 0.06 to 0.8 ppm. Concentrations in any reservoir can not be less than zero. Also, averaged Br/Cl ratios need to be within a reasonable range which may not be more than about three times larger than the average terrestrial (Morgan and Anders 1980, McDonough 2001) or solar system abundances (Palme et al. 2014, Lodders 2020) ratios. This, because there are no fractionation mechanisms that do not change the oxidation state of Cl and Br, that can increase the Br/Cl ratios beyond that level. Based upon these considerations, especially considering the calculated Br/Cl ratios, it appears that the most realistic chloride and bromide estimations for TE are from McDonough (2001), and the most realistic chloride and bromide estimates for BSE from Kendrick et al. (2017). These suggestions result in calculated chloride and bromide contents in the core that are respectively 181 and 0.77 ppm. The most realistic suggestions are presented in bold in Table 3.2. These data may suggest that the chloride and bromide concentrations in the mantle reservoirs as estimated by Guo and Korenaga (2021) are probably too high, as using their data in mass balance calculations for the global chloride and bromide inventory results in very low or even negative Cl contents (these calculations are not presented in Table 3.2) and extremely high Br/Cl ratios that appear to be unnatural. These data suggest that the chloride and bromide concentrations in the core and for the total Earth are well described by the estimations determined by McDonough (2001).

Within Table 3.2 a summary of the chloride and bromide contents in the different reservoirs is given. For the deeper reservoirs also a few options are given based on concentration estimates from different researchers. For BSE, the core and TE, where different options are given, the most realistic estimation is presented in bold. Interestingly, the most realistic estimate suggests that the overwhelming part of both chloride and bromide are present in the core. On the contrary, should we follow the much lower chloride and bromide concentrations as estimated for the TE as suggested by Morgan and Anders (1980), which are 19.9 and 0.106 ppm respectively, the calculations would suggest that the mantle might be the largest reservoir for both elements. This estimation was however rejected because from a mass balance calculation the Br/Cl ratio of the core would be unrealistically high with a value of 23.97 as can ce seen in Table 3.2. The large differences between the various chloride and bromide concentration estimates in the deeper reservoirs suggest however that further research on chloride and bromide concentrations from the mantle and core (and total Earth) is highly desirable.

A totally different approach to the chloride content of the mantle was suggested by Graedel and Keene (1996). They estimate that the total chlorine content of the mantle is $4.61*10^{21}$ kg, estimating that 99.6% of all chlorine is present in the mantle. Unfortunately however, their total chloride content in the mantle was based on a misinterpretation by a factor of 10 of the chondritic Cl/Si ratio as established by Cameron (1968). Their value would indicate an average concentration of 1150 ppm for the mantle, about 40 times higher than the estimate that was accepted here as the most realistic. Harben and Kužvart (1996) estimate the total amount of bromine in the mantle as $1.0471*10^{19}$ kg, estimating that 98.9% of all bromine is present in the mantle. This value would indicate that the average bromine concentration in the mantle is 2.6 ppm, several orders of magnitude more than the more realistic values discussed above.

3.2.2 Fluxes of Chlorine and Bromine Between the Major Reservoirs

Some of the chlorine and bromine fluxes between the reservoirs were estimated by Schilling et al. (1978). Some more data on the chlorine fluxes were published by Ito et al. (1983) and nicely visualised by Kaufmann (1984). As the sedimentary reservoir can be considered to exist of as two different reservoirs (solid chlorides in evaporites and dissolved chlorides in sediment brines, with clearly distinguishable characteristics) the figure presented by Kaufmann (1984) could better be presented with these reservoirs separated. Figure 3.2 shows a simplification of the chlorine cycle as presented by Jarrard (2003) with the addition of the evaporite reservoir and the flux of chloride from the continents split between chloride coming from evaporites and the rest of the continents (which is the sum of chloride coming from formation waters and chloride leached from rocks). The erosion of evaporites in this figure has been estimated from Hay et al. (2006). Most fluxes are only very roughly known, and it is still not known if a steady state exists between the mantle and surface reservoirs. Bromine fluxes are

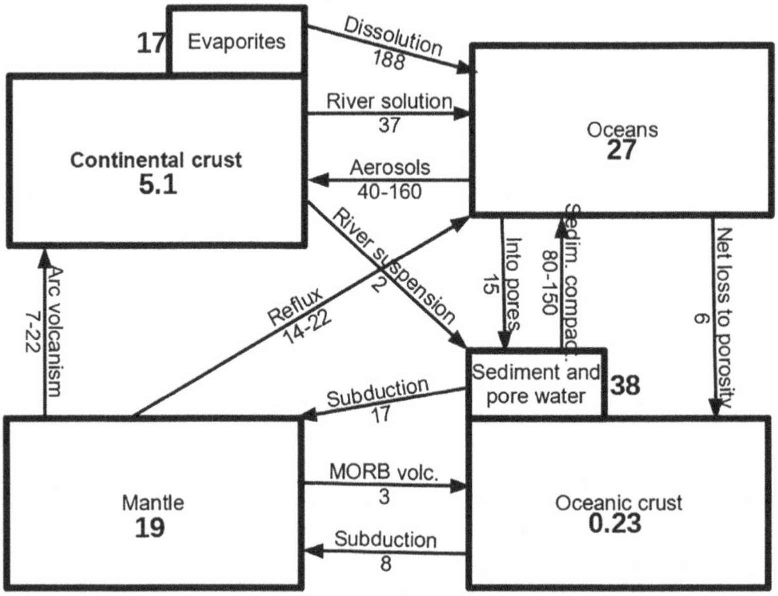

Fig. 3.2 The chlorine cycle of the earth. Chloride fluxes between reservoirs (simplified) are in 10^{12} g/year. The total amounts of chloride per reservoir (also simplified) are in 10^{21} g, with data taken from Table 3.2. Figure after Jarrard (2003) with additional data (evaporites) from Hay et al. (2006)

hardly studied and they have not been added to this figure as they are virtually unknown at the present day. Although estimates could theoretically be deducted from the chloride data and Br/Cl ratios of the reservoirs these were not incorporated into this figure. As a result all fluxes must be considered as very rough estimates.

References

Alcala FJ, Custodio E (2008) Using the Cl/Br ratio as a tracer to identify the origin of salinity in aquifers in Spain and Portugal. J Hydrol 359:189–207

Alexeeva LP, Alexeev SV, Kononov M, Ma T, Liu YD (2015) Chlorine and bromine isotops in brines of the Western Yakutia. Izvestia 13:19–30. (in Russian with English abstract)

Bagheri R, Nadri A, Raeisi E, Eggenkamp HGM, Kazemi GA, Montaseri A (2014) Hydrochemical and isotope (δ^{18}O, δ^{2}H, ^{87}Sr/^{86}Sr, δ^{37}Cl and δ^{81}Br) evidence for the origin of saline formation water in a gas reservoir. Chem Geol 384:62–75

Behne W (1953) Untersuchungen zur Geochemie des Chlor und Brom. Geochim Cosmochim Acta 3:186–214

Belluci JJ, Whitehouse MJ, John T, Nemchin AA, Snape JF, Bland PA, Benedix GK (2017) Halogen and Cl isotopic systematics in Martian phosphates: implications for the Cl cycle and surface halogen reservoirs on Mars. Earth Planet Sci Lett 458:192–202

Boeke HE (1908) Über das Krystallisationsschema der Chloride, Bromide, Jodide von Natrium, Kalium und Magnesium, sowie über das Vorkommen des Broms und das Fehlen von Jod in den Kalisalzlagerstätten. Z Kristallogr 45:346–391

Bonifacie M, Jendrzejewski N, Agrinier P, Humler E, Coleman M, Javoy M (2008) The chlorine isotope composition of earth's mantle. Science 319:1518–1520

Boschetti T, Toscani L, Shouakar-Stash O, Iacumin P, Venturelli G, Mucchino C, Frape SK (2011) Salt waters of the Northern Apennine Foredeep Basin (Italy): origin and evolution. Aquat Geochem 17:71–108

Braitsch O (1962) Entstehung und Stoffbestand der Salzlagerstätten. Springer-Verlag

Braitsch O (1971) Salt deposits, their origin and composition. Springer-Verlag

Braitsch O, Hermann AG (1963) Zur Geochemie des Broms in salinaren Sedimenten: Teil I: Experimentelle Bestimmung der Br-Verteilung in verschiedenen natürlichen Salzsystemen. Geochim Cosmochim Acta 27:361–391

Braunger S, Marks MAW, Wenzel T, Zaitsev AN, Markl G (2021) The petrology of the Tarosero volcanic complex: constraints on the formation of extrusive Agpaitic rocks. J Petrol 62:egab015

Cameron AGW (1968) A new table of abundances of the elements in the Solar System. In: Ahrens LH (ed) Origin and distribution of the elements. Pergamon Press, Oxford, pp 125–143

Chen LZ, Ma T, Du Y, Yang J, Liu L, Shan HM, Liu CF, Cai HH (2014) Origin and evolution of formation water in North China plain based on hydrochemistry and stable isotopes (^{2}H, ^{18}O, ^{37}Cl and ^{81}Br). J Geochem Explor 145:250–259

D'Ans and Kuhn (1940) Über den Bromgehalt von Salzgesteinen der Kalisalzlagerstätten. Z Kali 34:42–46. 59–64, 77–83

Dong P (2005) Halogen-element (F, Cl, and Br) behaviour in apatites, scapolite, and sodalite: an experimental investigation with field applications PhD Thesis University. Saskatchewan, Saskatoon

Eggenkamp HGM, Marks MAW, Bonifacie M, Wenzel T, Stochek G, Bardoux G, Agrinier P, Markl G (in prep) The behaviour of chlorine isotopes and halogens in sodalite and eudialyte group minerals from evolved alkaline magmatic systems

Eggenkamp HGM, Louvat P, Griffioen J, Agrinier P (2019a) Chlorine and bromine isotope evolution within a fully developed upper Permian natural salt sequence. Geochim Cosmochim Acta 245:316–326

Eggenkamp HGM, Louvat P, Agrinier P, Bonifacie M, Bekker A, Krupenik V, Griffioen J, Horita J, Brocks JJ, Bagheri R (2019b) The bromine and chlorine isotope composition of primary halite deposits and their significance for the secular isotope composition of seawater. Geochim Cosmochim Acta 264:13–29

Eggenkamp HGM, Marks MAW, Atanasova P, Wenzel T, Markl G (2020) Changes in halogen (F, Cl, Br, and I) and S ratios in rock-forming minerals as monitors for magmatic differentiation, volatile-loss, and hydrothermal overprint: the case for peralkaline systems. Minerals 10:995

Eggenkamp HGM, Marks MAW, Bonifacie M, Wenzel T, Stochek G, Bardous G, Agrinier P, Markl G (in prep) The behaviour of chlorine isotopes and halogens in sodalite and eudialyte group minerals from evolved alkaline magmatic systems

Eissa MA, Thomas JM, Pohll G, Shouakar-Stash O, Hershey RL, Dawoud M (2016) Groundwater recharge and salinization in the arid coastal plain aquifer of the Wadi Watir delta, Sinai, Egypt. Appl Geochem 71:48–62

Epp T, Marks MAW, Ludwig T, Kendrick MA, Eby N, Neidhardt H, Oelmann Y, Markl G (2019) Crystallographic and fluid compositional effects on the halogen (Cl, F, Br, I) incorporation in pyromorphite-group minerals. Am Mineral 104:1673–1688

Ercan HÜ, Karakaya MÇ, Bozdağ A, Karakaya N, Delikan A (2019) Origin and evolution of halite based on stable isotopes (δ^{37}Cl, δ^{81}Br, δ^{11}B and δ^{7}Li) and trace elements in Tuz Gölü Basin, Turkey. Appl Geochem 105:17–30

Ganapathy R, Anders E (1974) Bulk compositions of the moon and earth, estimated from meteorites. Geochim Cosmochim Acta, Suppl 5:1181–1206

Gast PW (1972) The chemical composition of the earth, the moon and chondritic meteorites. In: Robertson EC (ed) Nature of the solid earth. McGraw-Hill, New York, pp 19–40

Giunta T, Devauchelle O, Ader M, Locke R, Louvat P, Bonifacie M, Métivier F, Agrinier P (2017) The *gravitas* of gravitational isotope fractionation revealed in an isolated aquifer. Geochem Perspect Lett 4:53–58

Graedel TE, Keene WC (1996) The budget and cycle of Earth's natural chlorine. Pure Appl Chem 68:1689–1697

Guo M, Korenaga J (2021) A halogen budget of the bulk silicate earth points to a history of early halogen degassing followed by net regassing. PNAS 118:51

Gwynne R, Frape S, Shouakar-Stash O, Love A (2013) ^{81}Br, ^{37}Cl and ^{87}Sr studies to assess groundwater flow and solute sources in the southwestern Great Artesian Basin, Australia. Prog Earth Planet Sci 7:330–333

Han PY, Rudnick RL, He T, Marks MAW, Wang SJ, Gaschnig RM, Hu ZC (2023) Halogen (F, Cl, Br, and I) concentrations of the upper continental crust through time as recorded in ancient glacial diamictite composites. Geochim Cosmochim Acta 341:28–45

Hanlon C, Stotler R, Frape S, Gwynne R (2017) Comparison of $\delta^{81}Br$ and $\delta^{37}Cl$ composition of volatiles, salt precipitates, and associated water in terrestrial evaporative saline lake systems. Isot Environ Health Stud 53:446–465

Harben PW, Kužvart M (1996) Industrial minerals – a global geology. Industrials Information Ltd Metal Bulletin, PLC London

Hassan I, Grundy HD (1984) The crystal structures of sodalite-group minerals. Acta Crystallogr B Struct Sci 40:6–13

Hay WW, Migdisov A, Balukhovsky AN, Wold CN, Flögel S, Söding E (2006) Evaporites and the salinity of the ocean during the Phanerozoic: implications for climate, ocean circulation and life. Palaeogeogr Palaeoclimatol Palaeoecol 240:3–46

Henkemans E, Frape SJ, Ruskeeniemi T, Anderson NJ, Hobbs M (2018) A landscape-isotopic approach to the geochemical characterization of lakes in the Kangerlussuaq region, West Greenland. Arct Antarct Alp Res 50:1

Herrmann AG, Knake D, Schneider J, Peters H (1973) Geochemistry of modern seawater and brines from salt pans: main components and bromine distribution. Contrib Mineral Petrol 40:1–24

Ito E, Harris DM, Anderson Jr AT (1983) Alteration of oceanic crust and geologic cycling of chlorine and water. Geochim Cosmochim Acta 47:1613–1624

Jambon A, Déruelle B, Dreibus G, Pineau F (1995) Chlorine and bromine abundance in MORB: the contrasting behaviour of the Mid-Atlantic Ridge and East Pacific Rise and implications for chlorine geodynamic cycle. Chem Geol 126:101–117

Jarrard RD (2003) Subduction fluxes of water, carbon dioxide, chlorine, and potassium. Geochem Geophys Geosyst 4. https://doi.org/10.1029/2002GC000392

Kaufmann RS (1984) Chlorine in groundwater: stable isotope distribution. Ph.D. thesis. University of Arizona, Tucson, Ariz

Kendrick MA, Scambelluri M, Honda M, Phillips D (2011) High abundances of noble gas and chlorine delivered to the mantle by serpentinite subduction. Nat Geosci 4:807–812

Kendrick MA, Hémons C, Kamenetsky VS, Danyushevsky L, Devey CW, Rodemann T, Jackson MG, Perfit MR (2017) Seawater cycled throughout Earth's mantle in partially serpentinized lithosphere. Nat Geosci 10:222–228

Knauth LP (1998) Salinity history of the Earth's early ocean. Nature 395:554–555

Knauth LP (2005) Temperature and salinity history of the Precambrian Ocean: implications for the course of microbial evolution. Palaeogeogr Palaeoclimatol Palaeoecol 219:53–69

Kusebauch C, John T, Whitehouse MJ, Engvik AK (2015) Apatite as probe for the halogen composition of metamorphic fluids (Bamble sector, SE Norway). Contrib Mineral Petrol 170:34

Land LS (1995) The role of saline formation water in crustal cycling. Aquat Geochem 1:137–145

Lodders K (2020) Solar elemental abundances, in the Oxford research encyclopedia of planetary science. Oxford University Press

Louvat P, Bonifacie M, Giunta T, Michel A, Coleman M (2016) Determination of bromine stable isotope ratios from saline solutions by "wet plasma" MC-ICPMS including a comparison between high- and high-resolution modes, and three introduction systems. Anal Chem 88:3891–3898

McArthur JM, Sikdar PK, Hoque MA, Ghosal U (2012) Waste-water impacts on groundwater: Cl/Br ratios and implications for arsenic pollution of groundwater in the Bengal Basin and Red River Basin, Vietnam. Sci Tot Environ 437:390–402

McDonough WF (2001) The composition of the earth. In: Teissseyre R, Majewski E (eds) Earthquake thermodynamics and phase transformations in the Earth's interior, vol 76. Academic Press, San Diego, pp 3–23

McDonough WF (2014) Compositional model for the Earth's core. In: Treatise on geochemistry, vol 3.16. 2nd edn, pp 559–577

Morgan JW, Anders E (1980) Chemical composition of Earth, Venus, and Mercury. PNAS 77:6973–6977

Muramatsu Y, Doi T, Tomatu H, Fehn U, Takeuchi R, Matsumoto R (2007) Halogen concentrations in pore waters and sediments of the Nankai trough, Japan: implications for the origin of gas hydrates. Appl Geochem 22:534–556

Palme H, Lodders K, Jones A (2014) Solar system abundances of the elements. In: Davis AM (ed) Planets, asteriods, comets and the Solar System, Volume 2 of Treatise on geochemistry (2nd ed.). Elsevier, pp 15–36

Pinti DL, Castro MC, López-Hernández A, Hernández Hernández MA, Shouakar-Stash O, Richard L, Nuñez-Hernández S, Hall CM, Ramírez-Montes M (2019) Signature of ongoing brine reinjection on noble gas

isotopes and fluid chemistry at Las Tres Vírgenes geothermal field, Mexico. J Volcanol Geotherm Res 377:33–42
Pinti DL, Shouakar-Stash O, Castro MC, Lopez-Hernandez A, Hall CM, Rocher O, Shibata T, Ramírez-Montez M (2020) The bromine and chlorine isotopic composition of the mantle as revealed by deep geothermal fluids. Geochim Cosmochim Acta 276:14–30
Rittenhouse G (1967) Bromine in oil field waters and its use in determining possibilities of origin of these waters. AAPG Bull 17:1213–1228
Rudnick RL, Gao S (2003) Composition of the continental crust. In: Holland HD, Turekian KK (eds) Treatise on geochemistry. The crust, vol 3. Elsevier-Pergamon, Oxford, pp 1–64
Schilling JG, Unni CK, Bender ML (1978) Origin of chlorine and bromine in the oceans. Nature 273:631–636
Shouakar-Stash O (2008) Evaluation of stable chlorine and bromine isotopes in sedimentary formation fluids. PhD Thesis. University of Waterloo
Shouakar-Stash O, Alexeev SV, Frape SK, Alexeeva LP, Drimmie RJ (2007) Geochemistry and stable isotope signatures, including chlorine and bromine isotopes, of the deep groundwaters of the Siberian Platform, Russia. Appl Geochem 22:589–605
Stotler RL, Frape SK, Shouakar-Stash O (2010) An isotopic survey of δ^{81}Br and δ^{37}Cl of dissolved halides in the Canadian and Fennoscandian shields. Chem Geol 274:38–55
Stotler RL, Kirk MF, Newell KD, Goldstein RH, Frape SK, Gwynne R (2021) Stable bromine isotopic composition of coal bed methane (CBM) produced water, the occurrence of enriched ^{81}Br, and implications for fluid flow in the midcontinent, USA. Minerals 11:358
Strydom S, Sterpenich J, Grgic D, Richard A, Eggenkamp HGM, Agrinier P, Louvat P, Mosser-Ruck R, Gaire P, Gaucher EC (2020) Experimental study of chemical evolution and isotope fractionation of Cl and Br in pore water expelled during strong clay compaction. Appl Geochem 140:15274
Vallelonga P, de Gois JS, Borges DLG, Costas-Rodríguez M, Gkinis V, Lannuzel D, Spolaor A, Vanhaecke F (2021) Concentration and isotopic composition of bromine and chlorine in Antarctic Sea ice. Geochim Cosmochim Acta 293:18–27
Valyashko MG (1956) Geochemistry of bromide in the process of salt deposition and the use of the bromide content as a genetic and prospecting tool. Geochem Int 6:570–587
Valyashko MG, Mandrykina TV (1952) Brom v Solyanykh Otlozheniyakh kak Geneticheskii i Poiskovyi Priznak (Bromine in salt deposits as a genetic and prospecting guide). Trudy Vsesoyuznogo nauchno-issled, in-ta galurii, Vyp. 23, L Goskhimizdat (in Russian)
Wang LX, Marks MAW, Keller J, Markl G (2014) Halogen variations in alkaline rocks from the upper Rhine graben (SW Germany): insights into F, Cl and Br behavior during magmatic processes. Chem Geol 380:133–144
Wang LC, Hu WX, Wang XL, Cao J, Yao SP (2020) Halogens (Cl, Br, and I) geochemistry in middle Triassic carbonates: implications for salinity and diagenetic alteration of I/(Ca + Mg) ratios. Chem Geol 533:119444
Worden RH (1996) Controls on halogen concentrations in sedimentary formation waters. Mineral Mag 60:259–274
Yoshida M, Takahashi K, Yonehara N (1971) The fluorine, chlorine, bromine, and iodine contents of volcanic rocks in Japan. Bull Chem Soc Jpn 44:1844–1850
Yu HT, Ma T, Du Y, Chen LZ (2019) Genesis of formation water in the northern sedimentary basin of South China Sea: clues from hydrochemistry and stable isotopes (D, ^{18}O, ^{37}Cl and 81Br). J Geochem Explor 196:57–65
Zhang M, Frape SK, Love AJ, Herczeg AL, Lehmann BE, Beyerle U, Purtschert R (2007) Chlorine stable isotope studies of old groundwater, southwestern great Artesian Basin, Australia. Appl Geochem 22:557–574
Ziegler K, Coleman ML, Howarth RJ (2001) Palaeohydrodynamics of fluids in the Brent Group (Oseberg field, Norwegian North Sea) from chemical and isotopic compositions of formation waters. Appl Geochem 16:609–632

4 Chlorine and Bromine Isotopes, Measurement Scales and Reference Standards

Chlorine and bromine are the only halogen elements that have two stable isotopes, and from which it thus is possible to determine stable isotope ratio variations.

4.1 Isotopes of Chlorine and Bromine

4.1.1 Isotopes of Chlorine

4.1.1.1 Stable Isotopes of Chlorine

Chlorine is the second element of the halogen group (group 17 of the periodic table) and is positioned between sulphur and argon in period 3 of the table. Its nucleus consists of 17 protons. Its two stable isotopes have been discovered by Aston (1919), ^{35}Cl contains 18 neutrons and ^{37}Cl contains 20 neutrons. According to the latest report from the IUPAC (Meija et al. 2016) ^{35}Cl is the most common stable isotope with a representative isotopic abundance between 75.5 and 76.1% of all chlorine, while ^{37}Cl consists of the remaining 23.9 to 24.5%.

4.1.1.2 Radioactive Isotopes of Chlorine

Aside from the two stable isotopes 25 radioactive isotopes of chlorine are known. These isotopes have mass numbers 28 to 52, thus they contain 11 to 35 neutrons. The isotopes ^{28}Cl to ^{30}Cl are unbound and decay by proton decay with extremely low half-lives that are shorter than the time-of-flight of 170 ns. They are found beyond the so called proton drip-line, which can be considered as the edge of the "stable" radioactive isotopes. The isotopes ^{31}Cl to ^{34}Cl decay by electron capture or β^+ decay and the isotopes ^{38}Cl to ^{52}Cl decay by β^- decay. All these isotopes have (very) short half-lives, the longest living of these is ^{39}Cl with a half-life of 56.2 min. ^{36}Cl which can decay with both β^- decay to ^{36}Ar or with electron capture to ^{36}S has a much longer half-life of 301,300 years. This cosmogenic isotope is formed by cosmic radiation in the upper atmosphere and is present in environmental chlorine at a ratio of about $7*10^{-13}$ compared to ^{37}Cl. The relatively long half-life of ^{36}Cl makes it useful for radioactive dating, especially for groundwater samples up to about one million years old. Apart from ^{36}Cl also ^{34m}Cl, ^{38}Cl and ^{39}Cl are formed on Earth though the action of cosmic rays. Due to their short half-lives they have no application in geological studies. The discovery of the chlorine isotopes known to exist until 2010 is summarised in Thoenessen (2012). Table 4.1 gives a summary of the known stable and radioactive chlorine isotopes.

Interestingly, in early studies on the existence of different chlorine isotopes it was suggested that ^{39}Cl might be a naturally occurring isotope too. Although in his first publication on the existence of different chlorine isotopes only ^{35}Cl and ^{37}Cl were observed (Aston 1919), in a subsequent

H. Eggenkamp, *The Geochemistry of Stable Chlorine and Bromine Isotopes*, Advances in Isotope Geochemistry, https://doi.org/10.1007/978-3-031-75633-7_4

Table 4.1 Some properties of the ground states of chlorine isotopes. Data are taken from Kondev et al. (2021), Huang et al. (2021) and Wang et al. (2021). Discovery information is taken from Thoenessen (2012) and subsequent discoveries published on https://people.nscl.msu.edu/~thoennes/isotopes/. # in the Mass column indicates that the mass value and its uncertainty are not derived from purely experimental data, but at least partly from trends from the mass surface (Huang et al. 2021). # in the Half-life column means that the half-life is not experimentally determined but estimated from trends in neighbouring nuclei (Kondev et al. 2021). Half-life units are the following: zs (zeptosecond) = 10^{-21} s, ns (nanosecond) = 10^{-9} s, ms (millisecond) = 10^{-3} s, s (second), m (minute) = 60 s, ky (1000 years) = 3.1556926*10^{10} s (Kondev et al. 2021). The half-lives of ^{28}Cl and ^{30}Cl are unknown but a lower limit of 100 ns is given in Kondev et al. (2021), with the warning that this may be far shorter than the actual half-life. Half-lives for ^{48}Cl to ^{52}Cl are not known and presented values are estimated. In order to easier compare the half-lives of the different short living isotopes the half-life (without uncertainty) is also given as the logarithm of the half-life in seconds in column 6

Isotope	Z	N	Mass	Half-life	Half-life, ^{10}log(s)	Decay mode (and intensity)	Discovery
^{28}Cl	17	11	28.030350(470)#	? unc. > 100 ns	> −7	p (100%)	Mukha et al. (2018)
^{29}Cl	17	12	29.015050(200)#	5.4 ± 1.9 zs	−20,27	p (100%)	Mukha et al. (2018)
^{30}Cl	17	13	30.005018(26)	? unc. > 100 ns	> −7	p (100%)	Mukha et al. (2018)
^{31}Cl	17	14	30.992448(4)	190 ± 1 ms	−0.72	β^+ (100%)	Benenson et al. (1977)
^{32}Cl	17	15	31.9856846(6)	298 ± 1 ms	−0.53	β^+ (100%)	Glass and Jensen (1953)
^{33}Cl	17	16	32.9774520(4)	2.5038 ± 0.0022 s	0.40	β^+ (100%)	Hoag (1940)
^{34}Cl	17	17	33.97376249(5)	1.5267 ± 0.0004 s	0.18	β^+ (100%)	Frisch (1934)
^{35}Cl	17	18	34.96885269(4)	Stable			Aston (1919)
^{36}Cl	17	19	35.96830682(4)	301.3 ± 1.5 ky	12.98	β^- (98.1 ± 0.1%), β^+ (1.9 ± 0.1%)	Grahame and Walker (1941)
^{37}Cl	17	20	36.96590257(6)	Stable			Aston (1919)
^{38}Cl	17	21	37.96801041(11)	37.230 ± 0.014 m	3.35	β^- (100%)	Kennedy and Seaborg (1940)
^{39}Cl	17	22	38.9680082(19)	56.2 ± 0.6 m	3.53	β^- (100%)	Haslam et al. (1949)
^{40}Cl	17	23	39.970420(30)	1.35 ± 0.03 m	1.91	β^- (100%)	Morinaga (1956)
^{41}Cl	17	24	40.970680(70)	38.4 ± 0.8 s	1.58	β^- (100%)	Artukh et al. (1971)
^{42}Cl	17	25	41.973340(60)	6.8 ± 0.3 s	0.83	β^- (100%)	Artukh et al. (1971)
^{43}Cl	17	26	42.974060(70)	3.13 ± 0.09 s	0.50	β^- (100%)	Kashy et al. (1976)
^{44}Cl	17	27	43.978010(90)	562 ± 106 ms	−0.25	β^- (100%)	Westfall et al. (1979)
^{45}Cl	17	28	44.980390(150)	413 ± 24 ms	−0.38	β^- (100%)	Westfall et al. (1979)
^{46}Cl	17	29	45.985250(100)	232 ± 2 ms	−0.63	β^- (100%)	Guillemaud-Mueller et al. (1989)
^{47}Cl	17	30	46.989720(220)#	101 ± 5 ms	−1.00	β^- (100%)	Guillemaud-Mueller et al. (1989)
^{48}Cl	17	31	47.995410(540)#	30# ms (± > 200 ns)	−1.52	β^- (?)	Guillemaud-Mueller et al. (1989)
^{49}Cl	17	32	49.000790(430)#	35# ms (± > 200 ns)	−1.46	β^- (?)	Guillemaud-Mueller et al. (1989)
^{50}Cl	17	33	50.008270(430)#	10# ms (± > 620 ns)	−2.0	β^- (?)	Tarasov et al. (2009)
^{51}Cl	17	34	51.015340(750)#	5# ms (± > 200 ns)	−2.3	β^- (?)	Lewitowicz et al. (1990)
^{52}Cl	17	35	52.024000(750)#	2# ms (± > 200 ns)	−2.7	β^- (?)	Tarasov et al. (2018)

paper that summarises all results on the discoveries of isotopes obtained until the end of 1920, Aston (1920a) reported the provisional existence of ^{39}Cl as a natural chlorine isotope. This, because in all spectra taken from chlorine a faint line was observed that corresponded with mass 39. For that reason, in his summary on observed isotopes until the end of 1920 (Aston 1920b) ^{39}Cl was included as being a provisional isotope of chlorine pending confirmation in later studies. In the International Table of Isotopes, 1923 (Aston et al. 1923) ^{39}Cl was not included, presumably because no further observations were made. ^{39}Cl as a possible naturally occurring isotope of chlorine survived however for a significant time. During the 1930s an isotope ^{39}Cl was apparently observed regularly in the mass spectrographically obtained spectra of HCl, where a mass 40 (H^{39}Cl?) was observed in a ratio of about 1:1850 relative to ^{37}Cl and 1:6000 relative to ^{35}Cl (Becker 1930; Hettner and Böhme 1931; Kallman and Lasareff 1932). In later studies this provisional isotope was no longer found. Nier and Hanson (1936) studied the spectrum of HCl and found no observable peak in the 39 mass range. They refer to Ashley and Jenkins (1931) and Hardy and Sutherland (1932) who were also unable to determine the ^{39}Cl peak. Nier and Hanson (1936) concluded, that even if ^{39}Cl were present its ratio to ^{35}Cl must be less than 1:20,000. As shown in Table 4.1, ^{39}Cl was finally discovered in 1949 (Haslam et al. 1949) and it was shown that it has only a half-life of 56 min. Although this isotope is formed by way of cosmic rays in the atmosphere, the half-live is too low to ever acquire a concentration large enough to determine with mass spectrographs. From then on it was clear that ^{39}Cl could not be a naturally occurring isotope with a significant, and measurable, isotope abundance.

4.1.2 Isotopes of Bromine

4.1.2.1 Stable Isotopes of Bromine

Bromine is the third of the group 17 elements (halogens) and is positioned between selenium and krypton in period 4 of the periodic table. Its nucleus has 35 protons and the two stable isotopes, (^{79}Br and ^{81}Br) have 44 and 46 neutrons respectively. Both isotopes have about comparable abundances although the lighter ^{79}Br is slightly more common than the heavier ^{81}Br with a representative isotope abundance between 50.5 and 50.8%. The representative isotope abundance of ^{81}Br is between 49.2 and 49.5% (Meija et al. 2016).

4.1.2.2 Radioactive Isotopes of Bromine

Aside from the two stable isotopes 30 radioactive isotopes are known. These have mass numbers 68 to 101, thus containing 33 to 66 neutrons. The isotopes up to ^{69}Br decay by proton decay. These isotopes have extremely low half-lives. They are found beyond the proton drip-line, which can be considered as the edge of the "stable" (particle bound) radioactive isotopes. The isotopes ^{70}Br to ^{77}Br decay mostly by positron or β^+ decay and the isotopes ^{82}Br to ^{101}Br decay by electron or β^- decay. ^{78}Br and ^{80}Br can decay by either β^+ or β^- decay. All these isotopes have (very) short half-lives, the longest living of these is ^{77}Br with a half-life of only 57.04 h. Unlike chlorine there is no radioactive isotope with a half-life long enough to be of geological importance. Gross et al. (2012) described the discovery of the bromine isotopes known up to 2010. Table 4.2 gives a summary of the known stable and radioactive bromine isotopes.

Table 4.2 Some properties on the ground states of bromine isotopes. Data are taken from Kondev et al. (2021), Huang et al. (2021) and Wang et al. (2021). Discovery information is taken from Gross et al. (2012) and subsequent discoveries published on https://people.nscl.msu.edu/~thoennes/isotopes/. # in the Mass column indicates that the mass value and its uncertainty are not derived from purely experimental data, but at least partly from trends from the mass surface (Huang et al. 2021). # in the Half-life column means that the half-life is not experimentally determined but estimated from trends in neighbouring nuclei (Kondev et al. 2021). Half-life units are the following: ns (nanosecond) = 10^{-9} s, ms (millisecond) = 10^{-3} s, s (second), m (minute) = 60 s, h (hour) = 3600 s (Kondev et al. 2021). ^{65}Br to ^{67}Br have not been discovered yet, but estimated data are presented in Kondev et al. (2021) and Wang et al. (2021). In column 8 this is stated with n.d. (not discovered). Half-lives for ^{95}Br to ^{98}Br are not known and presented values are estimated. In order to easier compare the half-lives of the different short living isotopes the half-life (without uncertainty) is also given as the logarithm of the half-life in seconds in column 6

Isotope	Z	N	Mass	Half-life	Half-life, ^{10}log(s)	Decay mode (and intensity)	Discovery
^{65}Br	35	30	64.982300(540)#	? unc. < 410 ns		p (?)	n.d. [Blank et al. 2016]
^{66}Br	35	31	65.974700(430)#	? unc. < 410 ns		p (?)	n.d. [Blank et al. 2016]
^{67}Br	35	32	66.965080(320)#			p (?)	n.d.
^{68}Br	35	33	67.958360(280)#	50 ns (estimate)	−7.3	p (?)	Wimmer et al. (2019)
^{69}Br	35	34	68.950340(50)	< 24 ns	< −7.6	p (100%)	Rogers et al. (2011)
^{70}Br	35	35	69.944792(16)	78.8 ± 0.3 ms	−1.10	β^+ (100%)	Alburger (1978)
^{71}Br	35	36	70.939342(6)	21.4 ± 0.6 s	1.33	β^+ (100%)	Vosicki et al. (1981)
^{72}Br	35	37	71.9365946(11)	78.6 ± 2.4 s	1.90	β^+ (100%)	Nolte et al. (1970)
^{73}Br	35	38	72.931673(7)	3.4 ± 0.2 m	2.31	β^+ (100%)	Murray et al. (1970)
^{74}Br	35	39	73.929910(6)	25.4 ± 0.3 m	3.18	β^+ (100%)	Hollander (1953)
^{75}Br	35	40	74.925811(5)	96.7 ± 1.3 m	3.76	β^+ (100%)	Woodward et al. (1948)
^{76}Br	35	41	75.924542(10)	16.2 ± 0.2 h	4.77	β^+ (100%)	Fultz and Pool (1952)
^{77}Br	35	42	76.921379(3)	57.04 ± 0.12 h	5.31	β^+ (100%)	Woodward et al. (1948)
^{78}Br	35	43	77.921146(4)	6.45 ± 0.04 m	2.59	β^+ (~100%), β^- (<0.01%)	Bothe and Gentner (1937a)
^{79}Br	35	44	78.9183376(11)	Stable			Aston (1920a)
^{80}Br	35	45	79.9185298(11)	17.68 ± 0.02 m	3.03	β^+ (8.3 ± 0.2%), β^- (91.7 ± 0.2%)	Bothe and Gentner (1937b)
^{81}Br	35	46	80.9162882(10)	Stable			Aston (1920a)
^{82}Br	35	47	81.9168018(10)	35.282 ± 0.007 h	5.10	β^- (100%)	Snell (1937)
^{83}Br	35	48	82.915175(4)	2.374 ± 0.004 h	3.93	β^- (100%)	Snell (1937)
^{84}Br	35	49	83.916496(28)	31.76 ± 0.08 m	3.28	β^- (100%)	Born and Seelman-Eggebert (1943)
^{85}Br	35	50	84.915646(3)	2.90 ± 0.06 m	2.24	β^- (100%)	Born and Seelman-Eggebert (1943)
^{86}Br	35	51	85.918805(3)	55.1 ± 0.4 s	1.74	β^- (100%)	Stehney and Steinberg (1962)
^{87}Br	35	52	86.920674(3)	55.68 ± 0.12 s	1.75	β^- (100%)	Born and Seelman-Eggebert (1943)
^{88}Br	35	53	87.924083(3)	1.53 ± 0.06 s	0.18	β^- (100%)	Sugarman (1949)
^{89}Br	35	54	88.926705(4)	4.357 ± 0.022 s	0.64	β^- (100%)	Perlow and Stehney (1959)
^{90}Br	35	55	89.931293(4)	1.910 ± 0.010 s	0.28	β^- (100%)	Perlow and Stehney (1959)

(continued)

Table 4.2 (continued)

Isotope	Z	N	Mass	Half-life	Half-life, $^{10}\log(s)$	Decay mode (and intensity)	Discovery
^{91}Br	35	56	90.934399(4)	543 ± 4 ms	−0.27	β^- (100%)	Kratz and Herrmann (1974)
^{92}Br	35	57	91.939632(7)	314 ± 16 ms	−0.50	β^- (100%)	Kratz and Herrmann (1974)
^{93}Br	35	58	92.943220(460)	152 ± 8 ms	−0.82	β^- (100%)	Vosicki et al. (1981)
^{94}Br	35	59	93.948850(220)#	70 ± 20 ms	−1.15	β^- (100%)	Vosicki et al. (1981)
^{95}Br	35	60	94.952930(320)#	80# ms (± > 300 ns)	−1.1	β^- (?)	Bernas et al. (1997)
^{96}Br	35	61	95.958980(320)#	20# ms (± > 300 ns)	−1.7	β^- (?)	Bernas et al. (1997)
^{97}Br	35	62	96.963500(430)#	40# ms (± > 300 ns)	−1.4	β^- (?)	Bernas et al. (1997)
^{98}Br	35	63	97.969890(430)#	15# ms (± > 400 ns)	−1.8	β^- (?)	Ohnishi et al. (2010)
^{101}Br	35	66					Sumikama et al. (2021)

4.2 Measurement Scales Utilised for Reporting Chlorine and Bromine Isotope Data

There are several different methods with which stable isotope data can be reported. These reporting methods depend on the method how the isotope data were measured and importantly in a way that data can be compared to similarly determined data. In principle there are four methods that were used in the past to report chlorine or bromine isotope data. In the early years of isotope studies isotope variations ware reported through differences in the atomic mass of the different samples. This was in part done as at the very start of the isotope studies (Harkins and Wilson 1915) it was not even known if, and if so which, different isotopes would actually exist and knowledge on the isotope composition is not necessary to determine the atomic mass. With this technique differences in atomic masses are determined with the use of very delicate balances where exact amounts of matter are weighted. This technique was dominant in the period until approximately 1930. Important studies during this period were those of Curie (1921), Harkins and Hayes (1921), Dorenfeldt (1922) and Harkins and Stone (1926). No natural isotope variations were observed in these studies within the analytical precision, but with the application of diffusion techniques to separate isotopes both heavy and light fractions with measurably different atomic masses were actually obtained.

Using mass spectrography the existence of the two stable isotopes of chlorine (Aston 1919) and bromine (Aston 1920a) was established. Once it was established which isotopes did exist, and mass spectrography and mass spectrometry were further developed intensities of the individual isotopes could be measured. Since then the ratio of the individual isotopes could be used to report the data. Many studies were done to determine the exact isotope ratios of chlorine and bromine (Aston 1931; Nier and Hanson 1936; Owen and Schaeffer 1955; Cameron and Lippert 1955). Since about 1960 plain isotope ratios are no longer used to report their variations in geological samples and they have been replaced by the more precise delta notation that is determined from the ratio of the isotope ratios of a sample versus a standard, and for which (exact) actual ratios are irrelevant. Plain isotope ratios are still reported in astronomy. In astronomy data are mostly reported as isotope ratios. In that field it is done because the two isotopes are measured separately and then the intensities are divided upon each other. The precision of these measurements is not comparable to the precision that is obtained by mass spectrometry of geological samples.

Isotope data can also be reported as the atomic fraction of one (or both) of the isotopes (Langvad 1954; Boyd et al. 1955; Cameron et al. 1956). The atom fraction (or percentage) is mostly reported in studies where one of the isotopes is experimentally enriched, since a rather high enrichment was necessary in order to show a significant difference with the techniques in use during the 1950s.

Starting with Hoering and Parker (1961) the standard delta notation is mostly used to report isotope data from samples measured by mass spectrometry. Variations in isotope data are reported in the familiar delta notation. The definition for chlorine as originally expressed was (Hoering and Parker 1961, following the original proposal in McKinney et al. 1950):

$$\delta^{37}Cl = \frac{\left({}^{37}Cl / {}^{35}Cl\right)_{sample} - \left({}^{37}Cl / {}^{35}Cl\right)_{standard}}{\left({}^{37}Cl / {}^{35}Cl\right)_{standard}} * 1000 \quad (4.1)$$

and the definition for bromine as originally expressed was (Eggenkamp and Coleman 2000):

$$\delta^{81}Br = \frac{\left({}^{81}Br / {}^{79}Br\right)_{sample} - \left({}^{81}Br / {}^{79}Br\right)_{standard}}{\left({}^{81}Br / {}^{79}Br\right)_{standard}} * 1000 \quad (4.2)$$

However, although the above equations are still regularly encountered in literature, the extraneous factor 1000 should be omitted because the generally used unit per mil (‰) already indicates that the value is one part per thousand (Coplen 2011; Brand 2011). As a result the equations cited above should be avoided and the following delta equations are recommended:

$$\delta^{37}Cl = \frac{\left({}^{37}Cl / {}^{35}Cl\right)_{sample} - \left({}^{37}Cl / {}^{35}Cl\right)_{standard}}{\left({}^{37}Cl / {}^{35}Cl\right)_{standard}} \quad (4.3)$$

for chlorine, and

$$\delta^{81}Br = \frac{\left({}^{81}Br / {}^{79}Br\right)_{sample} - \left({}^{81}Br / {}^{79}Br\right)_{standard}}{\left({}^{81}Br / {}^{79}Br\right)_{standard}} \quad (4.4)$$

for bromine.

4.2.1 Conversion Factors Between the Various Measurement Scales for Chlorine

So, there are four possible scales to report chlorine isotope data, the relative atomic mass (A_r), the absolute isotope ratio, the atom percentage of either of the isotopes and the delta scale. The absolute atomic ratio can be reported as either $^{35}Cl/^{37}Cl$ or $^{37}Cl/^{35}Cl$. The atom percentage can be reported of either $\%^{35}Cl$ or $\%^{37}Cl$. Although in theory it is possible to calculate the $\delta^{35}Cl$ of a sample (based on $^{35}Cl/^{37}Cl$ ratios) in practice this has never been done, and only $\delta^{37}Cl$ is reported. Overall, this leads to 6 possible scales with which chlorine isotope numbers were reported in the various literature studies. The natural variation of chlorine isotope ratios on Earth is relatively small. The far majority of chlorine isotope values (reported as $\delta^{37}Cl$) are between −1 and + 1‰ vs. SMOC (Standard Mean Ocean Chloride), although the most extreme values for inorganic chlorides are between approximately −8 and + 20‰. Perchlorates can be more negative and values down to −15‰ have been observed. Within the solar system the observed variability is much larger, with for example Lunar data that show values between +4 and + 29‰ and from perchlorates from Mars the observed data range is between −81 and −1‰ (Liuzzi et al. 2021). While these values were reported in per mil versus SMOC, values reported on other scales can also vary significantly. Diffusion experiments can lead to large differences in isotope ratios. Harkins and Hayes (1921) reported the results from their diffusion experiments in differences from the standard relative atomic mass of chlorine. In the heavy fraction the relative atomic mass was increased by 0.03 to more than 0.04 units (Harkins and Ligget 1924), in the light fraction it was decreased by 0.04 mass units (Harkins 1925). In a series of diffusion experiments initiated by Alfred Klemm in the 1950s (e.g. Klemm and Lundén 1955; Herzog and Klemm 1958) the atom percentage of ^{37}Cl could be decreased to values as low as 8.4%. Chlorine isotope ratios ($^{35}Cl/^{37}Cl$) in stars and interstellar medium vary from as low as 1 to as high as 5, as compared to 3.13 for the solar system (Maas and Pilachowski 2018). The data reported on different measurement scales are difficult to compare, and for that reason it is useful to present conversion scales for the different the different scales.

Because variations in chlorine isotope ratios differ greatly, from very small variations in most terrestrial samples (only a few ‰ on the delta scale) to several units on the $^{35}Cl/^{37}Cl$ ratio scale for stars and interstellar medium here we present a few tables with conversions between the different scales from small to large variations. In order to make these conversions it is necessary to know the absolute atomic masses of ^{35}Cl and ^{37}Cl and their isotope ratio in seawater (SMOC). This is necessary to calculate the relative atomic weight of a sample and the $\delta^{37}Cl$. The absolute atomic masses were last evaluated by Wang et al. (2021). The $^{37}Cl/^{35}Cl$ isotope ratio was determined by Xiao et al. (2002) based on the comparison with NIST-SRM 975. In this study the $\delta^{37}Cl$ of NIST-SRM 975 was established as +0.34‰ vs. SMOC following extensive TIMS measurements. In this same study measurements done in Tucson, Arizona reported a $\delta^{37}Cl$ value of +0.43‰ for NIST-SRM 975, based on three measurements by IRMS. This last value is accepted in Brand et al. (2014) as the exact value defining the SMOC scale. Based on the much more extensive testing by TIMS in the Xining laboratory reported in Xiao et al. (2002) it may be better to accept that the $\delta^{37}Cl$ of NIST-SRM 975 was +0.34‰, although this discrepancy, that actually was reported in the same published article, shows clearly that new, and more precise determinations of the standard reference materials are necessary. In the conversion scales presented in this chapter the $^{37}Cl/^{35}Cl$ isotope ratio for mean seawater of 0.319409 as determined by Xiao et al. (2002) will be utilised.

In Table 4.3 conversion values for the chlorine isotope ratios, the atom percentages and the relative atomic weight are given for $\delta^{37}Cl$ values between −100 and + 100‰, spanning roughly the total range of $\delta^{37}Cl$ values known from solar system bodies.

In Table 4.4 conversions are given from the $^{35}Cl/^{37}Cl$ ratios to the delta scale, the relative atomic mass and the isotope's atom percentages for the full scale from almost no ^{35}Cl to almost no ^{37}Cl. That way it is possible to get an idea on how the different scales look like over the full possible range. For as far as is known in stars and interstellar matter $^{35}Cl/^{37}Cl$ values range from approximately 1 to approximately 5. Values beyond those values can be reached by isotope separation techniques.

4.2.2 Conversion Factors Between the Various Measurement Scales for Bromine

In principle from bromine the same measurement scales as for chlorine can be applied. However, since bromine isotopes are much less frequently measured in the past than chlorine isotopes, all of them have been less encountered than the comparative scales for chlorine. For as far as is known there are no papers describing variations in bromine isotope ratios with the use of the relative atomic mass. However, the atomic mass of bromine was determined in the nineteenth century, for example by Stas (1865; see Mallet 1893). Attempts to separate bromine isotopes were made by Ogawa (1936) who determined the separation of these isotopes in the exchange reaction between Br_2 and Br^- (as NaBr) and he managed to increase the atomic weight of bromine with 0.11 mass units in Br_2. The results from his experiments were presented as atomic weights. In the relatively small amount of papers on bromine isotopes that were published prior to 2000 isotope ratios and atom percentages were mostly encountered (Blewett

Table 4.3 Conversion between the $\delta^{37}Cl$ scale and the atomic weight, isotope ratio and isotope atom percentages for values between 100 and + 100‰. Relatively close to the average chlorine isotope ratio of the solar system, which is represented by a seawater $\delta^{37}Cl$ of 0‰. The values in all columns are rounded to actually measurable values (indicating that the last decimal is roughly of the order of magnitude of the error of measurement). The table shows that variations within the range of the most common $\delta^{37}Cl$ values (between −1 and + 1 on the delta scale) are hardly recognisable when measured with other techniques than with isotope ratio mass spectrometry. This explains why before the 1960s no significantly measurable chlorine isotope variations could be reported

$\delta^{37}Cl$	Atomic weight	$^{37}Cl/^{35}Cl$	$^{35}Cl/^{37}Cl$	Atom% ^{37}Cl	Atom% ^{35}Cl
100	35.4881	0.3514	2.846	26.00	74.00
80	35.4811	0.3450	2.899	25.65	74.35
60	35.4740	0.3386	2.945	25.29	74.71
40	35.4668	0.3322	3.010	24.94	75.06
20	35.4596	0.3258	3.069	24.57	75.43
10	35.4560	0.3226	3.100	24.39	75.61
8	35.4552	0.3220	3.106	24.35	75.65
6	35.4545	0.3213	3.112	24.32	75.68
4	35.4538	0.3207	3.118	24.28	75.72
2	35.4530	0.3200	3.125	24.25	75.75
1	35.4527	0.3197	3.128	24.23	75.77
0.8	35.4526	0.3197	3.128	24.22	75.78
0.6	35.4525	0.3196	3.129	24.22	75.78
0.4	35.4525	0.3195	3.130	24.22	75.78
0.2	35.4524	0.3195	3.130	24.21	75.79
0.0	**35.4523**	**0.3194**	**3.131**	**24.21**	**75.79**
−0.2	35.4522	0.3193	3.131	24.20	75.80
−0.4	35.4522	0.3193	3.132	24.20	75.80
−0.6	35.4521	0.3192	3.133	24.20	75.80
−0.8	35.4520	0.3192	3.133	24.19	75.81
−1	35.4519	0.3191	3.134	24.19	75.81
−2	35.4516	0.3188	3.137	24.17	75.83
−4	35.4508	0.3181	3.143	24.13	75.87
−6	35.4501	0.3175	3.150	24.10	75.90
−8	35.4494	0.3169	3.156	24.06	75.94
−10	35.4486	0.3162	3.162	24.02	75.98
−20	35.4449	0.3130	3.195	23.84	76.16
−40	35.4375	0.3066	3.261	23.47	76.53
−60	35.4300	0.3002	3.331	23.09	76.91
−80	35.4224	0.2939	3.403	22.71	77.29
−100	35.4148	0.2875	3.479	22.33	77.67

1936; Williams and Yuster 1946; Cameron and Lippert 1955; Cameron et al. 1956; Xiao et al. 1993). Only from 2000 onwards the delta scale was used (Eggenkamp and Coleman 2000) for bromine isotope measurements. Natural variations are very small, most values measured on terrestrial samples are between −1 and + 2‰ vs. SMOB (Standard Mean Ocean Bromide). No extraterrestrial bromine isotope ratio determinations are known, except from one that was measured on the comet 67P/Churyumov-Gerasimenko (Dhooghe et al. 2017, 2021), suggesting no significant variation form terrestrial values. In spite of this conversion tables are prepared to show the conversion for $\delta^{81}Br$ values between −100 and + 100‰, like for chlorine. These data are presented in Table 4.5.

The remaining range in $^{79}Br/^{81}Br$ ratios from nearly pure ^{81}Br to nearly pure ^{79}Br are then presented in Table 4.6. This table can be used when, in the future, values far beyond the terrestrial delta scale are encountered, or when, experimentally, larger isotope variations are encountered.

Table 4.4 Conversion between the $^{35}Cl/^{37}Cl$ ratio and the $\delta^{37}Cl$, atomic weight, $^{37}Cl/^{35}Cl$ ratio and isotope atom percentages for ratios between 0.01 and 100. $^{35}Cl/^{37}Cl$ ratios between 1 and 5 are encountered in stars and interstellar material. Values below and above these values can be reached though isotope separation processes, such as diffusion

$^{35}Cl/^{37}Cl$	$\delta^{37}Cl$	Atomic weight	$^{37}Cl/^{35}Cl$	Atom% ^{37}Cl	Atom% ^{35}Cl
0.01	312,078	36.946	100	99.01	0.99
0.05	61,616	36.871	20	95.24	4.76
0.1	30,308	36.784	10	90.91	9.09
0.2	14,654	36.633	5.0	83.33	16.67
0.4	6827	36.395	2.5	71.43	28.57
0.6	4218	36.217	1.67	62.50	37.50
0.8	2913	36.078	1.25	55.56	44.44
1.0	2131	35.967	1.00	50.00	50.00
1.5	1087	35.768	0.67	40.00	60.00
2.0	565.4	35.635	0.50	33.33	66.67
2.5	252.3	35.539	0.40	28.57	71.43
3.0	43.59	35.468	0.33	25.00	75.00
3.13	**0.00**	**35.452**	**0.32**	**24.21**	**75.79**
3.3	−51.28	35.433	0.30	23.26	76.74
3.5	−105.5	35.413	0.29	22.22	77.78
4.0	−217.3	35.368	0.25	20.00	80.00
4.5	−304.0	35.332	0.22	18.18	81.82
5.0	−373.8	35.302	0.20	16.67	83.33
6.0	−478.2	35.254	0.17	14.29	85.71
8.0	−608.7	35.191	0.13	11.11	88.89
10	−686.9	35.150	0.10	9.09	90.91
20	−843.5	35.064	0.05	4.76	95.24
100	−968.7	34.989	0.01	0.99	99.01

Table 4.5 Conversion between the $\delta^{81}Br$ scale and the atomic weight, isotope ratio and isotope atom percentages for values between 100 and + 100‰. Relatively close to the average bromine isotope ratio of the solar system, which is represented by a seawater $\delta^{81}Br$ of 0‰. The values in all columns are rounded to actually measurable values (indicating that the last decimal is roughly of the order of magnitude of the error of measurement). The table shows that variations within the range of the most common $\delta^{81}Br$ values (between − 1 and + 1 on the delta scale) are hardly recognisable when measured with other techniques than with isotope ratio mass spectrometry

$\delta^{81}Br$	Atomic weight	$^{81}Br/^{79}Br$	$^{79}Br/^{81}Br$	Atom% ^{81}Br	Atom% ^{79}Br
100	79.9517	1.0714	0.9334	51.72	48.28
80	79.9426	1.0519	0.9507	51.26	48.74
60	79.9332	1.0324	0.9686	50.80	49.20
40	79.9237	1.0129	0.9872	50.32	49.68
20	79.9140	0.9935	1.0066	49.84	50.16
10	79.9091	0.9837	1.0166	49.59	50.41
8	79.9081	0.9818	1.0186	49.54	50.46
6	79.9071	0.9798	1.0206	49.49	50.51
4	79.9061	0.9779	1.0226	49.44	50.56
2	79.9051	0.9759	1.0247	49.39	50.61
1	79.9046	0.9750	1.0257	49.37	50.63
0.8	79.9045	0.9748	1.0259	49.36	50.64
0.6	79.9044	0.9746	1.0261	49.36	50.64
0.4	79.9043	0.9744	1.0263	49.35	50.65
0.2	79.9042	0.9742	1.0265	49.35	50.65
0.0	**79.9041**	**0.9740**	**1.0267**	**49.34**	**50.66**
−0.2	79.9040	0.9738	1.0269	49.34	50.66

(continued)

Table 4.5 (continued)

δ^{81}Br	Atomic weight	^{81}Br/^{79}Br	^{79}Br/^{81}Br	Atom% ^{81}Br	Atom% ^{79}Br
−0.4	79.9039	0.9736	1.0271	49.33	50.67
−0.6	79.9038	0.9734	1.0273	49.33	50.67
−0.8	79.9037	0.9732	1.0275	49.32	50.68
−1	79.9036	0.9730	1.0277	49.32	50.68
−2	79.9031	0.9720	1.0288	49.29	50.71
−4	79.9021	0.9701	1.0308	49.24	50.76
−6	79.9011	0.9681	1.0329	49.19	50.81
−8	79.9001	0.9662	1.0350	49.14	50.86
−10	79.8991	0.9642	1.0371	49.09	50.91
−20	79.8941	0.9545	1.0477	48.84	51.16
−40	79.8838	0.9350	1.0695	48.32	51.68
−60	79.8733	0.9155	1.0923	47.80	52.20
−80	79.8626	0.8961	1.1160	47.26	52.74
−100	79.8516	0.8766	1.1408	46.71	53.29

Table 4.6 Conversions between the ^{79}Br/^{81}Br ratio and the δ^{81}Br, atomic weight, ^{81}Br/^{79}Br ratio and isotope atom percentages for ratios between 0.01 and 100. Values shown in this table have not yet been encountered in geological or astronomical samples, but can be reached though isotope separation processes, such as diffusion

^{79}Br/^{81}Br	δ^{81}Br	Atomic weight	^{81}Br/^{79}Br	Atom% ^{81}Br	Atom% ^{79}Br
0.01	101,671	80.897	100	99.01	0.99
0.05	19,534	80.821	20	95.24	4.76
0.10	9267	80.735	10	90.91	9.09
0.20	4134	80.583	5.0	83.33	16.67
0.50	1053	80.250	2.0	66.67	33.33
0.75	368.9	80.060	1.33	57.14	42.86
0.90	140.8	79.970	1.11	52.63	47.37
1.00	26.71	79.917	1.00	50.00	50.00
1.0267	**0.00**	**79.904**	**0.9740**	**49.34**	**50.66**
1.10	−66.63	79.870	0.91	47.62	52.38
1.25	−178.6	79.806	0.80	44.44	55.56
1.50	−315.5	79.718	0.67	40.00	60.00
2.0	−486.6	79.584	0.50	33.33	66.67
3.0	−657.8	79.418	0.33	25.00	75.00
5.0	−749.7	79.251	0.20	16.67	83.33
7.5	−863.1	79.153	0.13	11.67	88.24
10	−897.3	79.100	0.10	9.09	90.91
20	−948.7	79.013	0.05	4.76	95.24
100	−989.7	78.938	0.01	0.99	99.01

4.3 Reference Standards in Use for Chlorine and Bromine Isotope Measurements

4.3.1 Chlorine Isotope Standards

Although no standard has been formally defined it was already shown by Kaufmann (1984) that no significant variability was found between seawater samples from different locations and depths. For that reason he proposed to use ocean water as the international standard. As name for this standard he proposed SMOC (Standard Mean Ocean Chloride) which has been in informal use since then by most laboratories working on chlorine isotope geochemistry. The big advantage of the use of ocean chloride as isotope standard is that it can be obtained easily and can be used and exchanged between laboratories without any problem (Kaufmann 1984). Godon et al. (2004) did collect ocean water samples from 15 different locations worldwide, and from three of these locations they collected also a depth profile. No significant variation was observed between all measured samples and from these observations it was concluded that ocean water has an invariant δ^{37}Cl of 0.00‰ vs. SMOC within analytical error, and could thus be used as isotope standard. This study has later been confirmed by Liu et al. (2012), who analysed, as part of their study to test a new on-line method to analyse chlorine isotopes with the use of a GasBench to purify chloromethane gas, 10 different seawaters taken from the Chinese coast. The measurements on these samples were all within 0.1‰ of their ocean water standard, with an average and standard deviation of 0.00 ± 0.06‰.

In two studies it was suggested that variations in ocean water chlorine stable isotopes exist (Shirodkar et al. 2003, 2006). These variations were partly unexplainable, and partly explained as due to evaporation of the upper layer of ocean water. Considering these variations the authors suggested that it would be advisable to use the ISL354 standard (as developed by Xiao et al. 2002) instead of ocean chloride as standard for chlorine isotope research. The samples that showed isotope variations in ocean water were measured by a different technique than the samples that indicated no variation (TIMS versus IRMS) and they were measured in one laboratory only. Considering the fact that the divergent analyses were done in only one laboratory it would be advisable that more interlaboratory exchange and measurement of samples between different laboratories that measure chlorine isotopes with different techniques are organised so that the differences in results between the different techniques can be independently confirmed.

4.3.1.1 Chloride Isotope Reference Materials

The observations discussed in 4.3.1 at least suggested that under some circumstances chlorine isotope variations in seawater could exist, so that it may not in all cases be possible to consider a seawater sample by definition equal to SMOC. To address this a limited number of international reference materials are available. Brand et al. (2014) list three of these for chlorine. These are the above mentioned ISL-354, NIST-SRM 975 and NIST-SRM 975a. Of these NIST-SRM 975 is exhausted. NIST-SRM 975 has been used to define the SMOC scale. Its isotope composition has been determined by Shields et al. (1962) and its absolute ^{35}Cl/^{37}Cl ratio was $3.1272^{+0.0079}_{-0.0082}$. Its atomic weight was calculated to be (based on the ^{12}C scale) $35.45273^{+0.00092}_{-0.00097}$. Its chlorine isotope value is supposed to be exact +0.43‰ vs. SMOC (Coplen et al. 2002; Xiao et al. 2002), although in Coplen et al. (2002) an error range of ±0.04‰ is given. NIST-SRM 975 is exhausted and replaced by NIST-SRM 975a. Its absolute ^{35}Cl/^{37}Cl ratio is 3.1279 ± 0.0047 and its atomic weight is 35.45265 ± 0.00055 (NIST 2001). NIST-SRM 975a is available from the National Institute of Standards and Technology in Gaithersburg, MD, USA. Its chlorine isotope value is exact +0.2‰ (Brand et al. 2014), although Coplen et al. (2002) report an error of ±1.5‰. Böhlke et al. (2017) also analysed NIST-SRM 975a and found its δ^{37}Cl to be equal to SMOC with an average value of +0.01 ± 0.04‰. The chlorine isotope standard ISL-354, developed at the Qinghai Institute of Salt Lakes in China (Xiao et al. 2002), is available from the IAEA in Vienna and has a certified δ^{37}Cl of +0.05 ± 0.03‰ vs. SMOC and a ^{37}Cl/^{35}Cl ratio of 0.31964 ± 0.00092.

In spite of this, as most researchers consider the oceans as a reservoir with constant Cl isotope ratio, in most studies a defined or undefined sample of ocean water is presented as reference material. That this assumption is justifiable is shown from the measurement of an additional 86 seawater samples taken from the world's oceans (including the Baltic sea, Black sea etc.) at various depths that have been analysed at the IPGP in Paris. The average δ^{37}Cl of these 86 samples was +0.006 ± 0.023‰ (1SD) from the Atlantic Ocean internal reference sample in use at the IPGP. The standard deviation of 430 measurements of this reference samples is 0.026‰, indicating that seawater is really homogeneous at this scale of precision (P Agrinier, pers. comm).

4.3.1.2 Perchlorate Isotope Reference Materials

Böhlke et al. (2017) prepared three $KClO_4$ perchlorate isotope reference materials (USGS37, USGS38 and USGS39). USGS39 has a $\delta^{37}Cl$ value of 0.05 ± 0.03‰ which is close to SMOC. USGS37 has a $\delta^{37}Cl$ of +0.90 ± 0.04 which is a significantly positive value. USGS38 has an extremely negative $\delta^{37}Cl$ value at −87.90 ± 0.24‰ vs SMOC. This large range makes it possible to evaluate the chlorine isotope compositions of perchlorates in the full range of samples that are encountered permitting isotope evaluation of environmental ClO_4^- sources and natural attenuation.

4.3.2 Bromine Isotope Standards

No standard has yet been defined officially, and unlike the situation for chlorine there has not been yet an exhaustive study to examine the natural variation of bromide in the oceans. Despite this lack of knowledge Eggenkamp and Coleman (2000) proposed to use the isotope ratio of seawater as its international stable isotope standard. They considered this ratio as constant in the oceans as the residence time of bromide in the oceans (130,000,000 years; Broeker and Peng 1982) is even longer than the residence time of chloride in the oceans (87,000,000 years, Berner and Berner 1987). These long residence time values suggest that the oceans are a very well mixed and homogeneous bromide reservoir. They proposed to call this reference SMOB (Standard Mean Ocean Bromide) as analogue to SMOC for chlorine. The first possible confirmation of the assumption that the isotope composition of ocean water is constant came from Du et al. (2013) who analysed the bromine isotope composition of five seawater samples collected along the Chinese coast and observed that they all had bromine isotopes ratios which were within 0.1‰ of each other.

Only one reference material for bromine has been defined. This is NIST-SRM 977. The isotope composition of this material has been determined by Catanzaro et al. (1964). Its absolute $^{79}Br/^{81}Br$ isotope ratio is 1.02784 ± 0.00105, resulting in 50.686 ± 0.025 atom % ^{79}Br and 49.314 ± 0.025 atom % ^{81}Br. The $\delta^{81}Br$ of NIST-SRM 977 is −0.64 ± 0.06‰ relative to SMOB as determined by Shouakar-Stash et al. (2005).

References

Alburger DE (1978) Half-lives of ^{62}Ga, ^{66}As, and ^{70}Br. Phys Rev C 18:1875–1877

Artukh AG, Avdeichikov VV, Gridnev GF, Mikheev VL, Volkov VV, Wilczynski J (1971) New isotopes $^{29,30}Mg$, $^{31,32,33}Al$, $^{33,34,35,36}Si$, $^{35,36,37,38}P$, $^{39,40}S$ and $^{41,42}Cl$ produced in bombardment of a ^{232}Th target with 290 MeV ^{40}Ar ions. Nucl Phys A 176:284–288

Ashley MF, Jenkins FA (1931) Possibility of the existence of the chlorine isotope Cl^{39}. Phys Rev 42:438–440

Aston FW (1919) The constitution of the elements. Nature 104:393

Aston FW (1920a) The constitution of the elements. Nature 105:547

Aston FW (1920b) The mass spectra of the chemical elements. Philos Mag 39:611–625

Aston FW (1931) The isotopic constitution and atomic weights of selenium, bromine, boron, tungsten, antimony, osmium, ruthenium, tellurium, germanium, rhenium and chlorine. Proc R Soc Lond A 132:487–498

Aston FW, Baxter GP, Brauner B, Bebierne A, Leduc A, Richards TW, Soddy F, Urbain G (1923) Report of the international committee on chemical elements, 1923. J Am Chem Soc 45:867–874

Becker H (1930) Ein drittes Chlorisotop. Z Phys 59:601–602

Benenson W, Mueller D, Kashy E, Nann H, Robinson LW (1977) Mass of ^{27}P and ^{31}Cl. Phys Rev C 15:1187–1190

Bernas M, Engelmann C, Armbruster P, Czajkowski S, Ameil F, Böckstiegel C, Dessagne P, Donzaud C, Geissel H, Heinz A, Janas Z, Kozhuharov C, Miehé C, Münzenberg G, Pfützner M, Schwab W, Stéphan C, Sümmerer K, Tassan-Got L, Voss B (1997) Phys Lett B 415:111

Berner EK, Berner RA (1987) The global water cycle: geochemistry and environment. Prentice-Hall, Inc, Englewood Cliffs, NJ

Blank B, Goigoux T, Ascher P, Gerbaux M, Giovinazzo J, Grevy S, Kurtukian Nieto T, Magron C, Agramunt J, Algora A, Guadilla V, Montaner-Piza A, Morales AI, Orrigo SEA, Rubio B, Ahn DS, Doornenbal P, Fukuda N, Inabe N, Kiss G, Kubo T, Kubono S, Nishimura S, Phong VH, Sakurai H, Shimizu Y, Soderstrom

PA, Sumikama T, Suzuki H, Takeda H, Wu J, Fujita Y, Tanaka M, Gelletly W, Aguilera P, Molina F, Diel F, Lubos D, de Angelis G, Napoli D, Borcea C, Boso A, Cakirli RB, Ganioglu E, Chiba J, Nishimura D, Oikawa H, Takei Y, Yagi S, Wimmer K, de France G, Go S (2016) New neutron-deficient isotopes from ^{78}Kr fragmentation. Phys Rev C 93:061301

Blewett JP (1936) Mass spectrograph analysis of bromine. Phys Rev 49:900–903

Böhlke JK, Mroczkowski SJ, Sturchio NC, Heraty LJ, Richman KW, Sullivan DB, Griffith KN, Gu B, Hatzinger PB (2017) Stable isotope analyses of oxygen (^{18}O:^{17}O:^{16}O) and chlorine (^{37}Cl:^{35}Cl) in perchlorate: reference materials, calibrations, methods, and interferences. Rapid Commun Mass Spectrom 31:85–110

Born JH, Seelmann-Eggebert W (1943) Über die Identifizierung einiger Uranspaltprodukte mit entsprechenden durch (nα)- und (np)-Prozesse erhaltenen Isotopen. Naturwiss 31:86–89

Bothe W, Gentner W (1937a) Herstellung neuer Isotope durch Kernphotoeffekt. Naturwiss 25:126

Bothe W, Gentner W (1937b) Künstliche Radioaktivität durch γ-Strahlen. Naturwiss 25:90

Boyd AW, Brown F, Lounsbury M (1955) Mass spectrometric study of natural and neutron irradiated chlorine. Can J Phys 33:35–42

Brand WA (2011) New reporting guidelines for stable isotopes—an announcement to isotope users. Isot Environ Health Stud 47:535–536

Brand WA, Coplen TB, Vogl J, Rosner M, Prohaska T (2014) Assessment of international reference materials for isotope-ratio analysis (IUPAC technical report). Pure Appl Chem 86:425–467

Broeker WS, Peng TH (1982) Tracers in the sea. Eldigio Press Lamont Doherty Geological Observatory

Cameron AE, Lippert EL (1955) Isotope composition of bromine in nature. Science 121:136–137

Cameron AE, Herr W, Herzog W, Lundén A (1956) Isotopen-Anreicherung beim Brom durch electrolytische Überführung in geschmolzenem Bleibromid. Z Naturforschng 11a:203–205

Catanzaro EJ, Murphy TJ, Garner EL, Shields WR (1964) Absolute isotopic abundance ratio and the atomic weight of bromine. J Res Natl Bur Stand A 68A:593–599

Coplen TB (2011) Guidelines and recommended terms for expression of stable-isotope-ratio and gas-ratio measurement results. Rapid Commun Mass Spectrom 25:2538–2560

Coplen TB, Hopple JA, Böhlke JK, Peiser HS, Rieder SE, Krouse HR, Rosman KJR, Ding T, Vocke Jr RD, Révész KM, Lamberty A, Taylor PDP, De Bièvre P (2002) Compilation of minimum and maximum isotope ratios of selected elements in naturally occurring terrestrial materials and reagents. USGS Water-Resources Investigations Report 01-4222. https://pubs.usgs.gov/wri/wri014222/pdf/wri01-4222.pdf

Curie I (1921) Sur le poids atomique du chlore dans quelques mineraux. Compte Rend Sean 172:1025–1028

Dhooghe F, De Keyser J, Altwegg K, Briois C, Balsiger H, Berthelier JJ, Calmonte U, Cessateur G, Combi MR, Equeter E, Fiethe B, Fray N, Fuselier S, Gasc S, Gibbons A, Gombosi T, Gunell H, Hässig M, Hilchenbach M, Le Roy L, Maggiolo R, Mall U, Marty B, Neefs E, Rème H, Rubin M, Sémon T, Tzou CY, Wurz P (2017) Halogens as tracers of protosolar nebula material in comet 67P/Churyumov–Gerasimenko. Mon Not R Astron Soc 472:1336–1345

Dhooghe F, De Keyser J, Hänni N, Altwegg K, Cessateur G, Jehin E, Maggiolo R, Rubin M, Wurz P (2021) Chlorine-bearing species and the ^{37}Cl/^{35}Cl isotope ratio in the coma of comet 67P/Churyumov–Gerasimenko. Mon Not R Astron Soc 508:1020–1032

Dorenfeldt M (1922) Relative determination of the atomic weight of chlorine in Bamle apatite. J Am Chem Soc 45:1577–1579

Du Y, Ma T, Yang J, Liu L, Shan HM, Cai HS, Liu CF, Chen LZ (2013) A precise analytical method for bromine stable isotopes in natural waters by GasBench II-IRMS. Int J Mass Spectrom 338:50–56

Eggenkamp HGM, Coleman ML (2000) Rediscovery of classical methods and their application to the measurement of stable bromine isotopes in natural samples. Chem Geol 167:393–402

Frisch OR (1934) Induced radioactivity of sodium and phosphorus. Nature 133:721–722

Fultz SC, Pool ML (1952) Radioisotopes of bromine. Phys Rev 86:347–350

Glass NW, Jensen LK (1953) The short-lived radioisotopes P^{28} and Cl^{32}. Phys Rev 90:320

Godon A, Jendrzejewski N, Eggenkamp HGM, Banks DA, Ader M, Coleman ML, Pineau F (2004) A cross calibration of chlorine isotopic measurements and suitability of seawater as the international reference material. Chem Geol 207:1–12

Grahame DC, Walker HW (1941) Preparation and properties of long-lived radiochlorine. Phys Rev 60:909

Gross JL, Claes J, Kathawa J, Thoenessen M (2012) Discovery of zinc, selenium, bromine, and neodymium isotopes. At Data Nucl Data Tables 98:75–94

Guillemaud-Mueller D, Penionzhkevich YE, Anne R, Artukh AG, Bazin D, Borrel V, Détraz C, Guerreau D, Gvozdev BA, Jacmart JC, Jiang DX, Kalinin AM, Kamanin VV, Kutner VB, Lewitowicz M, Lukyanov SM, Mueller AC, Hoai Chau N, Pougheon F, Richard A, Saint-Laurent MG, Schmidt-Ott WD (1989) Z Phys A 332:189

Hardy JD, Sutherland GBBM (1932) Isotope separations in the infrared absorption Spectrum of HCl and the possible existence of a chlorine isotope of mass 39. Phys Rev 41:471–479

Harkins WD (1925) The separation of chlorine into isotopes (isotopic elements) and the whole number rule for atomic weights. PNAS 11:624–628

Harkins WD, Hayes A (1921) The separation of the element chlorine into isotopes (isotopic elements). The heavy fraction from diffusion. J Am Chem Soc 43:1803–1825

Harkins WD, Ligget TH (1924) The discovery and separation of the isotopes of chlorine and the whole number rule. J Phys Chem 28:74–82

Harkins WD, Stone SB (1926) The isotopic composition and atomic weight of chlorine from meteorites and from minerals of non-marine origin. J Am Chem Soc 48:938–949

Harkins WD, Wilson ED (1915) The changes of mass and weight involved in the formation of complex atoms. J Am Chem Soc 37:1367–1383

Haslam RNH, Katz L, Johns HE, Moody HJ (1949) Confirmation of Cl^{39} activity. Phys Rev 76:704

Herzog W, Klemm A (1958) Die Temperaturabhängigkeit des Isotopie-Effekts bei der elektrolytischen Wanderungen der Chlorionen in herschmolzenem Thallium(I)-chlorid. Z Naturforschg 13a:7–16

Hettner G, Böhme J (1931) Das Chlorisotop Cl^{39}. Z Phys 72:95–101

Hoag JB (1940) The production and half-life of chlorine 33. Phys Rev 57:937

Hoering TC, Parker PL (1961) The geochemistry of the stable isotopes of chlorine. Geochim Cosmochim Acta 23:186–199

Hollander JM (1953) Bromine isotopes produced by carbon-ion bombardment of copper. Phys Rev 92:518–519

Huang WJ, Wang M, Kondev FG, Audi G, Naimi S (2021) The Ame2020 atomic mass evaluation. Chin Phys C 45:030002

Kallman H, Lasareff W (1932) Über die Isotopenuntersuchungen (Sauerstoff, Neon und Chlor). Z Phys 80:237–241

Kashy E, Benenson W, Mueller D, Nann H, Robinson L (1976) Observation of highly neutron-rich ^{43}Cl and ^{59}Mn. Phys Rev C 14:1773–1775

Kaufmann RS (1984) Chlorine in groundwater: stable isotope distribution. Ph.D. thesis, University of Arizona, Tucson, Ariz

Kennedy JW, Seaborg GT (1940) Isotopic identification of induced radioactivity by bombardment of separated isotopes; 37-minute Cl^{38}. Phys Rev 57:843–844

Klemm A, Lundén A (1955) Isotopenanreicherung beim Chlor durch electrolytische Überfürung in geschmolzenem Bleichlorid. Z Naturforschg 10A:282–284

Kondev FG, Wang M, Huang WJ, Naimi S, Audi G (2021) The NUBASE2020 evaluation of nuclear physics properties. Chin Phys C 45:030001

Kratz KL, Herrmann G (1974) Delayed-neutron emission from short-lived Br and I isotopes. Nucl Phys A 229:179–188

Langvad T (1954) Separation of chlorine isotopes by ion-exchange chromatography. Acta Chem Scand 8:526–527

Lewitowicz M, Anne R, Artukh AG, Bazin D, Belozyorov AV, Bricault P, Détraz C, Guillemaud-Mueller D, Jacmart JC, Kashy E, Latimier A, Lukyanov SM, Mueller AC, Penionzhkevich YE, Pougheon F, Richard A, Schmidt-Ott WD, Zhang Y (1990) First observation of the neutron-rich nuclei ^{42}Si, $^{45,46}P$, ^{48}S, and ^{51}Cl from the interaction of 44 MeV/u^{48}Ca + ^{64}Ni. Z Phys A 335:117–118

Liu YD, Zhou AG, Gan YQ, Liu CF, Yu TT, Li XQ (2012) An online method to determine chlorine stable isotope composition by continuous flow isotope ratio mass spectrometry (CF-IRMS) coupled with a Gasbench II. J South Cent Univ 20:193–198

Liuzzi G, Villanueva GL, Viscardy S, Mège D, Crismani MMJ, Aoki S, Gurgurewicz J, Tesson PA, Mumma MJ, Smith MD, Faggi S, Kofman V, Knutsen EW, Daerden F, Neary L, Schmidt F, Trompet L, Erwin JT, Robert S, Thomas IR, Ristic B, Bellucci G, Lopez-Moreno JJ, Patel MR, Vandaele AC (2021) Probing the atmospheric Cl isotopic ratio on mars: implications for planetary evolution and atmospheric chemistry. Geophys Res Lett 48:e2021GL092650

Maas ZG, Pilachowski CA (2018) Chlorine isotope ratios in M giants. Astron J 156:2

Mallet JW (1893) Stas memorial lecture. J Chem Soc Trans 63:1–56

McKinney CR, McCrea JM, Epstein S, Allen HA, Urey HC (1950) Improvements in mass spectrometers for the measurement of small differences in isotope abundance ratios. Rev Sci Instrum 21:724–730

Meija J, Coplen TB, Berglund M, Brand WA, De Bièvre P, Gröning M, Holden NE, Irrgeher J, Loss RD, Walczyk T, Prohaska T (2016) Isotopic compositions of the elements 2013 (IUPAC technical report). Pure Appl Chem 88:293–306

Morinaga H (1956) Radioactive isotopes Cl^{40} and Ga^{74}. Phys Rev 103:504–505

Mukha I, Grigorenko LV, Kostyleva D, Acosta L, Casarejos E, Ciemny AA, Dominik W, Dueñas JA, Dunin V, Espino JM, Estradé A, Farinon F, Fomichev A, Geissel H, Gorshkov A, Janas Z, Kamínski G, Kiselev O, Knöbel R, Krupko S, Kuich M, Litvinov YA, Marquinez-Durán G, Martel I, Mazzocchi C, Nociforo C, Ordúz AK, Pfützner M, Pietri S, Pomorski M, Prochazka A, Rymzhanova S, Sánchez-Benítez AM, Scheidenberger C, Sharov P, Simon H, Sitar B, Slepnev R, Stanoiu M, Strmen R, Szarka I, Takechi M, Tanaka YK, Weick H, Winkler M, Winfield JS, Xu X, Zhukov MV (2018) Deep excursion beyond the proton dripline. I. Argon and chlorine isotope chains. Phys Rev 98:064308

Murray G, White WJK, Willmott JC, Entwistle RF (1970) The decay of ^{73}Br. Nucl Phys A 142:21–34

Nier AO, Hanson EE (1936) A mass-spectrographic analysis of the ions produced in HCl under electron impact. Phys Rev 50:722–726

NIST (2001) Certificate of analysis, standard reference material 975a isotopic standard for chlorine. Gaithersburg, Maryland, National Institute Standards and Technology, p 2

Nolte E, Kutschera W, Shida Y, Morinaga H (1970) Neutron deficient even-even isotopes of Kr, se, Ge and pseudo SU_3 coupling. Phys Lett B 33:294–296

Ogawa E (1936) The isotopic separation of oxygen, chlorine, bromine and nitrogen by the chemical methods. Bull Chem Soc Jpn 11:428–430

Ohnishi T, Kubo T, Kusaka K, Yoshida A, Yoshida K, Ohtake M, Fukuda N, Takeda H, Kameda D, Tanaka K, Inabe N, Yanagisawa Y, Gono Y, Watanabe H, Otsu H, Baba H, Ichihara T, Yamaguchi Y, Takechi M, Nishimura S, Ueno H, Yoshimi A, Sakurai H, Motobayashi T, Nakao T, Mizoi Y, Matsushita M, Ieki K, Kobayashi N, Tanaka K, Kawada Y, Tanaka N, Deguchi S, Satou Y, Kondo Y, Nakamura T, Yoshinaga K, Ishii C, Yoshii H, Miyashita Y, Uematsu N, Shiraki Y, Sumikama T, Chiba J, Ideguchi E, Saito A, Yamaguchi T, Hachiuma I, Suzuki T, Moriguchi T, Ozawa A, Ohtsubo T, Famiano MA, Geissel H, Nettleton AS, Tarasov OB, Bazin D, Sherrill BM, Manikonda SL, Nolen JA (2010) J Phys Soc Jpn 79:073201

Owen HR, Schaeffer OA (1955) The isotope abundances of chlorine from various sources. J Am Chem Soc 77:898–899

Perlow GJ, Stehney AF (1959) Halogen delayed-neutron activities. Phys Rev 113:1269–1276

Rogers AM, Famiano MA, Lynch WG, Wallace MS, Amorini F, Bazin D, Charity RJ, Delaunay F, de Souza RT, Elson J, Gade A, Galaviz D, van Goethem MJ, Hudan S, Lee J, Lobastov S, Lukyanov S, Matos M, Mocko M, Schatz H, Shapira D, Sobotka LG, Tsang MB, Verde G (2011) Ground-state proton decay of 69Br and implications for the 68Se astrophysical rapid proton-capture process waiting point. Phys Rev Lett 106:252503

Shields WR, Murphy TJ, Garner EL, Dibeler VH (1962) Absolute isotopic abundance ratios and the isotopic weight of chlorine. J Am Chem Soc 84:1519–1522

Shirodkar PV, Xiao YK, Hai L (2003) Boron and chlorine isotopic signatures of seawater in the Central Indian ridge. Curr Sci 85:313–320

Shirodkar PV, Xiao YK, Sarkar A, Dalal SG, Chivas AR (2006) Influence of air-sea fluxes on chlorine isotopic composition of ocean water: implications for constancy in $\delta^{37}Cl$ – a statistical inference. Environ Int 32:235–239

Shouakar-Stash O, Frape SK, Drimmie RJ (2005) Determination of bromine stable isotopes using continuous-flow isotope ratio mass spectrometry. Anal Chem 77:4027–4033

Snell AH (1937) The radioactive isotopes of bromine: isomeric forms of bromine 80. Phys Rev 52:1007–1022

Stas (1865) Nouvelles recherches sur les lois des proportions chimiques, sur les poids atomiques et leurs rapports mutuels. Mém Acad Roy Belg 35:3–26

Stehney AF, Steinberg EP (1962) Br^{86} – a new nuclide. Phys Rev 127:563–569

Sugarman N (1949) Short-lived halogen fission products. J Chem Phys 17:11–25

Sumikama T, Fukuda N, Inabe N, Kameda D, Kubo T, Shimizu Y, Suzuki H, Takeda H, Yoshida K, Baba H, Browne F, Bruce AM, Carroll R, Chiga N, Daido R, Didierjean F, Doornenbal P, Fang Y, Gey G, Ideguchi E, Isobe T, Lalkovski S, Li Z, Lorusso G, Lozeva R, Nishibata H, Nishimura S, Nishizuka I, Odahara A, Patel Z, Zs P, Regan PH, Rice S, Roberts OJ, Sakurai H, Simpson GS, Sinclair L, Söderström PA, Tanaka M, Taprogge J, Watanabe H, Werner V, Wieland O, Wu J, Xu ZY, Yagi A (2021) Observation of new neutron-rich isotopes in the vicinity of ^{110}Zr. Phys Rev C 103:014614

Tarasov OB, Morrissey DJ, Amthor AM, Baumann T, Bazin D, Gade A, Ginter TN, Hausmann M, Inabe N, Kubo T, Nettleton A, Pereira J, Portillo M, Sherrill BM, Stolz A, Thoennessen M (2009) Phys Rev Lett 102:142501

Tarasov OB, Ahn DS, Bazin D, Fukuda N, Gade A, Hausmann M, Inabe N, Ishikawa S, Iwasa N, Kawata K, Komatsubara T, Kubo T, Kusaka K, Morrissey DJ, Ohtake M, Otsu H, Portillo M, Sakakibara T, Sakurai H, Sato H, Sherrill BM, Shimizu Y, Stolz A, Sumikama T, Suzuki H, Takeda H, Thoennessen M, Ueno H, Yanagisawa Y, Yoshida K (2018) Discovery of ^{60}Ca and implications for the stability of ^{70}Ca. Phys Rev Lett 121:022501

Thoenessen M (2012) Discovery of the isotopes with $11 \leq Z \leq 19$. At Data Nucl Data Tables 98:933–959

Vosicki B, Björnstad T, Carraz LC, Heinemeier J, Ravn HL (1981) Intense beams of radioactive halogens produced by means of surface ionization. Nucl Instrum Methods 186:307–313

Wang M, Huang WJ, Kondev FG, Audi G, Naimi S (2021) The Ame2020 atomic mass evaluation. (II). Tables, graphs and references. Chin Phys C 45:030003

Westfall GD, Symons TJM, Greiner DE, Heckman HH, Lindstrom PJ, Mahoney J, Shotter AC, Scott DK, Crawford HJ, McParland C, Awes TC, Gelbke CK, Kidd JM (1979) Production of neutron-rich nuclides by fragmentation of 212-MeV/amu ^{48}Ca. Phys Rev Lett 43:1859–1862

Williams D, Yuster P (1946) Isotopic constitution of tellurium, silicon, tungsten, molybdenum, and bromine. Phys Rev 69:556–567

Wimmer K, Doornenbal P, Korten W, Aguilera P, Algora A, Ando T, Arici T, Baba H, Blank B, Boso A, Chen S, Corsi A, Davies P, de Angelis G, de France G, Doherty DT, Gerl J, Gernhäuser R, Jenkins DG, Koyama S, Motobayashi T, Nagamine S, Niikura M, Obertelli A, Lubos D, Rubio B, Sahin E, Saito TY, Sakurai H, Sinclair L, Steppenbeck D, Taniuchi R, Wadsworth R, Zielinska M (2019) Discovery of ^{68}Br in secondary reactions of radioactive beams. Phys Lett B 795:266–270

Woodward LL, Mccown DA, Pool ML (1948) Radioactive Br isotopes. Phys Rev 74:870–875

Xiao YK, Zhou YM, Wei HZ WQZ, Liu WG, Eastoe CJ (2002) A secondary isotopic reference material of chlorine from selected seawater. Chem Geol 182:655–661

Xiao YK, Liu WG, Qi HP, Zhang CG (1993) A new method for the high precision isotopic measurement of bromine by thermal ionization mass spectrometry. Int J Mass Spectrom Ion Processes 123:117–123

Part II

Preparation Techniques

5 Preparation Techniques for the Analysis of Chlorine Stable Isotopes

5.1 Classical Methods

Before the development of "modern" isotope ratio mass spectrometers (Nier 1947; McKinney et al. 1950) stable isotope ratios and their ratio variations were measured by techniques such as gravimetric determination (e.g. Richards and Wells 1905; Curie 1921) and mass spectrography (Aston 1921, 1931; Kallmann and Lasareff 1932; Nier and Hanson 1936). Although these techniques produced results that were not as precise as the more modern mass spectrometric techniques, surprisingly good results were already obtained with them and the atomic weight of chlorine was determined with good precision using these techniques. Isotope ratio variations in natural samples however were not yet observed using these techniques as natural isotope variations are smaller than the precisions of these techniques. Isotope ratio variations and differences in atomic masses were however observed in samples obtained by separation techniques such as diffusion (e.g. Harkins and Hayes 1921). A short description of these techniques therefore seems viable here as they facilitated the later development of more advanced isotope ratio analysis techniques.

5.1.1 Gravimetric Determination of the Atomic Weight of Chlorine

Using chemicals which were purified from commercially available reagents Richards and Wells (1905) were able to determine the atomic weights of sodium and chlorine by conversion of sodium chloride to silver chloride and silver to silver chloride and exactly weighing the reaction products. They observed that the atomic weight of chlorine is either 35.470 or 35.473 depending on the atomic weight of silver (107.920 in the former and 107.930 in the latter), because the atomic weight of silver was not yet known to the same precision as Richards and Wells (1905) were able to weigh their silver chloride precipitate samples. Using the same technique they were also able to determine the atomic weight of sodium as either 23.006 or 23.008. These values are only very slightly higher (0.05–0.07%) than the values accepted today (107.8682 for silver, 22.9898 for sodium and about 35.45 for chlorine (Prohaska et al. 2022). The error in the measurement of the chlorine atomic weight is almost perfectly explainable by the fact that the value for the atomic weight of silver that was used in their study was too high, indicating the great quality of their work.

H. Eggenkamp, *The Geochemistry of Stable Chlorine and Bromine Isotopes*, Advances in Isotope Geochemistry, https://doi.org/10.1007/978-3-031-75633-7_5

Gravimetric determination of the atomic weight was also used in a few more studies from the early twentieth century which were done to determine the atomic weight of chlorine from various natural rock samples. In these studies (Curie 1921; Gleditsch and Samdahl 1922; Dorenfelt 1922; Harkins and Stone 1926) small variations between different rock samples were found, however in all cases these variations were smaller than the precision of the determinations.

5.1.2 Mass Spectrographic Analyses

Very shortly after the development of the mass spectrograph (Aston 1919b) it was determined that chlorine was composed of only two stable isotopes with masses 35 and 37, from which it was clear that the isotope with mass 35 was the most common (Aston 1919a, 1920, 1921). In these early mass spectroscopic studies the isotopes were detected on photographic plates (Aston 1925) and the measurement of the actual ratio between ions was complex.

In the following years improved mass spectrographs were developed and it was possible to increase the quality if the isotope measurements (Aston 1931; Kallman and Lasareff 1932; Nier and Hanson 1936). The ratio between the two stable ions was determined in this period with good precision (about 3.1; Aston 1931), however, in spite of these improvements the precision of the measurements did not improve beyond the point where natural variations became measurable.

5.1.3 Early Mass Spectrometric Analyses

Only after the development of mass spectrometers with double ion collectors (Nier et al. 1946; Nier 1947; McKinney et al. 1950; Nier 1955) it became possible to determine variations beyond the analytical precision and at least it was finally possible to easily analyse isotope variations produced in chemical fractionation experiments (e.g. Bartholomew et al. 1954, Klemm and Lundén 1955, Lundén and Herzog 1956, Herzog and Klemm 1958, Hill and Fry 1962, Howald 1960). These measurements were of very good quality as could be concluded from the results obtained by Madorsky and Strauss (1948) who determined that, during the process of diffusion, the diffusion coefficient of ^{35}Cl was about 1.0012 to 1.0022 times that of ^{37}Cl. These values have been confirmed in almost all subsequent studies where the fractionation that is caused by diffusion of chloride in aqueous solutions was determined, both in experimental and natural systems.

5.2 General Methods to Analyse Stable Chlorine Isotopes

The information containing is this chapter is for a large part described earlier in my PhD thesis (Eggenkamp 1994), the chapter (Eggenkamp 2004) I wrote for the "Handbook of Stable Isotope Analytical Techniques" (De Groot 2004) and the first edition of this book (Eggenkamp 2014). It has been repeated here to fully describe the various methods that have been developed in search for the "best" method to analyse chlorine isotopes. The information is supplemented with methods that have been developed after the first edition of this book has been written.

5.2.1 Methods Using Other Gases than Chloromethane in Dual Inlet IRMS

5.2.1.1 Application of Hydrogen Chloride as Gas in Dual-Inlet IRMS

Hydrogen chloride was used in several early chlorine isotope studies (Nier and Hanson 1936; Madorsky and Strauss 1948; Johnston and Arnold 1953; Hoering and Parker 1961) as the gas that was introduced into the mass spectrometer. Among these studies was the first major chlorine isotope study of natural, geological samples (Hoering and Parker 1961). They measured the chlorine isotope composition of 81 samples in different geological samples ranging from aqueous samples to rock samples. Hoering and Parker (1961) were able to analyse the chlorine isotope ratio in hydrogen chloride gas with a precision of 0.8‰. The reason why hydrogen chloride gas was used as mass spectrometer gas was that it can be prepared quantitatively and because it produces a very simple mass scan with only four ion species formed (Table 5.1). The major disadvantage of hydrogen chloride as a mass spectrometer gas is however that it sticks to the walls of the vacuum system, and as a result it has a very large memory effect.

Chloride samples were precipitated as silver chloride (AgCl). For rock and mineral samples this was produced according to the method described by Kuroda and Sandell (1953). Samples were then dissolved in ammonium hydroxide, to which magnesium metal was added in excess. The magnesium displaces the silver, and the ammonium chloride solution that was formed was filtered. The remaining solution was evaporated to dryness in a vacuum oven. The solution from which the ammonium chloride was precipitated has a high pH, and care must be taken that it would not react with atmospheric carbon dioxide as this can not be separated from hydrogen chloride, and as such it will interfere during isotope analysis. The ammonium chloride residue was converted to hydrogen chloride by reaction with sulphuric acid under vacuum. The gaseous hydrogen chloride was trapped onto frozen phosphorus pentoxide to remove traces of water, and the dried hydrogen chloride was frozen into a sample container for introduction into the mass spectrometer.

5.2.1.2 Application of Chlorine Gas in Dual-Inlet IRMS

Chlorine gas has only been used rarely as mass spectrometer gas. It could be considered as a potentially useful mass spectrometer gas due to its simple mass scan which should be approximately as shown in Table 5.2. As shown in this table only ^{35}Cl and ^{37}Cl peaks are present near the atomic mass of chlorine so that no calculations have to be made to take interference into account. If the larger peaks around mass 70 to 74 are used to calculate the isotope ratio the presence of the mixed $^{35}Cl^{37}Cl^+$ ion must be taken into account, introducing a more complex calculation. A large disadvantage of this gas appears however the fact that it produces very large memory effects. These memory effects were so pronounced that Boyd et al. (1955) made sure that no chlorine gas had entered their mass spectrometer for 6(!) months between their experiments with non-radioactive chlorine and irradiated samples.

Bartholomew et al. (1954) used chlorine gas as their mass spectrometer gas to study the isotope effects in reactions of *tert*-butyl chloride (2-chloro-2-methyl-propane). Hydrogen chloride was prepared from silver chloride using a method comparable to Hoering and Parker (1961). In their preparation technique however, Bartholomew et al. (1954) heated ammonium chloride with concentrated sulphuric acid, and the hydrogen chloride that was produced was trapped in a bubbler containing a small amount of cold water by means of a stream of nitrogen gas. The aqueous hydrogen chloride was oxidised to chlorine gas by persulphate oxidation (Brown et al. 1953) in a stream of helium. The chlorine gas was frozen in a cold trap, and the helium pumped away. The chlorine gas was then transferred to the mass spectrometer and analysed.

Table 5.1 Mass scan for HCl as mass spectrometer gas (Hoering and Parker 1961)

Species	*m/z*	Relative intensity
$^{35}Cl^+$	35	17.0
$H^{35}Cl^+$	36	100
^{37}Cl	37	5.4
$H^{37}Cl^+$	38	32.5

Table 5.2 Mass scan for Cl_2 gas as mass spectrometer gas according to Eggenkamp (2004). The pattern is based on the abundances of the isotopes ^{35}Cl and ^{37}Cl and assumes that the total intensity of the monoatomic species is 10% of the total intensity of the diatomic species

Species	*m/z*	Relative intensity
$^{35}Cl^+$	35	13
$^{37}Cl^+$	37	4
$^{35}Cl_2^+$	70	100
$^{35}Cl^{37}Cl^+$	72	64
$^{37}Cl^+$	74	10

5.2.1.3 Application of Phosgene as Mass Spectrometer Gas

According to Kaufmann (1984) phosgene ($COCl_2$) has also been applied as mass spectrometer gas by Owen and Schaeffer (1955). This was however not the case, Owen and Schaeffer (1955) only referred to an earlier use of phosgene as mass spectrometer gas by Aston (1941), where it was observed that phosgene produced troublesome memory effects that might persist even after several days of pumping and baking. For that reason, and of course its extreme toxicity, phosgene was never again used as potential gas to measure chlorine isotope ratios.

5.2.1.4 Application of Chloroethane as Mass Spectrometer Gas

Very recently Ertl et al. (2024) suggested to use chloroethane (C_2H_5Cl) as the gas to measure $\delta^{37}Cl$ on. They developed a method using "Purge & Trap-gas chromatography-continuous flow-isotope ratio mass spectrometry" (P&T-GC-CF-IRMS). In this technique they react their aqueous chloride with phosphoric acid to obtain HCl in solution. This HCl is reacted with ethanol to form chloroethane. These two individual reactions leading to chloroethane are:

$$MCl + H_3PO_4 \leftrightarrow HCl_{(g)} + MH_2PO_4 \quad (5.1)$$

$$HCl_{(g)} + C_2H_5OH \leftrightarrow C_2H_5Cl + H_2O \quad (5.2)$$

The two reactions take place in a single vial so that the overall reaction is:

$$MCl + H_3PO_4 + C_2H_5OH \rightarrow C_2H_5Cl + MH_2PO_4 + H_2O\left(T = 120°C\right) \quad (5.3)$$

The water that is formed in this reaction is consumed by an excess amount of phosphoric acid. The result of this reaction is gaseous chloroethane that can be measured at m/z = 64 and m/z = 66. To make sure that the ion beams are about equal the resistor at mass 64 is set at $3*10^{10}$ Ω ant the one at mass 66 at $1*10^{11}$ Ω. Measurement of $\delta^{37}Cl$ on a Finnigan MAT Delta S mass spectrometer was within error equal to the measurement of the same samples with chloromethane gas. The standard deviation of repeated measurements (n = 2 or 4) of individual samples was better than 0.18‰. This suggests that the method is slightly less precise than dual inlet measurements with chloromethane, but still very acceptable for most samples, especially considering the fast preparation times.

5.2.2 Application of Chloromethane in Dual-Inlet IRMS

The most commonly used gas for both dual inlet and continuous flow isotope ratio mass spectrometry of chlorine at the moment is chloromethane (methyl chloride, CH_3Cl). At least three different methods have been proposed to produce quantitative yields of this gas. Chloromethane is much less reactive than either hydrogen chloride or chlorine gas and it shows no memory effects in the inlet and vacuum systems of mass spectrometers.

Chloromethane was first studied in a mass spectrometer by Bauer and Hogness (1935) who discovered the various primary ions that are formed during the process of ionization. Of these, CH_3Cl^+ had the highest intensity. In later studies, when mass separation of the mass spectrometers was improved especially this ion was shown to be of particular interest and use for the study of stable chlorine isotopes.

Chloromethane can be produced from several chloride compounds. In the chlorine stable isotope literature the most important of these are: chloromethane produced from ammonium chloride (Owen and Schaeffer 1955; Herzog and Dörnenburg 1958), chloromethane produced from silver chloride (Langvad 1954; Hill and Fry 1962; Taylor and Grimsrud 1969; Kaufmann 1984; Long et al. 1993; Eggenkamp 1994; Godon et al. 2004a), and chloromethane produced from copper chloride (Holt et al., 1997). This last technique was developed with the purpose of analysing organochlorine compounds and will be discussed in Sect. 5.6.1.3.

Chloromethane can be measured both as positive or negative ions. At this moment, most laboratories using chloromethane as their mass spectrometer gas measure chlorine isotope ratios on the positive CH_3Cl^+ ion. The positive ions are measured at *m/z* 50 and 52 at which positions the main peaks are found. Its mass scan is much more complex than that of hydrogen chloride or chlorine gas and shows several more peaks in its isotope spectrum as not only chlorine but also carbon has a minor isotope with a relatively large abundance, as is shown in Table 5.3.

In the case of negative ions the isotope ratio is determined on chloride ions formed from the decomposition of chloromethane molecules in the ion source. In the case of negative ions the mass scan is very simple with only two major peaks at *m/z* 35 and 37 (see Table 5.4) as shown in both major studies on chlorine isotope analyses using a negative ion source (Taylor and Grimsrud 1969; Pelc and Hałas 2008).

5.2.2.1 Preparation of Chloromethane from Ammonium Chloride

Chloromethane can be produced from ammonium chloride, and this technique was used regularly in chlorine isotope in studies from the 1950s (Owen and Schaeffer 1955; Herzog and Dörnenburg 1958). Silver chloride was precipitated from a liquid sample containing dissolved chloride. The silver chloride was subsequently dissolved in ammonia. Silver was removed from this solution as silver sulphide. After evaporation of the remaining solution pure ammonium chloride was left. Ammonium chloride can then be reacted into chloromethane following a procedure described by Blatt (1943). 240 mL of concentrated sulphuric acid is diluted with 40 mL of distilled water, to which is added 350 mL of methanol. During this procedure the temperature must be kept below 70 °C at all times. Approximately 2 mL of this solution is added to 50 mg of ammonium chloride under vacuum. The following reactions can potentially occur:

$$(CH_3)_2SO_4 + 2NH_4Cl > 2CH_3Cl + (NH_4)_2SO_4 \quad (5.4)$$

$$CH_3OH > (CH_3)_2O \left(\begin{array}{c}\text{Catalysed by}\\ \text{sulphuric acid}\end{array}\right) \quad (5.5)$$

$$2NH_4Cl + H_2SO_4 > 2HCl + (NH_4)2SO_4 \quad (5.6)$$

The second reaction can be minimised by keeping the temperature of the mixture below 50 °C. A weighed sample of ammonium chloride is placed into a bulb and evacuated. Approximately 2 mL sulphuric acid and methanol are added through a stopcock. The mixture is heated using an infrared lamp until the reaction ceases. The reaction tube is then chilled with dry ice and the chloromethane that has formed is distilled into an evacuated sample bulb. A potassium hydroxide

Table 5.3 Mass scan for chloromethane as mass spectrometer gas according to Owen and Schaeffer (1955) and Taylor and Grimsrud (1969). Owen and Schaeffer (1955) only published the mass scan of chloromethane in the region *m/z* = 47 to 52

		Relative intensity	
Species	*m/z*	Owen and Schaeffer (1955)	Taylor and Grimsrud (1969)
$^{35}Cl^+$	35		2.8
$H^{35}Cl^+$	36		1.2
$^{37}Cl^+$	37		1.0
$H^{37}Cl^+$	38		0.4
$^{12}C^{35}Cl^+$	47	6.0	7.7
$^{12}CH^{35}Cl^+$	48	3.2	3.2
$^{12}CH_2{}^{35}Cl^+$+ $^{12}C^{37}Cl^+$	49	9.0	9.6
$^{12}CH_3{}^{35}Cl^+$+ $^{12}CH^{37}Cl^+$	50	100	100
$^{13}CH_3{}^{35}Cl^+$+ $^{12}CH_2{}^{37}Cl^+$	51	3.7	3.4
$^{12}CH_3{}^{37}Cl^+$	52	31.7	31.4
$^{13}CH_3{}^{37}Cl^+$	53		0.5

Table 5.4 Mass scan of chloromethane when analysed as negative ions in dual-inlet mass spectrometry. The results obtained by Taylor and Grimsrud (1969) and Pelc and Hałas (2008) agree very well with respect to the most common ions in the mass spectrometer source

		Relative intensity	
Species	*m/z*	(Taylor and Grimsrud 1969)	(Pelc and Hałas 2008)
CN^-	26		0.05
CO^-	28		0.05
$^{35}Cl^-$	35	100	100
$^{37}Cl^-$	37	32.4	32
$^{12}C^{35}Cl^-$	47	0.25	
$^{12}CH^{35}Cl^-$	48	0.20	
$^{12}CH_2{}^{35}Cl^-$ + $^{12}C^{37}Cl^-$	49	0.50	
$^{12}CH^{37}Cl^-$	50	0.10	
$^{12}CH_2{}^{37}Cl^-$	51	0.14	

trap removes any hydrogen chloride that is formed during this reaction. The major disadvantage of this technique is that the yield will not be higher than approximately 35%. Several tests were done by Owen and Schaeffer (1955) to determine the potential errors caused because of this low yield and it was observed that data were reliable within an experimental error of 2‰. However, as it is currently known that the large majority of all measured $\delta^{37}Cl$ data are within 2‰ of SMOC, an experimental error with the size of ±2‰ is definitely not acceptable for natural chlorine isotope measurements.

5.2.2.2 Preparation of Chloromethane from Silver Chloride

The most commonly used technique in recent studies to produce chloromethane for dual inlet and continuous flow IRMS is the production of chloromethane from silver chloride that has been precipitated from an aqueous solution. The first to apply this technique was Langvad (1954) in a study on potential isotope fractionation of chlorine isotopes by ion chromatography. Over the next 50 years this method was improved considerably by Hill and Fry (1962), Taylor and Grimsrud (1969), Kaufmann (1984), Long et al. (1993), Eggenkamp (1994) and Godon (2000). The production of chloromethane from chloride in aqueous solution involves in principle three steps, precipitation of silver chloride, reaction of silver chloride with iodomethane in evacuated glass capsules to form chloromethane and finally either a gas chromatographic or cryogenic separation of chloromethane from iodomethane prior to isotope analysis.

5.2.2.2.1 Precipitation of Silver Chloride

Depending on the chloride concentration of the solution silver chloride is precipitated in slightly different ways. When the chloride concentration is lower than about 500 mg/L silver chloride can be precipitated after addition of dry chemicals to the sample solution, if the chloride concentration is higher by addition of these chemicals in dissolved form. Originally the amount of chloride estimated to be necessary for chlorine isotope measurements was 0.1 mmole (Kohnen 1988). As the result from improvements in sample preparation the minimum amount necessary for dual inlet mass spectrometric measurements is at present about 0.02 mmole, although 0.05 mmole seems a relatively ideal amount. The amount of sample containing this amount of chloride is put in a glass beaker. In the case of higher chloride concentrations in the sample 4 mL of a 1 M KNO_3 solution and 2 mL of a Na_2HPO_4-citric acid buffer solution is added.

The potassium nitrate solution is necessary to reach a high ionic strength. Taylor and Grimsrud (1969) discovered that if the ionic strength of the solution from which silver chloride is to be precipitated is less than 0.4 M the chloromethane yield fell below 100%. For example at an ionic strength of 0.2 M it was only 45%. This effect is probably caused because smaller silver chloride crystals are formed at higher ionic strengths which react completely to silver iodide while larger crystals precipitated at lower ionic strengths form a thicker layer of silver iodide which prevents the inner silver chloride from reacting with iodomethane so that the silver chloride does not react completely. This effect leads to isotope fractionation for which Taylor and Grimsrud (1969) found an average value of +0.43‰. The buffer solution is used to reach a constant pH of 2.2. This is necessary to remove small amounts of sulphide which precipitates with silver ions and forms silver sulphide (Kaufmann 1984), but also to prevent potential precipitation of other silver salts such as phosphates and carbonates (Vogel 1951). Kaufmann (1984) and Long et al. (1993) proposed to acidify the solution with HNO_3 while Eggenkamp (1994) followed Kohnen (1988) who proposed to use a pH 2.2 buffer solution after McIlvaine (1921) which is a solution containing 0.71 gram (0.004 mole) $Na_2HPO_4.2H_2O$ and 20.6 gram (0.098 mole) $HOC(CH_2CO_2H)_2CO_2H.H_2O$ (citric acid) per litre.

The solution, consisting of the sample plus pH buffer and ionic strength buffer, is placed on a boiling ring and heated to just below 100 °C. To this hot solution 1 mL of a 0.2 M silver nitrate solution is added. At the same moment that silver nitrate is added silver chloride starts to precipitate.

The solution will not be stirred using a magnetic stirrer during silver chloride precipitation because silver chloride may clot and it will be difficult to scrape it from the magnetic stirrer which would result in the loss of sample. The suspension is filtered over a Whatman® glass fibre filter, type GF/F with a retention of 0.7 μm and a standardized filter speed of 6 mL/s. During filtration the suspension is rinsed with a diluted (1 mL concentrated nitric acid in 500 mL water) nitric acid solution. This is done because when silver chloride is washed with distilled water it can become colloidal and run through the filter. For this reason the wash solution must contain an electrolyte. Nitric acid is chosen for this purpose as it is without action on the precipitate and is readily volatile (Vogel 1951). It is essential that glass fibre filters are used for filtering the silver chloride suspension as other filter types will react with iodomethane later in the procedure. If this is the case no (or little) chloromethane will be formed during the last step of its preparation.

As mentioned above, samples with low chloride concentrations are treated in a different way. As it would otherwise be necessary to add very large amounts of potassium nitrate and buffer solutions to these samples, and as such to avoid the handling of unnecessarily large sample volumes the reagents are added as dry chemicals. Per 100 mL of sample 6.00 g (0.06 mole) potassium nitrate, 2.06 g (0.0098 mole) citric acid and 0.07 g (0.0004 mole) $Na_2HPO_4.2H_2O$ is added after which the solution is heated to a temperature of just below 100 °C. Then 1 mL silver nitrate solution is added. Silver chloride will then start to precipitate and the suspension is treated as described above.

If a solution has a very low chloride concentration there is the risk that the chloride concentration is so low that the solubility product (approximately $1*10^{-8}$ at 80 °C, Eggenkamp 1994) for the formation of silver chloride is not reached. That means that, if the product of Cl^- and Ag^+ ions in solution is lower than the solubility product silver chloride will not precipitate. According to Weast and Astle (1982) the solubility of AgCl in cold water (10 °C) is 8.9 mg/L and in hot water (100 °C) it is 21 mg/L. That suggests that, when the chloride concentration is a sample is lower than 5 to 10 mg/L, larger quantities of $AgNO_3$ solution, to increase Ag^+ in the ion product, must be added, or the sample must be preconcentrated. This can be done by slowly evaporating part of the sample at temperatures just below the boiling point of water. This is however a relatively rough method that is time consuming and has the risk of contamination of the sample during the long duration of the process, as evaporation is generally done in the open air. Giunta et al. (2015) developed a different method to concentrate the chloride in a small amount of water. They use a 3 mL solid phase extraction (SPE) cartridge that is filled with 1 mL of Amberlite resin in the hydroxide form (100–200 mesh) and that is then rinsed with 50 mL of a 1 M NaOH solution. The water sample containing the low concentration of chloride is loaded through the column at a flow rate of about 1.5 mL/min. Because of the high partition coefficient of chloride between the resin and water (Kd > 5300) the exchange of chloride onto the resin is very efficient. Chloride can be removed efficiently from the resin with a potassium nitrate solution. Giunta et al. (2015) found that a volume as low as 5 mL of a 1 M KNO_3 solution is enough for the quantitative chloride recovery from 1 mL of resin. It was found that both the chloride recovery and the obtained $\delta^{37}Cl$ were very good with chloride recoveries that were on average 101 ± 7% of the expected amount and $\delta^{37}Cl$ values on average within 0.04‰ of the expected values, that is within the error of measurement. The final solution can be treated as described above for solutions with high chloride concentrations.

After filtration the filter containing the precipitate is dried overnight in an oven at a temperature of about 80 °C. During the drying process and after the filters were dried they need to be stored in a dark place to avoid the decomposition of silver chloride under the influence of light according to the reaction:

$$AgCl > Ag + Cl_2 \tag{5.7}$$

5.2.2.2.2 Treatment of Samples with Very High pH Values

Samples with a high pH will precipitate silver oxide when they are treated with silver nitrate. This will lead to reduced silver chloride yields, as well as other problems during the preparation of chloromethane and thus during measurement of the gas sample in the mass spectrometer. It is suggested that dimethyl ether can be formed during the procedure which might not be separated properly from chloromethane later in the procedure. If the pH is only moderately high, it will be lowered properly by the addition of the citric acid/phosphate buffer. However, in those cases where samples have a very high pH the buffer will not be efficient enough to decrease the pH to values close to 2. To prevent silver oxide from precipitating in this case nitric acid is added to the sample solution instead of the buffer solution. Enough nitric acid needs to be added to decrease the pH to a value below 2 after which the silver nitrate solution can be added and silver chloride will precipitate without problems.

5.2.2.2.3 Treatment of Samples Containing High Sulphide Concentrations

Samples that contain high sulphide concentrations will precipitate silver sulphide instead of silver chloride when silver nitrate is added to the solution. As a result the silver chloride yield will be lower but also dimethyl sulphide can be produced during the reaction with iodomethane and this will interfere during isotope measurement. If the sulphide content is low enough it will evaporate as hydrogen sulphide from the sample solution after adding either the citric acid/phosphate buffer or nitric acid (Kaufmann 1984). However, this is not sufficient for samples with higher sulphide concentrations. To remove sulphide from these samples it is oxidised through the addition of a few ml 30% hydrogen peroxide and heated to a temperature of approximately 80 °C for approximately 1 h. After this treatment silver nitrate can be added as if it were originally a normal sample.

5.2.2.2.4 Removal of Sulphate from Samples with Excessive Sulphate

Samples that also need to be analysed for ^{36}Cl by accelerator mass spectrometry need to be sulphate free to avoid the presence of the isotope ^{36}S which would completely swamp the small amount of ^{36}Cl in a sample. In samples that are only to be measured for chlorine stable isotopes excessive amounts of sulphate may need to be removed as this can be precipitated as silver sulphate leading to a miscalculation of the amount of silver chloride and subsequently lowering the apparent yield of chloromethane in the sample. Long et al. (1993) proposed to remove (excessive) sulphate by leaving the filter with silver chloride on the filtration set up. After emptying and cleaning the filtrate flask the silver chloride is dissolved in an ammonia solution which is filtered and collected in a small flask. Barium nitrate is added to precipitate any remaining sulphate. If it is necessary to keep silver chloride in solution some extra ammonia solution may be added. After all sulphate is precipitated as barium sulphate the solution is filtered. The filtrate should than be free of sulphate. Finally the pH is decreased to <2 by adding some nitric acid. Silver chloride should precipitate in this process because at low pH values the ammonia-ammonium equilibrium moves towards more ammonium so that less ammonia is available as ligand to the silver ion. To ensure that all silver chloride is precipitated a few extra drops of silver nitrate solution are added. The purified silver chloride can be treated further to produce chloromethane.

5.2.2.2.5 Preparation of Glass Reaction Capsules

The reaction from silver chloride to chloromethane takes place in glass capsules. From a length of pyrex tubing (with outer diameter of 12 mm) parts of about 38 cm are cut. After melting and separating this piece of tubing in the middle two tubes with a length of about 19 cm and sealed at one end are formed. After cooling and drying the glass tubes a filter containing silver chloride is put in the tube. A capillary is then drawn about 7 cm from the closed end. The tube is put onto a vacuum line which is pumped down to a pressure of less than 10^{-1} mbar. The tubes are filled with nitrogen gas to avoid the presence of oxygen that might react with iodomethane later in the procedure. Then, they are removed from the extraction line and closed with a rubber stopper to prevent leakage of air. 100 µl (1.6 mmole) iodomethane is added in a fume hood. The tube is returned to the vacuum line and the iodomethane is frozen on the silver chloride with liquid nitrogen and then pumped to a vacuum of less than $1*10^{-1}$ mbar). At the capillary the capsule is flame closed and molten from the line, keeping a glass ampoule containing solely a glass fibre filter with silver chloride and iodomethane. The capsule is placed in an oven for 48 h at a temperature of 80 °C. In the oven the reaction from silver chloride to chloromethane takes place according to:

$$AgCl + CH_3I \rightarrow AgI + CH_3Cl \quad (5.8)$$

As this reaction is an equilibrium reaction iodomethane needs to be added in excess. If the reaction temperature is too high iodomethane may decompose following:

$$CH_3I \rightarrow C_2H_6 + I_2 \quad (5.9)$$

as suggested by Eastoe et al. (1989). If iodomethane decomposes the normally colourless liquid will change to yellow or brown. Measured $\delta^{37}Cl$ values of samples that showed colourisation of the iodomethane have less accurate values (the standard deviation of repeatedly measured values is higher than normal). When a scan is made from a sample containing decomposed iodomethane it shows the presence of a peak with mass 29 which is not normally present. This peak may represent the ion $C_2H_5^+$. Except peak 29 also peaks with masses 45 and 46 which are not present in clean samples can be found in decomposed samples (see Fig. 5.1). This contamination will probably also affect peaks with m/z 50 and 52 and as such depreciate the quality of the analysis.

5.2.2.2.6 Cryogenic Separation of Chloromethane from Iodomethane

The two most common techniques to obtain pure chloromethane are cryogenic and gas chromatographic separation from iodomethane. In the oldest studies in which chloromethane was produced from the reaction of silver chloride with iodomethane chloromethane was separated from iodomethane with cryogenic separation (Langvad 1954; Hill and Fry 1962). In these studies separation was obtained by dry ice-acetone and dry ice-isopropyl alcohol (propan-2-ol) cryogenic mixtures respectively. In later studies the preferred method to separate chloromethane from iodomethane was gas chromatography (Taylor and Grimsrud 1969; Kaufmann 1984). More recently new methods for cryogenic separation of chloromethane from iodomethane were proposed by Holt et al. (1997) and Wu and Satake (2006).

Holt et al. (1997) used melting n-pentane (−130 °C) to separate chloromethane from iodomethane. The capsule containing chloromethane and iodomethane was frozen in dry ice-acetone slush (−79 °C), and two cold traps were frozen with n-pentane and liquid nitrogen respectively. After breaking the capsule that was frozen in the dry ice-acetone slush chloromethane and iodomethane slowly evaporate and move to the n-pentane cold trap. Iodomethane is trapped here while chloromethane moves to the second cold trap that is frozen with liquid nitrogen. This separation is done twice to ensure that chloromethane and iodomethane are perfectly separated. According to Holt et al. (1997) the iodomethane content in the resulting pure chloromethane must be lower than 0.2%. This threshold was reached after two cryogenic separations and further (gas chromatographic) purification was not necessary.

Wu and Satake (2006) considered using n-pentane a serious risk due to its flammability. To overcome this problem they proposed to seal the cold traps as proposed earlier by Kusakabe (2005) for the separation of sulphur dioxide from carbon dioxide for precise $\delta^{18}O$ and $\delta^{34}S$ measurements. They also found that at the temperature of the n-pentane trap neither iodomethane nor chloromethane vaporised, and that they both vaporise at higher temperatures, although of course chloromethane vaporises earlier than iodomethane. Wu and Satake (2006) searched for another compound that could be used for

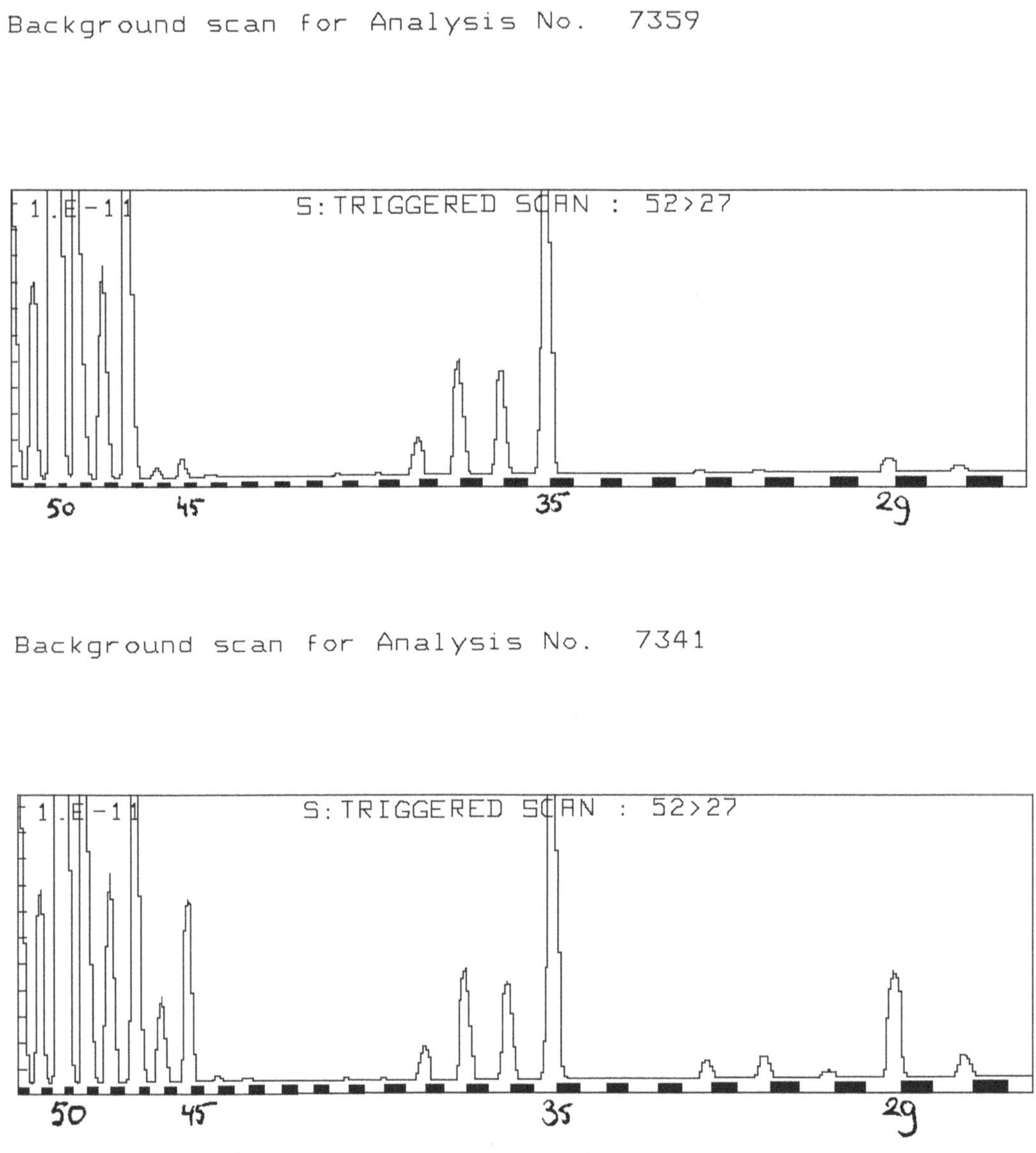

Fig. 5.1 Mass scans of chloromethane samples from the SIRA 24 mass spectrometer at Utrecht University taken using the original SIRA software running on a Hewlett Packard computer. Analysis 7359 shows a sample that was not overheated, analysis 7341 shows a sample that was overheated. Note that in analysis 7341 the intensities of the peaks at m/z 29, 45 and 46 have been increased significantly (Eggenkamp 1994)

cryogenic separation and found that 2,2,4-trimethylpentane (iso-octane) with a melting point of −107 °C was able to release chloromethane at its melting point, while iodomethane was only released after the complete melting of the 2,2,4-trimethylpentane. Their procedure is as follows. A scratched capsule with a mixture of chloromethane and iodomethane is put on-line in a tube cracker. The lowest part of the sample capsule is cooled to −72 °C with a dry ice-ethanol slush. On the vacuum line two sealed cold traps are installed containing 2,2,4-trimethylpentane. The line is evacuated after which the lower three quarters of the cold traps are frozen with liquid nitrogen. After braking the sample capsule the contents is slowly frozen to the first cold trap. After freezing all chloromethane and iodomethane to the first cold trap the liquid nitrogen from this trap is removed so that the temperature could increase. At the melting point of 2,2,4-trimethylpentane chloromethane vaporises and is frozen onto the second cold trap which is still frozen with liquid nitrogen. As long as both liquid and solid 2,2,4-trimethylpentane is present in the first trap only chloromethane is transferred to the second trap. After transferring all chloromethane, which is monitored from the pressure in the vacuum line the two cold traps are separated by closing a valve. To remove all remaining iodomethane from the chloromethane in the sample the separation is repeated by freezing the chloromethane to a third trap containing liquid nitrogen while the temperature of the second trap is increased to the melting point of 2,2,4-trimethylpentane. As long as both liquid and solid 2,2,4-trimethylpentane are present in the second trap only chloromethane is transferred to the third trap. Using this two step technique Wu and Satake (2006) managed to transfer chloromethane with a yield between 96 and 101%. The δ^{37}Cl of the transferred chloromethane was not significantly different before and after the separation. Analysis after only one separation step revealed still some iodomethane in the sample which led to δ^{37}Cl values in the samples which were higher than the original δ^{37}Cl values of the chloromethane by about 0.2 to 0.3‰, showing that a second separation step was necessary to implement on their extraction line.

A new and more simplified preparation line to separate chloromethane from iodomethane was recently described by Pelc (2024). The system to inject iodomethane into the preparation system, as well as the ampoules are equipped with Teflon valves. The proposed improvements resulted in significantly shorter preparation times, while the measurement uncertainty of 0.05 per mill is still comparable to other preparation and separation methods, including those with the use of gas chromatography.

5.2.2.2.7 Gas Chromatographic Separation of Chloromethane from Iodomethane

The second method to separate chloromethane from iodomethane for dual inlet mass spectrometry is by gas chromatography. Considering the large amount of chloromethane necessary for isotope analysis the gas chromatographic separation is done with the use of a packed column. Taylor and Grimsrud (1969) used Apiezon-L on Teflon T as stationary phase in a $^{5}/_{16}''$ glass capillary column for the separation of chloromethane from iodomethane. Kaufmann (1984) and Long et al. (1993) proposed the use of Porapack Q 80–100 mesh in a single column to separate chloromethane from iodomethane. Eggenkamp (1994) concluded that a two stage separation of chloromethane from iodomethane would reduce the risk of keeping any iodomethane in the chloromethane that is measured for isotope analysis. Godon (2000) improved the gas chromatographic separation by completely physically separating the two gas chromatography runs. In more detail the procedure developed is as follows (Eggenkamp 2004): chloromethane and iodomethane are separated first on a 75 cm, ¼" OD SS column, filled with Porapak-Q 80–100 mesh. Because there is a risk that this column is overloaded with the large excess of iodomethane, the gasses are separated in two runs. The carrier gas is helium, at a pressure of 3 bars and a gas flow of about 100 mL/min. The column temperature is kept at 140 °C. The procedure starts with backflushing the gas chromatograph in order to minimize iodomethane contamination of the column and detector. The Pyrex reaction tube is scratched by a glass cutting knife, and placed in a crushing tube. The tube cracker is evacuated and liquid nitrogen is placed around the first cold-trap. This cold trap is isolated from the rest of the gas chromatograph and the reaction capsule is broken. A helium flow is put on the cold trap to collect all chloromethane and iodomethane within this trap. After a short time, when all chloromethane and iodomethane are trapped the liquid nitrogen around the first cold-trap is replaced with warm water and the liquid nitrogen is now placed around the second cold-trap. The gas chromatography recorder is started and a chloromethane peak is detected after about 2 min. When the recorder is back on the base line, but before the iodomethane peak is detected (after about 4 min) valves are switched so that the first column is now back flushed. The chloromethane is at that moment trapped in the second cold trap, and only very little iodomethane should be mixed with it. Then the liquid nitrogen around the second cold-trap is replaced with warm water to release the chloromethane. Liquid nitrogen is placed around the sample cask. Chloromethane is swept over the second gas chromatographic column with the same characteristics as the first column. After the chloromethane has passed the column it is collected, frozen onto liquid nitrogen, in a sample vessel. The gas yield in the vessel is determined by measuring the chloromethane pressure in the vessel after heating the sample vessel with a hair dryer to evaporate the chloromethane. Finally the valves on the vessel are closed and the vessel can be removed from the gas chromatograph and subsequently attached to the mass spectrometer for isotopic analysis.

5.2.2.3 Measurement of Chloromethane as Positive Ions

Measurements of chlorine stable isotopes with most modern isotope ratio mass spectrometers, both dual inlet and continuous flow, are done on positive CH_3Cl^+ ions. These positive ions are produced from chloromethane when a positive acceleration voltage is applied in the ion source of the mass spectrometer (Taylor and Grimsrud 1969). For most modern mass spectrometers this is the normal mode of operation. It leads to a relatively high yield of positive ions. A potential disadvantage of the use of positive CH_3Cl^+ ions is however the relatively complex mass spectrum (see Table 5.3). This mass spectrum is so complex as both chlorine and carbon have minor isotopes (which is ^{13}C for carbon) and thus a large number of different ions can be formed from a chloromethane molecule in the ion source. Under normal conditions the isotope ratio is determined from the *m/z* mass ratio 52/50, where the major ions are $CH_3{}^{37}Cl^+$ and $CH_3{}^{35}Cl^+$. While $CH_3{}^{37}Cl$ is the only important ion at $m/z = 52$ at $m/z = 50$ also a significant fraction (about 1%) of the mass is from the ion $CH^{37}Cl^+$. As the isotope variation of chlorine is relatively modest (most data observed in natural samples are between +2.0 and −2.0‰ relative to SMOC) the error introduced by this small amount of $CH^{37}Cl^+$ is not normally measurable at the precision of the "best" possible measurements of chlorine isotopes, which are in the order of ±0.05‰ to ±0.1‰. For this reason most laboratories do not apply a correction factor to correct for this effect.

5.2.2.4 Measurement of Chloromethane as Negative Ions

In a few studies it is proposed to measure chlorine isotope compositions on Cl^- which is produced if the accelerating voltage in the ion source is switched to negative values. The advantage of negative ions is that a very simple mass spectrum is formed (Table 5.4) and the ions measured are $^{37}Cl^-$ and $^{35}Cl^-$ at $m/z = 37$ and $m/z = 35$. As these ions are also the only ions formed at these mass to charge ratios there is no necessary at all to apply correction factors to the measurement of chlorine isotopes this way.

Taylor and Grimsrud (1969) were the first to propose to analyse chlorine isotopes on negative ions. They compared positive and negative ions from chloromethane formed in their ion source by switching the acceleration voltage between +3000 and −3000 V. Although the ion yield for negative ions was about 30 times smaller than for positive ions the analytical error (standard deviation) was about ±0.22‰ and this was comparable to studies for positive ions. These results made that Taylor and Grimsrud (1969) recommended negative ions for future studies on chlorine isotopes. However, probably because chlorine isotopes in later studies were always measured on mass spectrometers that were also used for other isotope systems where positive ions were used, these recommendations were largely ignored.

A new ion source for the measurement of chlorine isotopes by negative ions was developed by Pelc and Hałas (2008). Based upon the original findings by Taylor and Grimsrud (1969) they built a new ion source specifically for negative ion measurements. As the yield in negative ions is relatively low compared to the yield in positive ions considerable effort was made to select the most appropriate filament material. It was found that the most suitable material to be used as a filament was a $Mo_{52.5}Re_{47.5}$ alloy as both the ion current intensity and the background were fully acceptable. As shown in Table 5.4 the mass spectrum for chlorine was comparable to that of Taylor and Grimsrud (1969). It was found however that small peaks of *m/z* 26 and 28 were found, probably consisting of the ions CN^- and CO^-. The development of this negative ion source also led to the development of a technique to analyse chlorine isotopes that was better adapted to the use of negative ions (Halas and Pelc 2009). The preparation of chloromethane from chloride is comparable to the method described above (precipitation of silver chloride and subsequent reaction of silver chloride with iodomethane. The reaction is done in glass ampoules with Teflon stopcocks and chloromethane and iodomethane are separated by cryogenic separation according to the method developed by Wu and Satake (2006). The analysis of a standard sample seems very promising with an internal precision better than 0.01‰. External precision was not determined yet, but seemed to be in the order of only few hundredths of a per mill (Dembska-Sięka et al. 2020).

5.2.3 Continuous Flow GC-IRMS

5.2.3.1 Gas Separation by Capillary Gas Chromatography

A large part of the currently measured chlorine stable isotope data are determined by continuous flow gas chromatography IRMS. Two techniques to establish this have been proposed in the literature. The first was proposed by Wassenaar and Koehler (2004). Their method included off line preparation of chloride bearing samples until the formation of chloromethane in stopper sealed glass vials. Silver chloride was produced from aqueous samples according to techniques described by Long et al. (1993) and Eggenkamp (1994). Conversion of silver chloride to chloromethane was done in 10 mL serum bottles crimp sealed with thick 13x20 mm butyl blue stoppers. Vials containing filters with silver chloride were evacuated through a hollow needle trough the stopper and subsequently iodomethane was injected through this stopper. The conversion to chloromethane is then done conventionally at 80 °C for 48 h.

To analyse the chlorine stable isotope ratio of a sample 150 to 700 uL of headspace is injected into a 2 meter Porapack-Q $^1/_4$ inch stainless steel packed column. The column was held at 160 °C with a helium flow of 100 mL/min and a column head pressure of about 150 kPa. At the beginning of a run a 30 s reference chloromethane pulse is introduced trough the dual-inlet bellows. The chloromethane peak followed at about 4 min. The rest of the gas introduced to the gas chromatograph (mainly iodomethane) was diverted to the waste through the open split. Due to the time necessary to get rid of the total iodomethane present in the sample the total analysis takes about 14.5 min. The standard deviation of a large series of measurements of the same sample is about 0.06‰.

Shouakar-Stash et al. (2005) proposed to analyse stable chlorine isotopes with continuous flow using a capillary gas chromatography column. The procedure also starts with precipitating silver chloride from an aqueous solution as described by Taylor and Grimsrud (1969) and Eggenkamp (1994). The conversion from silver chloride to chloromethane takes place in 20 mL amber crimp cap vials. As a capillary column is easily overloaded with large amounts of chloromethane and iodomethane small samples (0.2 mg) are weighted and added to the vials. Iodomethane is added in a glove bag and vials are sealed afterwards to avoid evaporation of iodomethane. The conversion reaction takes place in an oven at 80 °C for 48 h. After the conversion is complete the vials are loaded into a CombiPAL autosampler. From each sample 750 uL of the headspace is taken and injected into an Agilent 6890 gas chromatograph containing a 60 m, 0.32 mm ID DB-5 column with a 1 um film thickness. After injection the temperature is held at 35 °C for 7 min after that the temperature is increased quickly (at 35 °C per minute) to 250 °C and held at that temperature for 2 min. The advantage of this temperature program is that the separation between chloromethane and iodomethane is complete and iodomethane is flushed rather quickly reducing overall analysis time. From the gas chromatograph the gas is diverted into the mass spectrometer. The mass spectrometer program is set as such that only the chloromethane peak is diverted to the mass spectrometer, while the iodomethane is diverted to the flame ionization detector from the gas chromatograph, avoiding contamination of the mass spectrometer but also making it possible to monitor the development of the iodomethane peak. Before and after the chloromethane peak several reference gas peaks are measured, and each sample is measured at least in triplicate. Shouakar-Stash et al. (2005) report an internal precision better than ±0.04‰ and an external precision better than ±0.07‰ (1 standard deviation) for this method.

5.2.3.2 Separation of Chloromethane and Iodomethane on a GasBench

Liu et al. (2013) proposed to use a Gasbench II as the interface to transmit chloromethane onto the mass spectrometer. Chloromethane was produced from chloride in aqueous solution after precipitation with silver nitrate and subsequent reaction of the formed silver chloride with iodomethane in a glass vial. Vials used by Liu et al. (2013) were of the screw cap type. Unlike reaction vials for previous techniques they were flushed with helium before iodomethane was added and the reaction between silver chloride and chloromethane takes place in a helium atmosphere and not in vacuum as in previously developed methods. The separation and subsequent measurement of the chlorine isotope composition is carried out after a reaction time of 48 h at 80 °C. The gas is sampled through a special dual-hole needle which is connected to the CombiPAL autosampler. A helium flow (0.5 mL/min) is delivered to the vial trough the side hole and the samples is flushed into the GasBench system through the needle tip. The helium flow gently transfers chloromethane and other gas species from the vial into the fused-silica capillary within the needle tip. The gas mixture is transferred through a Nafion® guard trap to remove water and through a Valco valve. It is then set onto the GasBench gas chromatography (GC) PoraPlot Q column. This is done through a 100 µl loop in the loop injection system. Here chloromethane is separated from all other gas species, most importantly iodomethane. The flow rate on the gas chromatograph is set at 2 mL/min and the helium pressure at 12 psi (0.83 bar). The temperature of the gas chromatographic column is kept at 120 °C. After the chloromethane has been eluted from the column it is passed through a second Nafion® guard column and introduced into the isotope ratio mass spectrometer via the active open split. The reference pulse of chloromethane gas, which is of 99.95% purity, is automatically introduced into the mass spectrometer via a reference gas injection system. Two Faraday cup detectors monitor simultaneously and continuously the CH_3Cl^+ signals for the two major ions at *m/z* 50 ($CH_3{}^{35}Cl^+$) and 52 ($CH_3{}^{37}Cl^+$). The active open split is closed immediately after the chloromethane signal has passed to prevent iodomethane from flowing into the mass spectrometer. The chlorine isotope composition of the chloromethane from a single vial can be analysed several times by switching the contents of the loop injection system forth and back onto the gas chromatography column. The precision of this method, deducted from 12 measurements of the secondary isotope standard ISL 354 produced from seawater (Xiao et al. 2002), and seven measurements of the sodium chloride standard NIST SRM 975 was 0.07‰.

5.2.4 N-TIMS

In theory the measurement of chlorine isotopes by negative thermal ionization mass spectrometry is very simple. The method was for example used by Shields et al. (1962) in their study to establish the absolute isotopic abundance ratio and atomic weight of chlorine. They concluded that they were able to analyse stable chlorine isotopes with a precision of ±2‰. After the first studies of chlorine isotopes with "measurable" natural chlorine isotope variations (Kaufmann et al. 1984; Desaulniers et al. 1986), Vengosh et al. (1989) attempted to apply a new method to directly measure both boron and chlorine isotopes by negative thermal ionization mass spectrometry. They added droplets containing 1 to 500 μg chloride (and 50–200 ng boron) to rhenium filaments which were dried in an oven at about 50 °C. Fifteen of these filaments were loaded into a sample carousel of a Nuclide mass spectrometer and evacuated to 10^{-7} to 10^{-8} Torr. Applying a technique that was strictly repeated for each measurement they were able to analyse first the chlorine isotope ratio and subsequently the boron isotope ratio. Briefly the procedure to analyse chlorine isotopes was to slowly heat the filament to 1.7–2.2 A filament current, then the signal was optimised by focusing the ion beam and slowly increasing the filament current. When a $^{37}Cl^-$ ion intensity of 2 to 3 V was reached chlorine isotope data were collected for 30 to 45 min. During this period a total of 96 to 192 ratio measurements ($^{37}Cl/^{35}Cl$) were made. From 18 analyses of a seawater standard it was revealed that the analytical error was ±2.0‰, which is surprisingly similar to the value determined by Shields et al. (1962). More surprisingly however was the fact that (with analytical errors of 1.4 to 2.9‰) very large variations between different samples were found (from −7.2 to +24.7‰ relative to their seawater sample) without a clear relationship with the origin of the samples. Several of these samples were later remeasured by Xiao et al. (1995) and these measurements showed a much smaller, and perhaps more realistic, range in chlorine isotope compositions. One other study is known in which attempts were made to analyse chlorine isotopes by N-TIMS from this period (Gaudette 1990). In Antarctic ice very negative $\delta^{37}Cl$ values were found with values between −14 and −10‰ relative to their ocean chloride standard. These extreme variations were however never found with other analytical techniques and apparently the negative ion beams were not stable enough for precise chlorine isotope analysis due to the effect of isotope fractionation during the evaporation and ionisation process and large filament blanks (Magenheim et al. 1994; Fujitani et al. 2010).

Despite these analytical problems potential interest continued to exist in this technique as very small amounts of sample, only several hundreds nanograms of chloride or ten times less than necessary for P-TIMS, are necessary for analysis as was shown by Fujitani and Nakamura (2006) in their study to analyse very small chloride concentrations by isotope dilution mass spectrometry. Their approach to solve the analytical problems was to apply a total evaporation method. The total evaporation method indicates that during measurement of a sample the whole sample is evaporated from the filament and analysed. Because of this total evaporation the entire ion beam is integrated so that the effect of isotopic fractionation during data acquisition is minimised. The total evaporation technique was originally developed for analysing the isotope composition of uranium, plutonium (Callis and Abernathey 1991) and neodymium (Wakaki et al. 2007). Another application to analyse stable isotopes with this technique was in a study of rhenium isotopes (Suzuki et al. 2004). In order to properly measure the chlorine isotope composition Fujitani et al. (2010) added either 100, 200 or 500 ng chloride on the filaments and increased the ionization current to 1.8 A, corresponding to a filament temperature of about 1300 °C, while keeping the evaporation filament at 0 A throughout the analysis. Chloride was evaporated from the filament at such a rate that the complete sample was evaporated within 100 to 200 scans. Fujitani et al. (2010) compared results from these N-TIMS analyses with P-TIMS analyses from

Numata et al. (2001). The results showed that at 500 and 200 ng chloride samples could be measured with an error of ±0.9‰, while at smaller sample amounts the error increased to ±2.0‰. In comparison a sample containing 2000 ng chloride (seawater) showed an error of ±0.31‰. These results showed that there may be potential for N-TIMS for chlorine isotope analysis for very small samples, however only when a relatively large variation in chlorine isotope values is expected. Realising that there is only a relatively small natural variation in chlorine isotope ratios application of this technique will continue to be rather limited.

5.2.5 P-TIMS

The application of positive ion thermal ionization mass spectrometry for chlorine isotopes has been developed as a spin-off of the development of boron isotope analysis by the same technique. It was already known for some time (see the review by Xiao et al. 1991) that $M_2BO_2^+$ (where M = Li, Na, K, Rb or Cs) ions could be used to determine boron isotope compositions using thermal ionization mass spectrometry with good precision. Xiao et al. (1991) considered the possibility to extent this technique to chlorine isotope analyses and to measure the chlorine isotope ratios in the form of M_2Cl^+ ions. They realised that it would be big challenge as it is very difficult to obtain a strong and stable M_2Cl^+ ion beam. Xiao et al. (1988) had already discovered that a relatively strong Cs_2Cl^+ ion peak was formed when they analysed boron isotopes on $Cs_2BO_2^+$ ions if the filament was first covered with graphite, while no Cs_2Cl^+ ion peak was found in the absence of graphite. Xiao et al. (1991) considered this a good opportunity to study the possibility of analysing chlorine isotopes using TIMS in the presence of graphite on the filament. During experiments with rubidium chloride as the compound to study isotope ratios it was confirmed that in the presence of graphite Rb_2Cl^+ ions were emitted, while no Rb_2Cl^+ ions were emitted in the absence of graphite. In the absence of graphite just Rb^+ ions were emitted. It was concluded that in those experiments where the filament was covered with graphite the ionization potentials of the M_2Cl^+ ions and the temperatures at which a strong ion beam could be emitted were reduced. This effect resulted in increased M_2Cl^+ intensities. Studies on the formation of M_2Cl^+ ions from rubidium chloride and caesium chloride were performed by Xiao et al. (1991). The initial results were very promising and as caesium has only one stable isotope it was decided to continue the experiments with this element. Another very important consideration is the high mass of the Cs_2Cl^+ ion. This ion is measured at $m/z = 301$ and 303 and this high mass makes it more easily to control the mass spectrometric discrimination factor. As the mass spectrometric discrimination is small and reproducible this results in precise isotope measurements. Finally, addition of caesium chloride to a filament covered with graphite was ultimately shown to give a successful procedure to analyse chlorine isotopes (Xiao and Zhang 1992).

Xiao and Zhang (1992) applied this method using tantalum filaments. These filaments were treated with 3 μL of a graphite slurry that consisted of about 100 μg graphite in a mixture of 80% ethanol / 20% water (v/v). This slurry was allowed to dry almost completely. Then a sample solution, containing of 1 to 6 μg chloride was added to the filament. To prepare this sample solution the original chloride solution was transferred into a hydrochloric acid solution by cation exchange chromatography with resin in the Ba^{2+} form. This acid solution was neutralised with a caesium carbonate solution. The sample solution was dried for 2 min while passing a current of 2 A through the filament. When the filaments were dry they were loaded into the source. After evacuation of the ion source the analysis was started by increasing the filament current to 1.1 A. Cs_2Cl^+ ions that were emitted were monitored and data were collected by switching between the masses 301 and 303. Data were acquired for a period of 1.5 h. Seawater and samples from terrestrial salt lakes in China were measured in these original studies. Results obtained by this technique were reported to have errors better than 0.1‰ (Liu et al. 1997; Shirodkar et al. 2006).

The method developed by Xiao and Zhang (1992) was adapted in several laboratories. Magenheim et al. (1994) adopted the method to study chlorine stable isotope variations in small natural samples such as rocks and aerosols. They discovered that large quantities of sulphate, fluoride and organics in samples resulted in ionization suppression, so they developed techniques to remove them during sample preparation. They used ion exchange to remove sulphate, chemical precipitation to remove fluoride and extraction by activated carbon powder to remove organics. Magenheim et al. (1994) produced a stoichiometric (1:1) caesium chloride solution by eluting their acid solution over an ion-exchange resin in the Cs^+ form. This way they removed all acidity, which is important as acidity present as hydrochloric acid could cause isotope fractionation in the sample (Volpe and Spivack 1994). The precision of this technique was found to be better than ±0.2‰ in general. More recently comparable methods to analyse chlorine isotope compositions with positive thermal ionization mass spectrometry were proposed by Numata et al. (2001) and Nešković et al. (2002). Numata et al. (2001) prepared samples as silver chloride precipitate and converted to caesium chloride with metallic magnesium and ion exchange resin in its caesium form. Samples were loaded on tantalum filaments and the reported precision is about 0.1 to 0.2‰. Nešković et al. (2002) used rhenium filaments and they neutralised their hydrochloric acid solution with a caesium carbonate solution as Xiao and Zhang (1992) did. The preliminary result published in their report showed a very large difference in isotope ratio between the measurement of their seawater sample from a sodium chloride working standard (8.8 ± 1.2‰) and a measurement of the same material done by the method described by Long et al. (1993) in Arizona (0.39 ± 0.03‰). Because of this large difference it was concluded that more development of the method was necessary, however no more reports from this group have unfortunately appeared since then.

5.2.6 SIMS

A method to analyse chlorine isotopes using secondary ion mass spectrometry has been developed by Layne et al. (2004). The advantage of analysing chlorine isotopes by this technique is the fact that is able to analyse the composition on a very small spatial resolution (10 to 20 μm) so that in theory very small samples can be analysed, or that spatial variations can be determined in larger samples. The technique however is only viable in samples with relatively large variations as the overall reproducibility of this technique was originally only 1.5‰. In spite of this the first published results were encouraging, and later studies showed that the overall reproducibility decreased with gaining experience using SIMS to measure chlorine stable isotope variations.

Layne et al. (2004) used a Cameca IMS 1270 ion microprobe to analyse their samples. They tested the method for very small spots from rock samples and direct determinations of the contents of small dried samples of natural fluids such as fluid inclusions.

SIMS techniques for chlorine isotopes can only be successful if isobaric interferences are suppressed. Natural samples almost always contain sulphur which forms $^{34}SH^-$ ions that interfere with $^{35}Cl^-$. Modern ion microprobes however have very high mass resolution capabilities and they make it possible to analyse the proper ions ($^{35}Cl^-$ and $^{37}Cl^-$). In the IMS 1270 chlorine concentration and chlorine isotopes are analysed simultaneously. The sample is bombarded with a microbeam of Cs^+ ions focused on a 10 to 20 um diameter spot. This bombardment releases secondary ions which are accelerated into the mass spectrometer. Peaks for $^{30}Si^-$, $^{35}Cl^-$ and $^{37}Cl^-$ are obtained by cyclical magnetic peak switching. The isotope composition is determined from switching between $^{35}Cl^-$ and $^{37}Cl^-$ and the chlorine concentration from comparing $^{35}Cl^-/^{30}Si^-$ (normalised for SiO_2 content) for unknown and standard glasses (see e.g. Hauri et al. 2000). As chlorine has a very high ionization efficiency under Cs^+ bombardment high count rates are found. This implicates that high internal precisions are reached, while only a few ng of sample from the small sputtered crater is consumed. The measurements normally yield internal precisions which are better than ±0.9‰. Overall reproducibility generally is better than ±1.5‰ for samples with as little as 250 ppm chloride. Layne et al. (2004) state that their method is able to analyse samples with chloride concentrations down to 100 ppm, while IRMS and TIMS techniques become difficult at 100 to 200 ppm. More recent IRMS analysis however yielded reproducible results for rock samples with chloride concentrations as low as 50 ppm (Bonifacie et al. 2008), so that this no longer is a major advantage. The main advantage however will continue to be that very small samples can be analysed and that the distribution of chlorine isotopes in individual minerals in larger samples can be analysed. Later studies showed improvement in the methods for chlorine isotope measurement by SIMS and in most recent studies a reproducibility of 0.3‰ (2σ) for larger and 0.8‰ (2σ) for smaller chloride contents is reported (Bouvier et al. 2022).

5.2.6.1 Measurement of Apatite Samples with SIMS

One of the most interesting developments in the last 10 years is the application of SIMS to measure the chlorine isotope composition of apatite. Apatite is a relatively chloride rich mineral (see Chap. 3) and regularly present in terrestrial samples, meteorites and material returned from Solar System Bodies by various space missions. Due to the high chloride content in apatite it is possible to measure the chlorine isotope composition by SIMS in the very small crystals present in the precious little samples that are available for study.

As chlorine isotopes in apatite can be measured by any regular SIMS, in recent years several studies have been published for which the samples have been measured at a large range of SIMS facilities, such as Milton Keys (Hayden 2023, Darling et al. 2021), Beijing (Liu et al. 2022), Guangzhou (Ciu et al. 2021), Lausanne (Bouvier et al. 2022) and Potsdam (Wudarska et al. 2020).

Although slight differences exist in the analytic procedures from one to another facility in principle the procedure is comparable between the facilities. Here the method described by Wudarska et al. (2020) is summarised, which represents the method applied at the Helmholtz Centre in Potsdam (Germany). Small apatite crystals are mounted in the centre of 25-mm diameter acrylic discs using epoxy resin and polished to a < 5 μm surface quality. $\delta^{37}Cl$ values are determined using a Cameca 1280-HR large geometry SIMS. Prior to each measurement session, the sample mounts are cleaned in high-purity ethanol and are coated with a 35-nm thick high-purity gold film to assure electrical conductivity. A 2 nA, 10 keV $^{133}Cs^{+}$ primary ion beam is focused to a ~ 5 μm diameter spot to produce $^{35}Cl^{-}$ and $^{37}Cl^{-}$ ions following a modified protocol based on that of Kusebauch et al. (2015). A 25 μm × 25 μm area of the sample surface is sputtered for 60 s to remove the gold coat and to suppress surface contaminants. Then, the raster is reduced to a 15 μm × 15 μm analysis area. ^{35}Cl and ^{37}Cl secondary ions are collected simultaneously on Faraday cups over 20 cycles, with each cycle lasting for 4 s. The mass resolving power is set at $M/\Delta M \approx 2000$ (at 10%). A normal incidence, low-energy electron flood gun is used for charge compensation of the sputtered region. $^{37}Cl/^{35}Cl$ ratios measured are corrected for instrumental mass fractionation (IMF) based on reference materials of known chemical and isotopic composition. The repeatability of the measurements conducted on a reference sample is better than 0.09‰ (1σ) in all cases, confirming the high precision of the analyses (Wudarska et al. 2020).

5.2.6.2 Measurement of Other Chlorine Rich Minerals by SIMS

SIMS can also be used to measure chlorine isotope variations in other chlorine rich minerals. Eggenkamp et al. (2021) described chlorine isotope measurements of eudialyte and sodalite samples. Unfortunately the method they used was not described in their abstract.

5.2.7 Multi Collector ICP-MS

Measurement of chlorine stable isotopes by multi collector ICP-MS is challenging. This is mostly due to the existence of isobaric interferences by argon in the form of $^{36}ArH^+$ fragments which has almost the same mass as ^{37}Cl. Still, in a small number of studies chlorine isotopes ratios were determined with ICP-MS which reached satisfactory results.

De Gois et al. (2016) solved the challenging conditions by using a very high mass resolving resolution of about 10,000, the use of a 5 mM Na_4OH wash solution to minimise memory effects and high enough chloride concentrations of ≥70 mgL^{-1}. The high cation load of seawater samples (mostly Na and K) was removed efficiently prior to analyses with the use of ion exchange over Dowex 50WX8 resin. They concluded that taking these conditions into account the method was both accurate and precise.

5.2.7.1 Laser Ablation Multi Collector ICP-MS

Fietzke et al. (2008) proposed a relatively simple method to analyse chlorine isotopes from aqueous fluids and rocks using laser ablation multi collector ICP-MS. They extracted chloride from rocks with pyrohydrolysis methods as developed by Dreibus et al. (1979), Schnetger and Muramatsu (1996) and Bonifacie et al. (2007). Chloride is precipitated as silver chloride following precipitation by addition of 8 mL of a (1 part 65%) HNO_3 / (6 parts 0.1 N) $AgNO_3$ mixture to 25 mL of the sample containing enough chloride for analysis. An advantage of this procedure is the fact that the chloride in all samples is in a pure silver chloride matrix which largely reduces matrix effect problems during measurement of the sample.

A problem with analysing chlorine isotopes by ICP-MS is interference of $^{36}ArH^+$ from the plasma with $^{37}Cl^+$. This problem was by-passed through simultaneously monitoring masses 35 ($^{35}Cl^+$), 37 ($^{37}Cl^+ + {}^{36}ArH^+$) and 41 ($^{40}ArH^+$). The $^{36}Ar/^{40}Ar$ ratio is constant at 0.003 and this ratio is used to correct the peak at mass 37 for the ArH^+ peak. This correction is shown to be very efficient. Without the correction $^{37}Cl/^{35}Cl$ ratios of the SMOC standard range from 0.351 to 0.356, while with the correction (subtraction of 0.003*mass 41) this is reduced to a range from approximately 0.3558 to 0.3562, a factor of 12 smaller. Using this correction it was possible to analyse the chlorine isotope composition with a precision of ±0.06‰ (2 SE) with the use of less than 1 μg chloride.

A different technique to measure silver chloride by laser ablation MC-ICP-MS was proposed by Toyama et al. (2015). They used the $^{36}Ar^+/^{38}Ar^+$ ratio as an external standard for the mass bias corrections between $^{39}K^+$-$^{41}K^+$, $^{36}Ar^1H^+$-$^{38}Ar^1H^+$-$^{40}Ar^1H^+$, and $^{35}Cl^+$-$^{37}Cl^+$ with isobaric overlap corrections between K^+, ArH^+, and Cl^+ ions. Overlap of the sulphur ($^{36}S^+$) isobar with $^{36}Ar^+$ was indirectly monitored and corrected by baseline modelling using the $^{36}Ar^+/^{38}Ar^+$ ratio measurement. These procedures collectively helped to accomplish an accurate and high-precision measurement of the $^{37}Cl^+/^{35}Cl^+$ ratio. Toyama et al. (2015) were able to measure samples containing as little as 4 μg chlorine with 2σ accuracies of 0.2‰ for halite and of 0.4‰ for rock samples.

5.2.7.2 Possible Sequential Analysis of Cl, Br and S Isotopes after Ion Chromatography

After methods were developed to analyse bromine isotopes by multi collector ICP-MS (Gelman and Halicz 2010), Zakon et al. (2014) coupled their ICP-MS to an Ion Chromatograph. In samples with sufficient sulphur, bromide and chloride present they were able to measure the stable isotope composition of all three elements, in principle, in a single run.

This method is based on online separation and purification of ions by ion chromatography before isotope analysis by MC-ICP-MS. The proposed technique can simplify $\delta^{37}Cl$, $\delta^{81}Br$, and $\delta^{34}S$ analysis in environmental samples significantly. It is also possible that, in samples with different anionic species of the same element (such as Cl^- and ClO_3^-), that the chlorine isotope composition of both species can also be analysed in a single analytical run. Major isobaric interferences, which are especially present with chlorine, are reduced by the use of "dry" plasma conditions and to apply sufficient mass resolution power. The sample-standard bracketing technique is used for instrumental drift correction. Precisions of 0.2‰ (1σ) have been obtained for $\delta^{37}Cl$ with analytes containing 0.7 μmol of Cl. For S and Br precisions up to 0.15‰ (1σ) and 0.1‰ (1σ) respectively have been achieved for analytes containing down to 5 nmol for S and 0.6 nmol of Br. Robustness of this analytical method, as well as its high precision and accuracy, was demonstrated in laboratory standard solutions and for a small number of environmental samples (Zakon et al. 2014).

5.2.8 Measurement of Chlorine Isotopes with Neutron Activation Analyses

The samples obtained during the diffusion experiments with molten salts (which are to be described in Sect. 8.1.2) were analysed after irradiation with thermal neutrons of silver chloride prepared from these samples. The procedure that was followed is described by Klemm and Lundén (1955). The salt obtained from the experiments was dissolved and the chloride was precipitated as silver chloride after addition of a silver nitrate solution. The silver chloride is then irradiated with neutrons and the chlorine isotopes are partly changed into the isotopes with one extra neutron (n,γ-process). ^{36}Cl and ^{38}Cl have a considerable different half-life, and from the β-activity in the sample the percentage ^{37}Cl could be determined. This method is not very precise, and the percentage ^{37}Cl in the samples could only be determined with a precision of about 1%. For that reason the chlorine isotope composition in later studies were also determined using the then newly developed chloromethane (IRMS) method (Herzog and Dörnenburg 1958).

5.2.9 Measurement of Chlorine Isotopes Using Fast Atom Bombardment Mass Spectrometry

Westaway et al. (1998) proposed a method to analyse the chlorine isotope composition of silver chloride precipitated from a chloride solution by Fast Atom Bombardment using a modified, hybrid model MI 1201E FAB isotope ratio mass spectrometer (PO Electron, Ukraine) equipped with three collectors: two Faraday cups with fixed focus and an electron multiplier with pulse analyser and a single-ion counter. Using this method silver chloride was put on a silver filament and melted. This was then mounted into the mass spectrometer and bombarded with xenon atoms. Negative ions ($^{35}Cl^-$ and $^{37}Cl^-$ were formed of which the ratio could be determined in 30 to 50 different determinations. The precision of this method was about ±1‰ and its main advantage was to avoid the time-consuming preparation, purification, and recovery of the gaseous chloromethane.

5.2.10 Measurement of Very Small Samples with the Use of AMS

In order to measure chlorine isotopes in "astromaterial", usually consisting of very small amounts of meteorites or return samples from space missions, a high quality isotope analysis of very small amounts of sample must preferably be made. The measurement of such small samples is hugely important in order to understand the chlorine isotope composition of the different parts of the solar system. Chlorine isotope measurements on small samples are regularly determined using SIMS, which delivers spot measurements from very small sample amounts, but that can vary significantly on nanoscale distances (Darling et al. 2021), and is only measurable in minerals that have high enough chloride contents such as apatite. That means that SIMS is not easy to use to measure bulk chlorine isotope ratios of chlorine poor and heterogeneous samples that are comprised of multiple phases. These bulk samples are mostly measured using IRMS, but for that technique 0.2 to 1 mg chloride is necessary for a proper, reliable measurement, and with "astromaterial" this is rarely available. For that reason Anderson et al. (2022) developed a method to measure extremely small amounts of chlorine (down to 1 μg). They produced AgCl from their small bulk sample through mixing the chloride bearing solution with Nb powder in a centrifuge tube and added Milli-Q water, KNO_3 and $AgNO_3$ to precipitate AgCl. This mixture was centrifuged and the precipitate put in a small quartz vial and dried overnight at 90 °C. The dried AgCl (plus Nb) was loaded into steel cathodes prior to insertion into a modified Middleton source of an accelerator mass spectrometer (AMS). Ions were sputtered using a cathode voltage of 4 kV and accelerated to an energy of 40 keV for samples >2 μg chloride and a higher voltage of 6 kV for smaller samples to produce sufficient current for a measurement. Typical $^{35}Cl^-$ currents for a target containing 25 μg Cl chloride standard were generally between 1 and 2 mA $^{35}Cl^-$, while samples with the lowest Cl content at 0.5 μg produced ~60 nA of $^{35}Cl^-$. Blanks appeared to be very good (only 0.01–0.015 μg chloride). 2σ errors of the measurement were between approximately 1 and 4‰, although relatively high with respect to chlorine isotope variations observed on Earth, these errors are well suited to understand the generally larger chlorine isotope variations as observed on other Inner Solar System bodies than Earth.

5.2.11 Comparisons Between Different Measurement Techniques

As chlorine isotopes are analysed by so many different techniques it is essential to know that they are mutually comparable, and that a sample analysed with one technique will have the same value when it is measured with a different technique. Only a few studies are known where samples measured with different techniques to analyse chlorine isotope compositions are compared. Two of these studies compare P-TIMS with IRMS techniques and two other studies compare IRMS with SIMS techniques.

Rosenbaum et al. (2000) compared P-TIMS with IRMS measurements of five samples with a $\delta^{37}Cl$ range between −4.4 and + 0.3‰ relative to SMOC. They found that within analytical precision the two techniques yielded similar results. It was concluded that P-TIMS would be the most appropriate technique for small samples with an optimal chloride amount of 0.1 to 0.3 μmol per analysis and an analytical uncertainty of ≤0.2‰, while IRMS would be more appropriate for larger samples with an optimal chloride amount of >10 μmol per analysis, but with a significantly better analytical uncertainty of ≤0.1‰. Godon et al. (2004a) extended this study by comparing both techniques for 6 samples with a total $\delta^{37}Cl$ range from −3‰ to +6‰. They also found a very good agreement between the two techniques.

Godon et al. (2004b) compared the results from analysis of chlorine isotopes from SIMS and IRMS techniques. They compared eight synthetic aluminosilicate glasses. They found that the raw results of the measurements from the two techniques did not agree very well but discovered that this was the result of a very systematic instrumental mass fractionation. It was found that the instrumental mass fractionation had a very good relation with the normalised major element composition of the analysed synthetic glasses. When the SIMS data are corrected for this mass fractionation the chlorine isotope data do agree very well within the errors of the measurements. This indicates that it is possible to compare results of the two techniques, but it also means that the instrumental mass fractionation needs to be determined. To obtain this information it is necessary that also IRMS measurements are produced. The combination of the two techniques on the same sample set is very useful as using SIMS it is possible to analyse very small areas (e.g. within individual crystals), so that very detailed variations in specimens can be determined. With IRMS the bulk isotope composition for the same mineral or rock can be determined. So, the two techniques can be complementary in their possible use to study chlorine isotope variations in rock and mineral samples.

Very important to that respect is the comparison of SIMS and IRMS data for apatites as recently the chlorine isotope composition of apatite in both terrestrial and extraterrestrial studies has been used extensively. When samples are measured by SIMS well characterised reference materials need to be used. Because apatite is increasingly used to characterise the chlorine isotope variation in a very large range of environments, from terrestrial samples of all ages, starting from the oldest rocks (e.g. Wudarska et al. 2020), to meteorites (e.g. Barrett et al. 2019) and potentially return samples from space missions (e.g. Morin et al. 2021). To facilitate future application of apatite in chlorine isotope studies, both by SIMS and IRMS Wudarska et al. (2021) made an inter-laboratory characterisation and comparison of six apatite reference materials to be used as future reference materials. The samples were selected for homogeneity (by SIMS) of their chlorine isotope composition and contained different chloride concentrations (0.26 to 6.09%) so that they can be applied over the full chloride range that is observed in apatite studies. Samples were measured by SIMS at the GeoForschungsZentrum in Potsdam (Germany) and compared to IRMS data determined at the Institut de physique du globe de Paris (France), at the University of Delaware (USA) and at the University of Texas (Austin, USA). Measurements showed to be challenging and show relative large $\delta^{37}Cl$ differences between the three IRMS laboratories. The results (see Table 5.5) show relatively large variations in $\delta^{37}Cl$ values of apatite depending on the method to extract chlorine from the apatite samples. The data suggest that for a better understanding and intercomparison of chlorine isotope data from apatite samples and considering the small number of samples considered in this study, that further research is definitely necessary.

Table 5.5 $\delta^{37}Cl$ values obtained by IRMS from 7 apatite samples and ISL-354 from three laboratories with different methods to dissolve the apatite, and, $\delta^{37}Cl$ values from the 7 apatite samples measured by SIMS in Potsdam. Full method descriptions and details on the analytical techniques are described in Wudarska et al. (2021)

Sample:	MGMH #133648	TUBAF #38	TUBAF #40	MGMH #128441A	TUBAF #50	TUBAF #37	Dur	ISL-354
B202IPGP, IRMS in dual-inlet mode, apatite dissolved in 4.4 M HNO_3	−0.01 ± 0.03 (n = 2)	−0.02 ± 0.06 (n = 3)	−0.01 ± 0.06 (n = 5)	+0.23 ± 0.06 (n = 3)	+0.01 ± 0.11 (n = 3)	−0.01 ± 0.05 (n = 2)		−0.01 ± 0.10 (n = 9)
Delaware, IRMS in dual-inlet mode, apatite dissolved in 0.5 M HNO_3	+0.09 ± 0.08 (n = 8)	+0.04 ± 0.04 (n = 6)	+0.22 ± 0.04 (n = 6)	+0.43 ± 0.06 (n = 6)	+0.36 ± 0.02 (n = 5)	+0.22 ± 0.04 (n = 6)	+0.20 ± 0.03 (n = 8)	+0.05 ± 0.04 (n = 13)
Austin, IRMS in continuous flow mode, chlorine extracted by pyrohydrolysis	+0.23 ± 0.40 (n = 2)	+0.43 ± 0.20 (n = 2)	+0.66 ± 0.08 (n = 2)	+0.68 ± 0.32 (n = 2)	+0.66 ± 0.15 (n = s)	+0.34 ± 0.09 (n = 2)	+0.17 ± 0.10 (n = 4)	−0.05 ± 0.14 (n = 2)
Potzdam, SIMS measurements randomised on random crystals	+0.10 ± 0.02 (n = 33)	+0.11 ± 0.05 (n = 34)	+0.23 ± 0.05 (n = 36)	+0.36 ± 0.06 (n = 30)	+0.26 ± 0.10 (n = 50)	+0.14 ± 0.19 (n = 30	+0.34 ± 0.10 (n = 43)	

5.3 Extraction of Chlorine from Silicate Rocks for Isotope Analyses

5.3.1 Alkali (NaOH) Fusion Method

Eggenkamp (1994) developed a method which was largely derived from the sodium hydroxide fusion method as used by Behne (1953). This procedure was described by Eggenkamp (1994, 2004) as follows: X grams of powdered rock, in which X is an amount of rock containing enough chloride to obtain one or more $\delta^{37}Cl$ analyses, is heated together with 10X grams of sodium hydroxide pellets for about 30 min in a nickel crucible. The rock dissolves in the molten sodium hydroxide and Si-O bonds are partly destroyed. The temperature must not be too high since sodium hydroxide will evaporate at high temperatures. The vapour pressure of sodium hydroxide is 1 mmHg (1.33 mbar) at 739 °C and 10 mmHg (13.3 mbar) at 897 °C (Stull 1947). After cooling the sample is dissolved in 35X ml distilled water. This is achieved by putting the nickel crucible in a beaker with water on a magnetic stirrer. After 2 h of stirring the content of the crucible is either dissolved or suspended in the water. This suspension is exposed to air for some time (overnight) to oxidise Fe^{2+} and other reduced ions. The solution has a very high pH so that addition of Ag^+ ions would result in precipitation of silver oxide; therefore the solution must be acidified. This is done by adding 17.5X mL 65% HNO_3 which produces a colloidal "solution" of silica gel according to:

$$SiO_4^{4-} + 4HNO_3 \rightarrow H_4SiO_4 + 4NO_3^- \quad (5.10)$$

This colloidal solution can unfortunately not be filtered but to make this possible 3X mL hydrofluoric acid 40% is added. The silica gel reacts with the hydrofluoric acid to form a combined silica oxyfluoride. This reaction does not go to completion. The oxyfluoride precipitates slowly (overnight) to the bottom of the glass container. Then 10.9X mL of a 2.5 mol.kg^{-1} magnesium nitrate ($Mg(NO_3)_2.6H_2O$) solution is added (64.1 gram $Mg(NO_3)_2.6H_2O$ and 35.9 gram H_2O which is a solution with a density 1.25 gr.ml^{-1}). Excess toxic HF will react to poorly soluble magnesium fluoride. The resulting suspension is filtered. The residue contains reacted silica gel, magnesium fluoride, insoluble hydroxides and rock components that were not dissolved in the sodium hydroxide melt. Chloride is concentrated in the filtrate. The filtrate has a low pH and a high ionic strength, so that the silver chloride can be precipitated directly from this solution without adding any buffer or potassium nitrate.

Interestingly when this method was repeated during a stay of the author at the Institut de physique du globe de Paris in early 2013 it appeared that the hydrofluoric acid was not removed completely from the solution after the addition of the magnesium nitrate solution. The result was that the final solution still contained hydrofluoric acid which made it impossible to filtrate the suspension as the remaining hydrofluoric acid dissolved the glass fibre filter during filtration of this suspension. While the procedure described above was successfully applied during the PhD study of the author (see Eggenkamp 1994, Chap. 11 and Eggenkamp and Schuiling 1995) it apparently didn't work anymore during a later test. The only differences between the procedures as described above and as applied during the stay at the IPGP was the fact that during the PhD studies all solutions were kept in glass beakers while, due to the potential to dissolve glass by hydrofluoric acid, in Paris they were all kept in plastic beakers. What perhaps actually happened was that the hydrofluoric acid dissolved part of the glass and neutralised itself in this action, something that did obviously not happen with the plastic containers in Paris. This interesting effect should be studied in the future to explain what really occurs during the procedure and perhaps to be able to develop a more convenient method to extract chloride from silicate rock samples.

The method proposed above was later slightly modified in Godon et al. (2004b) who decided to centrifuge the solution after acidification with HNO_3 to remove the silica gel that has formed. After centrifuging a clear solution could be decanted containing the chloride. This way they were able to avoid the addition of hydrofluoric acid to their sample.

5.3.2 Hydrogen Fluoride Decomposition Method

Musashi et al. (1998) pointed to the fact that a weakness in the technique described in Sect. 5.3.1 is the addition of large amounts of chemicals to the samples, resulting in risk of contamination. The changes proposed later by Godon et al. (2004b) addressed this matter to a large part, but it has to be realised that addition of large amounts of chemicals has the risk that the sample is contaminated with chloride that may be present in those chemicals. However, tests done by Eggenkamp (1994) showed no measurable chloride in a blank (chloride free sea-sand) that was treated in the same way as actual samples, and it was also shown that no fractionation of chlorine was found due to the complex chemistry. However, Musashi et al. (1998) considered the drawback such that they decided to develop a different technique to extract chloride from rocks.

Musashi et al. (1998) proposed a method to extract chloride from rock by dissolving it in a concentrated hydrofluoric acid solution. They added approximately 1 gram of rock, 1 ml distilled water and 4 ml of 17 M hydrofluoric acid solution into a teflon centrifuge tube. This tube was closed and shaken overnight (or longer) to dissolve the silica rock. The next day 10 mL distilled water was added and 5 mL of a slurry containing 2 to 3 gram calcium hydroxide. The calcium in this slurry did precipitate the fluoride in the solution as calcium fluoride immediately. Again the solution was shaken overnight to assure that as much fluoride as possible has reacted to calcium fluoride. Then the tube was centrifuged and the suspension was filtered over a glass fibre filter. This treatment was repeated twice, and the filtrated solutions we added together. To remove the remaining fluoride from the solution it was passed over a strongly acidic cation exchanger and to increase the chloride in the final solution this was passed over a strongly basic anion-exchange resin. This resin has a higher affinity for chloride compared to fluoride so that it separates them effectively. The chloride is then removed by rinsing a 20 to 25 mL 0.5 M potassium nitrate solution over the column. As the affinity for nitrate is larger than for chloride this is exchanged and eluted from the column. Chlorine isotopes in the eluted solution can be analysed as described earlier by whatever technique is available in the laboratory of the researcher.

5.3.3 Dissolution of Chloride Rich Silicates (Sodalite, Tugtupite and Eudialyte) in Nitric Acid

Some silicate minerals are soluble in nitric acid. Eudialyte [$Na_{15}Ca_6Fe_3Zr_3Si_{26}O_{72}(O,OH,H_2O)_4(Cl,F,OH)_2$] a cyclosilicate consisting of [Si_3O_9]$^{6-}$ and [Si_9O_{27}]$^{18-}$ rings (Golyshev et al. 1971, Giuseppetti et al. 1971), even received its name due to this characteristic. Stromeyer (1819) described that this new mineral, discovered in Greenland by Prof. von Giesecke, looks very much like garnet, but that it is easily gelatinised by acids. For that reason the name eudialyte, from Greek *"eu dialytos"*, that means "well decomposable, was proposed for this mineral (Stromeyer 1819). For that reason, Eggenkamp et al. (2022) dissolved eudialyte in diluted (1:5) nitric acid in liberate chloride ions for chlorine isotope measurement. During this process it was discovered also that chloride rich zeolite minerals (sodalite ($Na_8Al_6Si_6O_{24}Cl_2$; up to 7.3% Cl) and tugtupite ($Na_8Be_2Al_2Si_8O_{24}Cl_2$; up to 7.6% Cl) were also easily soluble in dilute nitric acid. For chlorine isotope studies these minerals can thus be dissolved in nitric acid (Eggenkamp et al. 2022). If possible, the amount of sample is chosen to contain 1 mg of chloride. So, for sodalite and tugtupite this is approximately 15 mg and for eudialyte between 70 and 100 mg, depending on the Cl content. If enough material is present, each sample should be dissolved twice to obtain duplicate measurements. Mineral separates are to be powdered as fine as possible. Sodalite and tugtupite can be dissolved in 1 mL of a solution containing 750 μL H_2O, 50 μL 30% H_2O_2 and 200 μL 69% HNO_3. H_2O_2 is added to oxidise any sulphide present that will precipitate as unwanted Ag_2S later in the procedure. Since eudialyte contains no sulphide no H_2O_2 needs to be added. But, due to larger sample amounts, eudialyte samples must dissolved in a mixture of 4 mL H_2O and 1 mL 69% HNO_3. All mineral samples can be dissolved at room temperature. Silica present in dissolved silicate minerals starts to gellify immediately after dissolution and within a few hours the solution can no longer be filtered effectively. Thus, AgCl must be precipitated as quickly as possible after dissolution of the minerals. For most samples this can be done after 2 to 4 h. When larger sample quantities are dissolved, for example due to low chloride contents, gel formation is much faster. When this happens filtration of the AgCl suspension becomes very difficult, and results in glass fibre filters that contain significant amounts of hygroscopic silica gel together with the AgCl precipitate. This causes trouble at later stages in the analytical procedure, because after reaction with CH_3I, the gaseous CH_3Cl to measure the chlorine isotope composition at, is contaminated with water. Although the water peak arrives earlier than the CH_3Cl peak during gas chromatography, the water peak has a very long tail that overlaps heavily with the CH_3Cl peak. At high water contents it is unavoidable that significant amounts of water vapour are present in the gas that is introduced into the mass spectrometer. This results in low quality Cl isotope data, that are characterised by very large standard deviations (up to 0.5‰) for repeated measurements of the same sample. Therefore, it is important that the dissolution time for silica-rich nitric acid soluble minerals is kept as short as possible, preferentially at less than 1 h. Unfortunately this may result in samples that are not fully dissolved, and a compromise needs to be found between full dissolution of the sample (that may take more time) and difficulty of filtration of the sample (that can be avoided to take the time between dissolution and filtration as short as possible). A possible suggestion may also be that the mineral grains are crushed to smaller particles to facilitate faster dissolution (P Agrinier, pers. comm). However, what would be the risk of faster silicagel formation is not known and would by necessity the subject of a possible future chlorine isotope research of mineral species.

5.3.4 Pyrohydrolysis Methods

The principle of pyrohydrolysis is the extraction of chloride and other volatile compounds from molten rock by a stream of water vapour. These volatile compounds are then trapped into distilled water. From this solution chloride ions can be analysed by mass spectrometry either directly or after subsequent clean-up.

The first pyrohydrolysis method for chlorine isotope measurements was proposed by Magenheim et al. (1994). They applied the method that was originally developed by Dreibus et al. (1979). They used an induction furnace to heat a sample to 1300 to 1400 °C in a pre-cleaned platinum boat. During extraction water vapour carrier gas was applied at a flow rate of 0.3 g water per minute for 50 min. After this period the induction furnace was turned off and the system was flushed with 2.5 g water per minute for another 20 min. Finally the interior of the combustion tube was also rinsed with distilled water. Sample was collected by condensation in a water cooled fused silica condenser and ultimately collected in a polyethylene bottle. Chlorine isotopes of the extracted chloride were analysed following the positive TIMS method developed by Xiao and Zhang (1992). Before using this method however fluoride and sulphate were removed from the solution by precipitation as calcium fluoride and cation exchange as barium sulphate respectively.

Pyrohydrolysis for the analyses of chlorine isotopes has also been applied at the University of New Mexico (Barnes et al. 2006; Sharp et al. 2007) and the Institut de physique du plobe de Paris (Bonifacie et al. 2007). Barnes et al. (2006) used the method as described by Magenheim et al. (1994) while Bonifacie et al. (2007) developed their method after Whitehead and Thomas (1985). In this technique the sample is mixed with 5 to 6 times its mass of vanadium pentoxide as flux. The mixture is heated (in a platinum boat) to a temperature of 1200 °C and extraction is done with a water vapour flow of about 1.5 mL/min for 25 min, after which it was purged with pure air to dry the system. The sample was collected by condensation in a water cooled fused silica condenser and finally collected in a Nalgene bottle. After extraction all glassware is rinsed with distilled water to ensure that the recovery of the chloride from the sample is as complete as possible. The main differences between the techniques developed by Magenheim et al. (1994) and Bonifacie et al. (2007) are the addition of vanadium pentoxide as flux to optimise the release of halogens from the sample and to lower the melting temperature of the samples, as well as keeping the whole system above 300 °C by the use of a resistance furnace and heating coil. The advantage of this system is that no condensation of water takes place before the collection flask.

After pyrohydrolysis both groups applied continuous flow or dual inlet isotope ratio mass spectrometry on chloromethane gas which was prepared from the dissolved chloride according to the techniques described in Long et al. (1993), Eggenkamp (1994) and Godon (2000).

5.4 Extraction of Chloride from Non-silicate Rocks

Eggenkamp (1994, 2004) described a few methods to extract chloride for chlorine isotope analysis from non-silicate rocks and minerals, such as evaporites and carbonatite samples, samples with high sulphide concentrations and samples with (very) high pH values.

5.4.1 Evaporites

Evaporites can simply be prepared by dissolution a known amount of sample in distilled water (Eggenkamp et al. 1995). This solution can be treated as any aqueous chloride solution for isotope analysis.

5.4.2 Carbonatites

Most carbonatites, with the exception of natrocarbonatites, contain little chloride. Eggenkamp and Koster van Groos (1997) extracted chloride from the carbonate fraction of carbonatites by the following method. Five gram of carbonatite was suspended in 100 mL distilled water and stirred in a glass beaker on a magnetic stirrer/heating plate at a temperature of 50 °C. The pH of this suspension was monitored. Concentrated nitric acid was added drop by drop from a burette until the suspension stopped effervescing which happened at the moment that the pH was as low as 0.5. Next, the solution was filtered. Only chloride present in carbonates and apatites is liberated with this method. The sample can further be treated as a dilute water sample. In the case of natro-carbonatites, which contain much more chloride (Keller and Krafft 1990), a significantly smaller amount of sample is required (Eggenkamp and Koster van Groos 1997).

5.4.3 Other Non-silicate and Non-carbonate Minerals

Other non-silicate minerals such as phosphates and hydroxide chlorides can generally be dissolved in dilute (1:5) nitric acid (Hintze 1915). Contrary to the nitric acid soluble silicate minerals described in Sect. 5.3.3 there is no risk of gel formation in these minerals as no silica is present in these minerals. That means there is no need to precipitate AgCl as quickly as possible so that better control to the total dissolution of the sample is possible. Some minerals of this type, especially those originating from oxidised sulphide ores have been shown to have extreme $\delta^{37}Cl$ values (Eggenkamp and Schuiling 1995).

5.5 Chlorine Isotopes in Chlorate and Perchlorate

Although Hoering and Parker (1961) already measured chlorine isotopes on two samples of Chilean perchlorates they did not specify how they reduced the perchlorate to chloride for isotopic analysis. Renewed interest in the chlorine isotope composition of chlorates and perchlorates developed at the end of the twentieth century due to potential applications of their study as environmental pollutants (Ader et al. 2001). Only later it was realised that perchlorates are also produced under natural conditions other than those existing in the ultra-dry Chilean Atacama desert and that their isotope composition can be studied to trace natural and anthropogenic origins of perchlorates in groundwaters (Sturchio et al. 2003, 2006, 2007; Gu et al. 2007; Bohlke et al. 2005). The isotope composition of chlorates and perchlorates is measured after reduction to chloride which is than reacted with iodomethane to form chloromethane for isotope ratio analysis. The methods applied are described in the following sections.

5.5.1 Chlorate

Until relatively recent it was understood that chlorate (ClO_3^-) was not normally naturally produced. Its anthropogenic production was however well known and as such it is used for example as herbicide, or in combination with perchlorate as oxidant in explosives and solid fuel rockets. It is also present in many drinking waters that are treated with chlorine for disinfection purposes. Considering its potentially toxic properties its isotopic composition might be useful to trace for forensic reasons. The first study in which the chlorine isotope composition of chlorate (ClO_3^-) has been conducted was Ader et al. (2001). Ader et al. (2001) developed a method to analyse the chlorine isotope composition in relatively large amounts of chlorate and analysed a set of chemical reagents (sodium chlorate; $NaClO_3$ and potassium chlorate; $KClO_3$). To analyse the chlorine isotope composition, the chlorate must be reduced to chloride which can be analysed by classic methods. Chlorate can be quantitatively reduced by heating it with excess iron(II) in the presence of a relatively high concentration of sulphuric acid, following the procedure for the quantitative analysis of chlorate by Vogel (1989). For the method proposed by Ader et al. (2001) an amount of 2.5 mL of a 0.02 M chlorate solution is reduced by the addition of 2.5 mL of a 0.2 M ammonium iron(II) sulphate solution in 2 M sulphuric acid and 1 mL of concentrated sulphuric acid. This mixture was heated to 80 °C for 10 min and cooled to ambient temperature afterwards. Chlorine isotope composition then can be determined after transformation of chloride to chloromethane as described by Eggenkamp (1994).

In more recent years it became obvious that chlorate is much more widespread in the environment due, not only to anthropogenic sources but also to natural processes (Gordon and Tachiyashiki 1991, Pisarenko et al. 2010, Rao et al. 2010, Jackson et al. 2016). The measurement of the chlorine isotope composition of environmental chlorate samples is quite complex, because the concentration of chlorate is always low and other ions (such as chloride, nitrate and sulphate) are often present in concentrations that can be several orders of magnitude higher (Rao et al. 2010; Jackson et al. 2012, 2016). To further complex the possibility to measure the isotope composition of chlorate, Jackson et al. (2021) state that chlorate is susceptible to reactions that can change its isotope composition, and that methods used previously to concentrate perchlorate in samples (Gu et al. 2011b, Hatzinger et al. 2011) cannot be used to concentrate chlorate as they require very low pH values, where chlorate is not stable, and the ion-exchange resins used do not sufficiently separate chlorate from nitrate. Jackson et al. (2021) developed purification methods to extract chlorate of sufficient purity for isotope analysis. The purification contained of different procedures including precipitation of co-occurring salts by evaporative concentration; extraction of ClO_3^- using acetone; removal of SO_4^{2-} and Cl^- using the addition of Ba and Ag, respectively; bacterial NO_3^- reduction; ion exchange to remove ClO_4^-; and removal of residual organic compounds by adsorption and/or oxidation using hydrogen peroxide (H_2O_2) and/or adsorption. After all steps, from test samples >95% of the initial chlorate was recovered, in natural samples however this could be anywhere between 6 and 33%. The main target of their study was that chlorate of sufficient purity would be obtained for isotope analysis, not that all chlorate was recovered (Jackson et al. 2021). Chlorate was recovered as $CsClO_3$ or $KClO_3$ because these salts have a relatively low solubility and because they are suitable for isotope analysis. Following the procedure it is assumed that only little isotope fractionation occurred, so that the loss of chlorate is not problematic. Stable isotope analyses on the purified chlorate samples were performed using the methods as described by Böhlke et al. (2017) for perchlorate as described below.

5.5.2 Perchlorate

Perchlorate is more common in natural samples than chlorate. It is found in quantities of up to 0.6% in the ultra-dry Atacama Desert in Northern Chile (Ericksen 1981). It was long assumed that these were the only natural occurrences of perchlorates in the world, but recently indigenous perchlorate has been detected in the drier regions of the United States (Plummer et al. 2006; Jackson et al. 2005; Rajagopalan et al. 2006, 2009; Rao et al. 2007; Parker et al. 2008). Also in Antarctic and Martian soils perchlorates have been detected recently (Hecht et al. 2009; Ming et al. 2010; Catling et al. 2010; Kounaves et al. 2010). Perchlorates from the Atacama Desert have been transported in large quantities to the United States as the major component of the soils they originated from is nitrate and as such this was worldwide used as fertilizer. Besides these natural perchlorates very large amounts have been artificially produced to be used as rocket propeller, and for purposes such as explosives, fire works, road flares and electroplating solutions. These discoveries have triggered interest in the measurement of the stable chlorine and oxygen isotope composition of the perchlorates as the isotope composition can supply information on the origin and behaviour of this potentially harmful ion in the environment.

5.5.2.1 Chemical Reduction of Perchlorate to Chloride

The first method to analyse the chlorine isotope composition of perchlorates was proposed by Ader et al. (2001). They applied two different methods to reduce perchlorate to chloride for subsequent isotope analysis following classical techniques. The first method they proposed applied the reduction of perchlorate with titanium(III)sulphate according to Von Rothmund (1909) a method that was originally developed to determine the perchlorate concentration in solution by back-titration of the excess of titanium(III) in solution. Ader et al. (2001) found that the perchlorate conversion yields of this method were very irregular and varied somewhere between 0 and 90% proving this method useless for isotope analysis for which a yield as closely as possible to 100% is necessary. The second method that was applied by Ader et al. (2001) was developed by Joan and Reedy (1940). For this method the perchlorate solution is made alkaline using potassium carbonate (approximately 300 mg of potassium carbonate for each 50 μmol of perchlorate) and evaporated to dryness. In a muffle furnace the resulting slid is heated for 45 min to a temperature of 600 °C and subsequently cooled to ambient temperature. Distilled water is added until the potassium carbonate is dissolved completely and the solution is acidified with nitric acid until the pH has decreased to a value lower than 3. At that moment all perchlorate has been reduced to chloride and the solution can be treated according to classic methods for chlorine isotope analysis.

5.5.2.2 In Vacuo Decomposition of Perchlorate to Chloride

In their first study on microbial reduction of perchlorate Sturchio et al. (2003) developed a different method to reduce perchlorate for isotope analyses. The samples they used in their experiments contained both chloride and perchlorate. The chloride was removed by adding a 1 M silver nitrate solution. The chloride precipitates as silver chloride which is insoluble while silver perchlorate is very soluble so that the perchlorate stays in solution. After the chloride is removed the resulting solution containing both perchlorate ions and silver nitrate is evaporated until near dryness. This substance is put into a Pyrex combustion tube and dried completely. The tube then is frozen by liquid nitrogen, evacuated and sealed off. Then the evacuated tube is combusted for 1 h at 550 °C. During this process the perchlorate is reduced to chloride which is combined with silver present in the tube to form silver chloride. If this silver is present in excess the formation of silver chloride is complete. The silver chloride that is produced is subsequently washed with dilute nitric acid to remove excess nitrates. The purified silver chloride is dried and iodomethane is added. The procedure to form chloromethane is the same as for common chlorine (chloride) isotope analysis, i.e. a short reaction time at high temperature and cryogenic separation of chloromethane and iodomethane as described by Holt et al. (1997). In a later paper Sturchio et al. (2007) changed the procedure to separate perchlorate and chloride from a mixture containing both anions through the use of a highly selective bifunctional anion exchange resin (Purolite A530E). Perchlorate is precipitated from the solution by the addition of caesium chloride, as the solubility of caesium perchlorate is relatively low. These anion exchange techniques have been developed by Gu et al. (2001) and Bao and Gu (2004) and they facilitate the collection of purified perchlorate for the combined analysis of chlorine and oxygen isotopes.

5.5.2.3 High Temperature Dissociation of Perchlorate to Chloride and Oxygen

Böhlke et al. (2017) presented methods for stable isotope analyses of both oxygen (including ^{17}O) and chlorine in perchlorate. They prepared and tested large quantities of $KClO_4$ to be used as isotope reference materials (USGS37, USGS38, USGS39). In order to determine the chlorine isotope composition of the their samples they took aliquots of $KClO_4^-$ containing 2.5 μmol of perchlorate. These were weighed into 9 mm outer diameter quartz glass tubes that were pre-baked at 850 °C. The tubes were evacuated and sealed with a H_2-O_2 torch, and batches of sealed tubes were then baked for 20 min in a pre-heated furnace at 650 °C to decompose the perchlorate. The chloride salts produced after decomposition of the $KClO_4$ are dissolved in warm 18.2 MΩ distilled water. This solution can be treated by conventional methods to measure the chlorine isotope composition of the chloride present in solution. Jackson et al. (2021) showed that this method can also be used to decompose chlorate for chlorine isotope analysis.

5.5.3 Collection of Perchlorate from Natural Samples Containing Low Perchlorate Concentrations

In natural groundwater samples perchlorate is always a trace component at concentrations which are virtually always lower than 10 μg/L and in most cases even lower than 1 μg/L. The low concentrations make it very difficult to extract perchlorate from natural samples. Sturchio et al. (2011) summarised the methods and results of the chlorine and oxygen isotope determinations on perchlorates, most of which had been done in one single laboratory. For the purpose of the large scale water treatment necessary to extract the small amounts of perchlorate from the samples a special anion-exchange resin was developed to extract perchlorate from aqueous solutions (Gu et al. 2001, 2007, 2011a, 2011b). This ion-exchange resin, commercially available as A530E, is suitable for the direct preconcentration of perchlorates from aqueous solutions. The sample from which perchlorate has to be removed is passed through a column packed with this resin. As common with ion exchange resins it is also (but less) sensitive for ions such as chloride, nitrate and sulphate. These interfering ions decrease the perchlorate efficiency, so that a larger amount of sample should normally be passed through the column than would expected from the known concentration of perchlorate in this sample. This also indicates that after the passage of the sample trough the ion-exchange column besides perchlorate several other ions are present that need to be removed. The recovery of perchlorate is done by flushing the column firstly with hydrochloric acid to remove most nitrate, sulphate, carbonate and organic anions from the column. As a result the amount of chloride on the column increases to substantial amounts. The perchlorate is then eluted from the column through elution with a 1 M iron(III) chloride solution that displaces the perchlorate ions with $FeCl_4^-$ ions on the column. The eluted perchlorate can be purified further following techniques described by Gu et al. (2011a, 2011b). Perchlorate concentrations above 1 μg/L can generally be recovered fairly well, while lower concentrations are considerably more challenging (Sturchio et al. 2011). The purified perchlorate is normally precipitated as caesium perchlorate and its purity is checked by ion chromatography or Raman spectrometry. The caesium perchlorate is treated for isotope analysis as described by Sturchio et al. (2007, 2011).

5.6 Chlorine Isotopes in Organic Molecules

There are several techniques developed to analyse the chlorine isotope composition of chlorinated organic compounds. Originally they were measured by extracting, through reduction, the chlorine from the compounds, often in the form of an inorganic chloride, and analyse this according to the techniques available for inorganic chloride isotope analyses. More recently on-line techniques have been developed that directly introduce organic compounds into isotope ratio or quadrupole mass spectrometers.

Chlorine isotope measurements of organic compounds were already done before it was known that stable isotope variations also exist in natural samples. These measurements were done on samples that were obtained from experiments to determine kinetic chlorine isotope effects. These experiments (e.g. Bartholomew et al. 1954; Howald 1960; Hill and Fry 1962; Turnquist et al. 1973; Williams and Taylor 1973) conducted S_N1 and S_N2 reactions in which the chloride ion was one of the reaction products. This free chloride could be analysed according to methods described above (such as the methods developed and described by Langvad (1954), Hill and Fry (1962) and Taylor and Grimsrud (1969). Based on the development of the chlorine isotope data of the free chloride the kinetic fractionation in these processes could be determined.

After the development of more precise methods to analyse inorganic chloride in natural samples studies were started to analyse chlorine isotopes in organic samples too. As it was known from experiments as cited above that during the processes to produce anthropogenic organic compounds large kinetic fractionation occurred (e.g. Hill and Fry, 1962; Turnquist et al. 1973) it was expected that chlorine isotope variations in organic compounds would be larger than in inorganic compounds, and thus it was considered feasible to search for these variations.

5.6.1 Methods that Apply Reduction of Chlorine to Chloride

Between 1990 and 2000 several laboratories developed their own methods to analyse chlorine isotope variations in organic compounds. These early techniques focussed on the development of methods to analyse pure compounds, only later methods evolved such that it was possible to concentrate and analyse the very low concentrations that are usually found in natural environments.

5.6.1.1 Reduction of Organochlorine Compounds with Metallic Lithium

The first direct measurements of stable chlorine isotopes from organochlorine compounds were made by Tanaka and Rye (1991). They analysed a set of six compounds to determine whether the variations in chlorine isotope compositions in organic compounds would be large enough to be determined in naturally occurring (stratosphere) environments.

To extract the chlorine from the organochlorine compounds Tanaka and Rye (1991) heated these compounds with lithium metal in evacuated sealed quartz tubes. The lithium reduces the chlorine in the organochlorine compounds to chloride which is recovered from the tube as lithium chloride. The chlorine isotope composition of this lithium chloride was determined after dissolution in water and following the procedure as described by Taylor and Grimsrud (1969). The reported analytical precision of this technique, based on one or two determinations per organochlorine compound was 0.15‰.

5.6.1.2 Parr Bomb Technique

Van Warmerdam et al. (1995) were the first to realise the strength of analysing both carbon and chlorine isotopes on the same sample. Although they did not analyse both isotopes on the same aliquot of sample it showed that for fingerprinting purposes analysing both isotopes in organic compounds is very useful. To analyse the chlorine isotopes in their samples they combusted the organochlorine compounds according to ASTM method D808–91. This method is developed for the analysis of chlorine in combustible solids and liquids and applies a Parr 1901 Oxygen Bomb. In short the method works as follows. 5 mL of a 5% calcium carbonate solution is injected to the bomb for absorption of the combustion products. Then the sample capsule is sealed with cellophane tape to prevent loss of the sample by volatilization. The sample is injected through the tape seal with the use of a gas tight syringe. Not more than 50 μL of chlorinated compound (sample) is added to avoid overpressure of the bomb during combustion. Then the fuse wire loop is put in contact with the centre of the tape cover after which the bomb is sealed and a maximum pressure of 20 atmosphere of oxygen is added to the bomb. The bomb is then placed in an ice-water bath and the fuse is ignited. After combustion the bomb is de-pressurised and rinsed with ultrapure water to collect the combustion products. Among these is the chlorine which is now present as chloride and which is analysed for chlorine isotopes according to the technique described by Long et al. (1993). The chlorine yield of the Parr bomb technique was within a narrow range between 65 and 75% and no isotope differences were observed in cases when lower or higher yields were obtained. The isotope measurements had an analytical reproducibility of ±0.29‰ (standard deviation), based on 21 analyses of a trichloroethylene sample, while for the other measured compounds the standard deviation of three to ten measurements was between 0.10 and 0.68‰. The carbon isotope ratios of the same compounds were determined according to the technique described by Boutton et al. (1983). Interestingly the combination of both chlorine and carbon isotope analyses of the various samples made it possible to better characterise the individual samples as it was observed that samples with comparable chlorine isotope ratios could have very different carbon isotopes, while samples with comparable carbons isotope ratios could have clearly different chloride isotope ratios.

5.6.1.3 Non-aqueous High Temperature Conversion Methods for both Chlorine and Carbon Isotopes

A non-aqueous method to convert organochlorine compounds such that both the carbon and chlorine isotope ratios could be analysed in the same aliquot of sample was developed by Holt et al. (1997). An aliquot of sample is injected into a glass ampoule that contains about 1 gram copper(II) oxide wire, frozen with liquid nitrogen and evacuated. The ampoule is sealed at a restriction and heated for 2 h at 550 °C. The copper(II) oxide wire oxidises carbon in the organic compound to CO_2 and reduces the chlorine to chloride that is retained as copper(I) chloride. During this process one part of the ampoule is frozen into a beaker containing liquid nitrogen to trap carbon dioxide and water, probably to avoid the pressure from getting too high. After the reaction has finished the ampoule is put into a tube cracker attached to a vacuum line and at first the carbon dioxide is transferred and cleaned for carbon isotope analysis. In the remaining part of the reaction ampoule the original chlorine is still present as copper(I) chloride. This is reattached to the vacuum line and after evacuation 35 μL iodomethane is frozen onto the copper(I) chloride using a syringe trough a septum. The reaction ampoule is sealed again and heated for 2 h at a temperature of 300 °C. During this reaction copper(I) chloride is reacted to copper(I) iodide and iodomethane to chloromethane. Finally the chloromethane was separated by cryogenic separation from iodomethane following the method described in Sect. 5.2.2.2.6. The chlorine yield from different chlorinated compounds with this method varied between 80 and 100%, and generally it appeared that the more complex the chlorinated hydrocarbon was the lower the yield was. The chlorine isotope data obtained by this method showed to be very precise. Most duplicate measurements of a single compound for chlorine isotopes were within 0.05‰ from each other and it was concluded that chlorine isotope ratios could be measured with a precision that is generally better than 0.1‰.

At about the same time Holt et al. (1997) developed their method a group in Reading (UK) independently developed a method which was comparable to the method developed by Holt et al. (1997). Jendrzejewski et al. (1997) also added one to three μL of a chlorinated hydrocarbon to a glass capsule containing about 1 gram copper(I) oxide. The capsule then was evacuated and sealed and heated in a furnace at 400 °C for at least 1 h. After this time the capsules are further heated to the final reaction temperature, which is 620 °C for borosilicate capsules. After the reaction has finished the capsule was put into a tube cracker and attached to a vacuum-line. After cracking condensible gases are transferred, and carbon dioxide is separated out and cleaned for carbon isotope measurement. All solid residues (glass, remaining copper oxide and newly formed copper chloride were then recovered into a glass beaker, rinsed with and dissolved in distilled water to dissolve the chloride into the water. After dissolution of all chloride it was analysed for the isotope composition according to the methods described by Long et al. (1993) and Eggenkamp (1994). Both the chlorine yields and the precision of the chlorine isotope measurements were comparable to the method developed by Holt et al. (1997) showing chloride yields between 86 and 119% and a 0.08‰ reproducibility for the stable chlorine isotope measurements.

5.6.1.4 On-Line IRMS Technique for the Analysis of Total Organic Chlorine in Environmental Samples

The first on-line technique to analyse the chlorine isotope composition of the total organic chlorine in environmental samples was developed by Wassenaar and Koehler (2004). The first stage of their method still concerned off-line preparation of chloromethane according to techniques described by Long et al. (1993) and Eggenkamp (1994), while organic samples were converted through the Parr Bomb technique (Van Warmerdam et al. 1995). The on-line part of the technique consisted of injection of the chloromethane and iodomethane mixture onto a 2 m 1/4" GC column packed with Porapack Q, where the two gases were separated and from which the chloromethane was directly introduced onto the mass spectrometer, while iodomethane was directed through a waste line. The chlorine isotope composition was determined through comparison of the sample peaks with reference gas peaks and it was found that the repeatability of the technique reached about 0.06‰ (1σ). The main advantage of this method was the time saving because of the direct injection of the chloromethane and iodomethane mixture onto the gas chromatograph and subsequent introduction to the MS.

5.6.2 Measurement of Organochlorine Compounds Using TIMS

Numata et al. (2002) and Holmstrand et al. (2004) have developed methods to make it possible to analyse the chlorine isotope composition of organic compounds by thermal ionization mass spectrometry. The advantage of this technique is that very small amounts of organochlorine compounds can be measured for their chlorine isotopes which is important because in natural environments the concentrations of organochlorine compounds are, in most cases, very small.

Numata et al. (2002) liberated chlorine as chloride from the chlorinated compounds using the sodiumbiphenyl reagent following Liggett (1954) with some modifications after Numata et al. (2001). The method had the organochlorine compound (containing about 1.0 mg chlorine) reacted with some organic solvent (toluene, n-decane or di-iso-propylether) in a test tube (sealed with rubber stopper in a nitrogen atmosphere) to which 0.2 mL sodiumbiphenyl reagent is added. Sodiumbiphenyl reagent is a 1 M sodiumbiphenyl solution in dimethoxiethane. After the reaction has finished (which takes 30 min) 0.5 mL water is added to extract the chloride ions from the organic mixture. The chloride that is extracted is converted to caesium chloride according to the method described by Numata et al. (2001) for measurement on a Finnigan MAT262 thermal ionization mass spectrometer.

Holmstrand et al. (2004) used a different method to extract chloride from the chlorinated solvents. They extracted chloride from the solvent by combustion using copper(II) oxide comparable to methods established by Holt et al. (1997) and Jendrzejewski et al. (1997). The chloride extracted this way was purified and converted to caesium chloride according to the technique described by Numata et al. (2002) and analysed on a Finnigan MAT261 thermal ionization mass spectrometer.

5.6.3 Techniques to Introduce Organic Molecules into the Mass Spectrometer Directly

The first study in which it was attempted to analyse chlorine isotopes by directly introducing organic molecules and fragments into the mass spectrometer was done by Shouakar-Stash et al. (2006). This followed the application of using of mass fragments to explore stable isotope analyses as was done for the first time by Rees (1978) who directly introduced SF_5^+ fragments into the mass spectrometer that were obtained from SF_6 for sulphur isotope studies. The aim of this study for chlorine isotopes was to attempt to analyse them using less preparative steps and improve sensitivity and precision.

Shouakar-Stash et al. (2006) used an Agilent 6890 gas chromatograph coupled to an Isoprime isotope ratio mass spectrometer with collectors for masses m/z 94 and 96. The procedure was based on the fact that fragments with masses 94 and 96, 95 and 97 and 96 and 98 were produced from PCE (tetrachloroethene), TCE (trichloroethene) and DCE (dichloroethene) respectively. Chlorinated solvents were introduced directly onto the gas chromatograph using a gas-tight syringe. The samples injected were either pure chemicals or the headspace sampled from above aqueous solutions. Separation of the gaseous phases was performed on a DB-5 column and the purified compounds were introduced into the mass spectrometer whereby outside the window of interest the flow was directed to the FID to avoid introduction of unwanted compounds to the mass spectrometer. The isotope ratios were analysed from the masses 94 and 96. Shouakar-Stash et al. (2006) discovered that the reported δ^{37}Cl values of the organochlorine compounds were highly non-linear. For example in the case of tetrachloroethene where depending on the peak area values between −4.0 and + 3.0‰ were found while the actual values should have been 0.0‰. As the non-linearity was very systematic this could be corrected for. Thanks to this systematic correction it is possible to obtain precisions as low as ±0.06 to 0.12‰ depending on the compound studied. One of the biggest improvements of this techniques was the possibility to analyse very small sample amounts. It was found that concentrations as low as 5 to 20 μg/L, equivalent to amounts of 6 to 9 nmole of chlorine and approximately 80 times lower than off-line methods developed earlier, could be analysed using this technique.

5.6.4 Analysis of Chlorine Isotopes in Organic Molecules by ICP-MS

At about the same time experiments were done to analyse chlorine isotope compositions on organochlorine molecules using a multi collector ICP-MS. Van Acker et al. (2006) were the first to describe a successful method to do so. Using their method the individual organochlorine compounds are separated using gas chromatography as in the earlier mentioned method, and the organochlorine compounds are introduced separately into a multi collector ICP-MS. Potential difficulties with this method are the low ionization yield for chlorine due to its high first ionization potential and the fact that helium carrier gas is mixed with argon. Argon produces $^{1}H^{36}Ar^{+}$ ions in the plasma which has a large interference with $^{37}Cl^{+}$. This problem was solved by using a very high mass resolving power of about 10,000. This is enough to find a small shoulder on the mass 37 peak which is from $^{37}Cl^{+}$, due to a small mass difference between $^{1}H^{36}Ar^{+}$ (m/z = 36.97537) and $^{37}Cl^{+}$ (m/z = 36.96590). The width of this shoulder is only about 0.0004 amu, but enough to be able to analyse the $\delta^{37}Cl$ with a precision of about 0.12‰. The $\delta^{37}Cl$ was determined in a way that was comparable to continuous flow GC-IRMS, thus by comparing the unknown sample with a known standard. With this technique it was possible to analyse very small amounts of sample. The precision of 0.12‰ was reached with samples sizes between 200 nmole to 1.3 μmole, while worse precisions were found with smaller sample sizes. This analytical method applied a more direct method to measure the chlorine isotope composition by ICP-MS as compared to the previously described method by Fietzke et al. (2008) where the $^{37}Cl^{+}$ signal was corrected for the presence of $^{1}H^{36}Ar^{+}$ in the sample. Different techniques to analyse chlorine isotopes by gas chromatography multi-collector ICP-MS were described in later years, such as by Horst et al. (2017) and Renpenning et al. (2018). These methods yielded comparable or slightly better errors, but were able to measure samples as small as 2 to 3 nmol of chlorine. Recent research suggests that the amount of HAr^{+} ions can de decreased significantly by the use of collision cells so that the quality of chlorine isotope measurements will increase significantly in the future (P Agrinier, pers. comm).

5.6.5 Analysis of Chlorine Isotopes by Quadrupole Mass Spectrometry

When the chlorine isotope composition of chlorinated organic compounds is measured by gas chromatography quadrupole mass spectrometry (GC-qMS) a mixture of chlorinated solvents is separated into individual compounds using a gas chromatograph and subsequently introduced into a quadrupole mass spectrometer. The isotope composition of the individual compounds is then calculated from the peak areas of a number of specifically selected molecular ions and fragment ions of the analysed substances (Sakaguchi-Söder et al. 2007, Sakaguchi-Söder 2010). Although the method is straightforward, calculation of the isotope composition is done using a complex set of mathematical equations. In their original study Sakaguchi-Söder et al. (2007) used an Agilent 6890 N gas chromatograph coupled to an Agilent 5973 N quadrupole mass selective detector. They separated the compounds using a DB-624 capillary column, which is sensitive to chlorinated ethenes and vinylchloride. To determine the isotope composition of an individual compound a selection of the most common mass fragments of this compound were measured. Unlike previously described methods where the isotope composition is determined on mass fragments containing only one chlorine atom, the isotope composition with quadrupole mass spectrometry is determined on all major fragments which may contain more chlorine atoms and an individual mass may contain either both isotopes or only one isotope as described in detail by Sakaguchi-Söder et al. (2007). The set of mathematical formulas can be solved and in their preliminary study they managed to get an internal precision of 0.4 to 1.1‰ on samples with a size of between 10 to 1000 pmoles. The theoretical basis for the calculation of chlorine isotope ratios was improved by Elsner and Hunkeler (2008). The method was further improved by Aeppli et al. (2010) and Jin et al. (2011). Their methods showed that even smaller samples, down to 1.6 pmole could be analysed with comparable precisions as found by Sakaguchi-Söder et al. (2007).

The method is currently in use in several, mostly German, laboratories. In a large intercomparison study Bernstein et al. (2011) studied the results from several of these laboratories and compared them to results obtained by GC-IRMS in other laboratories. After sometimes complex correction methods it was found that the results of the various direct injection methods were in good agreement with each other even when it was considered that the precision for the IRMS methods was much better (±0.1‰) compared to the qMS methods (0.2 to 0.5‰ or 0.2 to 0.9‰, depending on the type of quadrupole mass spectrometer used).

In recent years more methods to measure the chlorine isotope compositions of organic molecules by qMS were published (e.g. Asfaw et al. 2020, Ponsin et al. 2019). In general errors obtained in these methods are still between 0.5 and 1‰. It must be noted however that in organic compounds chlorine isotope variations are generally rather large and errors this size may well be acceptable, especially considering the small amounts of sample available that will not be measurable by more precise techniques.

Zhang et al. (2021) developed a method where a negative chemical ionization source was fitted between the gas chromatograph and the quadrupole MS. In negative chemical ionization (NCI), the sample molecules do not directly interact with electrons emitted from the filament, but reaction gas molecules collide with electrons to generate hot electrons, which are trapped by the sample molecules to generate negative ions, achieving ionization. NCI can detect compounds with strong electronegativity, such as organochlorines, that readily capture electrons, resulting in high sensitivity and selectivity (Munson 2006). The result is that only $^{35}Cl^-$ and $^{37}Cl^-$ ions need to be detected in order to calculate the chlorine isotope composition. The errors given by this method are still relatively large and range from 0.37 to 2.15‰ (Zhang et al. 2021), but the main advantage is that no complex calculations need to be done on ion fragments to calculate the $\delta^{37}Cl$.

5.6.6 Measurement of Chlorine Isotopes in Organic Molecules after High Temperature Conversion

Hitzfeld et al. (2011) proposed a different method to analyse chlorine isotopes in organic molecules after simple high temperature conversion to hydrogen chloride and subsequent measurement in a Balzers GAM 445 Analyzer. A mixture of organic molecules are separated by capillary gas chromatography and then led into a high temperature conversion oven. Using hydrogen as make-up gas chlorine in the organic molecules is converted to hydrogen chloride at temperatures between 1300 and 1400 °C. The isotope composition of the hydrogen chloride is determined directly in the isotope analyser at m/z = 36 and 38. The precision of the method is not very good at just under 1‰, but the authors saw possibilities for the improvement of the method in the future.

5.6.7 A Method for the Measurement of Chlorine Stable Isotopes in Polyvinylchloride

Plastics such as polyvinylchloride can not be introduced into gas chromatographic systems as they are polymerised solids. Musashi et al. (2010) developed a method to analyse the chlorine isotope composition of this type of compounds with the use of supercritical water. Supercritical water, at pressures above 30 MPa and temperatures above 600 °C, is able to extract about 80% of the chlorine from polyvinylchloride. Chlorine isotopes were analysed from the extracted chloride following the method described by Eggenkamp (1994) and preliminary results indicated that no obvious isotope fractionation appeared. Unfortunately this promising method was not developed further.

5.7 Extraction of Chlorinated Solvents from (Ground) Water Samples

After the establishment of the first techniques to analyse chlorine isotope compositions of organochlorine compounds it was realised that it would be essential to be able to extract these compounds from the natural environment. As the concentrations of these compounds in natural environments, especially ground waters, always are relatively low these methods are important to be developed as otherwise it would never be possible to study their behaviour in natural environments.

Holt et al. (1997) were the first to develop a method to extract organochlorine compounds from water samples. To extract organochlorine compounds from groundwater they have helium bubbled through a sample and trapped the organochlorine compounds into a cold trap cooled by a dry ice-acetone slush after the sample was led trough a cold water condenser. Through this procedure the amount of water in the cold trap is minimised. The procedure is repeated to a second cold trap which was also cooled with a dry ice-acetone slush and the organochlorine compounds are finally trapped in a cold trap containing liquid nitrogen. The water free organochlorine compounds are collected in a combustion tube and subsequently analysed as described by Holt et al. (1997). This method has later been adapted by Gan et al. (2013) who used a comparable technique to extract carbon and chlorine from organochlorine compounds using copper(II) oxide combustion, but used a GasBench II analogue to Liu et al. (2013) to introduce the produced chloromethane into the mass spectrometer.

5.8 Extraction of Chloromethane from the Atmosphere for Isotope Analysis

Thornton et al. (2013) developed a cryosampler and sample purification system that allowed them to extract large enough quantities of chloromethane from the atmosphere for isotope analyses. As their method was primarily developed to be able to extract and analyse bromomethane from the atmosphere it is described in more detail in the following Chap. 6.

5.9 In Situ Determination of Chlorine Isotope Ratios on Solar System Bodies

There have been several chlorine isotope studies on extraterrestrial material. The first of these studies were on meteorites and later studies were applied to Apollo samples that were returned from the Moon to Earth. These samples were analysed by the same conventional techniques that are also used to measure samples from Earth that had comparable compositions. In the last decades space craft equipped with mass spectrometers have visited several Solar System Bodies (see e.g. Vazques et al. 2020). On two of these bodies in situ chlorine isotope measurements have been made, the planet Mars and the comet 67 P/Churyumov-Gerasimenko.

5.9.1 Chlorine Isotopes in Mars Surface Samples

Farley et al. (2016) reported chlorine isotopes in rocks from Gale Crater on Mars. The samples were measured on HCl that was evolved from unknown phases of sedimentary rocks and sand. Chlorine isotope ratios can be measured on HCl evolved from the rock sample in the SAM (Sample Analysis at Mars instrument, Mahaffy et al. 2012) quadrupole mass spectrometer (QMS). Due to corrections for isobaric species it is not possible to determine high precision isotope ratios. HCl is produced during EGA (Evolved Gas Analysis) at high temperature in the SAM most probably from the decomposition of an oxyanion species such as perchlorate (Glavin et al. 2013). Chlorine isotope ratios are determined on the produced HCl on masses 38 and 36 and must be corrected for isobaric corrections. These interferences include, at mass 38, the C_3H_2 hydrocarbon fragment and a few minor constituents such as ^{38}Ar, double-ionized CS_2, and fragments of acetonitrile (CH_3CN). The interfering species at mass 36 include $H_2{}^{34}S$, ^{36}Ar, $^{18}O^{18}O$ and the C_3 hydrocarbon fragment (Farley et al. 2016). Farley et al. (2016) used three approaches to characterise the chlorine isotopic ratio of the samples. First, they computed the ratio and its uncertainty in each few-second duration sweep of the mass spectrum, yielding a time series over the entire course of the EGA run. Second, they computed an uncertainty weighted and HCl weighted mean from the time series data. And third, they computed the Cl isotopic ratio for the specific time interval of the EGA run that minimizes uncertainty in the isotopic ratio. The latter occurs when the isobaric corrections are a minimum. They observed that the three methods yield consistent results but provide different ways to view and evaluate the isotope ratios. It was observed that the SAM had a background HCl signal too, that increased over the course of the mission. Farley et al. (2016) concluded that this might be a memory effect from previously processed samples. For that reason, and because isobaric interferences during EGA are highest at lower evolving temperatures, they evaluate the chlorine isotope ratio in their samples at higher evolving temperatures (above 200 °C) only. At these high temperatures the memory effect is small enough to be neglected. After correction for isobaric interferences the chlorine isotope ratio is determined by dividing the ^{37}Cl and the ^{35}Cl signals and the ratio is converted to the delta scale with the assumption that the $^{37}Cl/^{35}Cl$ ratio is the true ratio. The ratio is converted to the SMOC scale using a $^{37}Cl/^{35}Cl_{(SMOC)}$ ratio of 0.31955 as determined by Wei et al. (2012). Although the errors of the measurement are large, with standard deviations between 3 and 37‰ the obtained values are useful due to the large range obtained in the dataset. Most interestingly is that the data showed that the SAM instrument was able to measure chlorine isotope ratios, an application that was not known before and that was not considered when developing the instrument.

5.9.2 Determination of Chlorine Isotopes in Samples from the Atmosphere of Mars

Chlorine isotopes in the Mars atmosphere were measured on HCl present in the Mars atmosphere with the use of the NOMAD (Nadir Occultation for MArs Discovery) spectrometer present at the ExoMars Trace Gas Orbiter (Vandaele et al. 2018). The presence of HCl was first detected in the Mars atmosphere by Aoki et al. (2021) and Korablev et al. (2021). Liuzzi et al. (2021) were able to determine the chlorine isotope ratios in this HCl in occultations with high HCl abundances. Because in most cases the $H^{37}Cl$ line has an intensity comparable to (or even below) the single-spectrum noise level, the data are used to derive an average, global value of $\delta^{37}Cl$ in two steps: first, single spectra are analysed to derive the $H^{35}Cl$ abundance by only probing the lines of the main isotopologue. Second, an average of all spectra where HCl is detected is calculated, weighing them with the signal to noise ratio of the $H^{35}Cl$ detection, and retrieve a single value for the $^{37}Cl/^{35}Cl$ isotopic ratio from this "grand average" spectrum. These analyses were all done by taking into account known instrument systematics, such as temperature-induced frequency shifts and resolving power variability, which were modelled and accounted for, according to the calibration of the instrument (Liuzzi et al., 2019). Due to the low ^{37}Cl concentrations the error of the single $\delta^{37}Cl$ determination that could be calculated by Liuzzi et al. (2021) was very large (at ±78‰). By taking into account a larger number of observations, Trokhimovskiy et al. (2021) could decrease the error significantly to ±20‰.

5.9.3 Determination of Chlorine Isotopes in 67P/Churyumov-Gerasimenko's Coma

Halogen compounds and chlorine isotopes can be determined by the Double Focusing Mass Spectrometer (DFMS) of ROSINA, the Rosetta

Orbiter Spectrometer for Ion and Neutral Analysis (Balsiger et al. 2007), onboard the European Space Agency's Rosetta mission (Dhooghe et al. 2017), that visited the comet 67P/Churyumov-Gerasimenko. For HCl detection, the DFMS operated in neutral mode so that only neutral cometary gas is detected (Calmonte et al. 2016). Part of the gas is ionised and/or fragmented by electron impact in the ion source. Ion fragments from a narrow, chosen *m/z* range, pass through the analyser and impact on a microchannel plate (MCP), creating an electron avalanche that is recorded by a Linear Electron Detector Array chip (LEDA) with two rows of 512 pixels each. Data are obtained as Analog-to-Digital Converter (ADC) counts as a function of LEDA pixel number. This method, described in Le Roy et al. (2015), is used to convert DFMS data into fragment ion intensities as a function of *m/z*. Due to fragmentation of molecules in the DFMS ion source, hydrogen halides can be detected in the form of singly charged fragmentation products (F, HF, ^{35}Cl, ^{37}Cl, $H^{35}Cl$, $H^{37}Cl$ ^{79}Br, ^{81}Br, $H^{79}Br$ and $H^{81}Br$, Dhooghe et al. (2017).

Accurate isotope ratios are obtained because fragmentation and instrument sensitivity is essentially the same for both stable isotopes. Systematic errors associated with instrument sensitivities can be removed by considering the ratio of the abundances of both isotopes. If the measurements of two stable isotopes are interpolated on the same timescale, the isotopic ratio can be obtained from the measured ion intensities minus their background correction. The isotope ratio is then obtained from the sum of all time-correlated measurements of ^{37}Cl divided by the sum of all time-correlated measurements of ^{35}Cl. The uncertainty is obtained by error propagation (Dhooghe et al. 2017).

5.10 Measurement of Extraterrestrial Chlorine Isotope Ratios from Earth

Chlorine isotope compositions can not only be determined from measurements on actual samples that were found on Earth, brought back to Earth or in situ at Solar System bodies. It is also possible, under certain circumstances, to determine the chlorine isotope ratio directly from Earth with the use of ground-based telescopes. Since, for example, the HCl content in the atmosphere of Venus is so much higher than in Earth's atmosphere, its concentration can be determined from Earth with ground-based telescopes. HCl contents in the Venus atmosphere could be measured as the concentrations are in the order of ppm instead of ppb as in the Earth's atmosphere, probably because Venus has no ocean that will dissolve excess HCl from the atmosphere (Iwagami et al. 2008). Because the actual measurement is done by spectroscopy both isotopes, ^{35}Cl and ^{37}Cl could be measured separately and an isotope ratio can be determined (Connes et al. 1967, Young 1972) by simply dividing both concentrations. Iwagami et al. (2008) describe the most recent approach to measure not only the HCl concentration but also the $^{35}Cl/^{37}Cl$ ratio of the measured HCl in the atmosphere of Venus. They measured the HCl concentrations with two telescopes, IRTF (Infrared Telescope Facility, Hawaii) with the CSHELL spectrometer (Greene et al. 1993) and at OAO (Okayama Astrophysical Observatory, Japan) with the OASIS spectrometer (Okumura et al. 2000). HCl radiance is determined around band R3 lines of $H^{35}Cl$ at 5739.3 cm^{-1} and of $H^{37}Cl$ at 5735.1 cm-1, and P24 line of CO^2 at 5837.2 cm^{-1}, respectively. The measured radiance is corrected for the calculated transmission of Earth's atmosphere and solar lines, as described in detail in Iwagami et al. (2008). As the concentrations of the individual isotopologues are measured with large errors, the calculated isotope ratio also has a large error. In this case it is about 0.4 ratio units which translates to approximately 180‰, so that it could only be concluded that the chlorine isotope ratio in Venus' atmosphere does not deviate from the average value in the Solar System.

There are several papers that present $^{35}Cl/^{37}Cl$ ratios in stars and interstellar matter. The chlorine isotope ratio is measured as summarised in Maas and Pilachowski (2018). Chlorine abundance measurements are difficult in stellar spectra due to low abundances and the absence of strong optical absorption features. ^{35}Cl abundances in stars have been measured using the $H^{35}Cl$ feature located at 3.7 μm in stars with T < 4000 K (Maas et al. 2016). Stars with temperatures above ~4000 K do not have HCl features in their spectra due to the molecule's low dissociation energy. Both $H^{35}Cl$ and $H^{37}Cl$ have been detected in the star RZ Ari, and a $^{35}Cl/^{37}Cl$ isotope ratio of 2.2 ± 0.4 was derived (Maas et al. 2016). As this is a cool star with an effective temperature of 3340 K (McDonald et al. 2012) it was possible to derive the chlorine isotope ratio. Maas and Pilachowski (2018) made their observations with the use of the Gemini South telescope using the Phoenix instrument (Hinkle et al. 1998). The method used to measure ^{35}Cl and ^{37}Cl abundances is described in detail in Maas and Pilachowski (2018). Both HCl lines have nearly identical excitation potentials (Rothman et al. 2013), which are listed in Maas et al. (2016). The chlorine isotope ratio in RZ Ari was determined by taking the ratio of the equivalent widths of the $H^{35}Cl$ line and the $H^{37}Cl$ line (Maas et al. 2016). Both HCl features in the star had weak line strengths on the linear portion of the curve of growth (COG) and were compared with six other stars to COG models for the two HCl lines. It was found that small deviations from the linear approximation existed. The full procedure is described in detail in Maas (2020), Maas et al. (2016) and Maas and Pilachowski (2018).

5.11 Recapitulation of Analytical Methods

As shown in this chapter the development of the analytical methods to analyse chlorine isotopes has a considerable history. Most laboratories that are able to analyse chlorine isotopes use their own method, that they also published when this was developed to such a level that the isotopes could be measured with good (or acceptable) precision. Striking in this respect is that in several cases a method was published, but little or no case studies were later published, so that, for these laboratories, it can be concluded that they have abandoned the field of chlorine isotope geochemistry research shortly after their method was developed and published.

It is clear from the method descriptions in this chapter that chlorine isotopes have been analysed by virtually all techniques and instruments that have been developed to measure stable isotopes in the past. Considering the relatively small number of laboratories that still actively analyse chlorine stable isotopes routinely it is expected that in the near future in inorganic chlorine isotope research preferably IRMS measurements with chloromethane as mass spectrometer gas (both dual-inlet and continuous flow; according to the techniques as developed by Long et al. 1993; Eggenkamp 1994 and Shouakar-Stash et al. 2005) and P-TIMS measurements (as developed by Xiao and Zhang 1992) will be used in chlorine isotope studies. Other techniques, such as ICP-MS will remain rarely used considering the difficulty to separate $^{36}ArH^{+}$ from $^{37}Cl^{+}$, although future developments with the use of collision cells may improve the quality of the measurements (especially with regards to the measurement of smaller quantities of chlorine) considerably, resulting in the increased opportunity to measure chlorine isotopes by ICP-MS. Although potential errors are still relatively high, SIMS is developing rapidly as the method of choice for small chlorine rich minerals such as apatite. This is so because apatite is a common mineral in a large array of samples, including meteorites and potentially in samples returned from Solar System Bodies. Due to the resources put in the development of SIMS for chlorine isotope measurements these measurements also improve in quality, including a trend to smaller errors. Other techniques that have relatively large errors of measurement, such as N-TIMS will only be applied rarely because the natural variation of chlorine isotope ratios is smaller than their routine measurement errors. These techniques will only be applied for specific applications such as extremely small samples or samples with large isotope ratio variations and not as routine chlorine isotope techniques.

Few new techniques to analyse organochlorine compounds have been developed in the past 10 years. Most analysis are still done using GC-MS or GC-qMS after preconcentration the organic compounds in the sample. In spite of all the efforts, the precision of the measurements is still not optimal. The main problem will continue to be that concentrations of chlorinated compounds in environmental samples are always low. On the upside however, since chlorine isotope variations in chloro-organic samples are in general relatively large, in most cases these relatively large errors are not a big problem in interpreting the obtained data.

References

Ader M, Coleman ML, Doyle SP, Stroud M, Wakelin D (2001) Methods for the stable isotopic analysis of chlorine in chlorate and perchlorate compounds. Anal Chem 73:4946–4950

Aeppli C, Holmstrand H, Andersson P, Gustafsson Ö (2010) Direct compound-specific stable chlorine isotope analysis of organic compounds with quadrupole GC/MS using standard isotope bracketing. Anal Chem 82:420–426

Anderson TS, Hidy AJ, Boyce JW, McCubbin FM, Tumey S, Dudley JM, Haney NC, Bardoux G, Bonifacie M (2022) Development towards stable chlorine isotope measurements of astromaterials using the modified Middleton source of an accelerator mass spectrometer. Int J Mass Spectrom 477:116849

Aoki S, Daerden F, Viscardy S, Thomas IR, Erwin JT, Robert S, Trompet L, Neary LG, Villanueva GL, Liuzzi G, MMJ C, Clancy RT, Whiteway J, Schmidt F, Lopez-Valverde MA, Ristic B, Patel MR, Bellucci G, Lopez-Moreno JJ, Olsen KS, Lefèvre F, Montmessin F, Trokhimovskiy A, Fedorova AA, Korablev O, Vandaele AC (2021) Annual appearance of hydrogen chloride on Mars and a striking similarity with the water vapor vertical distribution observed by TGO/NOMAD. Geophys Res Lett 48:e2021GL092506

Asfaw BA, Sakaguchi-Söder K, Bernstein A, Siebner H, Schüth C (2020) Optimization of compound-specific chlorine stable isotope analysis of chloroform using the Taguchi design of experiments. Rapid Commun Mass Spectrom 34:e8922

Aston FW (1919a) The constitution of the elements. Nature 104:393

Aston FW (1919b) A positive ray spectrograph. Philos Mag 38:707–714

Aston FW (1920) The mass spectra of the chemical elements. Philos Mag 39:611–625

Aston FW (1921) The mass spectra of the chemical elements—part 3. Philos Mag 42:140–144

Aston FW (1925) Photographic plates for the detection of mass rays. Math Proc Cambridge Philos Soc 22:548–554

Aston FW (1931) The isotopic constitution and atomic weights of selenium, bromine, boron, tungsten, antimony, osmium, ruthenium, tellurium, germanium, rhenium and chlorine. Proc R Soc Lond A 132:487–498

Aston FW (1941) Mass spectra and isotopes. Longman, Green and Co, New York

Balsiger H, Altwegg K, Bochsler PP, Eberhardt P, Fischer J, Graf S, Jäckel A, Kopp E, Langer U, Mildner M, Müller J, Riesen T, Rubin M, Scherer S, Wurz P, Wüthrich S, Arijs E, Delanoye S, De Keyser J, Neefs E, Nevejans D, Rème H, Aoustin C, Mazelle C, Médale JL, Sauvaud JA, Berthelier JJ, Bertaux JL, Duvet L, Illiano JM, Fuselier SA, Ghielmetti AG, Magoncelli T, Shelley EG, Korth A, Heerlein K, Lauche H, Livi S, Loose A, Mall U, Wilken B, Gliem F, Fiethe B, Gombosi TI, Block B, Carignan GR, Fisk LA, Waite JH, Young DT, Wollnik H (2007) Rosina-Rosetta orbiter spectrometer for ion and neutral analysis. Space Sci Rev 128:745–801

Bao H, Gu B (2004) Natural perchlorate has its unique oxygen isotope signature. Environ Sci Technol 38:5073–5077

Barnes JD, Selverstone J, Sharp ZD (2006) Chlorine isotope chemistry of serpentinites from Elba, Italy, as an indicator of fluid source and subsequent tectonic history. Geochem Geophys Geosyst 7:Q08015

Barrett TJ, Barnes JJ, Anand M, Franchi IA, Greenwood RC, Charlier BLA, Zhao X, Moynier F, Grady MM (2019) Investigating magmatic processes in the early solar system using the Cl isotopic systematics of eucrites. Geochim Cosmochim Acta 266:582–597

Bartholomew RM, Brown F, Lounsbury M (1954) Chlorine isotope effect in reactions of *tert*-butyl chloride. Can J Chem 32:979–983

Bauer SH, Hogness TR (1935) Products and processes of ionization in methyl chloride as determined by a mass spectrometer. J Chem Phys 3:687–692

Behne W (1953) Untersuchungen zur Geochemie des Chlor und Brom. Geochim Cosmochim Acta 3:186–214

Bernstein A, Shouakar-Stash O, Ebert K, Laskov C, Hunkeler D, Jeannottat S, Sakaguchi-Söder K, Laaks J, Jochmann MA, Cretnik S, Jager J, Haderlein SB, Schmidt TC, Aravena R, Elsner M (2011) Compound-specific chlorine isotope analysis: a comparison of gas chromatography / isotope ratio mass spectrometry and gas chromatography / quadrupole mass spectrometry methods in an interlaboratory study. Anal Chem 83:7624–7634

Blatt AH (1943) Organic synthesis, vol II. John Wiley & Sons, p 251

Böhlke JK, Sturchio NC, Gu B, Horita J, Brown GM, Jackson WA, Batista JR, Hatzinger PB (2005) Perchlorate isotope forensics. Anal Chem 77:7838–7842

Böhlke JK, Mroczkowski SJ, Sturchio NC, Heraty LJ, Richman KW, Sullivan DB, Griffith KN, Gu B, Hatzinger PB (2017) Stable isotope analyses of oxygen (^{18}O,^{17}O,^{16}O) and chlorine (^{37}Cl,^{35}Cl) in perchlorate: reference materials, calibrations, methods,

and interferences. Rapid Commun Mass Spectrom 31:85–110
Bonifacie M, Jendrzejewski N, Agrinier P, Coleman M, Pineau F, Javoy M (2007) Pyrohydrolysis-IRMS determination of silicate chlorine stable isotope compositions. Application to oceanic crust and meteorite samples. Chem Geol 242:187–201
Bonifacie M, Jendrzejewski N, Agrinier P, Humler E, Coleman M, Javoy M (2008) The chlorine isotope composition of earth's mantle. Science 319:1518–1520
Boutton TW, Wong WW, Hachey DL, Lee LS, Cabrera MP, Klein PD (1983) Comparison of quartz and pyrex tubes for combustion of organic samples for stable carbon isotope analysis. Anal Chem 55:1832–1833
Bouvier AS, Portnyagin MV, Flemetakis S, Hoernle K, Klemme S, Berndt J, Mironov NL, John T (2022) Chlorine isotope behavior in subduction zone settings revealed by olivine-hosted melt inclusions from the Central America volcanic arc. Earth Planet Sci Lett 581:117414
Boyd AW, Brown F, Lounsbury M (1955) Mass spectrometric study of natural and neutron-irradiated chlorine. Can J Phys 33:35–42
Brown F, Gillies A, Stevens WH (1953) A note on the preparation of chlorine gas containing 36Cl. Can J Chem:31–768
Callis EL, Abernathey RM (1991) High-precision analyses of uranium and plutonium by total sample volatilization and signal integration. Int J Mass Spectrom Ion Proc 103:93–105
Calmonte U, Altwegg K, Balsiger H, Berthelier JJ, Bieler A, Cessateur G, Dhooghe F, van Dishoeck EF, Fiethe B, Fuselier SA, Gasc S, Gombosi TI, Hässig M, Le Roy L, Rubin M, Sémon T, Tzou CY, Wampfler SF (2016) Sulphur-bearing species in the coma of comet 67P/Churyumov–Gerasimenko. Mon Not R Astron Soc 462(Suppl 1):S253–S273
Catling DC, Clair MW, Zahnle KJ, Quinn RC, Clarc BC, Hecht MH, Kounaves SP (2010) Atmospheric origins of perchlorate on Mars and in the Atacama. J Geophys Res 115:E00E11
Ciu ZX, Yang Q, Xia XP, Wang R, Bonifacie M, Lai CK, Zhang WF, Zhang YQ, Xu J (2021) High-precision apatite δ^{37}Cl measurement by SIMS with a 10^{12} Ω amplifier faraday cup. J Anal At Spectrom 37:222–238
Connes P, Connes J, Benedict J, Kaplan LD (1967) Traces of HCl and HF in the atmosphere of Venus. Astrophys J 147:1230–1237
Curie I (1921) Sur le poids atomique du chlore dans quelques mineraux. Compte Rend Sean 172:1025–1028
Darling JR, White LF, Kizovski T, Černok A, Moser DE, Tait KT, Dunlop J, Langelier B, Douglas JO, Zhao X, Franchi IA, Anand M (2021) The shocking state of apatite and merrillite in shergottite Northwest Africa 5298 and extreme nanoscale chlorine isotope variability revealed by atom probe tomography. Geochim Cosmochim Acta 293:422–437
De Gois JS, Costas-Rodriguez M, Vallelonga P, Borges DLG, Van Haecke F (2016) A simple method for high-precision isotopic analysis of chlorine via pneumatic nebulization multi-collector inductively coupled plasma-mass spectrometry. J Anal At Spectrom 31:537–542
De Groot PA (2004) Handbook of stable isotope analytical techniques, vol I. Elsevier
Dembska-Sięka P, Pelc A, Rajchel L (2020) The origin and chemical composition of Carpathians chloride CO_2-rich waters in the light of stable chlorine isotope studies (^{37}Cl and ^{35}Cl). Geol Quart 64:197–204
Desaulniers DE, Kaufmann RS, Cherry JA, Bentley HW (1986) ^{37}Cl-^{35}Cl variations in a diffusion controlled groundwater system. Geochim Cosmochim Acta 50:1757–1764
Dhooghe F, De Keyser J, Altwegg K, Briois C, Balsiger H, Berthelier JJ, Calmonte U, Cessateur G, Combi MR, Equeter E, Fiethe B, Fray N, Fuselier S, Gasc S, Gibbons A, Gombosi T, Gunell H, Hässig M, Hilchenbach M, Le Roy L, Maggiolo R, Mall U, Marty B, Neefs E, Rème H, Rubin M, Sémon T, Tzou CY, Wurz P (2017) Halogens as tracers of protosolar nebula material in comet 67P/Churyumov–Gerasimenko. Mon Not R Astron Soc 472:1336–1345
Dorenfeldt M (1922) Relative determination of the atomic weight of chlorine in Bamle apatite. J Am Chem Soc 45:1577–1579
Dreibus C, Spettel B, Wänke H (1979) Halogens in meteorites and their primordial abundances. In: Ahrens LH (ed) Origin and distribution of the elements, vol 34. Pergamon Press, pp 33–38
Eastoe CJ, Guilbert JM, Kaufmann RS (1989) Preliminary evidence for fractionation of stable chlorine isotopes in ore forming hydrothermal systems. Geology 17:285–288
Eggenkamp HGM (1994) δ^{37}Cl; the geochemistry of chlorine isotopes. Geol Ultraiect 116:1–150. Thesis Utrecht University
Eggenkamp HGM (2004) Summary of methods for determining the stable isotope composition of chlorine and bromine in natural materials. In: de Groot PA (ed) Chapter 28 in: handbook of stable isotope analytical techniques. Elsevier, Amsterdam, pp 604–622
Eggenkamp HGM (2014) The geochemistry of stable chlorine and bromine isotopes. Springer
Eggenkamp HGM, Koster van Groos AF (1997) Chlorine stable isotopes in carbonatites: evidence for isotopic heterogeneity in the mantle. Chem Geol 140:137–143
Eggenkamp HGM, Schuiling RD (1995) δ^{37}Cl variations in selected minerals: a possible tool for exploration. J Geochem Explor 55:249–255
Eggenkamp HGM, Kreulen R, Koster van Groos AF (1995) Chlorine stable isotope fractionation in evaporites. Geochim Cosmochim Acta 59:5169–5175
Eggenkamp HGM, Marks MAW, Ludwig T, Markl G (2021) Intragranular halogen (F, Cl, Br), S and δ^{37}Cl variability as determined by SIMS in soda-

lite and eudialyte from the Ilímaussaq intrusion, South Greenland, GeoKarlsruhe 2021. https://doi.org/10.48380/dggv-hcj1-5702
Eggenkamp HGM, Marks MAW, Bonifacie M, Bardoux G, Agrinier P, Markl G (2022) Cl isotope fractionation in magmatic and hydrothermal eudialyte, sodalite and tugtupite (Ilímaussaq intrusion, South Greenland). Chem Geol 604:120932
Elsner M, Hunkeler D (2008) Evaluating chlorine isotope effects from isotope ratios and mass spectra of polychlorinated molecules. Anal Chem 80:4731–4740
Ericksen GE (1981) Geology and origin of the Chilean nitrate deposits. USGS Prof. Pap. 1188:37
Ertl S, Eichinger F, Pichlmayer J, Voropaev A (2024) A new method to determine δ^{37}Cl values of chloride in aquatic systems. Grundwasser 29:63–71
Farley KA, Martin P, Archer Jr PD, Atreya SK, Conrad PG, Eigenbrode JL, Fairén AG, Franz HB, Freissinet C, Glavin DP, Mahaffy PR, Malespin C, Ming DW, Navarro-Gonzalez R, Sutter B (2016) Light and variable 37Cl/35Cl ratios in rocks from gale crater, Mars: possible signature of perchlorate. Earth Planet Sci Lett 438:14–24
Fietzke J, Frische M, Hansteen TH, Eisenhauer A (2008) A simplified procedure for the determination of stable chlorine isotope ratios (δ^{37}Cl) using LA-MC-ICP-MS. J Anal At Spectrom 23:769–772
Fujitani T, Nakamura N (2006) Determination of chlorine in nine rock reference materials by isotope dilution mass spectrometry. Geostand Geoanal Res 30:113–120
Fujitani T, Yamashita K, Numata M, Kanazawa N, Nakamura N (2010) Measurement of chlorine stable isotopic composition by negative thermal ionization mass spectrometry using total evaporation technique. Geochem J 44:241–246
Gan YQ, Yu TT, Zhou AG, Liu YD, Liu CF (2013) A technique for carbon and chlorine isotope analyses of chlorinated aliphatic hydrocarbons in groundwater. J Earth Sci 24:274–281
Gaudette HE (1990) Chlorine and boron isotopic analyses of Antarctic ice and snow: indicators of marine and volcanic atmospheric inputs. Geol Soc Am Annu Meeting 1990:173
Gelman F, Halicz L (2010) High precision determination of bromine isotope ratio by GC-MC-ICPMS. Int J Mass Spectrom 289:167–169
Giunta T, Ader M, Bonifacie M, Agrinier P, Coleman M (2015) Pre-concentration of chloride in dilute water-samples for precise δ^{37}Cl determination using a strong ion-exchange resin: application to rainwaters. Chem Geol 413:86–93
Giuseppetti G, Mazzi F, Tadini C (1971) The crystal structure of eudialyte. Tschermak mineral. Petrogr Mitt 16:105–127
Glavin DP, Freissinet C, Miller KE, Eigenbrode JL, Brunner AE, Buch A, Sutter B, Archer PD, Atreya SK, Brinckerhoff WB, Cabane M, Coll P, Conrad PG, Coscia D, Dworkin JP, Franz HB, Grotzinger JP, Leshin LA, Martin MG, McKay C, Ming DW, Navarro-Gonzalez R, Pavlov A, Steele A, Summons RE, Szopa C, Teinturier S, Mahaffy PR (2013) Evidence for perchlorates and the origin of chlorinated hydrocarbons detected by SAM at the Rocknest aeolian deposit in gale crater. J Geophys Res Planets 118:1955–1973
Gleditsch E, Samdahl B (1922) Radioactivité sur le poids atomique de chlore dans un mineral ancien, l'apatede Balme. Compte Rend Sean 174:746–748
Godon A (2000) La subduction des Petites Antilles: apports de la géochimie isotopique du chlore. PhD thesis. Paris VII Univ, p 300. (Abstract in English)
Godon A, Jendrzejewski N, Eggenkamp HGM, Banks DA, Ader M, Coleman ML, Pineau F (2004a) A cross calibration of chlorine isotopic measurements and suitability of seawater as the international reference material. Chem Geol 207:1–12
Godon A, Webster JD, Layne GD, Pineau F (2004b) Secondary ion mass spectrometry for the determination of δ^{37}Cl. Part II. Intercalibration of SIMS and IRMS for aluminosilicate glasses. Chem Geol 207:291–303
Golyshev VM, Simonov VI, Belov NV (1971) Crystal structure of eudialyte. Sov Phys Crystallogr 16:70–74
Gordon G, Tachiyashiki S (1991) Kinetics and mechanism of formation of chlorate ion from the hypochlorous acid/chlorite ion reaction at pH 6–10. Environ Sci Technol 25:468–474
Greene TP, Tokunaga AT, Toomey DW, Carr JS (1993) CSHELL: a high spectral resolution 1–5 mm cryogenic echelle spectrograph for the IRTF. Proc SPIE 1946:313
Gu B, Brown GM, Maya L, Lance MJ, Moyer BA (2001) Regeneration of perchlorate (ClO_4^-)-loaded anion exchange resins by novel tetrachloroferrate ($FeCl_4^-$) displacement technique. Environ Sci Technol 35:3363–3368
Gu B, Brown GM, Chiang CC (2007) Treatment of perchlorate contaminated groundwater using highly selective, regenerable ion-exchange technologies. Environ Sci Technol 41:6277–6282
Gu B, Böhlke JK, Sturchio NC, Hatzinger PB, Jackson WA, Beloso Jr AD, Heraty LJ, Bian Y, Brown GM (2011a) Removal, recovery and fingerprinting of perchlorate by ion exchange processes. In: SenGupta AK (ed) Ion exchange and solvent extraction: a series of advances, vol 20. CRC, Boca Raton, FL
Gu B, Böhlke JK, Sturchio NC, Hatzinger PB, Jackson WA, Beloso AD, Heraty LJ, Bian Y, Jiang X, Brown GM (2011b) Applications of selective ion exchange for perchlorate removal, recovery, and environmental forensics. In: SenGupta AK (ed) Ion exchange and solvent extraction: a series of advances: 20. Taylor & Francis, pp 117–144
Halas P, Pelc A (2009) New isotope ratio mass spectrometric method of precise δ Cl determinations. Rapid Commun Mass Spectrom 23:1061–1064

Harkins WD, Hayes A (1921) The separation of the element chlorine into isotopes (isotopic elements). J Am Chem Soc xx:1803–1825
Harkins WD, Stone SB (1926) The isotopic composition and atomic weight of chlorine from meteorites and from minerals of non-marine origin. J Am Chem Soc 48:938–949
Hatzinger PB, Böhlke JK, Sturchio NC, Gu B (2011) Guidance manual for forensic analysis of perchlorate in groundwater using chlorine and oxygen isotopic analyses. U.S. Department of Defense, Environmental Security Technology Certification Program (ESTCP Project ER-200509)
Hauri E, Wang J, Dixon JE, King PE, Mandeville C, Newman S (2000) SIMS analysis of volatiles in silicate glasses: 1. Calibration, matrix effects and comparisons with FTIR. Chem Geol 183:99–114
Hayden TS (2023) Assessing the volatile inventory and history of the moon using lunar meteorites. PhD thesis. Open University, Milton Keynes, UK
Hecht MH, Kounaves SP, Quinn RC, West SJ, Young SMM, Ming DW, Catling DC, Clark BC, Boynton WV, Hoffman J, DeFlores LP, Gospodinova K, Kapit J, Smith PH (2009) Detection of perchlorate and the soluble chemistry of Martian soil at the Phoenix Lander site. Science 325:64–67
Herzog W, Dörnenburg E (1958) Die Verwendung van Methylchlorid zur massenspektrometrichen Isotopenanalyse von Chlor. Z Naturforschg 13a:51–52
Herzog W, Klemm A (1958) Die Temperaturabhängigkeit des Isotopie-Effekts bei der elektrolytischen Wanderungen der Chlorionen in geschmolzenem Thallium(I)-chlorid. Z Naturforschg 13a:7–16
Hill JW, Fry A (1962) Chlorine isotope effects in the reactions of benzyl and substituted benzyl chlorides with various nucleophiles. J Am Chem Soc 84:2763–2769
Hinkle KH, Cuberly RW, Gaughan NA, Heynssens JB, Joyce RR, Ridgway ST, Schmitt P, Simmons JE (1998) Phoenix: a cryogenic high-resolution 1- to 5-um infrared spectrograph. Proc SPIE 3354:810–821
Hintze C (1915) Handbuch der Mineralogie. Verlag von Veit & Comp, Leipzig
Hitzfeld KL, Gehre M, Richnow HH (2011) A novel online approach to the determination of isotopic ratios for organically bound chlorine, bromine and Sulphur. Rapid Commun Mass Spectrom 25:3114–3122
Hoering TC, Parker PL (1961) The geochemistry of the stable isotopes of chlorine. Geochim Cosmochim Acta 23:186–199
Holmstrand H, Andersson P, Gustafsson Ö (2004) Chlorine isotope analysis of submicromole organochlorine samples by sealed tube combustion and thermal ionization mass spectrometry. Anal Chem 76:2336–2342
Holt BD, Sturchio NC, Abrajano TA, Heraty LJ (1997) Conversion of chlorinated volatile organic compounds to carbon dioxide and methyl chloride for isotopic analysis of carbon and chlorine. Anal Chem 69:2727–2733
Horst A, Renpenning J, Richnow HH, Gehre M (2017) Compound specific stable chlorine isotopic analysis of volatile aliphatic compounds using gas chromatography hyphenated with multiple collector inductively coupled plasma mass spectrometry. Anal Chem 89:9131–9138
Howald RA (1960) Ion pairs. I. Isotope effects shown by chloride solutions in glacial acetic acid. J Am Chem Soc 82:20–24
Iwagami N, Ohtsuki S, Tokuda K, Ohira N, Kasaba Y, Imamura T, Sagawa H, Hashimoto GL, Takeuchi S, Ueno M, Okumura S (2008) Hemispheric distributions of HCl above and below the Venus' clouds by ground-based 1.7 mm spectroscopy. Planet Space Sci 56:1424–1434
Jackson WA, Anandam SK, Anderson T, Lehman T, Rainwater K, Rajagopalan S, Ridley M, Tock R (2005) Perchlorate occurrence in the Texas southern high plains aquifer system. Ground Water Monit Remidiat 25:137–149
Jackson WA, Davila A, Estrada N, Lyons WB, Coates JD, Priscu J (2012) Perchlorate and chlorate biogeochemistry in ice-covered lakes of the McMurdo dry valleys, Antarctica. Geochim Cosmochim Acta 98:19–30
Jackson WA, Davila AF, Böhlke JK, Sturchio NC, Sevanthi R, Estrada N, Brundrett M, Lacelle D, McKay CP, Poghosyan A, Pollard W, Zacny K (2016) Deposition, accumulation, and alteration of Cl^-, NO_3^-, ClO_4^- and ClO_3^- salts in a hyper-arid polar environment: mass balance and isotopic constraints. Geochim Cosmochim Acta 182:197–215
Jackson WA, Brundrett M, Böhlke JK, Hatzinger PB, Mroczkowski SJ, Sturchio NC (2021) Isotopic composition of natural and synthetic chlorate ($\delta^{18}O$, $\Delta^{17}O$, $\delta^{37}Cl$, $^{36}Cl/Cl$): methods and initial results. Chemosphere 274:129586
Jendrzejewski N, Eggenkamp HGM, Coleman ML (1997) Sequential determination of chlorine and carbon isotopic composition in single microliter samples of chlorinated solvent. Anal Chem 69:4259–4266
Jin B, Laskov C, Rolle M, Haderlein SB (2011) Chlorine isotope analysis of organic contaminants using GCqMS: method optimization and comparison of different evaluation schemes. Environ Sci Technol 45:5279–5286
Joan M, Reedy JH (1940) The detection of oxy-halogen anions. Trans Ill State Acad Sci 33:123–125
Johnston WH, Arnold JR (1953) The existence of multiply charged molecular ions of HBr, HCl, and DCl. J Chem Phys 21:1499–1502
Kallman H, Lasareff W (1932) Über die Isotopenuntersuchungen (Sauerstoff, Neon und Chlor). Z Phys 80:237–241
Kaufmann RS (1984) Chlorine in groundwater: stable isotope distribution. Ph.D. thesis. University of Arizona, Tucson, Ariz
Kaufmann RS, Long A, Bentley H, Davis S (1984) Natural chlorine isotope variations. Nature 309:338–340

Keller J, Krafft M (1990) Effusive natrocarbonatite activity of Oldoinyo Lengai, June 1988. Bull Volcanol 52:629–645

Klemm A, Lundén A (1955) Isotopenanreicherung beim Chlor durch electrolytische Überfürung in geschmolzenem Bleichlorid. Z Naturforschg 10A:282–284

Kohnen MEL (1988) Stabiele chloorisotopen onderzoek. Internal report. University of Utrecht, p 17

Korablev O, Olsen KS, Trokhimovskiy A, Lefèvre F, Montmessin F, Fedorova AA et al (2021) Transient HCl in the atmosphere of Mars. Sci Adv 7:eabe4386

Kounaves SP, Stroble ST, Anderson RM, Moore Q, Catling DC, Douglas S, McKay CP, Ming DW, Smith PH, Tamppari LK, Zent AP (2010) Discovery of natural perchlorate in the Antarctic dry valleys and its global implications. Environ Sci Technol 44:2360–2364

Kuroda PK, Sandell EB (1953) Chlorine in igneous rocks. Bull Geol Soc Am 64:879–896

Kusakabe M (2005) A closed pentane trap for separation of SO_2 from CO_2 for precise $\delta^{18}O$ and $\delta^{34}S$ measurements. Geochem J 39:285–287

Kusebauch C, John T, Barnes JD, Klügel A, Austrheim HO (2015) Halogen element and stable chlorine isotope fractionation caused by fluid-rock interaction (Bamble sector, SE Norway). J Petrol 56:299–324

Langvad T (1954) Separation of chlorine isotopes by ion-exchange chromatography. Acta Chem Scand 8:526–527

Layne GD, Godon A, Webster JD, Bach W (2004) Secondary ion mass spectrometry for the determination of $\delta^{37}Cl$. Part I. Ion microprobe analysis of gasses and fluids. Chem Geol 207:277–289

Le Roy L, Altwegg K, Balsiger H, Berthelier JJ, Bieler A, Briois C, Calmonte U, Combi MR, De Keyser J, Dhooghe F, Fiethe B, Fuselier SA, Gasc S, Gombosi TI, Hässig M, Jäckel A, Martin Rubin M, Chia-Yu Tzou CY (2015) Inventory of the volatiles on comet 67P/Churyumov-Gerasimenko from Rosetta/ROSINA. Astron Astrophys 583:A1

Liggett LM (1954) Determination of organic halogen with sodium biphenyl reagent. Anal Chem 26:748–750

Liu WG, Xiao YK, Wang QZ, Qi HP, Wang YH, Zhou YM, Shirodkar PV (1997) Chlorine isotopic geochemistry of salt lakes in the Qaidam Basin, China. Chem Geol 136:271–279

Liu YD, Zhou AG, Gan YQ, Liu CF, Yu TT, Li XQ (2013) An online method to determine chlorine stable isotope composition by continuous flow isotope ratio mass spectrometry (CF-IRMS) coupled with a Gasbench II. J South Cent Univ 20:193–198

Liu Y, Li QL, Tang GQ, Li XH (2022) Simultaneous determination of $\delta^{18}O$ and $\delta^{37}Cl$ isotopic compositions and the concentrations of F, Cl, and S in apatite using SIMS. Crystals 12:383

Liuzzi G, Villanueva GL, Mumma MJ, Smith MD, Daerden F, Ristic B, Thomas I, Vandaele AC, Patel MR, Lopez-Moreno JJ, Belluci G, The NOMAD Team (2019) Methane on Mars: new insights into the sensitivity of CH_4 with the NOMAD/ExoMars spectrometer through its first in-flight calibration. Icarus 321:671–690

Liuzzi G, Villanueva GL, Viscardy S, Mège D, Crismani MMJ, Aoki S, Gurgurewicz J, Tesson PA, Mumma MJ, Smith MD, Faggi S, Kofman V, Knutsen EW, Daerden F, Neary L, Schmidt F, Trompet L, Erwin JT, Robert S, Thomas IR, Ristic B, Bellucci G, Lopez-Moreno JJ, Patel MR, Vandaele AC (2021) Probing the atmospheric Cl isotopic ratio on Mars: implications for planetary evolution and atmospheric chemistry. Geophys Res Lett 48:e2021GL092650

Long A, Eastoe CJ, Kaufmann RS, Martin JG, Wirt L, Finley JB (1993) High-precision measurement of chlorine stable isotope ratios. Geochim Cosmochim Acta 57:2907–2912

Lundén A, Herzog W (1956) Isotopenanreicherung bei Chlor durch electrolytische Überführung in geschmolzenem Zinkchlorid. Z Naturforschg 11a:520

Maas ZG (2020) The galactic chemical evolution of chlorine and phosphorus. PhD. Indiana Univ

Maas ZG, Pilachowski CA (2018) Chlorine isotope ratios in M giants. Astron J 156:2

Maas ZG, Pilachowski CA, Hinkle K (2016) Chlorine abundances in cool stars. Astron J 152:196

Madorsky SL, Strauss S (1948) Concentration of isotopes of chlorine by the counter-current electromigration method. J Res Natl Bur Stand 38:185–189

Magenheim AJ, Spivack AJ, Volpe C, Ransom B (1994) Precise determination of stable chlorine isotopic ratios in low-concentration natural samples. Geochim Cosmochim Acta 58:3117–3121

Mahaffy PR, Webster CR, Cabane M, Conrad PG, Coll P, Atreya SK, Arvey R, Barciniak M, Benna M, Bleacher L, Brinckerhoff WB, Eigenbrode JL, Carignan D, Cascia M, Chalmers RA, Dworkin JP, Errigo T, Everson P, Franz H, Farley R, Feng S, Frazier G, Freissinet C, Glavin DP, Harpold DN, Hawk D, Holmes V, Johnson CS, Jones A, Jordan P, Kellogg J, Lewis J, Lyness E, Malespin CA, Martin DK, Maurer J, McAdam AC, McLennan D, Nolan TJ, Noriega M, Pavlov AA, Prats B, Raaen E, Sheinman O, Sheppard D, Smith J, Stern JC, Tan F, Trainer M, Ming DW, Morris RV, Jones J, Gundersen C, Steele A, Wray J, Botta O, Leshin LA, Owen T, Battel S, Jakosky BM, Manning H, Squyres S, Navarro-Gonzalez R, McKay CP, Raulin F, Sternberg R, Buch A, Sorensen P, Kline-Schoder R, Coscia D, Szopa C, Teinturier S, Baffes C, Feldman J, Flesch G, Forouhar S, Garcia R, Keymeulen D, Wood-ward S, Block BP, Arnett K, Miller R, Edmonson C, Gorevan S, Mumm E (2012) The sample analysis at Mars investigation and instrument suite. Space Sci Rev 170:401–478

McDonald I, Zijlstra AA, Boyer MI (2012) Fundamental parameters and infrared excesses of Hipparcos stars. Mon Not R Astron Soc 427:343–357

McIlvaine TC (1921) A buffer solution for colorimetric comparison. J Biol Chem 49:183–186

McKinney CR, McCrea JM, Epstein S, Allen HA, Urey HC (1950) Improvements in mass spectrometers for the measurement of small differences in isotope abundance ratios. Rev Sci Instrum 21:724–730

Ming DW, Smith PH, Tamppari LK, Zent AP (2010) Discovery of natural perchlorate in the Antarctic dry valleys and its global implications. Environ Sci Technol 44:2360–2364

Morin SM, Barnes JJ, Wilbur ZE, Stadermann AC, Domanik K, McCubbin FM (2021) Assessing the volatile inventory of basaltic fragments in Luna soils. In: 84th Ann meet meteor Soc, vol 2609. LPI Contrib, p 6229

Munson B (2006) Chemical ionization mass spectrometry: theory and applications. In: Encyclopedia of analytical chemistry. John Wiley & Sons, Hoboken, NJ, USA

Musashi M, Markl G, Kreulen R (1998) Stable chlorine-isotope analysis of rock samples: new aspects of chlorine extraction. Anal Chem Acta 362:261–269

Musashi M, Eggenkamp HGM, Van Cappellen P (2010) Experiment on chlorine extraction from chlorinated pollutants by supercritical water and their Cl isotope analysis. Bull Sci Technol Shibaura Inst Technol 54:57–61. (In Japanese with English abstract)

Nešković OM, Veljković MV, Veličković SR, Derić AJ, Miljević NR, Golobočanin DD (2002) Precise measurement of chlorine isotopes by thermal ionization mass spectrometry. Nukleonika 47(supp. 1):s85–s87

Nier AO (1947) A mass spectrometer for isotope and gas analysis. Rev Sci Instrum 18:398–411. [4]

Nier AO (1955) Determination of isotopic masses and abundances by mass spectrometry. Science 121:737–744

Nier AO, Hanson EE (1936) A mass-spectrographic analysis of the ions produced in HCl under electron impact. Phys Rev 50:722–726

Nier AO, Ney EP, Inghram MG (1946) A null method for the comparison of two ion currents in a mass spectrometer. Phys Rev 70:116–117

Numata M, Nakamura N, Gamo T (2001) Precise measurement of chlorine stable isotopic ratios by thermal ionization mass spectrometry. Geochem J 35:89–100

Numata M, Nakamura N, Koshikawa H, Terashima Y (2002) Chlorine stable isotope measurements of chlorinated aliphatic hydrocarbons by thermal ionization mass spectrometry. Anal Chim Acta 455:1–9

Okumura S, Nishihara E, Watanabe E, Mori A, Kataza H, Yamashita T (2000) OASIS: a multi-purpose near-infrared camera and spectrograph. Publ Astron Soc Jpn 52:931

Owen HR, Schaeffer OA (1955) The isotope abundances of chlorine from various sources. J A Chem Soc 77:898–899

Parker DR, Seyfferth AL, Reese BK (2008) Perchlorate in groundwater: a synoptic survey of "pristine" sites in the conterminous Unites States. Environ Sci Technol 42:1465–1471

Pelc A (2024) New chloromethane preparation line for the stable chlorine isotope composition analysis. Isot Environ Health Stud:1–9. https://doi.org/10.1080/10256016.2024.2379414

Pelc A, Hałas S (2008) Negative ion source for chlorine isotope ratio measurements. Rapid Commun Mass Spectrom 22:3977–3982

Pisarenko AN, Stanford BD, Quinones O, Pacey GE, Gordon G, Snyder S (2010) Rapid analysis of perchlorate, chlorate and bromate ions in concentrated sodium hypochlorite solutions. Anal Chim Acta 659:216–223

Plummer LN, Böhlke JK, Doughten MW (2006) Perchlorate in Pleistocene and Holocene groundwater in north-Central New Mexico. Environ Sci Technol 40:1757–1763

Ponsin V, Torrentó C, Lihl C, Elsner M, Hunkeler D (2019) Compound-specific chlorine isotope analysis of the herbicides atrazine, acetochlor, and metolachlor. Anal Chem 91:14290–14298

Prohaska T, Irrgeher J, Benefield J, Böhlke JK, Chesson LA, Coplen TB, Ding TP, Dunn PJH, Gröning M, Holden NE, Meijer HAJ, Moossen H, Possolo A, Takahashi Y, Volg J, Walczyk T, Wang J, Wieser ME, Yoneda S, Zhu XK, Meija J (2022) Standard atomic weights of the elements 2021 (IUPAC technical report). Pure Appl Chem 94:573–600

Rajagopalan S, Anderson TA, Fahlquist L, Rainwater KA, Ridley M, Jackson WA (2006) Widespread presence of naturally occurring perchlorate in high plains of Texas and New Mexico. Environ Sci Technol 40:3156–3162

Rajagopalan S, Anderson TA, Cox S, Harvey G, Cheng Q, Jackson WA (2009) Perchlorate in wet deposition across North America. Environ Sci Technol 43:616–622

Rao B, Anderson TA, Orris GJ, Rainwater KA, Rajagopalan S, Sandvig RM, Scanlon BR, Stonestrom DA, Walvoord MA, Jackson WA (2007) Widespread natural perchlorate in unsaturated zones of the Southwest United States. Environ Sci Technol 41:4522–4528

Rao B, Anderson TA, Redder A, Jackson WA (2010) Perchlorate formation by ozone oxidation of aqueous chlorine/oxy-chlorine species: role of Cl_xO_y radicals. Environ Sci Technol 44:2961–2967

Rees CE (1978) Sulphur isotope measurements using SO2 and SF6. Geochim Cosmochim Acta 42:383–389

Renpenning J, Horst A, Schmidt M, Gehre M (2018) Online isotope analysis of $^{37}Cl/^{35}Cl$ universally applied for semi-volatile organic compounds using GC-MC-ICPMS. J Anal At Spectrom 33:314–321

Richards TW, Wells RC (1905) A revision of the atomic weights of sodium and chlorine. J Am Chem Soc 27:459–529

Rosenbaum JM, Cliff RA, Coleman ML (2000) Chlorine stable isotopes: a comparison of dual inlet and thermal ionization mass spectrometric measurements. Anal Chem 72:2261–2264

Rothman LS, Gordon IE, Babikov Y, Barbe A, Chris Benner D, Bernath PF, Birk M, Bizzocchi L, Boudon V, Brown LR, Campargue A, Chance K, Cohen EA, Coudert LH, Devi VM, Drouin BJ, Fayt A, Flaud JM, Gamache RR, Harrison JJ, Hartmann JM, Hill C, Hodges JT, Jacquemart D, Jolly A, Lamouroux J, Le Roy RJ, Li G, Long DA, Lyulin OM, Mackie CJ,

Massie ST, Mikhailenko S, Müller HSP, Naumenko OV, Nikitin AV, Orphal J, Perevalov V, Perrin A, Polovtseva ER, Richard C, Smith MAH, Starikova E, Sung K, Tashkun S, Tennyson J, Toon GC, Tyuterev VG, Wagner G (2013) The HITRAN2012 molecular spectroscopic database. J Quant Spectrosc Radiat Transfer 130:4–50

Sakaguchi-Söder KA (2010) New method for compound-specific stable chlorine isotope analysis: basics and application. Ph.D. Thesis. Technische Universität, Darmstadt, Germany

Sakaguchi-Söder K, Jager J, Grund H, Matthäus F, Schüth C (2007) Monitoring and evaluation of dechlorination processes using compound-specific chlorine isotope analysis. Rapid Commun Mass Spectrom 21:3077–3084

Schnetger B, Muramatsu Y (1996) Determination of halogens, with special reference to, iodine, in geological and biological samples using pyrohydrolysis for preparation and inductively coupled plasma mass spectrometry and ion chromatography for measurement. Analyst 121:1627–1631

Sharp ZD, Barnes JD, Brearly AJ, Chaussidon M, Fisher TP, Kamenetsky VS (2007) Chlorine isotope homogeneity of the mantle, crust and carbonaceous chondrites. Nature 446:1062–1065

Shields WR, Murphy TJ, Garner EL, Dibeler VH (1962) Absolute isotopic abundance ratios and the isotopic weight of chlorine. J Am Chem Soc 84:1519–1522

Shirodkar PV, Xiao YK, Sarkar A, Dalal SG, Chivas AR (2006) Influence of air-sea fluxes on chlorine isotopic composition of ocean water: implications for constancy in $\delta^{37}Cl$—a statistical inference. Environ Int 32:235–239

Shouakar-Stash O, Drimmie RJ, Frape SK (2005) Determination of inorganic chlorine stable isotopes by continuous flow isotope ratio mass spectrometry. Rapid Commun Mass Spectrom 19:121–127

Shouakar-Stash O, Drimmie RJ, Zhang M, Frape SK (2006) Compound-specific chlorine isotope ratios of TCE, PCE and DCE isomers by direct injection using CF-IRMS. Appl Geochem 21:766–781

Stromeyer F (1819) Analyse einiger von dem Prof. von Giesecke in Grönland entdeckten Fossilien: Giesekit, Saphirin, Apophyllit, Dichroit, Arragonit und Eudialit. Ann Phys 63:372–381

Stull DR (1947) Vapour pressure of pure substances. Inorganic compounds. Ind Eng Chem 39:540–550

Sturchio NC, Hatzinger PB, Arkins MD, Suh C, Heraty LJ (2003) Chlorine isotope fractionation during microbial reduction of perchlorate. Environ Sci Technol 37:3859–3863

Sturchio NC, Böhlke JK, Gu B, Horita J, Brown GM, Beloso Jr AD, Hatzinger PB, Jackson WA, Batista JR (2006) Stable isotopic compositions of chlorine and oxygen in synthetic and natural perchlorates. In: Gu B, Coates JD (eds) Perchlorate environmental occurrences, interactions, and treatment. Springer, New York, pp 93–109

Sturchio NC, Böhlke JK, Beloso Jr AD, Streger SH, Heraty LJ, Hatzinger PB (2007) Oxygen and chlorine isotopic fractionation during perchlorate biodegradation: laboratory results and implications for forensics and natural attenuation studies. Environ Sci Technol 41:2796–2802

Sturchio NC, Böhlke JK, Gu BH, Hatzinger PB, Jackson WA (2011) Isotopic tracing of perchlorate in the environment. In: Baskaran M (ed) Handbook of environmental isotope geochemistry, advances in isotope geochemistry. Springer-Verlag, Berlin Heidelberg

Suzuki K, Miyata Y, Kanazawa N (2004) Precise Re isotope ratio measurements by negative thermal ionization mass spectrometry (NTI-MS) using total evaporation technique. Int J Mass Spectrom 235:97–101

Tanaka N, Rye DM (1991) Chlorine in the stratosphere. Nature 353:707

Taylor JW, Grimsrud EP (1969) Chlorine isotopic ratios by negative ion mass spectrometry. Anal Chem 41:805–810

Thornton B, Horst A, Carrizo D, Holmstrand H, Andersson P, Crill PM, Gustafsson Ö (2013) A high-volume cryosampler and sample purification system for bromine isotope studies of methyl bromide. J Atmos Ocean Technol 30:2095–2107

Toyama C, Kimura JI, Chang Q, Vaglarov BS, Kuroda J (2015) A new high-precision method for determining stable chlorine isotopes in halite and igneous rock samples using UV-femtosecond laser ablation multiple faraday collector inductively coupled plasma mass spectrometry. J Anal At Spectrom 30:2194–2207

Trokhimovskiy A, Fedorova AA, Olsen KS, Alday J, Korablev O, Montmessin F, Lefèvre F, Patrakeev A, Belyaev D, Shakun AV (2021) Isotopes of chlorine from HCl in the Martian atmosphere. Astron Astrophys 651:A32

Turnquist CR, Taylor JW, Grimsrud EP, Williams RC (1973) Temperature dependence of chlorine kinetic isotope effects for aliphatic chlorides. J Am Chem Soc 95:4133–4138

Van Acker MRMD, Shahar A, Young ED, Coleman ML (2006) GC/multiple collector-ICPMS method for chlorine stable isotope analysis of chlorinated aliphatic hydrocarbons. Anal Chem 78:4663–4667

Van Warmerdam EM, Frape SK, Aravena R, Drimmie RJ, Flatt H, Cherry JA (1995) Stable chlorine and carbon isotope measurements of selected chlorinated organic solvents. Appl Geochem 10:547–552

Vandaele AC, Lopez-Moreno JJ, Patel MR, Bellucci G, Daerden F, Ristic B, Robert S, Thomas IR, Wilquet V, Allen M, Alonso-Rodrigo G, Altieri F, Aoki S, Bolsée D, Clancy T, Cloutis E, Depiesse C, Drummond R, Fedorova A, Formisano V, Funke B, González-Galindo F, Geminale A, Gérard JC, Giuranna M, Hetey L, Ignatiev N, Kaminski J, Karatekin O, Kasaba Y, Leese M, Lefèvre F, Lewis SR, López-Puertas M, López-Valverde M, Mahieux A, Mason J, McConnell J, Mumma M, Neary L, Neefs E, Renotte E, Rodriguez-Gomez J, Sindoni G, Smith M, Stiepen

A, Trokhimovsky A, Vander Auwera J, Villanueva G, Viscardy S, Whiteway J, Willame Y, Wolff M, The NOMAD Team (2018) NOMAD, an integrated suite of three spectrometers for the ExoMars trace gas mission: technical description, science objectives and expected performance. Space Sci Rev 214:80

Vazques T, Vuppaia S, Ayodeji I, Song LX, Grimes N, Evans-Nguyes T (2020) In situ mass spectrometers for applications in space. Mass Spectrom Rev 40: 670–691

Vengosh A, Chivas AR, McCulloch MT (1989) Direct determination of boron and chlorine isotopic compositions in geological materials by negative thermal-ionization mass spectrometry. Chem Geol (Isot Geosci Sect) 79:333–343

Vogel AI (1951) A textbook of quantitative inorganic analysis, theory and practice. Longmans, Green and Co, London, p 918

Vogel AI (1989) Jeffrey GH et al (eds) A textbook of quantitative inorganic analysis, theory and practice, 5th ed. Longmans, Green and Co, London

Volpe C, Spivack AJ (1994) Stable chlorine isotopic composition of marine aerosol-particles in the Western Atlantic Ocean. Geophys Res Lett 21:1161–1164

Von Rothmund V (1909) Concerning the reduction and determination of perchlorate. Z Anorg Chem 62:108–113

Wakaki S, Shibata S, Tanaka T (2007) Isotope ratio measurements of trace Nd by the total evaporation normalization (TEN) method in thermal ionization mass spectrometry. Int J Mass Spectrom 264:157–163

Wassenaar LI, Koehler G (2004) On-line technique for the determination of the $\delta^{37}Cl$ of inorganic and total organic Cl in environmental samples. Anal Chem 76:6384–6388

Weast RC, Astle MJ (1982) Handbook of chemistry and physics.63rd edn. CRC Press, Inc, Boca Raton, FL

Wei HZ, Jiang SY, Xiao YK, Wang J, Lu H, Wu B, Wu HP, Li Q, Luo CG (2012) Precise determination of the absolute isotopic abundance ratio and the atomic weight of chlorine in three international reference materials by the positive thermal ionization mass spectrometer-Cs_2Cl^+-graphite method. Anal Chem 84:10350–10358

Westaway KC, Koerner T, Fang YR, Rudziñski J, Paneth P (1998) A new method of determining chlorine kinetic isotope effects. Anal Chem 70:3548–3552

Whitehead D, Thomas JE (1985) Use a nebulizer in pyrohydrolytic decomposition of silicate materials for determination of fluorine and chlorine. Anal Chem 57:2421–2423

Williams RC, Taylor JW (1973) Chlorine kinetic isotope-effect models. 1. Isotopic dependence in nominal C-Cl stretching vibrations of aliphatic chlorides and vibrational analysis of *tert*-butyl chloride ground-state. J Am Chem Soc 95:1710–1714

Wu JH, Satake H (2006) Purification of CH_3Cl from CH_3I using cold trap with sealed 2,2,4-trimethylpentane for $\delta^{37}Cl$ measurement. Anal Chim Acta 555:41–46

Wudarska A, Słaby E, Wiedenbeck M, Birski Ł, Wirth R, Götze J, Lepland A, Kusebauch C, Kocjan I (2020) Chlorine isotope composition of apatite from the >3.7 Ga Isua Supracrustal Belt, SW Greenland. Fortschr Mineral 10:27

Wudarska A, Słaby E, Wiedenbeck M, Barnes JD, Bonifacie M, Sturchio NC, Bardoux G, Couffignal F, Glodny J, Heraty L, John T, Kusebauch C, Mayanna S, Wilke FDH, Deput E (2021) Inter-laboratory characterisation of apatite reference materials for chlorine isotope analysis. Geostand Geoanal Res 45:121–142

Xiao YK, Zhang CG (1992) High precision isotopic measurement of chlorine by thermal ionization mass spectrometry of the Cs_2Cl^+ ion. Int J Mass Spectrom Ion Proc 116:183–192

Xiao YK, Beary ES, Fassett JD (1988) An improved method for the high-precision isotopic measurement of boron by thermal ionization mass-spectrometry. Int J Mass Spectrom Ion Proc 85:203–213

Xiao YK, Jin L, Qi HP (1991) Investigation of thermal ion emission characteristics of graphite. Int J Mass Spectrom Ion Proc 107:205–213

Xiao YK, Zhou YM, Liu WG (1995) Precise measurement of chlorine isotopes based on Cs_2Cl^+ by thermal ionization mass-spectrometry. Anal Lett 28:1295–1304

Xiao YK, Zhou YM, Wang QZ, Wei HZ, Liu WG, Eastoe CJ (2002) A secondary isotopic reference material of chlorine from selected seawater. Chem Geol 182:655–661

Young LDG (1972) High resolution spectra of Venus—a review. Icarus 17:632–658

Zakon Y, Halicz L, Gelman F (2014) Isotope analysis of sulfur, bromine, and chlorine in individual anionic species by ion chromatography / multicollector—ICPMS. Anal Chem 86:6495–6500

Zhang J, Liu SH, Gui JY, Li ZY, Qi GC (2021) Compound-specific chlorine isotope analysis of organochlorine pesticides by gas chromatography-negative chemical ionization mass spectrometry. Hindawi 2021:8874679

6 Preparation Techniques for the Analysis of Stable Bromine Isotopes

6.1 Early Attempts to Analyse Bromine Isotopes

Shortly after the development of the mass spectrograph it was discovered that bromine consists of two stable isotopes (Aston 1920a, 1920b). Later it was confirmed that the two isotopes have abundances which are bout equal (Aston 1931). More precise measurements on the ratio between the two stable bromine isotopes were done by Blewett (1936) who was able to analyse variations with a precision of about 25‰. After the development of a new type of mass spectrometers by Nier (1940), Williams and Yuster (1946) were able to measure the bromine isotope composition with a precision of about 4‰. The first negative ion thermal ionization mass spectrometry study (Cameron and Lippert 1955), in which attempts were made to observe natural bromine variations with a precision of (also) approximately 4‰, did not find natural bromine isotope variations. Techniques improved to about a precision of 1.8‰ in the early 1960s (Catanzaro et al. 1964), while the first study showing that very precise measurements are possible was published in 1993 (Xiao et al. 1993), using positive ion thermal ionization mass spectrometry on the Cs_2Br^+ ion. Since Willey and Taylor (1978), who were the first to test bromomethane as mass spectrometer gas, it is expected that this may be the most promising gas for gas source isotope ratio mass spectrometry (both for dual inlet and continuous flow techniques) and it was using this gas that the first natural variations were finally determined by Eggenkamp and Coleman (2000). The techniques that were used by Eggenkamp and Coleman (2000) in their first study to determine natural bromine isotope variations are also described by Eggenkamp (2004) in De Groot (2004). They are repeated here due to the importance of these methods for the development of bromine isotope analysis techniques. Especially the method to separate bromide from a large excess of chloride (Dechan 1886; Friedheim and Meyer 1892; Boeke 1908) is still almost universally used in laboratories that study bromine isotope variations in groundwaters.

6.2 Techniques to Separate Bromine from Chlorine in Natural Samples

As chlorine and bromine are chemically very similar they are difficult to separate in samples where they are both present. Considering the importance to analyse the composition of samples containing multiple halogens (chlorine, bromine and iodine) during the nineteenth century many attempts have been made to physically separate the halogens from mixtures to analyse them individually by gravimetric means. It took until the end of the century before fast and quantitative methods were developed (Dechan 1886).

H. Eggenkamp, *The Geochemistry of Stable Chlorine and Bromine Isotopes*, Advances in Isotope Geochemistry, https://doi.org/10.1007/978-3-031-75633-7_6

Friedheim and Meyer (1892) summarised the previous attempts and described the method which was later adopted by Eggenkamp and Coleman (2000) to extract bromide from saline samples in their first study reporting natural bromine isotope variations. Friedheim and Meyer (1892) mixed known amounts of chloride, bromide and iodide salts and separated them very efficiently, all with less than 0.5% contamination by the other halogens. Boeke (1908) finally adopted the method for geological (evaporite) samples, which are normally characterised by a large excess of chlorine over bromine. Typical Cl/Br ratios (w/w) are 50 to 5000, for example in seawater it is 292 (Rankama and Sahama 1950). Methods currently in use for the isotopic ratio analyses of bromine still use this technique, although small variations in the method exist between the different laboratories.

6.2.1 Oxidative Separation of Bromine from Chlorine by a $K_2Cr_2O_7/H_2SO_4$ Solution

The technique to separate bromine from an excess of chlorine used by Eggenkamp and Coleman (2000) followed very closely the procedure developed by Friedheim and Meyer (1892). The method used is based upon the well known (see e.g. already Löwich 1829) difference in oxidation-reduction behaviour of the halogen elements. With increasing atomic number the halide ions are increasingly easier to oxidise to the native elements. Thus, bromine can be separated from chlorine with the use of an oxidising agent that is strong enough to oxidise bromide to bromine, while it must not be strong enough to oxidise chloride to chlorine. It is possible to oxidise bromide to bromine by a boiling solution containing potassium dichromate ($K_2Cr_2O_7$) and sulphuric acid (H_2SO_4), while chloride is not. Separation of bromine from chlorine is taken place in a simple distillation apparatus (Fig. 6.1).

The left flask (a) is filled with the sample (either in solution or as a water-soluble salt), containing approximately 2 to 10 mg bromide. Then 10 gram potassium dichromate ($K_2Cr_2O_7$) and subsequently distilled water is added until the flask contains approximately 100 mL solution in total. To facilitate gentle boiling an amount of anti-bumping granules are added to this flask. Through the separating funnel (b) 20 mL of a 1:1 H_2SO_4:H_2O mixture is added, after which the

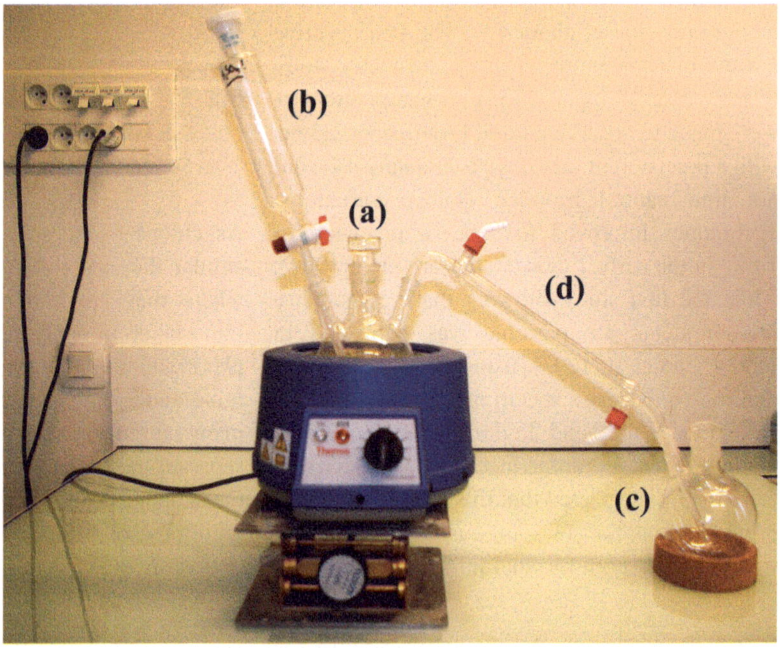

Fig. 6.1 Distillation equipment used for the separation of bromine from chlorine. (**a**) and (**c**) are 500 mL round bottom flasks, (**b**) is a separating funnel and (**d**) is the connection between the round-bottom flasks (**a**) and (**c**). For operation of this equipment see text. Photograph taken by Thomas Giunta and used by permission

fluid in flask (a) is diluted to approximately 200 mL with distilled water. The right flask (c) is filled with 100 mL of a solution containing 1 gr potassium hydroxide. This flask is put into a tray containing cooling water. The water is stagnant, so that it slowly increases in temperature during the cause of the distillation. The left flask (a) is then heated. When the contents of this flask starts to boil a yellow-brown, very pale, cloud is distilled over to flask (b) through connection (d). This is bromine gas. The chloride in flask (a) is transferred for a small part (<1%) to flask (b) in the form of hydrogen chloride, and a small portion is reacted with the potassium dichromate to form chromyl chloride (CrO_2Cl_2), which is not carried over in the distillation (Dechan 1886). The bromine will react with the hydroxide in flask (b) to form hypobromite (BrO^-) and bromide (Br^-) ions. While the solution slowly increases in temperature due to the distillation of boiling water into flask (b) the hypobromite ions slowly decompose to bromide ions and dissolved oxygen (O_2, Bremner 1965). Although the bulk of bromine is transferred over in the first few minutes, the distillation is proceeded for 20 min to facilitate a 100% yield of bromine. After 20 min the boiling is stopped and the distillation set-up is disassembled and the contents of flask (b) is transferred into a 500 mL Erlenmeyer flask to which approximately 2 gr of zinc dust is added, and the mixture is boiled for 10 min. Any residual hypobromite left in solution will then be reduced to bromide which can be processed further. As a few tenths of a percent of the originally present chloride is also transferred the procedure should be repeated for samples with high (>1000) initial Cl/Br ratios. The resulting solutions can be stored until they are analysed, for example in polyethylene bottles.

6.2.2 Oxidative Separation of Bromine from Chlorine by a HNO_3 Solution

While oxidation with potassium dichromate and reduction with zinc metal is perfectly acceptable for IRMS measurements it is not when the resulting solution needs to be analysed by MC-ICPMS. With MC-ICPMS also chromium ($\delta^{53}Cr$) and zinc ($\delta^{66}Zn$) isotopes can be measured, generally on very low concentrations of the metals in solution. As a consequence even the smallest amounts of chromium and zinc present in the mass spectrometer solution can and will cause unacceptable large memory effects in the mass spectrometer source of the ICP-MS. For that reason, for ICP-MS applications a method to separate bromine from chlorine without the use of metal ions needs to be used.

For that reason, oxidation with a HNO_3 solution was considered (Eggenkamp and Louvat 2018). It has long been known that a dilute HNO_3 solution will oxidise Br^-, but not Cl^- (Kebler 1892, Benedict and Snell 1903). Murphy et al. (1954) developed the method to quantitatively separate Br from Cl. During oxidation of Br^-, Br_2 is formed and in alkaline solutions the Br_2 is disproportionated into Br^- and BrO^-, the latter of which subsequently disproportionates to Br^- and BrO^{3-} following:

$$3Br_2 + 3OH^- \rightarrow 3Br^- + 3BrO^- + 3H^+ \quad (6.1)$$

$$3BrO^- \rightarrow 2Br^- + BrO_3^- \quad (6.2)$$

The final solution thus contains Br^- and BrO_3^- in a 5:1 ratio. When bromine is trapped in an ammonia solution, which is both alkaline and a strong reducing agent, the BrO_3^- that is formed is reduced to Br^-. The resulting solution could thus be directly introduced into the inlet system of an MC-ICP-MS instrument. The method developed by Eggenkamp and Louvat (2018) for the analysis of Br isotopes follows the procedure described by Murphy et al. (1954) with several modifications:

Br isotopes are routinely measured in a 0.2 M NH_4NO_3 solution (Louvat et al. 2016). This medium (0.2 M NH_4NO_3) was kept for the oxidative separation for bromine form chlorine as it has proven to produce reproducible measurements of Br isotope ratios and fast washouts (Louvat et al. 2016). The use of the same medium for both analytical procedures (distillation and ion exchange) in use in the same laboratory also allows for immediate comparison between the two methods when they are used to extract Br from geological samples. For that reason Eggenkamp and Louvat (2018) decided to trap the distilled Br_2 in an ammonia solution. Murphy et al. (1954) used a 300 mL distillation flask, but as the size of samples for bromine isotope measurements is mostly between 1 and 25 mL, such a large flask is very inconvenient because the volume of the distilled solution should be as small as possible (about 2 mL) to obtain a good signal level during analyses.

In short the method described by Eggenkamp and Louvat (2018) is a simple distillation that separates bromide from chloride. The sample is mixed with dilute (1:5) nitric acid in a distillation flask and heated over a candle flame for 10 min. The distillate (bromine) is trapped in an ammonia solution and reduced to bromide. Chloride is only distilled to a very small extent. The obtained solution can be measured directly by MC-ICP-MS for stable Br isotopes.

6.2.3 Separation of Bromine from Chlorine over an Ion Exchange Resin

In order to measure bromine isotope compositions by multi collector ICP-MS, Louvat et al. (2016) proposed to separate bromine from chlorine using ion exchange techniques. Following several tests with different eluents for bromide extraction it was concluded that the most successful protocol to extract bromide from a solution with a Cl/Br ratio equivalent to that in seawater consisted of the use of 1 mL of AG 1-X4 200–400 mesh resin in OH form that was put in Kartell funnels (162/4, 4 mm inner diameter) with PTFE frits. Elution was done by gravity, with a rate of 0.2 mL/min for H_2O, but much lower for NaOH solutions. The resin was washed with 10 mL 0.5 M NH_4NO_3 solution, 5 mL H_2O and 20 mL 4 M NaOH to remove all bromide from the resin. 1 to 2 mL of sample volume was introduced at a pH of 9. Then the resin is rinsed with two times 1 mL of H_2O, 5 mL 4 M NaOH, again 2 times 1 mL H_2O and finally 2 mL of 0.2 M NH_4NO_3 solution to remove as much as possible chloride and other ions while keeping all bromide on the resin. Finally bromide is collected from the resin in solution by eluting with 5 mL 0.2 M NH_4NO_3 solution. This elution protocol was successfully applied to natural seawater, with extraction yields of bromide between 96 and 103% (within the 5% HPLC analytical error), and final Cl/Br ratios from 10 to 50. This solution could be measured successfully for bromine isotope compositions by ICP-MS. However, for initial solutions with a Cl/Br ratio significantly above that of seawater this protocol resulted in final Cl/Br ratios too high to be measured by ICP-MS, and for those samples the protocol as described in Sect. 6.2.2 needs to be applied. A different ion exchange resin and approach was used by De Gois et al. (2016). They used Dowex 50WX8 resin and used solutions (including sample solutions) containing 5 mM NH_4NO_3 to counteract memory effects in the bromine isotope determination. They cleaned a polypropylene column with ultrapure water and filled the column with 1 mL of precleaned resin. 2 mL sample solution was loaded onto the column, and the bromide fraction was eluted with 2 mL of ultrapure water. The 2 mL of eluent leaving the column during sample loading and the 2 mL of eluent leaving the column during bromide elution were collected in a single PTFE vessel. The bromide concentration in the solution obtained after ion exchange chromatography was, after addition of an aliquot of NH_4OH, increased by evaporation on a hot plate at 90 °C until (after about 5 h) a total volume of about 1 mL was reached. No loss of bromine was observed during this process.

6.3 Analysis of Stable Bromine Isotopes from Inorganic Bromides

6.3.1 N-TIMS

Cameron and Lippert (1955) used negative thermal ionization mass spectrometry in their first attempt to find natural bromine isotope variations. Cameron and Lippert (1955) obtained bromine samples which were acquired from several suppliers who prepared their bromine from bromide obtained from different salt deposits as well as Pacific ocean water. The bromine was treated with ammonium hydroxide, evaporated till dryness and treated with sodium hydroxide to produce sodium bromide. This sodium bromide was added to a nickel filament as aqueous solution, dried and installed in the ion source of the mass spectrometer. The bromide was evaporated in the source and the ratio of $^{79}Br/^{81}Br$ was measured with a precision of approximately 4‰. The five samples analysed had $^{79}Br/^{81}Br$ ratios between 1.0206 and 1.0231, well within the precision of the measurements.

Catanzaro et al. (1964) improved the absolute $^{79}Br/^{81}Br$ ratio to 1.02784 ± 0.00190 with calibration for bias by the use of samples of known isotopic composition prepared from nearly pure separated bromine isotope samples. All isotopic measurements were done with solid-source single-focusing mass spectrometers equipped with 68° analyser tubes, 60° magnet pole pieces and 12 inch radii of curvature. Triple-filament rhenium ribbon sources were used and ion currents were measured by means of a vibrating reed electrometer with an expanded-scale recorder. Ratios were measured by varying the magnetic field at constant ion-accelerating voltage. Bromide was deposited on the sample filaments in the form of ammoniacal solutions of silver bromide. Preliminary experiments showed that silver bromide was far superior to sodium bromide with respect to signal stability during mass spectrometric analyses. One drop of a silver bromide solution, containing 3 mg Br/ml (~0.06 mg Br), was deposited on each filament and dried under a heat lamp. To minimize the variability of discrimination, all analyses were made identical. A strict pattern of filament heating was followed and a set of 10 measurements was made on a growing signal (3 to $5*10^{-13}$ A) between 38 and 52 min after the filaments were turned on. Analyses which did not follow the normal signal-growth pattern were discarded (Catanzaro et al. 1964).

6.3.2 P-TIMS

Xiao et al. (1993) developed a method to analyse bromine isotope compositions using positive ion thermal ionization mass spectrometry applying a method that is equivalent to the method they developed earlier for chlorine isotopes. They analysed their samples on a VG 354 thermal ionization mass spectrometer. As an analogue to the method to measure chlorine isotopes it involved transferring bromide to caesium bromide which is attached to the filament and which is analysed as Cs_2Br^+ ions. Xiao et al. (1993) tested the method with pure potassium bromide and copper(II) bromide reagents from various sources. These salts were converted to a hydrogen bromide solution (hydrobromic acid) using a cation exchange resin. This acid solution was neutralised to a pH of 3 to 4 using solid caesium carbonate. Filaments in the mass spectrometer were at first treated with about 3 μL of a graphite slurry containing about 100 μg graphite. Then the sample solution was added to the filament and dried for 2 min with a current of 1.1 A. When the filaments were dry the samples were loaded into the instrument and were analysed after evacuation. The Cs_2Br^+ ions were monitored and its intensity was adjusted to 3 to $4*10^{-12}$ A by adjusting the filament current. Data were collected by switching between the masses 345 ($Cs_2{}^{79}Br^+$) and 347 ($Cs_2{}^{81}Br^+$). Several sample sizes were tested (2 to 32 μg) and it appeared that at sample sizes below 4 μg both the intensity and the stability degraded, so that preferentially the size of a sample should be above this limit. The method appeared to be very precise with a relative standard deviation (at the 95% confidence limit) of 0.11‰. Relatively small amounts of chloride in the samples seemed not to have large effects on the measured bromine isotope measurements. Amounts of up to 51.5 μg Cl per 24 μg Br did not change the isotope measurements significantly.

Unfortunately no natural samples were analysed in this study. The reason for this was that co-existing chloride produced a wide interference peak, two peaks at $Cs_2{}^{79}Br^+$ (m/z = 345) and $Cs_2{}^{81}Br^+$ (m/z = 347) were just set on top of this wide peak. This seriously affected the accuracy of the determination of the bromine isotope composition and restricted the wide application of the method (Ma et al. 2016a) in spite of what was stated in Xiao et al. (1993). Another problem that arose was that the newer generation of mass spectrometer used in the laboratory, the Thermo Fisher TRITON mass spectrometer, could not measure masses above $m/z = 310$, so that Cs_2Br^+ ions (with $m/z = 345$ and 347) cannot normally be measured. In order to be able to measure bromine stable isotopes on a TRITON thermal ionisation mass spectrometer Ma et al. (2016b) modified the power unit of the TRITON. They reduced the acceleration voltage from 10 kV to 8 kV, whereby the mass range was extended from $m/z = 310$ to $m/z = 350$. A mass calibration curve was established on the acceleration voltage of 8 kV and stable isotope compositions of bromine were successfully measured on masses $m/z = 345$ and 347. Ma et al. (2016a) then measured bromine isotopes using the same procedure as described in Xiao et al. (1993). However the caesium bromide ions were collected in a different way. Ion ratios were determined using static multicollection, unlike peak jumping mass spectrometry in previous studies. Multicollection mass spectrometry has advantages in comparison to peak jumping mass spectrometry, the data acquisition time is shorter so that a smaller sample size can be analysed before ion beam intensity decays and time drift correction for ion beam decay is not necessary, something that can help to improve the accuracy of the measurement. Using this technique $\delta^{81}Br$ values could be measured with errors between 0.03 and 0.24‰ which is considered of good enough precision because bromine isotope variations in nature show a range that is considerably larger (Ma et al. 2016a).

6.3.3 Dual Inlet IRMS

The first natural bromine isotope variations were found using dual inlet mass spectrometry. This was done on samples from which chloride and bromide were separated from each other using the classical technique where bromide was oxidised to bromine that could be evaporated from the sample solution while chloride was not (Friedheim and Meyer 1892; Boeke 1908; Eggenkamp and Coleman 2000; see Sect. 6.2).

After the separation of bromide from chloride according to the technique described above, an amount of the solution containing 2 to 8 mg bromide is poured into a glass beaker. As the pH of this solution, being effectively a potassium hydroxide solution, is high nitric acid is added until the pH is decreased to a value less than 2. This procedure is followed as otherwise not silver bromide, but silver oxide will precipitate when silver nitrate is added to the solution later in the procedure. Unlike the method described by Eggenkamp (1994) for the analysis of chlorine isotopes no pH buffer is added, as after the addition of nitric acid the pH will remain low enough to prevent precipitating of silver oxide. An ionic strength buffer (6 gr potassium nitrate per 100 mL sample solution) is added to facilitate small silver bromide crystals which will react with more ease later on in the sample preparation procedure (after Taylor and Grimsrud 1969). The remainder of the procedure, such as precipitation of silver bromide, and the reaction to form bromomethane after addition of iodomethane in a glass ampoule is entirely comparable to the method for chloride samples as described in Sect. 5.2.2.2. The separation of bromomethane from iodomethane needs to be done more carefully than the separation of chloromethane from iodomethane as the two peaks are much closer to each other when they are separated by gas chromatography (actually the bromomethane peak is found between chloromethane and iodomethane on the Porapack Q column), and as a small amount of chloride is always transferred in the procedure to separate bromide from chloride always a small amount of chloromethane is present in the gas mixture and this needs to be removed during the gas-chromatographic separation too. This means that the timing of the gas-chromatographic separation is much more important than in the chloromethane separation for chlorine isotopes. The separation through two gas chromatograph runs is in the case of bromine isotope analyses even more important than in the case of chlorine isotopes. Eggenkamp and Coleman (2000) suggest that the distillation is done twice when the Cl/Br ratio is above 1000. Data obtained by Eggenkamp and Coleman (2000) indicate that if the intensity of the chloromethane peak in the mass spectrometer is more than 10% of the intensity of the bromomethane peak the measured $\delta^{81}Br$ is in error, with a value that is more positive than would otherwise be measured in a clean sample.

6.3.4 Continuous Flow GC-IRMS

Considering the fact that the bromide concentration in most samples is relativity low, it has been considered very important that methods would be developed that use much less sample than the 2 to 8 mg necessary for dual inlet mass spectrometry. Shouakar-Stash et al. (2005a) proposed to use continuous flow mass spectrometry to obtain this goal. The initial step in their procedure is the separation of bromine from chlorine according to the technique described in Sect. 6.2. Then silver bromide is precipitated from the solution and this is reacted with iodomethane in a procedure comparable to the method described earlier by Shouakar-Stash et al. (2005b) for chlorine isotopes. The reaction from silver bromide to bromomethane is done in crimp cap vials that can be used in a gas chromatograph auto sampler.

For the final step, the isotope analysis of the bromomethane, the CH_3Br/CH_3I gas mixture is injected into an Agilent 8690 GC through a CombiPal autosampler. During this procedure the injection needle is thoroughly cleaned with air and helium, to avoid any cross contamination between samples. The gas mixture is separated on a DB-5 capillary column. The temperature program used is the same as for the measurement of chlorine isotopes (Shouakar-Stash et al. 2005b). To avoid the entrance of iodomethane into the mass spectrometer this is diverted to a flame ionisation detector (FID). As the measurement of a sample takes a relatively long time (about 15 min) both before and after the bromomethane peak four reference peaks are measured. The precision of this method is found to be about 0.06‰. In spite of the fact that in a continuous flow process much smaller samples can be analysed than in a dual-inlet set up the minimum sample size that can be analysed is still about 1 mg bromide. The reason for this is that the initial step, the separation of the bromide from the large excess of chloride, still has the same dimensions as in the original development of the method by Eggenkamp and Coleman (2000).

6.3.5 Continuous Flow Analysis After GasBench II Chromatography

Du et al. (2013) developed a method to analyse bromine isotopes after separation of bromomethane and iodomethane on a GasBench II system as an analogue to the chlorine isotope determination from Liu et al. (2013). Bromide is separated from chloride following a combination of the methods described by Eggenkamp and Coleman (2000) and Shouakar-Stash (2005a), but maintaining the temperature in the oxidation (left) flask well below the boiling temperature at about 60 °C. At this temperature the oxidation of bromide already takes place and bromine gas is transferred to the second flask as a gas but because the solution doesn't boil the amount of chloride that is transferred is reduced considerably. Silver bromide is produced as described in earlier sections and the reaction between silver bromide and iodomethane is now done in 8 mL screw-cap vials which are filled with helium before adding iodomethane. The reaction to form bromomethane takes place for 56 h at 80 °C. The vials are sampled via a CombiPAl autosampler. The gas is dried over a perfluorinated sulfonic acid resin trap, and then introduced on a CP-PoraPlot-Q chromatographic column (27.5 m × 0.32 mm) to separate bromomethane from iodomethane and introduced into the mass spectrometer. During each run three reference peaks, three sample peaks and tree reference peaks are measured from which the isotope composition is determined.

6.3.6 ICP-MS

Gelman and Halicz (2011) developed a method to analyse the stable bromine isotope composition in very small inorganic samples by miniaturizing the oxidation/distillation set up in which bromide is separated from chloride in combination with MC-ICP-MS analyses of the molecular bromine formed in the oxidative separation. Gelman and Halicz (2011) decreased the size of the oxidation vessel such that only 1 mL of sample (containing only about 0.02 mg bromide), 2 mL of a 1 M potassium dichromate solution and 3 mL of a 2:1 H_2SO_4:H_2O mixture are used for the separation procedure. The reaction vessel is continuously purged with helium which transfers the produced bromine to a wash bottle containing anhydrous sulphuric acid to remove any moisture. Then the helium/bromine mixture is combined with an argon stream and transferred quantitatively into the MC-ICP-MS. Because the sample is introduced into the mass spectrometer as a dry and hydrogen free gas mixture the measurement can be made in low resolution (~300) mode without risk of producing Ar_2H^+. The instrumental mass bias is corrected using a strontium external spike (Gelman and Halicz 2010). Isotope analyses are done by simultaneously analysing $^{79}Br^+$ and $^{81}Br^+$ ions. Thanks to the mass bias correction with Sr they managed to increase the precision of the method from about 0.6‰ to about 0.1‰ (1σ), about equal to other bromine isotope techniques, but with the added benefit of being able to analyse samples containing about 50 times less bromide than could be measured with other techniques. At a later stage the ICP-MS was coupled to an ion chromatograph (Zakon et al. 2016). The samples are nebulized into an Apex-Q desolvator to achieve dry plasma conditions. Using this set up they were able to measure chlorine, bromine and sulphur isotopes in principle in a single run (see also Sect. 5.2.7.2), and were also able to measure different ions of the same element in a single run. Zakon et al. (2016) were able to reach precisions of 0.1‰ (1σ) for bromine isotopes for sample amounts down to 0.6 nmol.

Based on the bromide eluate that they produced with their ion exchange chromatography method Louvat et al. (2016) checked various methods of introduction such a sample into an ICP-MS source. They checked sample introduction by direct injection (d-DIHEN), desolvation (APEX) and through a small spray chamber and observed that this last method provided the best sensitivity and stability of the Br signal. Despite Ar_2H^+ spectral interference on $^{81}Br^+$ it was observed that low-resolution mode (with appropriate tuning of the $Ar_2H^+/^{81}Br^+$ sensitivity) showed higher precision bromine isotope measurements than high-resolution (HR) mode, due to the narrowness of the $^{81}Br^+$ plateau in HR mode and to slight mass drifting with time. Louvat et al. (2016) measured their samples in a wet plasma. This can cause possible Br memory effects in acid solutions and interference of Ar_2H^+ on $^{81}Br^+$. It was found however that high-pH solutions can efficiently remove these memory effects and give much shorter washout times between samples. Long term reproducibility of the method was between ±0.11 and 0.27‰ (2σ) and samples as small as 1 μg bromide could be measured.

While Louvat et al. (2016) used low resolution mode to analyse the bromine isotope composition Wei et al. (2015) developed a method to measure bromine isotopes by ICM-MS in high resolution mode. They were able to measure the bromine isotope composition with a mass resolution of about 8000 as available from the Neptune ICP-MS, while they observed that a resolution of about 5000 would be sufficient. The precision of their method was about (1σ) 0.1 to 0.15‰. Wei et al. (2015) did not measure any natural samples. Also De Gois et al. (2016) applied high resolution mode. With the resolving power of about 8000 they were able to measure bromine isotope compositions with a comparable precision of about 0.1‰. De Gois et al. (2016) measured the bromine isotope composition in seawater and NIST SRM 977 after ion exchange chromatography as described above in Sect. 6.2.3.

6.3.7 Measurement of Bromine Isotopes with Neutron Activation Analyses

The samples obtained during diffusion experiments with molten bromide salts (see Sect. 8.1.2) were analysed after irradiation with thermal neutrons of the silver bromide prepared from these samples. The procedure that was followed is described by Cameron et al. (1956). The salts obtained from the experiments were dissolved in water and bromide was precipitated as silver bromide after addition of a silver nitrate solution. Then the silver bromide was irradiated with neutrons and the bromine isotopes are partly changed into isotopes with one extra neutron (n,γ-process). Radioactive isotopes ^{80}Br and ^{82}Br are formed in this process and from their decay energies the percentage ^{79}Br in the original sample can be determined. This method is not very precise, and the percentage ^{79}Br in the samples could only be determined with a precision of about 1%. For that reason the bromine isotope composition was also determined by the more precise N-TIMS method as described by Cameron and Lippert (1955).

6.4 Measurements of Bromine Isotopes in Organic Molecules

6.4.1 Measurements by ICP-MS

Isotope techniques to analyse the stable bromine isotope composition in organic molecules have been developed since about 2007. Most techniques apply direct introduction of individual organic compounds of bromine into an MC-ICP-MS after separation the compounds by gas chromatography. Sylva et al. (2007) were the first to analyse organobromine compounds by this technique. They separated a mixture of three brominated benzenes on a DB-1 capillary column in a Hewlett-Packhard 5890 gas chromatograph. The individual compounds were then introduced into a Thermo Neptune ICP-MS. In the plasma $^{79}Br^{+}$ and $^{81}Br^{+}$ ions are formed from which the isotope ratio could be determined with a precision of about 0.3‰. Holmstrand et al. (2010) extended the range of compounds to be analysed for bromine isotopes. They used the same type of gas chromatograph but applied a Factor Four VF-5 ms column (equivalent to a DB-5 ms column) to separate their phases. The GC was coupled to a VG Isoprobe ICP-MS. They constructed three different torch assemblies to test different set-ups for bromine analyses. They observed that the precision of the measurements for brominated diphenyl ethers was only 1.4 to 1.8‰, as compared to a precision of 0.4‰ for monobromobenzene (that was also used by Sylva et al. 2007). It appeared that the precision of the bromine isotope measurements is dependent on the compounds that are analysed, apparently due to a different behaviour in the torch where they are ionised and where the bromine ions are formed.

Horst et al. (2011) extended this system for the measurement of extremely small amounts of bromomethane that is extracted from air samples. Their system shows a precision of better than 0.1‰ for samples as small as 40 ng and still better than 0.7‰ for samples in the range between 5 and 20 ng. Linearity of the measured isotope ratios was shown over the entire range of tested concentrations. The long-term reproducibility was determined by measuring three batches of commercially available bromomethane gases against each other over a time period of 3 months and their long-term precision was 0.4‰ or better.

Gelman and Halicz (2010) managed to improve the precision of ICP-MS measurements considerably in a comparable set up through the simultaneous introduction of strontium and bromine into the ICP-MS. Strontium is used as a spike to correct for mass bias and matrix effects. Strontium can be used for this purpose as its *m/z* is close to that of bromine, in spite of their large differences in first ionization energy. It was proposed to use strontium as it is well known as a spike to correct for mass bias and matrix effects in $^{87}Sr/^{86}Sr$ analyses (Ehrlich et al. 2001a, 2001b). Using this correction they were able to improve the precision of bromine isotope measurements to about 0.2‰ for samples containing about 1 nmole (80 ng) of bromine injected to 0.5‰ for samples from which only 0.02 nmole (1.6 ng) was injected into the mass spectrometer. As was also observed by Holmstrand et al. (2010) they also found that different compounds showed different precisions. The precision of tribromophenol measurements were in all different sample amounts that were measured much better than for bromotoluene for example.

6.4.2 High Temperature Conversion of Organic Molecules

Hitzfeld et al. (2011) described a method to directly analyse chlorine, bromine and sulphur isotopes after gas chromatographic separation and subsequent high temperature conversion with hydrogen as make-up gas (see Sect. 5.6.6 for the procedure for the measurement of chlorine isotopes using this technique). For the determination of bromine isotopes the conversion is to hydrogen bromide gas, which is measured at *m/z* values of 80 and 82. Just as was observed for chlorine the precision of this method is about 1‰.

6.4.3 Measurement of Bromine Isotope Compositions by Quadrupole Mass Spectrometry

The method to analyse the chlorine isotope composition of organic molecules by direct introduction into a quadrupole mass spectrometer as developed by Aeppli et al. (2010) can also be used as a method to analyse bromine isotopes in organic compounds as was indicated by Thornton et al. (2013) when they described their method to extract bromomethane from the atmosphere.

The GC/qMS method was optimised and validated on bromoform, 3-bromophenol and 4-bromotoluene by Zakon et al. (2016). They also compared results obtained by GC/qMS and GC/MC-ICP-MS for 1,2-dibromoethane and 3-bromophenol samples with different bromine isotope compositions to evaluate the analytical performance of their method. The precisions of the method is 0.2–0.3‰ for sample amounts from tens to thousands pmol. They also obtained good correlations between the results from GC/qMS and GC/MC-ICPMS for 1,2-dibromoethane and 3-bromophenol.

6.5 Extraction of Bromomethane from the Atmosphere for Isotope Analysis

A system for the collection of the methyl halides bromomethane and chloromethane from the ambient atmosphere in quantities sufficient for bromine and chlorine isotope analysis was developed and described in detail by Thornton et al. (2013) and shorter by Horst (2013). Their system consists of a sampler that features a stainless steel cryotrap with a volume of approximately 1 L which is immersed in liquid nitrogen and filled with glass beads to increase the condensation surface. A vacuum pump is used to pull air through the trap at flow rates between 10 and 20 L/min making it possible to sample the nearly 40 ng required for subsequent $\delta^{81}Br\text{-}CH_3Br$ analyses from 4 to 5 m^3 of air in 4 h. Water is removed with a condenser and a Drierite® trap at the inlet. Carbon dioxide is separated from the sample with a Carbosorb® trap during transfer to a storage canister. In the laboratory, the air sample is extracted from the canister again and the collected amounts of bromomethane are quantified and purified from co-trapped interfering compounds (such as nitrous oxide) using a packed column mounted into a GC-MS. Thornton et al. (2013) were able to show that neither the cryogenic trapping in parts of the system nor the complete sampling and purification method caused any isotope fractionation, which definitely was the most crucial requirement for their trapping system intended to collect samples for isotope analysis. The study by Thornton et al. (2013) demonstrated the capability of the cryosampler by quantifying the chloromethane and bromomethane collected from atmospheric samples and the non-fractionating bromine isotope fingerprint of bromomethane from synthetic air samples of controlled compositions. The sample purification system was designed around a packed column gas chromatography-quadropole-mass spectrometry (GCqMS) system with three additional cryotraps and backflushing capacity. The system's suitability was tested by observing both the mass recovery and the lack of δ^{81}Br isotope fractionation induced during sample purification under varying flowrates and loading scenarios. To demonstrate that the entire system samples and subsequently delivers bromomethane to the isotope analysis system without inducing isotope fractionation, diluted synthetic air mixtures were prepared from standard gases with known isotope compositions and were processed through the entire system, showing a δ^{81}Br-CH_3Br of +0.03 ± 0.10‰ relative to the original starting isotope composition. Finally, the combined cryosampler-purification and analysis system was applied to demonstrate the first-ever δ^{81}Br-CH_3Br measurements in the ambient atmosphere with two samples that were collected in the autumn of 2011, yielding δ^{81}Br values of −0.08 ± 0.43‰ and + 1.75 ± 0.13‰ respectively versus standard mean ocean bromide for samples that were collected at a suburban Stockholm, Sweden, site.

6.6 Extraterrestrial Bromine

Bromine isotope ratios in extraterrestrial material are only known from comet 67P/Churyumov–Gerasimenko. Dhooghe et al. (2017) measured ^{79}Br and ^{81}Br abundances in the coma of 67P/Churyumov–Gerasimenko with the Double Focusing Mass Spectrometer (DFMS) of ROSINA as they also did for chlorine too (Dhooghe et al. 2017). Bromine abundances are about 100 times less than chlorine abundance and thus much more difficult to measure. The bromine isotope ratio (^{79}Br/^{81}Br) was determined through division of the individual ion counts.

6.7 Recapitulation and Expectations of the Application of Bromine Isotopes in the Future

Techniques to analyse stable bromine isotopes are still in significant development. The number of laboratories where currently bromine isotopes are analysed is still very limited, so that this development is unfortunately slow. For example, the far majority of all published data on inorganic bromide are measured in only a single laboratory. Also for the isotope analyses of organobromine compounds the majority of the published data are from only few laboratories.

More and more it is shown that stable bromine isotope ratios are preferably analysed with different techniques than chlorine isotopes. For the originally developed duel inlet technique the amount of sample that is necessary is in most cases so large that only deep saline formation waters can successfully be analysed. Even the on-line method developed by Shouakar-Stash et al. (2005b) needs more than 1 mg of bromine and in many cases the available amount of sample is (much) less. Measurement with ICP-MS seems the most promising for the near future as only very small sample amounts are necessary to measure and these measurements show very good results. ICP-MS techniques for bromine are more comfortable than for chlorine as there are less problems with interference from argon, which are expected to become even less in the future due to the use of collision cells (P Agrinier, pers. comm). In principle also P-TIMS could be a very good method to analyse bromine isotopes, but unfortunately it has not yet shown its full potential. The advantage of these methods is the significantly smaller sample consumption than for IRMS techniques and for that reason it is expected it will see a larger number of studies where natural samples are measured in the future.

In the case of organobromine compounds it also seems that ICP-MS techniques are the techniques that are developing as the most common method to analyse these compounds for bromine isotopes. Several laboratories have developed techniques using ICP-MS (Sylva et al. 2007; Holmstrand et al. 2010, Gelman and Halicz 2010), and the impression is that, until the present day, most organobromine isotope studies use these techniques. It is however not to be excluded that other techniques like direct injection techniques with IRMS for larger sample quantities and quadrupole mass spectrometry techniques as developed by Zakon et al. (2016) will show to be very convenient for future bromine isotope studies.

References

Aeppli C, Holmstrand H, Andersson P, Gustafsson Ö (2010) Direct compound-specific stable chlorine isotope analysis of organic compounds with quadrupole GC/MS using standard isotope bracketing. Anal Chem 82:420–426

Aston FW (1920a) The constitution of the elements. Nature 105:547

Aston FW (1920b) The mass spectra of the chemical elements - part 2. Philos Mag 40:628–634

Aston FW (1931) The isotopic constitution and atomic weights of selenium, bromine, boron, tungsten, antimony, osmium, ruthenium, tellurium, germanium, rhenium and chlorine. Proc R Soc Lond A 132:487–498

Benedict S, Snell JF (1903) A method for the detection of chlorides, bromides and iodides. J Am Chem Soc 25:809–814

Blewett JP (1936) Mass spectrograph analysis of bromine. Phys Rev 49:900–903

Boeke HE (1908) Über das Krystallisationsschema der Chloride, Bromide, Jodide von Natrium, Kalium und Magnesium, sowie über das Vorkommen des Broms und das Fehlen von Jod in den Kalisalzlagerstätten. Z Kristallogr 45:346–391

Bremner JM (1965) Inorganic forms of nitrogen. In: Black CA et al (eds) Methods of soil analysis, part 2. Agronomy 9. American Society of Agronomy Inc, Madison, WI, pp 1179–1237

Cameron AE, Lippert EL (1955) Isotope composition of bromine in nature. Science 121:136–137

Cameron AE, Herr W, Herzog W, Lundén A (1956) Isotopen-Anreicherung beim Brom durch electrolytische Überführung in geschmolzenem Bleibromid. Z Naturforschng 11a:203–205

Catanzaro EJ, Murphy TJ, Garner EL, Shields WR (1964) Absolute isotopic abundance ratio and the atomic weight of bromine. J Res Natl Bur Stand (US) 68A:593–599

De Gois JS, Vallelonga P, Spolaor A, Devulder V, Borges DLG, Vanhaecke F (2016) Bromine isotope ratio measurements in seawater by multi-collector inductively coupled plasma-mass spectrometry with a conventional sample introduction system. Anal Bioanal Chem 408:409–416

De Groot PA (2004) Handbook of stable isotope analytical techniques, vol Volume I. Elsevier, Amsterdam

Dechan M (1886) Detection and estimation of iodine, bromine, and chlorine. J Chem Soc 49:682–685

Dhooghe F, De Keyser J, Altwegg K, Briois C, Balsiger H, Berthelier JJ, Calmonte U, Cessateur G, Combi MR, Equeter E, Fiethe B, Fray N, Fuselier S, Gasc S, Gibbons A, Gombosi T, Gunell H, Hässig M, Hilchenbach M, Le Roy L, Maggiolo R, Mall U, Marty B, Neefs E, Rème H, Rubin M, Sémon T, Tzou CY, Wurz P (2017) Halogens as tracers of protosolar nebula material in comet 67P/Churyumov–Gerasimenko. Monthly Not R Astron Soc 472:1336–1345

Du Y, Ma T, Yang J, Liu L, Shan HM, Cai HS, Liu CF, Chen LZ (2013) A precise analytical method for bromine stable isotopes in natural waters by GasBench II-IRMS. Int J Mass Spectrom 338:50–56

Eggenkamp HGM (1994) $\delta^{37}Cl$; the geochemistry of chlorine isotopes. Geol Ultraiect 116:1–150

Eggenkamp HGM (2004) Summary of methods for determining the stable isotope composition of chlorine and bromine in natural materials. Chapter 28. In: de Groot PA (ed) Handbook of stable isotope analytical techniques. Elsevier, Amsterdam, pp 604–622

Eggenkamp HGM, Coleman ML (2000) Rediscovery of classical methods and their application to the measurement of stable bromine isotopes in natural samples. Chem Geol 167:393–402

Eggenkamp HGM, Louvat P (2018) A simple distillation method to extract bromine from natural water and salt samples for isotope analysis by multi-collector inductively coupled plasma mass spectrometry. Rapid Commun Mass Spectrom 32:612–618

Ehrlich S, Karpas Z, Ben-Dor L, Halicz L (2001a) High precision lead isotope ratio measurements by multicollector-ICP-MS in variable matrices. J Anal At Spectrom 16:975–977

Ehrlich S, Gavrieli I, Ben-Dor L, Halicz L (2001b) Direct high-precision measurements of the $^{87}Sr/^{86}Sr$ isotope ratio in natural water, carbonates and related materials by multiple collector inductively coupled plasma mass spectrometry (MC-ICP-MS). J Anal At Spectrom 16:1389–1392

Friedheim C, Meyer RJ (1892) Über die quantitative Trennung und Bestimmung von Chlor, Brom und Jod. Z Anorg Chem 1:407–422

Gelman F, Halicz L (2010) High precision determination of bromine isotope ratio by GC-MC-ICPMS. Int J Mass Spectrom 289:167–169

Gelman F, Halicz L (2011) High-precision isotope ratio analysis of inorganic bromide by continuous flow MC-ICPMS. Int J Mass Spectrom 307:211–213

Hitzfeld KL, Gehre M, Richnow HH (2011) A novel online approach to the determination of isotopic ratios for organically bound chlorine, bromine and Sulphur. Rapid Commun Mass Spectrom 25:3114–3122

Holmstrand H, Unger M, Carrizo D, Andersson P, Gustafsson Ö (2010) Compound-specific bromine isotope analysis of brominated diphenyl ethers using GC-ICP-multicollector-MS. Rapid Commun Mass Spectrom 24:2135–2142

Horst A (2013) Stable bromine isotopic composition of methyl bromide. In: Method development and applications. PhD thesis Stockholm University

Horst A, Holmstrand H, Andersson P, Andersson A, Carrizo D, Thornton BJ, Gustafsson Ö (2011) Compound-specific bromine isotope analysis of methyl bromide using gas chromatography hyphenated with inductively coupled plasma multiple-collector mass spectrometry. Rapid Commun Mass Spectrom 25:2425–2432

Kebler LF (1892) A method for detecting chlorine, bromine, and iodine in the same mixture. J Anal Appl Chem 6:569–572

Liu YD, Zhou AG, Gan YQ, Liu CF, Yu TT, Li XQ (2013) An online method to determine chlorine stable isotope composition by continuous flow isotope ratio mass spectrometry (CF-IRMS) coupled with a Gasbench II. J South Cent Univ 20:193–198

Louvat P, Bonifacie M, Giunta T, Michel A, Coleman M (2016) Determination of bromine stable isotope ratios from saline solutions by 'wet plasma' MC-ICPMS including a comparison between high and low-resolution modes, and three introduction systems. Anal Chem 88:3891–3898

Löwich C (1829) Das Brom und seine chemischen Verhaltnisse. Heidelberg, p 175

Ma YQ, Peng ZK, Chen YJ, Yang J, Xiao YK, Zhang YL (2016a) High precise determination of bromine isotopic ratios by positive thermal ionization mass spectrometry using static multicollection based on Cs_2Br^+ ions. Chin J Anal Chem 44:186–191

Ma YQ, Peng ZK, Han FQ, Liu WG, Yang J, Xiao YK, Zhang YL (2016b) Extension of mass range and application in the determination of bromine isotope by TIMS. J Chin Mass Spectrom Soc 37:465–470. (in Chinese with English abstract)

Murphy TJ, Clabaugh WS, Gilchrist R (1954) Separation of iodide, bromide, and chloride from one another and their subsequent determination. J Res Natl Bur Stand 53:13–18

Nier AO (1940) A mass spectrometer for routine isotope abundance measurements. Rev Sci Instrum 11:212–216

Rankama K, Sahama TG (1950) Geochemistry. The University of Chicago Press, Chicago

Shouakar-Stash O, Frape SK, Drimmie RJ (2005a) Determination of bromine stable isotopes using continuous-flow isotope ratio mass spectrometry. Anal Chem 77:4027–4033

Shouakar-Stash O, Drimmie RJ, Frape SK (2005b) Determination of inorganic chlorine stable isotopes by continuous flow isotope ratio mass spectrometry. Rapid Commun Mass Spectrom 19:121–127

Sylva SP, Ball L, Nelson RK, Reddy CM (2007) Compound-specific $^{81}Br/^{79}Br$ analysis by capillary gas chromatography/multicollector inductively coupled plasma mass spectrometry. Rapid Commun Mass Spectrom 21:3301–3305

Taylor JW, Grimsrud EP (1969) Chlorine isotopic ratios by negative ion mass spectrometry. Anal Chem 41:805–810

Thornton B, Horst A, Carrizo D, Holmstrand H, Andersson P, Crill PM, Gustafsson Ö (2013) A high-volume cryosampler and sample purification system for bromine isotope studies of methyl bromide. J Atmos Ocean Techn 30:2095–2107

Wei HZ, Jiang SY, Zhu ZY, Yang T, Yang JH, Yan X, Wu HP, Yang TL (2015) Improvements on high-precision measurement of bromine isotope ratios by multicollector inductively coupled plasma mass spectrometry. Talanta 143:302–306

Willey JF, Taylor JW (1978) Capacitive integration to produce high precision isotope ratio measurements on methyl chloride and bromide. Anal Chem 50:1930–1933

Williams D, Yuster P (1946) Isotopic constitution of tellurium, silicon, tungsten, molybdenum, and bromine. Phys Rev 69:556–567

Xiao YK, Liu WG, Qi HP, Zhang CG (1993) A new method for the high precision isotopic measurement of bromine by thermal ionization mass spectrometry. Int J Mass Spectrom Ion Proc 123:117–123

Zakon Y, Halicz L, Lev O, Gelman F (2016) Compound-specific bromine isotope ratio analysis using gas chromatography/quadrupole mass spectrometry. Rapid Commun Mass Spectrom 30:1951–1956

Part III

Theoretical and Experimental Estimates of Chlorine and Bromine Isotope Fractionation

7 Isotope Fractionation

7.1 Stable Isotope Fractionation

Variations in light stable isotopes (such as hydrogen, carbon, nitrogen, oxygen, sulphur, chlorine and bromine) are caused by processes that are called isotope fractionation. Fractionation is the process where differences in physical or chemical properties, mostly the result from differences in the weight of the isotopes result in different enrichments of one of the isotopes of an element. Variations in the isotope composition of an element can also have its origin from radioactive decay of mother isotopes which may be present in different quantities in the original environment. In this case these daughter isotopes are called radiogenic. The isotope geochemistry of strontium is a good example of an element where isotope variations are mainly caused by radioactive decay of a mother isotope, in this case the β-decay of ^{87}Rb to stable ^{87}Sr. Chlorine and bromine do not have any radiogenic isotopes and all terrestrial isotope variations observed for chlorine and bromine are caused by fractionation processes of the stable isotopes.

In isotope geochemistry two major types of fractionation are distinguished. These are equilibrium fractionation and kinetic fractionation. Equilibrium fractionation is the isotope effect that is observed if two phases are found in thermodynamic equilibrium. Due to the mass difference of the isotopes of an element the zero-point energy of the isotopes are different. The heavier isotope has a lower zero-point energy and as a result the energy necessary to break a bond with the heavier isotope is higher. As a consequence the heavier isotope is concentrated in the phase in which it is most strongly bound. As the difference between the zero-point energies of the lighter and heavier isotopes decreases with temperature increases equilibrium fractionation has the tendency to decrease with increasing temperature. In theory at infinitely high temperatures it reduces to zero. A very important effect that has been observed is that oxidation and reduction processes are associated with very large equilibrium fractionation.

Kinetic fractionation on the contrary is a non-equilibrium process. It is found in systems where variations arise as a result of a velocity difference between a heavier and a lighter isotopes, such as diffusion, or if one phase is removed preferentially from a system that is formed in equilibrium, so that the reaction moves to one side. A very well known example of this effect is the removal of water from a cloud in the form of rain or snow which ultimately leads to very low oxygen and hydrogen isotope values in precipitation. This process can be treated as a fractional distillation under equilibrium conditions as is expressed by the Rayleigh (1896) fractionation. A short but clear introduction in the different fractionation processes is given in Hoefs (2021).

H. Eggenkamp, *The Geochemistry of Stable Chlorine and Bromine Isotopes*, Advances in Isotope Geochemistry, https://doi.org/10.1007/978-3-031-75633-7_7

7.2 Chlorine and Bromine Stable Isotope Fractionation

On Earth chlorine and bromine are primarily found as chloride and bromide ions. This means that redox reactions are not normally playing a large role in the chemistry of these elements. This results in isotope variations that are in most cases rather modest. Equilibrium fractionation between phases where chlorine or bromine are present in the chloride or bromide form in both phases is generally not very large. As a consequence, kinetic effects are in many cases responsible for most of the observed isotope variations. In the following chapters the various process that cause variations in the chlorine and bromine isotope ratios will be described. Results obtained in theoretical, experimental and field studies are reported and compared. Even before it was possible to measure natural variations of chlorine and bromine isotopes attempts were made to estimate, both theoretically and experimentally, the fractionation that might be expected if it would ever become possible to measure the actual variations. When natural variations became measurable in the 1980s the observations confirmed for a large part the results of these early experimental studies. Given the still fragmentary knowledge of the chlorine and bromine isotope geochemistry some of these early data still have not been either confirmed or re-analysed in modern studies. Especially with respect to the isotope fractionation of bromine isotopes considerable future work is needed to obtain a complete knowledge of its isotope behaviour.

Detailed introductions to the theoretical backgrounds can be found in classic works such as Urey and Greiff (1935), Bigeleisen and Mayer (1947), Urey (1947), Melander (1960), Bigeleisen (1965) and Richet et al. (1977). Schauble et al. (2003) estimated the theoretical equilibrium isotope fractionation for different chlorine isotope systems. Theoretical estimates for both chlorine and bromine equilibrium fractionation in gas-solution systems have been calculated by Czarnacki and Hałas (2012). In recent years various estimates of equilibrium fractionation in magma systems (Balan et al. 2019, Liu et al. 2021) have been published and these studies suggest that surprisingly large chlorine isotope fractionation can be expected in high temperature systems. Also, recent calculations for bromine isotope fractionation in salts (Gao and Liu 2021), suggest that bromine isotope fractionation is closely related to the Cl/Br ratio in minerals. This effect can explain the observations that bromine isotope variations in nature are regularly larger than chlorine isotope variations in natural settings (e.g. Eggenkamp 2015, Eggenkamp et al. 2019).

In the following chapters various isotope fractionation processes that are relevant to chlorine and bromine stable isotope variations are described. These processes are not always described as being either equilibrium or kinetic fractionation as such as in many cases the variations that are observed are the result of a combination of both types of fractionation. In many cases the size of equilibrium fractionation is relatively modest and a large part of the observed fractionation is caused by kinetic effects. A common example is the isotope variation caused by the removal of one of the phases in a system. Relatively large equilibrium fractionation is only expected during reduction/oxidation reactions between phases of different oxidation states, which are, as already cited above, relatively rare processes in the earth's surface reservoirs.

7.3 Expression of Fractionation Quantities

In the case of equilibrium fractionation the quantity α is the isotope fractionation factor and describes the ratio of $^{37}Cl/^{35}Cl$ (or $^{81}Br/^{79}Br$) in the two phases that are in equilibrium (Coplen 2011). For example, the isotope fractionation factor for the precipitation of halite from a saturated NaCl solution is (Eggenkamp et al. 2016):

$$\alpha_{halite-Cl_{aq}} = \frac{\left({}^{37}Cl / {}^{35}Cl \right)_{halite}}{\left({}^{37}Cl / {}^{35}Cl \right)_{Cl_{aq}}} = 1.00035 \quad (7.1)$$

It is difficult to determine the isotope fractionation factor directly and in experimental isotope studies it is generally determined from the difference in $\delta^{37}Cl$ or $\delta^{81}Br$ between the two phases. The isotope fractionation can be determined this way because the isotope value difference between the two phases is (within the error of the measurements) equal to 1000lnα (in this case 0.35‰) and is defined as the per mil isotope fractionation $\varepsilon^{37}Cl$ or $\varepsilon^{81}Br$. ε is defined as α-1, and in the case of the per mil isotope fractionation ε in this particular example can be written down as $\varepsilon^{37}Cl_{(halite\text{-}Cl(aq))}$ (e.g. Szaran 1998, Coplen 2011). Written this way it indicates that with an isotope fractionation of 0.35‰ $\delta^{37}Cl$ of halite is 0.35‰ higher than $\delta^{37}Cl$ of the brine.

In the case of kinetic fractionation a different approach is needed. In the case of diffusion the fractionation is defined as the ratio of the diffusion coefficients of the lighter over the heavier isotope. Thus, in the case of chlorine, the fractionation due to diffusion is D_{35Cl}/D_{37Cl}. The advantage of using this ratio and not the ratio D_{37Cl}/D_{35Cl} is that the fractionation ratio than has a number larger than 1 which is for the general reader easier to interpret (a larger number defines a larger fractionation). Other kinetic fractionation factors, for example due to different reaction rates can be defined the same way, such as k_{35Cl}/k_{37Cl}, again, first the isotope ^{35}Cl because reaction rates of the lighter isotope are in general higher than that of the heavier isotope due to the fact that the heavier isotope normally has a stronger bond.

References

Balan E, Créon L, Sanloup C, Aléon J, Blanchard M, Paulotto F, Bureau H (2019) First-principles modeling of chlorine isotope fractionation between chloride bearing molecules and minerals. Chem Geol 525:424–434

Bigeleisen J (1965) Chemistry of isotopes. Science 147:463–471

Bigeleisen J, Mayer MG (1947) Calculation of equilibrium constants for isotopic exchange reactions. J Chem Phys 15:261–267

Coplen TB (2011) Guidelines and recommended terms for expression of stable-isotope-ratio and gas-ratio measurement results. Rapid Commun Mass Spectrom 25:2538–2560

Czarnacki M, Hałas S (2012) Isotope fractionation in aqua-gas systems: Cl_2-HCl-Cl^-, Br_2-HBr-Br^- and H_2S-S^{2-}. Isot Environ Health Stud 48:55–64

Eggenkamp HGM (2015) Why are variations in bromine isotope compositions in the Earth's history larger than chlorine isotope compositions? Ann UMCS, Sectio AAA 70:183–194

Eggenkamp HGM, Bonifacie M, Ader M, Agrinier P (2016) Experimental determination of stable chlorine and bromine isotope fractionation during precipitation of salt from a saturated solution. Chem Geol 433:46–56

Eggenkamp HGM, Louvat P, Agrinier P, Bonifacie M, Bekker A, Krupenik V, Griffioen J, Horita J, Brocks JJ, Bagheri R (2019) The bromine and chlorine isotope composition of primary halite deposits and their significance for the secular isotope composition of seawater. Geochim Cosmochim Acta 264:13–29

Gao CH, Liu Y (2021) First-principles calculations of equilibrium bromine isotope fractionations. Geochim Cosmochim Acta 297:65–81

Hoefs J (2021) Stable isotope geochemistry.9th edn. Springer Nature Switzerland

Liu X, Wei HZ, Li YC, Williams-Jones AE, Lu JJ, Jiang SY, Dong G, Ma J, Eastoe CJ (2021) Chlorine isotope mantle heterogeneity: constraints from theoretical first-principles calculations. Chem Geol 572:120193

Melander L (1960) Isotope effects on reaction rates. Ronald, New York

Rayleigh L (1896) Theoretical considerations respecting the separation of gases by diffusion and similar processes. Philos Mag Ser 5(42):493–498

Richet P, Bottinga Y, Javoy M (1977) A review of hydrogen, carbon, nitrogen, oxygen, Sulphur, and chlorine stable isotope fractionation among gaseous molecules. Annu Rev Earth Planet Sci 5:65–110

Schauble EA, Rossman GR, Taylor HP (2003) Theoretical estimates of equilibrium chlorine-isotope fractionations. Geochim Cosmochim Acta 67:3267–3281

Szaran J (1998) Carbon isotope fractionation between dissolved and gaseous carbon dioxide. Chem Geol 150:331–337

Urey HC (1947) The thermodynamic properties of isotopic substances. J Chem Soc 1947:562–581

Urey HC, Greiff LJ (1935) Isotope exchange equilibria. J Chem Soc Am 57:321–327

8 Theoretical and Experimental Fractionation Studies Involving Chloride and Bromide Ions Only

8.1 Diffusion

Diffusion probably is the most well known and most well understood process responsible for variations in chlorine and bromine isotope compositions. Molecular diffusion is the process in which matter is transported from one part of a system to another as a result of arbitrary molecular movements (Crank 1956). It was first described by Fick (1855). His work is now referred to as Fick's First and Second Laws, and was published even before quantitative experimental measurements had ever been done. Although Lindemann (1921) discussed the possibility that, due to the mass difference of isotopes from a single element isotopes could theoretically be separated electrolytically, it must be realised that diffusion experiments to separate potential (because still undiscovered) isotopes of chlorine were started already in 1915 (Harkins and Hall 1916, p. 221) even before the actual discovery to the chlorine isotopes by Aston (1919). Experimental separation of chlorine isotopes by diffusion was shown to exist in 1920 (Harkins 1920a, 1920b, Harkins and Hayes 1921; Harkins and Ligget 1924; Harkins and Jenkins 1926). In these studies, set up to confirm that chlorine exists of several isotopes, no attempts were made yet to quantify the fractionation factor of diffusion, although estimates were made to calculate on what scale diffusion would separate the chlorine isotopes as compared to neon. The first study in which fractionation due to diffusion was well quantified was published in the 1940s (Madorsky and Straus 1948) and a variety of studies on diffusion of chlorine (and bromine) isotopes have been published since then, consisting of theoretical, experimental and field studies.

8.1.1 Diffusion of Chlorine and Bromine in Aqueous Solutions

8.1.1.1 Theoretical Fractionation During Aqueous Diffusion

The chlorine isotope fractionation factor due to diffusion can be estimated in several ways. Firstly, by analogy to ideal gases the isotopic fractionation factor in aqueous solution can be estimated from the ratio of the square root of their masses (Graham's Law):

$$\alpha = \sqrt{\frac{m_L}{m_H}} \tag{8.1}$$

Where m_L is the molecular mass of the lighter diffusing isotope and m_H that of the heavier isotope. There are, however, two problems when applying this approach: (A) Collisions between gas molecules are elastic, whereas those between hydrated molecules and water molecules are inelastic. (B) The masses of actual diffusing molecules in solution must be known. This requires knowledge of the degree of chloride hydration, and the resi-

H. Eggenkamp, *The Geochemistry of Stable Chlorine and Bromine Isotopes*, Advances in Isotope Geochemistry, https://doi.org/10.1007/978-3-031-75633-7_8

dence time of a water molecule next to a chloride ion relative to the residence time of a water molecule next to another water molecule. There is much confusion about the degree of hydration of the chloride ion. The hydration number is generally assumed to be ≤6 (Powell et al. 1988). Impey et al. (1983) defined a dynamic hydration number which is 2.6 at 287 K. Marchese and Beveridge (1984) suggested a nine fold hydration. Samoilov (1957) even assumed a negative hydration number on the basis of his calculations which showed that a water molecule stays for a shorter period next to a chloride ion than next to another water molecule. On the basis of a commonly assumed hydration number of 6, the diffusion coefficient ratio (D_{35Cl}/D_{37Cl}) would be 1.0080, much higher than found in both experimental and field studies.

Considering the fact that dissolved species in a solution do not behave like an ideal gas a different approach is needed. Within solutions it was already realised by Lindemann (1921) that different isotopes of an element had different velocities which would mean that isotopes (in ionic form) can be separated electrolytically. This effect can be described using the reduced masses of the colliding species in solution. The reduced mass is defined as:

$$\mu_i = \frac{m_i M}{m_i + M} \tag{8.2}$$

Where m_i and M are the masses of the colliding species (in an aqueous solution of chloride, m_i is a hydrated chloride ion and M is a water molecule). This kinetic theory was described for gases (Tabor 1991; Bird et al. 2002) but it was shown that it is also valid for species in aqueous solution (e.g. Jähne et al. 1987). The definition of the fractionation as a result of diffusion ($^{35/37}\alpha$) for chlorine is:

$$^{35/37}\alpha = \frac{D_{35}}{D_{37}} = \sqrt{\frac{\mu_{37}}{\mu_{35}}} = \sqrt{\frac{m_{37}\left(m_{35}+M\right)}{m_{35}\left(m_{37}+M\right)}} \tag{8.3}$$

Where D_{35} and D_{37} are the diffusion coefficients of the light and heavy isotopes of chlorine (Richter et al. 2006). Calculation of $^{35/37}\alpha$ requires knowledge of the hydration number n of the diffusing chloride ion $[Cl(H_2O)_n^-]$, as would also be the case if Graham's Law was applicable. Based on the Debye-Hückel å parameter (Appelo and Postma 2005) the hydration number n of Cl^- would be 3: $[Cl(H_2O)_3^-]$.

A problem with the use of either Graham's Law or the reduced masses theory to determine fractionation of the diffusing species became apparent when the diffusion fractionation of divalent species such as Mg^{2+} and Ca^{2+} was studied (Richter et al. 2006). They observed that the experimentally determined fractionation was much lower than expected based upon the reduced masses of the species in solution. For this reason Richter et al. (2006) proposed that fractionation due to diffusion is better described by an empirical inverse power relation:

$$^{35/37}\alpha = \frac{D_{35}}{D_{37}} \propto \left(\frac{m_{37}}{m_{35}}\right)^{-\beta} \tag{8.4}$$

Where m is the mass of the light or heavy isotope without considering any water of hydration. Molecular dynamics simulations (Alder et al. 1974; Impey et al. 1983; Nuevo et al. 1995; Willeke 2003) as applied by Bourg and Sposito (2007) produced results on the potential isotope fractionation due to diffusion for different ions in water which were similar to those experimentally determined by Richter et al. (2006). Further investigations by Bourg and Sposito (2007) and Bourg et al. (2010) revealed β-values between 0 and 0.2 for different noble gas atoms and mono- and divalent ions in solution. The numerical value of β was found to be dependent on the strength of the solute-solvent interaction which is defined as the average residence time of water molecules in the first solvation shell of ions. The value of β varies from ~0.00 to ~0.01 for divalent ions; from ~0.01 to ~0.06 for monovalent ions and from ~0.05 to ~0.19 for noble gases (Bourg et al. 2010). For chloride in solution it was simulated to be 0.034 ± 0.018 by Bourg and Sposito (2007), indicating a fractionation factor $^{35/37}\alpha$ for chloride of 1.0019 ± 0.0010. For bromide (Bourg et al. 2010) did not apply molecular dynamics simulations, but calculated a β-value of 0.0320 ± 0.0097 based on experimentally obtained data published by Eggenkamp and Coleman (2009). This β-value agrees to a $^{79/81}\alpha$ of 1.0008 ± 0.0002 (see Sect. 8.1.1.2).

These data on fractionation due to diffusion can be used to predict possible chloride concentration and chlorine isotope profiles (and of course of other species) in systems where solely simple diffusion takes place. With the aim to study diffusion of radioisotopes in sediments Duursma en Hoede (1967) presented mathematical solutions for various diffusion models. Three of these were applied to chlorine isotope fractionation systems by Eggenkamp (1994). Eggenkamp (1994) showed examples of what kind of chloride concentration and chlorine isotope profiles could be expected depending on the source of the chloride: diffusion from a source with a constant concentration, diffusion after a momentary release of chloride and diffusion from a source with a constant inflow.

8.1.1.1.1 Diffusion from a Source with a Constant Concentration

Diffusion from a source with a constant concentration can be described as the situation where an infinite amount of matter with high chloride or bromide content diffuses into an infinite amount of matter with low chloride or bromide content. The chloride or bromide concentration at the boundary between these two parts will be constant during diffusion. The solution for the concentration as a function of x (distance from the boundary) and t (time) is (see e.g. Carslaw and Jaeger 1959):

$$c_{(x,t)} = c_0 erfc \frac{x}{2\sqrt{Dt}} \quad (8.5)$$

Where *erfc* is the complementary error function. Characteristic of this diffusion model is that in parts of the system with a low original chloride or bromide concentration δ^{37}Cl or δ^{81}Br can reach very low values and also that δ^{37}Cl or δ^{81}Br depends very much on the chloride or bromide concentration ratio between the maximum and minimum original chloride or bromide concentrations in the system. Calculations from Eggenkamp (1994) on chloride showed that, when the chloride concentration in the low concentration part of the experiment is negligible very large, negative δ^{37}Cl values can be obtained. Later experiments reported by Eggenkamp and Coleman (2009) showed however that these low δ^{37}Cl values can not normally be measured because the chloride concentrations in the part with the low δ^{37}Cl values are generally too low to facilitate chlorine isotope measurements. In the part of the system with the higher original chloride or bromide concentrations δ^{37}Cl or δ^{81}Br increases due to the faster diffusion of the isotope ^{35}Cl or ^{79}Br. As the chloride or bromide concentration in this part of the system is relatively high the isotope effect is much less than in the low concentration part of the system (Fig. 8.1).

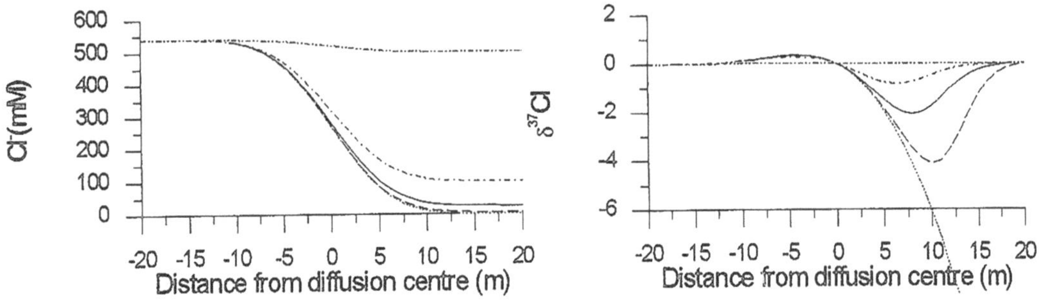

Fig. 8.1 Chloride concentration (left) and chlorine isotope variation (right) profiles for a system with diffusion from a constant source for different chloride concentrations in the low concentration part of the system. Profiles are developed from Eq. (8.5), and drawn with a diffusion time of 250 years, a diffusion coefficient of 15*10–10 m^{2}s^{-1}, a diffusion coefficient ratio ($^{35/37}\alpha$) of 1.00245, a chloride concentration in high concentration part of the system of 540 mM, and in the low concentration part of 0, 5, 25, 100 and 500 mM (from Eggenkamp 1994)

8.1.1.1.2 Diffusion After a Momentary Release of Chloride/Bromide

In this model, at time t = 0 a fixed amount of chloride or bromide ions is released at an injection point. From this injection point (at location x = 0) the chloride/bromide will start to diffuse. In the following it is assumed that the original concentration in the region around the injection point is lower than the concentration after injection of the fixed amount of chloride/bromide at the injection point, the moment it is injected. As the release of chloride or bromide is momentarily the concentration at x = 0 will decrease from the start as t increases. The equation for linear diffusion after a momentary release of chloride or bromide is:

$$c_{(x,t)} = \frac{s}{\sqrt{4\pi Dt}} \exp - \frac{x^2}{4\mathrm{Dt}} \tag{8.6}$$

Where s is the amount of the original momentary release of chloride/bromide. In this system the concentration of chloride or bromide in the diffusion centre decreases rapidly, while the isotope ratio in the centre increases due to the higher diffusion coefficient of the isotope ^{35}Cl or ^{79}Br relative to ^{37}Cl or ^{81}Br. The isotope ratio in the transported chloride or bromide in this scenario can also reach very low values, where, just as in the former model, the lowest values depend on the original chloride or bromide concentration in the environment where the momentary release diffuses into. The diffusion profile seems rather similar to the profile in the first model, however, as the chloride or bromide concentration in the diffusion centre decreases the isotope effect also decreases as the chloride or bromide concentration near the diffusion centre approaches the original chloride or bromide concentration in the environment. Figure 8.2 shows the effect on the chloride or bromide concentration and the chlorine or bromine isotope composition for a system where after the initial release the chloride or bromide diffuses in one dimension (linear) away from the location of the initial momentary release.

8.1.1.1.3 Diffusion from a Source with a Constant Inflow

In this model, a constant inflow of chloride with an infinitely small volume is introduced from a source at a constant rate. It can be considered as an extension of the former model. The diffusion solution of this model can be found by integrating Eq. (8.6) over time, replacing *s* by *q*d*t*, (see Carslaw and Jaeger 1959). In the mathematical solutions that describe the diffusion from a constant inflow *q* defines the amount of chloride added per unit of time.

For linear diffusion the solution is given by (Carslaw and Jaeger 1959):

$$c(x,t) = \frac{qx}{2\mathrm{D}\sqrt{\pi}} \left[\frac{2\sqrt{Dt}}{x} \exp\left(\frac{-x2}{4\mathrm{Dt}} - \sqrt{\pi}\, erfc \frac{x}{2\sqrt{Dt}} \right) \right] \tag{8.7}$$

In this system a continuously increasing chloride concentration at the location of the inflow is observed that decreases at larger distances from

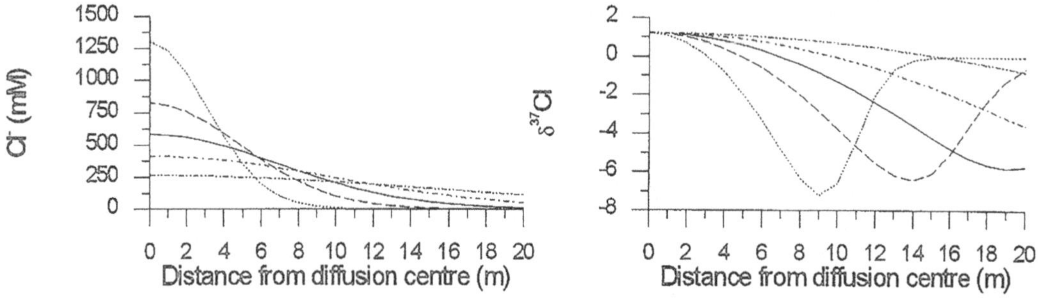

Fig. 8.2 Chloride concentration (left) and chlorine isotope variation (right) profiles for a system with diffusion after a momentary release of chlorine. The figure shows the effects for subsequent linear diffusion after Eq. (8.6), and is drawn for an initial release of chloride of 10 moles, a diffusion coefficient of 15*10–10 m^2s^{-1}, a diffusion coefficient ratio ($^{35/37}\alpha$) of 1.00245, and diffusion times of 100, 250, 500, 1000 and 2500 years. The initial chloride concentration in the area around the initial release was 5 mM (from Eggenkamp 1994)

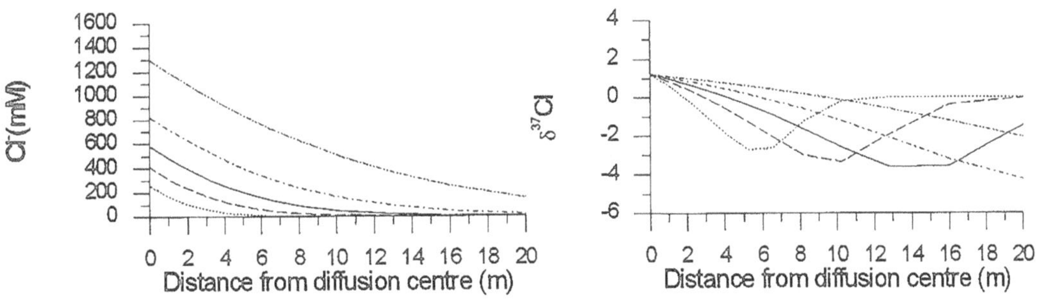

Fig. 8.3 Chloride concentration (left) and chlorine isotope variation (right) profiles for a system with diffusion from a constant inflow of chloride. The figure shows the effects for subsequent linear diffusion after Eq. (8.7), and is drawn for an influx of chloride equal to 10 mmole/year, a diffusion coefficient of 15*10–10 m^2s^{-1}, a diffusion coefficient ratio ($^{35/37}\alpha$) of 1.00245, and diffusion times of 100, 250, 500, 1000 and 2500 years. The initial chloride concentration in the area around the initial release was 5 mM (from Eggenkamp 1994)

the inflow. The chlorine or bromine isotope ratios of the transported chloride or bromide reaches a minimum which moves slowly away from the source as time progresses. As the chloride or bromide concentration at the source increases the δ^{37}Cl or δ^{81}Br value of the minimum decreases due to the larger ratio between the chloride or bromide concentration between the source and the background. Figure 8.3 shows how the chloride or bromide concentration and chlorine or bromine isotope compositions evolve over time in the case of one dimensional diffusion where chloride or bromide is continuously added at the source.

8.1.1.2 Diffusion Experiments to Study Separation of Possible Multiple Chlorine Isotopes

From the earliest days that the existence of stable isotopes in the light elements was known or suggested (Harkins and Hall 1916) experimental work was applied to use diffusion as a technique to separate these isotopes. The result of these experiments was that both heavy (Harkins and Hayes 1921) and light (Harkins and Jenkins 1926) fractions were produced. The actual chlorine isotope fractionation as a result of diffusion was however not yet determined in these experiments as the main purpose of these experiments was just to prove that more than one isotope did exist and that it was possible to separate them through a diffusion process.

William Harkins had hypothesised already in in 1915 (Harkins and Wilson 1915) that the light elements, chlorine, magnesium and silicon, just like neon as was discovered some years before (Thomson 1913) might be mixtures of isotopes because their atomic masses deviated too much from a whole number. Harkins (Harkins and Hayes 1921) decided that it was important to demonstrate the existence of different isotopes of at least one of these light elements by making an actual separation into at least two fractions which would differ in their atomic weights, as determined by some chemical method. The prediction of the existence of different isotopes was made on the basis of their chemical or mean atomic weights, and at that time the atomic weight of silicon was not known with a high degree of precision, while the atomic weights of magnesium and of chlorine, especially the latter, were considered to be very exact. For this reason and also because the deviation of the atomic weight of chlorine from a whole number is thc largest known for any light element, Harkins and Hall (1916) decided to attempt to separate chlorine into its possible isotopes.

It was decided that hydrogen chloride would be the most usable compound to study diffusion of potential chlorine isotopes because it is gaseous and the molecule consists of only one chlorine atom, so that no contamination with molecules consisting of both potentially existing isotopes would exist. In order to make sure that the separation of the isotopes was detectable a very large diffusion instrument was built consisting of 12.2 meters of porous pipe stems to let the hydrogen chloride diffuse through. The whole equipment and procedure is described in detail in Harkins and Hayes (1921). In 1920 several preliminary reports on the separation of chlorine in isotopes were published (Harkins 1920a, 1920b, 1920c). Some discussion arose whether it was possible to separate chlorine into its isotopes (e.g. Aston 1920a, 1920b, Chapman 1920; Soddy 1920; Brönsted and Hevesy 1921) but the results obtained by the very thorough experiments from Harkins and his co-workers showed that indeed chlorine could be separated into its isotopes and these experiments clearly confirmed the observations by Aston (1919), with the use of a spectrograph, that (at last) two isotopes of chlorine exist.

The separation of the chlorine isotopes was reported in the form of the atomic weight of the fraction. The normal atomic weight of chlorine was determined to be 35.460 in the early twentieth century (Harkins and Hayes 1921, 1922). Harkins and Hayes (1921) reported the following increases in the atomic weight of chlorine. In the first series of experiments, with Mr. CE Broeker as research assistant and obtained in 1920, the measured increase of the atomic weight of chlorine was 0.052, 0.059, 0.057, 0.055 and 0.053, thus on average a 0.055 ± 0.003 increase in the atomic weight of chlorine. Converted to the delta scale this fraction had a $\delta^{37}Cl$ value of 165 ± 10‰, a very significant increase. Later experiments gave slightly smaller increases of the atomic weight of chlorine of 0.034 ± 0.003 and 0.038 ± 0.002. These increases agree with values on the delta scale of 96 ± 8‰ and 105 ± 3‰. In a later report Harkins and Jenkins (1926) reported the chlorine atomic weight of the light fractions that were obtained in a different set of experiments. The most extreme of these experiments yielded a fraction that showed a decrease of 0.042 ± 0.001 of the atomic weight of chlorine. Three other experiments decreased the atomic weight of chlorine by 0.036 ± 0.003, 0.035 ± 0.001 and 0.029 (single determination only) respectively. This translates to values on the delta scale of −111 ± 3‰, 95 ± 9‰, −92 ± 3‰ and −76‰. Harkins and Jenkins (1926) conclude that overall the separation of the isotopes of chlorine reached in their laboratory is about 0.097, approximately one tenth of an atomic weight unit, or in modern delta values 276‰. At the moment that Harkins and Jenkins (1926) published their results this was by far the greatest relative difference produced in the atomic weight of any element. These experiments showed clearly the great potential of using diffusion in purifying individual isotopes of an element, a process that was later applied as one of the processes to produce sufficient ^{235}U to build atomic bombs for the Manhattan project (Columbia University 1942; Hewlett and Anderson 1962).

8.1.1.3 Experimental Studies to Determine Fractionation Due to Diffusion for Chlorine and Bromine

When methods to measure individual isotopes improved it became easier to study actual isotope fractionation that resulted from diffusion. Fink et al. (1934) observed actual fractionation of hydrogen isotopes as the result of diffusion. In their experiment, where they diffused hydrogen through a thin palladium foil, they observed that protium, the isotope ^{1}H, was diffusing more readily than the heavy isotope deuterium (^{2}H). With further improvement of analytical methods to determine the isotope composition of chemical species more of these diffusion studies were done. Klemm (1943) studied for example how the isotope composition of copper changes when it diffuses into silver sulphide. Only a few years later the first experimental study in which the difference in velocity, and thus the fractionation of chlorine isotopes as a result of diffusion was determined, was applied by Madorsky and Strauss (1948). They determined how the two isotopes of chloride in solution can be separated by the then recently developed counter current electromigration method (Brewer et al. 1947). As the result of this technique ^{35}Cl is concentrated in an anode compartment, while the isotope ratio is kept constant at a cathode compartment. In a 1:30 sodium chloride solution which was used as the electrolyte the concentration of ^{35}Cl could be increased to almost 81% after a migration time of almost 500 h (20–21 days). The initial enrichment factor of ^{35}Cl compared to ^{37}Cl in the five experiments that Madorsky and Strauss (1948) conducted was between 1.00115 and 1.00207. During all experiments the enrichment factors dropped during the course of the experiments due to the formation of silica gel as the result of interaction with the sodium hydroxide anode restituent.

A second study in which the relative mobilities of ^{35}Cl and ^{37}Cl in solution were determined was conducted by Konstantinov and Bakulin (1965). They determined the isotope separation of lithium chloride, sodium chloride and hydrochloric acid in a system where solutions of the chloride salts could exchange with potassium dichromate through a quartz filled capillary at a temperature of 50 °C. The ions could move under an electric current and after some time (6 to 12 h) the isotopic composition had changed enough to be analysed. It was observed that the chlorine isotope separation for lithium chloride and sodium chloride was about twice as large as for hydrochloric acid, and it was also shown that the isotope separation was larger if higher concentrations were applied. For lithium chloride it increased from 1.0012 in a 1 N solution to 1.0018 in a 7.5 N solution, for sodium chloride it increased from 1.0010 in a 1 N solution to 1.0014 in a 5.4 N solution. Contrary to these observed large fractionation in salt solutions, in a hydrochloric acid solution the isotope separation increased from 1.00022 in a 2 N solution to 1.00068 in a 10 N solution.

Richter et al. (2006) conducted experiments to determine the diffusion fractionation factors for KCl, LiCl, $MgCl_2$ and $CaCl_2$. Although the main focus was on the determination of the fractionation factors for the cations, in the case of the lithium and magnesium experiments they also measured the chlorine isotope composition of their samples. In their experiments diffusion run from a small (0.5 to 0.7 cm^3) container with a high concentration solution into a large (200 mL) container with only water. Diffusion took place through a tube with an inner diameter of 1 mm and a length of 0.8 or 1.2 cm. Diffusion of the concentrated solution into the larger fresh water value was allowed and samples were taken at regular intervals. Both the concentrations and isotope composition of the cations and in a few aliquots the chloride were measured. For chlorine the D_{35Cl}/D_{37Cl} ratio Richter et al. (2006) observed was 1.00143 ± 0.00080 for experiments conducted at 75 °C.

With the improvement of analytical techniques it also became possible to conduct diffusion experiments without applying electrical currents to increase isotope separation. Eggenkamp and Coleman (2009) determined isotope fractionation as a result of diffusion for both chloride and bromide ions in a system that would approach natural diffusion rather closely. Their diffusion experiments were similar in concept to those developed to analyse the diffusion of actinides in clay (Schreiner et al. 1982). Eggenkamp and Coleman (2009) filled a piece of glass tubing halfway with a polyacrylamide gel to which little or no sodium chloride or bromide solution was added, while the other half was filled with a gel with a high sodium chloride or bromide concentration. This system was left to diffuse for some time. Several experiments were prepared with different chloride or bromide concentrations, and with different diffusion temperatures. In these experiments is was observed that the fractionation due to diffusion for chloride seemed to increase at higher temperatures, while no variations depending on the concentration were observed. In the case of bromide however no significant variations in fractionation due to diffusion were observed for either the temperature or the concentration. The results of these experiments indicate that for chlorine the ratio D_{35Cl}/D_{37Cl} seems to increase from 1.00128 ± 0.00017 (1σ) at 2 °C to 1.00192 ± 0.00015 at 80 °C. For bromine the ratio D_{79Br} /D_{81Br} do not suggest a temperature dependency with the observations varying from 1.00098 ± 0.00009 at 2 °C to 1.0064 ± 0.00013 at 21 °C and 1.00078 ± 0.00018 (1σ) at 80 °C.

Diffusion of chlorine through porous rock has been studied by Hasegawa and Nakata (2018). They measured the chlorine isotope composition trough sandstones, granite and ceramic disks which were placed between two tanks, one filled with water with a high chloride concentration and one with a low concentration. The solution in both tanks was stirred permanently. At regular time intervals, depending on the diffusion coefficient of chloride trough the rock samples, a water sample was taken from both tanks and the chloride concentration and chlorine isotope composition were measured. From the evolution of the chloride concentrations and $\delta^{37}Cl$ values D_{35Cl}/D_{37Cl} ratios could be determined. In the various experiments the D_{35Cl}/D_{37Cl} ratios were between 1.0015 and 1.0016 for the ceramic disc, between 1.0020 to 1.0022 for granite and between 1.0018 and 1.0021 for sandstone. These values are in good agreement with those of previous studies, which ranged between 1.0013 and 1.0030 (Weaver 2000; Eggenkamp 2014). The measured D_{35Cl}/D_{37Cl} ratios for the ceramic disks are nearly equal to those in free water (1.0013 to 1.0019; Eggenkamp and Coleman 2009). The ceramic disk is considered to be inert for ion transport so that the fractionation is the same as in free water. The D_{35Cl}/D_{37Cl} ratios for geological material, ranging from 1.0018 to 1.0022 for granite and sandstone, are higher than in free water. These materials may have higher D_{35Cl}/D_{37Cl} ratios due to interaction between dissolved ions the sample surface.

8.1.1.4 Observations on Aqueous Natural Systems to Determine the Isotope Fractionation

The first study in which chlorine isotope diffusion was detected in a natural system was conducted by Desaulniers et al. (1986) in groundwater sampled from Quaternary glacial deposits in Ontario, Canada. This natural system has a low permeability so that even after long diffusion times chloride concentrations and chlorine isotope signals have not been levelled out. Based on observations that chloride in these deposits must have its origin from underlying saline formations (Desaulniers et al. 1981) it was possible to estimate the diffusion of chloride from these underlying sediments. Assuming simple diffusion it was possible, after chlorine isotope compositions were measured from these groundwater samples, to determine the fractionation between the light and heavy isotope of chlorine in an environmental setting for the first time. As diffusion of chloride in this system was upward from saline deposits into younger deposits containing freshwater it was observed that $\delta^{37}Cl$ increased with depth. Desaulniers et al. (1981) were able to model the chloride concentrations at their sites assuming diffusion times between 10,000 and 16,000 years. They calculated possible isotope fractionation factors depending on the groundwater flow during this period, which was modelled to be between 0 and 0.4 cm per year. Applying their model with current chloride concentrations in the system Desaulniers et al. (1986) could conclude that diffusion times from 12,000 to 16,000 years and a groundwater flow close to 0 cm per year would fit the chloride concentration data best. These data were used to calculate a best fit for the fractionation of chlorine isotopes due to diffusion and the results showed that in this system the diffusion coefficient of ^{35}Cl was about 1.0012 times larger than the diffusion coefficient of ^{37}Cl, actually just within the range determined by both Madorsky and Strauss (1948) and Konstantinov and Bakulin (1965) in earlier experiments.

Eggenkamp et al. (1994) determined the chlorine isotope fractionation in a diffusion system where saline ocean water diffused downwards into sediments containing relatively freshwater. Currently Kau Bay on the Indonesian Island of Halmahera is connected to the ocean through a shallow sill that, during the last glacial period, was above sea-level so that fresh and brackish sediments were deposited in the then lake. After the glacial period sea-level rose and when it increased to a level above the sill the lake changed into a bay and the water turned saline. Since then downward diffusion of chloride took place in the sediments of Kau Bay (Middelburg and de Lange 1989). The diffusion was also clear from chlorine isotope ratios that were analysed in the sediment pore water, which show a decrease in $\delta^{37}Cl$ with depth. After correction for advective porewater flow the fractionation could be determined in two different sediment cores. They were reported as 1.0023 and 1.0030. Unfortunately no errors were determined, but probably these were rather large given the small total variation in $\delta^{37}Cl$ measurements. The D_{35Cl}/D_{37Cl} values obtained are at the upper limit when compared to the experimental data from Madorsky and Strauss (1948), and a bit outside the limits obtained by both Konstantinov and Bakulin (1965) and later by Eggenkamp and Coleman (2009).

Later, two more studies have determined the isotope fractionation from natural settings. Groen et al. (2000) determined it from sediment porewater profiles sampled along the Surinam coast where salt water diffuses into sediments containing fresh porewater from both overlying younger saline sediments and underlying older saline sediments. Although they only analysed the chlorine isotope composition of a rather limited number of samples they were able to estimate the ratio in diffusion coefficient between ^{35}Cl and ^{37}Cl as being 1.0027.

The most recent study in which the difference in diffusion coefficient between ^{35}Cl and ^{37}Cl is determined described one of the more complex systems in which this fractionation is determined (Beekman et al. 2011). The Dutch IJsselmeer area has a fairly complex history. The area was a relative freshwater lake until medieval times after which its salinity slowly increased. During this period the area was also prone to storm surges and during the very short period a storm surge lasted salt water from the North Sea was introduced into the sediment. After 1932 the area was closed from the sea by a dam, and the area turned into a fresh water lake. The only possibility to model both chloride concentration as well as chlorine isotope composition in this system was to take the whole (known) history into account and to apply numerical modelling. A very satisfactory model was developed and a best fit D_{35Cl}/D_{37Cl} value of 1.00165 was found, which agrees very well with experimental data obtained by Eggenkamp and Coleman (2009).

Overall, five studies in which the D_{35Cl}/D_{37Cl} ratio in aqueous solutions was determined experimentally, four in which it was determined from observations in natural systems, plus one in which it was determined through molecular dynamics simulations are known for chlorine, while only one such study is known for bromine. Within reasonable limits the results of the diffusion studies in aqueous systems agree well with each other. In the different studies D_{35Cl}/D_{37Cl} has been determined as between 1.0010 and 1.0030 for chlorine, while the majority of the data fall between 1.0015 and 1.0020. Although some experimental results suggest that the D_{35Cl}/D_{37Cl} ratio is dependent on the temperature of the experiments the results from the environmental studies show a larger range in the D_{35Cl}/D_{37Cl} ratios than the experiments from (Eggenkamp and Coleman 2009) indicate, so that further research in obviously needed to get a better understanding of the potential temperature dependency of the D_{35Cl}/D_{37Cl} ratio.

Currently only one study is known in which the fractionation due to diffusion for bromide is determined. The results of this study (Eggenkamp and Coleman 2009) are also included in Fig. 8.4. It is hoped that, when more of these studies are done, they are conducted on bromine isotopes too. The results obtained for bromine up to now agree well with the data as they were expected with respect to chlorine considering the smaller relative mass difference between both stable bromine isotopes as compared to both chlorine isotopes (Eggenkamp and Coleman 1997).

8.1.2 Diffusion in Melt Systems

During the 1950s several experimental studies have been published in which the isotope fractionation due to diffusion of chloride and bromide in molten salts was determined by researchers from the Max Planck Institute in Mainz and the Higher Technical Institute in Göteborg. Chloride diffusion was determined in lead chloride ($PbCl_2$, Klemm and Lundén 1955), zinc chloride ($ZnCl_2$, Lundén and Herzog 1956) and thallium chloride (TlCl, Herzog and Klemm 1958). From this last system also the temperature dependence has been determined. Bromide diffusion was determined on lead bromide ($PbBr_2$, Cameron et al. 1956) and on zinc bromide ($ZnBr_2$, Lundén and Lodding 1960). In these experiments the isotopes ^{35}Cl or ^{79}Br were enriched by the application of an electric current on the molten

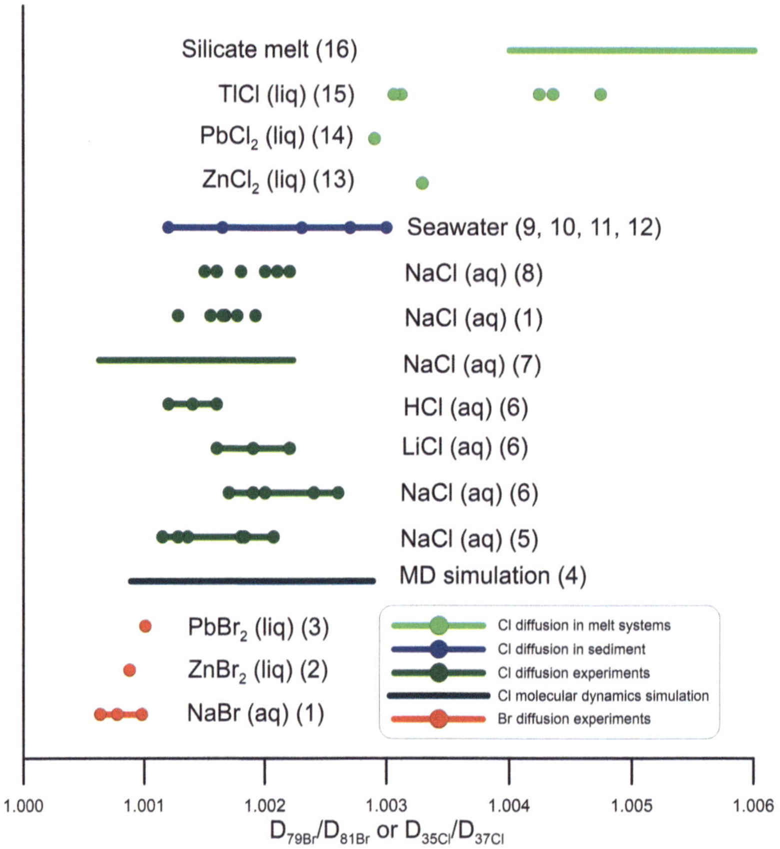

Fig. 8.4 Fractionation factors due to diffusion for bromine and chlorine. The figure summarises all published data, including molecular dynamics simulations, experiments and estimates from natural environments. It shows clearly that fractionation for bromine is relatively low, while for chlorine there is a clear difference between fractionation due to diffusion in aqueous systems and in melt systems. In aqueous systems the chlorine isotope fractionation varies, depending on the study between less than 1.001 and 1.003. In high temperature melt systems reported values are between approximately 1.003 and 1.005. In the (few) studies on bromine isotopes this difference is not visible. References cited: (1) Eggenkamp and Coleman (2009), (2) Lundén and Lodding (1960), (3) Cameron et al. (1956), (4) Bourg and Sposito (2007), (5) Madorsky and Strauss (1948), (6) Konstantinov and Bakulin (1965), (7) Richter et al. (2006), (8) Hasegawa and Nakata (2018), (9) Desaulniers et al. (1986), (10) Eggenkamp et al. (1994), (11) Groen et al. (2000), (12) Beekman et al. (2011), (13) Lundén and Herzog (1956), (14) Klemm and Lundén (1955), Herzog and Klemm (1958), (16) Fortin et al. (2017)

salts. The difference in diffusion coefficient for the two isotopes was determined in each case.

The data obtained in these experiments are either comparable to the maximum values or slightly larger than the data obtained from the chloride and bromide diffusion experiments and observations in aqueous solution. Konstantinov and Bakulin (1965) observed that their fractionation data increased with the electrolyte concentration and concluded that the highest values they observed approached in absolute magnitude the values obtained in these earlier experiments with molten salts.

Diffusion of chlorine isotopes has also been studied in silicate melt environments. Fortin et al. (2017) characterised chlorine isotope fractionation due to diffusion in dacite melts at 1200 to 1350 °C. They used a piston-cylinder apparatus, to generate silicate melts with a dacitic composition that were enriched in chloride. From these melts they fabricated diffusion couples in which chlorine atoms could diffuse in chemical gradients at temperatures of 1200 to 1350 °C and a pressure of 1 GPa. The run products were analysed by SIMS for the chlorine isotope compositions along the diffusion profiles. The diffusion profiles are rather short, with lengths of up to 3 mm. The profiles look very comparable to the theoretical diffusion profiles described in Eggenkamp (1994) and experimental profiles from Eggenkamp and Coleman (2009). The diffusion coefficient ratio ($^{35}Cl/^{37}Cl$) is 1.005 ± 0.001, which equals to a β of 0.09 ± 0.02. No temperature effect on the diffusivity ratio was recognisable over the 150 °C temperature range applied in their experiments. The fractionation due to diffusion in melts thus is significantly higher than the fractionation due to diffusion in aqueous solutions as determined in both experimental and field studies, at least under specific conditions and length scales. Interestingly, Fortin et al. (2017) state that the diffusion fractionation for chloride in melts is comparable to the diffusion fractionation of Ca and Mg in silicate melts, which are 1.004 ($^{40}Ca/^{44}Ca$) and 1.006 ($^{24}Mg/^{26}Mg$) respectively (Richter et al. 1999, 2003, 2008).

8.1.3 Overall Summary of Fractionation Estimates Due to Diffusion

Figure 8.4 shows a summary of all estimations, calculated, experimental and from environmental measurements of values for the D_{35Cl}/D_{37Cl}, and D_{79Br}/D_{81Br} that have been studied in the past.

Fractionation due to diffusion has been studied in a large array of systems, in aqueous solutions with more or less interaction with the medium the solution diffuses and in various melt systems. The results show that a relative large range in fractionation factors ($\alpha = D_{35Cl}/D_{37Cl}$ or D_{79Br}/D_{81Br}) is found, especially for chlorine. For bromine the fractionation factor has been determined in aqueous solution and the melt of $PbBr_2$ and $ZnBr_2$. The fractionation for bromine is relatively modest and values are comparable in low temperature aqueous systems and melt systems. The fractionation for chlorine is very variable and there appears a significant difference between fractionation in aqueous systems and in melt systems. The values (D_{35Cl}/D_{37Cl}) in aqueous systems are mostly between 1.001 and 1.003. Because the studies have been done in a large series of environments, temperatures and even with different co-diffusion cations it is not very clear what are the determining conditions that govern the exact fractionation. This is for example clear from the four studies of seawater diffusing in various sedimentary systems, where D_{35Cl}/D_{37Cl} ratios between 1.001 and 1.003 are found, spanning the whole range of observed D_{35Cl}/D_{37Cl} values for aqueous chlorine diffusion. Interestingly however is that in high temperature melt systems the D_{35Cl}/D_{37Cl} ratios are mostly above 1.003. At modest temperatures (500 to 700 °C) in molten $ZnCl_2$, $PbCl_2$ and TlCl ratios are mostly lower than in high (1100 to 1350 °C) silicate melts. This would suggest that the fractionation ratio increases with increasing temperature. However, in the experiments with molten thallium chloride lower D_{35Cl}/D_{37Cl} ratios were observed at the higher experiment temperatures (about 700 °C) and higher D_{35Cl}/D_{37Cl} ratios at lower temperatures (about 550 °C). Further research to understand the effects that rule the fractionation due to diffusion of chloride and bromide ions in different environments is thus necessary to fully understand the effects.

8.1.4 Gravitational Isotope Fractionation

Because the effects of gravitation on Earth, in a motionless and chemically homogeneous water column, dissolved isotopic substances must be distributed according to their masses. Giunta et al. (2017) observed this characteristic first in a motionless and chemically homogeneous water column and observed that both $\delta^{37}Cl$ and $\delta^{81}Br$ increased downward. While the chemical composition of the brine in this 180 m thick Cambrian aquifer was constant $\delta^{37}Cl$ increased from −0.5 at the top to +0.2‰ at the base, and $\delta^{81}Br$ increased from 0.0 at the top to +1.2‰ at the base. These increases were explained by the fact that the heavier isotope moves downward faster as a result of gravity. Giunta et al. (2017) estimated that in a stagnant aquifer of Cambrian age both $\delta^{37}Cl$ and $\delta^{81}Br$ would be enriched by ~0.78‰ per 100 meter. Giunta et al. (2017) observed a good correlation between $\delta^{37}Cl$ and $\delta^{81}Br$ with a slope slightly less than the theoretical equilibrium between diffusion and gravitational setting. Although it was expected that both ions (chloride and bromide) would show the same effect it appears that in the case of bromide it was slightly larger than for chloride, possibly due to the higher mass of the bromide ion compared to the chloride ion. Overall however the observed isotope variations could be well explained by gravitational fractionation.

8.2 Ion-Filtration/Ion Exclusion

When a solution is forced through a semi-permeable membrane the effect can be that the solvent passes though the membrane with more ease than the solute. The reason for this effect can be for example the difference in size between the ions and molecules of the solvent and the solute or a difference in charge. As a result of this effect, which is called ion-filtration, isotope fractionation can occur. During the development of chlorine isotope techniques at the University of Arizona already a few studies were done in which the isotope fractionation of chlorine as a result of ion filtration effects were investigated.

8.2.1 Theory

Phillips and Bentley (1987) described the theory on ion-filtration with respect to chlorine isotopes in some detail. During the ion-filtration process the solvent is able to, while the dissolved ions are unable to move through a membrane. Ion-filtration is a process which is similar to reverse osmosis. Following the description by Kaufmann (1984) osmosis is the effect that appears when two reservoirs, one with water and one with an aqueous salt solution are in connection to each other through a semi-permeable membrane. During this process water will be able to pass the membrane while the dissolved salt will not. As a result of this process the contents of the reservoirs will interact through the membrane and will try to balance the chemical potential between the two reservoirs. That means that water flows through the semi-permeable membrane in the direction of the reservoir with the higher chemical potential (concentration of salts) in an attempt to equalise the potentials, effectively diluting the contents of the reservoir with the higher concentration. In the case of reverse osmosis the process is comparable (two reservoirs with aqueous solutions of different chemical potential), but now a pressure is applied on the reservoir with the higher concentration. Because of this pressure water is forced through the membrane in the direction of the freshwater side. As a result the salinity increases on the saline side of the membrane. This effect may have a consequence on the isotope composition of the chlorine in the water. In many cases semi-permeable membranes are consisting of clay-minerals. These minerals have negative charges at their surfaces, which means that negative ions are repelled from the membrane. The effect is comparable to diffusion as the lighter ion will be repelled more efficiently than the heavier ion, so that isotope fractionation will take place.

Phillips and Bentley (1987) describe a different model to explain chlorine isotope fractionation occurring from ion-filtration. In their model actually ions move through the membrane, but because the lighter isotope moves away from the membrane more efficiently than the heavier isotope the result is that more of the heavier isotopes are actually transferred through the membrane, increasing the $\delta^{37}Cl$ in the sedimentary layer behind the membrane. In their simplified model development of the chloride concentration and chlorine isotopes is described as the result of compaction of the sedimentary layers. At the start the chloride concentration is highest in the lower and lowest in the shallowest sedimentary layer, and the chlorine isotope composition is chosen to be zero in the upper two layers and + 0.25‰ in the lowest. When pressure is applied it is assumed that the compaction started in the lower and progressed to the shallower layers. In step 1 the lowest layer is compacted. As a result of this both water as well as chloride with an increased $\delta^{37}Cl$ moves to the second layer, and in this layer both the chloride and the chlorine isotope composition increase. When the lowest layer is fully compacted the second layer starts to compact. During this process again water and isotopically heavy chloride move to the third layer. As a result the chloride concentration increases and the isotope ratio decreases in this second layer, while in the upper layer both the concentration and the isotope composition increase. When the second layer is fully compacted the third starts to compact resulting in higher concentrations and a lower isotope composition in this layer. The effect of this process is that considerably positive $\delta^{37}Cl$ values can theoretically be observed in transported chloride as opposed to diffusion which tends to lead to more negative $\delta^{37}Cl$ values in transported chloride.

8.2.2 Experimental

The effects of ion-filtration on chlorine isotope fractionation have been experimentally established by Campbell (1985). In one of her experiments she forced a chloride solution with a $\delta^{37}Cl$ value of 0.04‰ through a thick (2.5 cm) plug of bentonite clay and measured the $\delta^{37}Cl$ of the effluent solution. It was found that the $\delta^{37}Cl$ of the effluent solution had significantly increased $\delta^{37}Cl$ values (+1.70, +1.60, +1.51 and + 0.51‰ respectively). Also results from various other experiments in which chloride was transported through different types of man-made semipermeable membranes showed elevated $\delta^{37}Cl$ values in the effluent solutions, ranging from no significant increase (+0.01 ± 0.12‰) to 1.26 ± 0.20‰.

Unfortunately all of these early experiments were conducted in the period that the technique to analyse chlorine isotope variations was still under development at the University of Arizona. For that reason the results were in general considered to be not very conclusive, in the sense that it could not be explained how the changes in $\delta^{37}Cl$ during the tests related to the different membranes that were applied, chloride concentrations and diffusion time. It is unfortunate that the experiments conducted in these studies were never repeated after the analytical techniques to analyse chlorine isotopes had been improved, so that it is still difficult to understand the effects of chlorine isotope fractionation as a result of ion-filtration in natural samples.

Strydom et al. (2022) did some clay compaction experiments to test if chlorine and bromine isotopes fractionate during this process and if the results were to relate to ion-filtration processes. They compacted a smectite-rich bentonite, a swelling clay, and an illite-rich marl, a non-swelling clay, that were saturated with ocean water to pressures up to a high fluid pressure of 45 MPa, a mechanical stress of up to 150 MPa and temperatures up to 150 °C. Strydom et al. (2022) observed a general decrease of the cation and anion concentrations in the expelled water for the swelling clay while the concentrations slightly increase in the remaining pore fluid. This was not the case for non-swelling clay. More freshening of the expelled water occurred during compaction at higher temperature. Strydom et al. (2022) also observed a larger range of isotopic variation for bromine (the $\delta^{81}Br$ of the expelled water ranged from +0.9‰ up to +1.5‰) than for chlorine (the $\delta^{37}Cl$ of the expelled water ranged from −0.5‰ to −0.1‰) in the compaction experiments. During the compaction of the illite-rich marl, no significant variation in the Cl isotope values was observed (the $\delta^{37}Cl$ was close to 0‰), while the $\delta^{81}Br$ value of the expelled water showed a comparable general increase (from +0.9‰ up to +1.5‰) as during the compaction of smectite-rich bentonite. Strydom et al. (2022) concluded that the surface chemistry of the clay, in combination with decreasing porosity, in part, drive the anion and isotope evolution. Strydom et al. (2022) proposed that the significant retention of chloride and bromide in pore water of compacted smectite-rich bentonite (a swelling clay) is indicative of ultrafiltration and that chlorine and bromine isotopes are promising tracers to consider when tracking the origin of low-salinity formation waters in sedimentary basins, although unfortunately, it was still not possible to quantify the exact effects.

8.2.3 Low δ^{37}Cl Values in Deep-Sea Porewater Originate from Ion-Filtration

It has long been known that deep-sea sediment pore waters can show very negative δ^{37}Cl values (Ransom et al. 1995; Spivack et al. 2002; Deyhle et al. 2003; Godon et al. 2004; Wei et al. 2008; Agrinier et al. 2019). Although it was not possible to conclusively explain these negative values by ion-exchange it was suspected that this effect must be the origin of the negative values as they could not be satisfactorily explained by other processes that can potentially fractionate chlorine isotopes. Recently however, it was shown by Agrinier et al. (2021) that compaction-induced ion filtration of chlorides through clay-rich membranes can produce the observed pore fluid δ^{37}Cl depletions with isotope fractionation factors ranging from 1.000 to 1.008 between chloride of the expelled fluid and chloride of the residual fluid. Agrinier et al. (2021) found that smectite-rich sediments were associated with higher isotopic fractionation factors. Illite and chlorite-rich sediments were associated with intermediate isotope fractionation factors and clay-poor sediments were associated with the lowest isotope fractionation factors. These data suggested that ion-filtration induced chlorine isotope fractionation is controlled by the surface charge that is associated with specific clay minerals.

8.3 Fractionation During Ion-Exchange Chromatography of Chloride

Since it was shown that halides in mixtures could be separated using ion-exchange methods (Atteberry and Boyd 1950; Riemann and Lindenbaum 1952) using a column filled with an ion-exchange resin of a strong base type, Langvad (1954) attempted to apply this method with a pure chloride solution with the aim to test if it was possible also to separate the two stable isotopes of this element. For his test he filled two columns with an inner diameter of 2.7 cm and a length of 150 cm with Amberlite IRA-400 in the nitrate form. The isotope separation was effectively applied by moving a potassium chloride solution continuously from column A into column B until it had virtually moved for 43 meters. At that moment the solution was sampled into ten fractions of which five were analysed for chlorine isotopes. In order to measure the chlorine isotope composition Langvad (1954) developed the analytical method using chloromethane gas which is still, after several subsequent improvements, the most precise method to measure chlorine isotopes to date. The results he obtained (see Fig. 8.5), recalculated to δ^{37}Cl values, showed that the first fractions were enriched in ^{37}Cl and the later fractions in ^{35}Cl. In spite of the very large errors (mostly 2 to 4‰, in one case almost 10‰!) it is obvious that the total variation of about 44‰ between the different fractions that was found by Langvad (1954) was a very significant isotope fractionation. Unfortunately however as the isotope data obtained were not very precise it was not possible to calculate an enrichment factor from these data.

Heumann and Hoffmann (1976) conducted a similar test using an 80 cm long column with an internal diameter of 1.5 cm filled with the strongly basic anion exchanger AG1-X 10 (nitrate form, 200 to 400 mesh). They analysed small chloride samples (200 mg) which they eluted with nitrate at three different concentrations, 0.01, 0.1 and 1 M. Their samples were analysed using thermal ionization mass spectrometry. They found the same characteristic isotope profile as Langvad (1954), thus it was observed that the first fractions were enriched in ^{37}Cl, and they were able to obtain enrichment factors (α) which were shown to be different for the different nitrate concentrations. They found that for a 0.01 M nitrate eluent α had a value of 1.00054, for a 0.1 M nitrate eluent a value of 1.00030 and for a 1.0 M nitrate eluent a value of 1.00019.

Musashi et al. (2004, 2007) were the latest to determine chlorine isotope fractionation during the passage of chloride over anion exchange col-

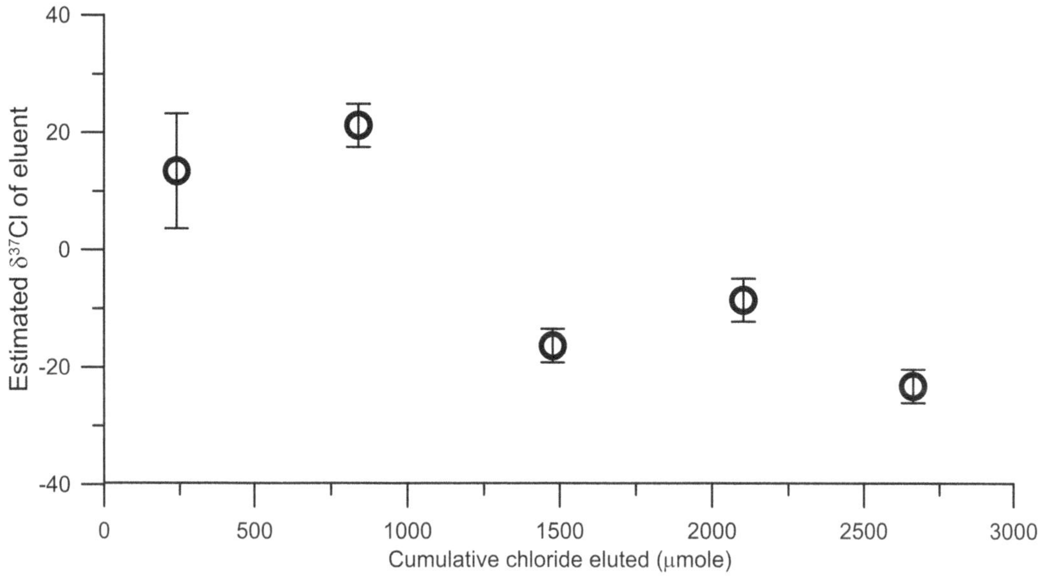

Fig. 8.5 Chlorine isotope variations as obtained by Langvad (1954) in his ion-exchange experiment. Data recalculated to the familiar delta notation

umns. In both studies a strongly basic anion exchanger, Muromac 1 × 8 analytical grade was used (mesh size, 200–400 mesh; ionic form, Cl^-; total exchange capacity, 1.2 eq/l; cross linkage, 8% divinylbenzene (DVB); moisture, 45–55%). The resin was packed in the column and was conditioned to the hydroxide form with a 2 mol/L sodium hydroxide solution fed into the column by a high pressure pomp (Nihon Seimitsu). Chloride-free distilled water was fed into the column in order to remove sodium ions from the column. During these experiments at first the column was fed with a 0.1 M hydrochloric acid solution and subsequently with a 0.1 M nitric acid solution. During this process the column was first saturated with chloride which in the second phase was replaced with nitrate and removed from the column. The isotope compositions could be measured from the fractions containing chloride. It was observed that in the two experiments the isotope fractionation of chlorine between the stationary phase and the solution was 1.00030 and 1.00034 respectively, values which are not only in good agreement with each other but also in very good agreement with the values obtained by Heumann and Hoffmann (1976) when their experiment with a nitrate concentration of 0.1 M is compared. It was concluded that on this ion-exchange resin ^{35}Cl is concentrated on the resin, while ^{37}Cl prefers to concentrate in solution, as was actually already also observed by Langvad (1954) and Heumann and Hoffmann (1976). As normally the heavier isotope concentrates in the most strongly bound phase it was concluded by Musashi et al. (2007) that this effect represents a hydration effect and the results were discussed in comparison with isotope fractionation between sodium chloride brine and halite precipitate where the heavier isotope is concentrated in the precipitate and not the solution.

8.4 Salt Precipitation

Attempts to determine isotope fractionation of chloride ions between a saturated solution and its co-precipitating solid have been made in theoretical, experimental as well as field studies. Most of these studies were done with sodium chloride (halite) solutions, which is not surprising as sodium is the most common cation in most saline solutions and chloride the most common anion. As potassium and magnesium are also common cations in salt deposits salts from these elements are studied as well especially in relationship to potash deposits in China. More recently also some studies have been published in which fractionation between bromide solutions and precipitates are determined, as well as studies were both chlorine and bromine isotopes were measured so that the influence of much higher chloride concentration in solution or precipitate on the bromine isotope fractionation could be studied.

8.4.1 Experimental Determination of Chlorine and Bromine Isotope Fractionation During Precipitation of Salt

Hoering and Parker (1961) were the first to study the chlorine isotope fractionation between sodium chloride (halite) and a saturated sodium chloride solution. The fractionation factor α obtained in their experiment:

$$Na^{37}Cl_{(s)} + {}^{35}Cl^-_{(aq)} \leftrightarrow Na^{35}Cl + {}^{37}Cl^-_{(aq)} \quad (8.8)$$

was 1.0002 ± 0.0003 ($\varepsilon^{37}Cl$ = 0.2 ± 0.3‰), suggesting that the chlorine isotope composition of the precipitated sodium chloride was probably, but not significantly lower than that of the saturated solution. This is different from the later observed isotope fractionation between halite and brine, where it was shown that the precipitated halite has a higher $\delta^{37}Cl$ than the brine it precipitates from. Unfortunately, as the analytical precision obtained by Hoering and Parker (1961) was still not good enough to determine small variations they could not conclude that this result indicated any significant fractionation.

As part of their study on Zechstein evaporites Eggenkamp et al. (1995) determined equilibrium fractionation factors between sodium chloride, potassium chloride and magnesium chloride saturated solutions and their precipitating salts. They observed that the chlorine isotope fractionation is dependent on the cation that precipitates together with the chloride. Near-saturated solutions of reagent grade sodium chloride, potassium chloride, and magnesium chloride were allowed to evaporate at ambient temperature. After precipitates of the respective salts were formed, the solutions and the precipitates were separated by decanting the brine and filtering the precipitate. To remove any adhering solution the precipitate was rinsed with acetone. Both the brine and the salt were analysed for chlorine isotopes composition and the fractionation was defined as the difference between the $\delta^{37}Cl$ of the precipitate and of the remaining brine. For sodium chloride they found that the precipitate was 0.26 ± 0.07‰ heavier than the brine, while for potassium chloride and magnesium chloride it was respectively 0.09 ± 0.07‰ and 0.06 ± 0.10‰ lighter. These fractionation factors were interpreted to explain the lack of variations in the studied Zechstein evaporites. As the sign of the isotope fractionation reverses from sodium chloride to potassium chloride this would mean that at a certain moment a minimum $\delta^{37}Cl$ would be reached, after which the $\delta^{37}Cl$ would increase again due to the domination of potassium and magnesium salt precipitate in the last stages of salt precipitation. The results obtained by Eggenkamp et al. (1995) were, at least qualitatively, confirmed in an experiment reported by Eastoe et al. (1999) who evaporated a batch of seawater and analysed the $\delta^{37}Cl$ of the remaining water and the precipitated halite at certain stages. The $\delta^{37}Cl$ of the brine decreased during the early stages, but increased again during the later. The observation that in all stages the halite that was analysed was considerably enriched in $\delta^{37}Cl$ compared to the brine was also in good agreement with earlier measurements.

Studies done in the Chinese Tarim and Western Qaidam Basins (Tan et al. 2005, 2006, 2009) indicated that during the latest stages of salt precipitation in these basins $\delta^{37}Cl$ continued to decrease, and this effect was used locally as a proxy for the exploration of potassium deposits. It was shown however in an earlier study by Xiao et al. (2000) that equilibrium isotope fractionation in potassium and magnesium chlorides was not as large as chlorine isotope fractionation during halite precipitation. A study by Luo et al. (2012) confirmed these results. The effect that the chlorine isotope composition continues to decrease during precipitation of magnesium chlorides might however be the result of the specific character of these deposits, which are terrestrial rather than marine, or perhaps of kinetic effects.

Later, Luo et al. (2014) published the results of chlorine isotope fractionation studies during precipitation of sodium, potassium and magnesium chloride from a saturated solution in order to better understand chlorine isotopic variations during brine evolution. They observed fractionation factors that were significantly higher than the fractionation factors obtained by Eggenkamp et al. (1995). The fractionation factors they obtained were 1.00055 for NaCl, 1.00025 for KCl and 1.00012 for $MgCl_2$. No errors were given, but the data were used the explain the large negative $\delta^{37}Cl$ values found in potash deposits in the Chinese Qaidam Basin. Luo et al. (2014) recalculated the Rayleigh fractionation for the evaporation of seawater to dryness and concluded that, based on their new fractionation factors indeed the $\delta^{37}Cl$ of potash deposits must be highly negative (lower than 1‰). The samples obtained from these experiments were measured with TIMS on CsCl as proposed by Xiao and Zhang (1992). Unfortunately, the large differences in fractionation factors as compared to those from Eggenkamp et al. (1995) were not discussed, except that was stated that the chlorine isotope fractionation in salt deposits was underestimated, and also the relatively large errors that were obtained were not discussed. This led to a discussion on the results between Eggenkamp (2015a) and Luo et al. (2015). Eggenkamp (2015a) stated that when the errors of the $\delta^{37}Cl$ measurements were taken into account the results are absolutely not as conclusive as was suggested by Luo et al. (2014). When errors are taken into account the fractionation factors could be given as 1.00055 ± 0.00046 for NaCl, 1.00025 ± 0.00010 for KCl and 1.00012 ± 0.00050 for $MgCl_2$. These errors are so high that within the margin of error the data obtained by Eggenkamp et al. (1995) and those obtained by Luo et al. (2014) are not significantly different from each other. Luo et al. (2015) disagree with this statement based on different viewpoints on how fractionation factors were determined and that the large errors were less relevant because more negative $\delta^{37}Cl$ values were clearly observed in the data presented in both Luo et al. (2012) and Luo et al. (2014).

Chlorine and bromine isotope fractionation in a larger set of salts precipitated from saturated solutions were determined by Eggenkamp et al. (2016). They measured, in a way similar to Eggenkamp et al. (1995) and at room temperature (22 ± 2 °C), the fractionation ($10^3\ln\alpha$) of NaCl (+0.35 ± 0.08), KCl (−0.12 ± 0.05), $MgCl_2$ (−0.02 ± 0.02), LiCl (+0.03 ± 0.05), NH_4Cl (+0.09 ± 0.03), RbCl (−0.31 ± 0.05), CsCl (−0.23 ± 0.02), $CaCl_2$ (+0.04 ± 0.02), $SrCl_2$ (+0.15 ± 0.02), $BaCl_2$ (+0.41 ± 0.03), $FeCl_3$ (−0.23 ± 0.05), LiBr (−0.03 ± 0.09), NaBr (−0.07 ± 0.05), Kbr (0.00 ± 0.01), NH_4Br (+0.11 ± 0.05), $MgBr_2$ (+0.06 ± 0.06), $CaBr_2$ (−0.06 ± 0.04) and $SrBr_2$ (+0.05 ± 0.06). Interesting to note is that the results for NaCl, KCl and $MgCl_2$, although slightly different form the data measured by Eggenkamp et al. (1995) confirm the sign of the fractionation as determined in that study. And, while the values are different, they are still closer in value to Eggenkamp et al. (1995) than the values measured by Luo et al. (2014). It is interesting to note that large differences in isotope fractionation factors exist depending on the cation that co-precipitates with chlorine and bromine. It is also interesting to notice that the sign of the fractionation can both be positive or negative. Further it is striking that isotope fractionation during precipitation of pure bromide salts is always very little. This is unexpected, as observations on natural samples suggested that bromine isotope variation in nature may be larger than that of chlorine (Eggenkamp 2015b).

Natural salt deposits show in general relatively high $\delta^{81}Br$ values (Eggenkamp et al. 2019a, 2019b). This effect had already been suggested in earlier observations where the bromine isotope composition in brines that originated from dissolved evaporites showed significantly increased values. Eggenkamp (2014) summarised $\delta^{81}Br$ data from four studies on formation waters originating from dissolved evaporites and found that the average $\delta^{81}Br$ of these samples was

+0.60 ± 0.29‰. In order to understand chlorine and bromine isotope fractionation in salt with natural chloride/bromide ratios Eggenkamp et al. (2019c) presented the results of experiments where chlorine and bromine isotope measurements were done on synthetic ocean water that was evaporated until dryness. In these experiments it was shown that while the chlorine isotope measurements of the precipitated salt closely followed the expected fractionation as discussed above, thus the $\delta^{37}Cl$ of the first precipitated NaCl is about +0.35‰, the $\delta^{81}Br$ of the first precipitated salt is about 0.6‰ more positive than the $\delta^{81}Br$ of the brine. This confirmed that bromine isotope fractionation during precipitation of natural salt indeed is even larger than chlorine isotope fractionation.

8.4.2 Theoretical Chlorine and Bromine Isotope Fractionation in Salt Minerals

Schauble et al. (2003) presented a study updating the theoretical knowledge on chlorine isotope fractionation. They also studied chlorine isotope fractionation of inorganic chloride that is distributed between different phases. One of the main discoveries they made was that chloride would concentrate in phases where it is bonded to +II cations relative to phases where it is bonded to +I cations. They attempted to calculate isotope fractionation for salts that precipitate from solution, however this appeared very difficult to calculate. Schauble et al. (2003) were able to estimate the differences between sodium chloride and potassium chloride in case these salts would coexist. Their calculation indicated that chloride in sodium chloride would be about 0.8‰ heavier than chloride in potassium chloride. Considering all uncertainties they concluded that these results are at least qualitatively in agreement with the experimental results reported by Eggenkamp et al. (1995).

In order to better understand the large bromine isotope fractionation that is found during the precipitation of salts from saturated solutions Gao and Liu (2021) conducted first-principles molecular dynamics simulations to better understand the origin of the large bromine isotope fractionations in these salt minerals. This first-principles molecular dynamics (FPMD) method is used to evaluate the solvation effects. That way the configurational effects of aqueous Br-bearing species could be included. For crystalline compounds, they used the static first-principles periodic DFT methods. Gao and Liu (2021) found that the β-factors generally increased as the oxidation state increase, indicating that ^{81}Br tend to concentrate in substances with higher Br oxidation states. Aqueous bromide bearing species in brines have detectable but small Br isotope fractionations with each other. However, fractionation between gaseous molecules and $NaBr_{(aq)}$ is significantly larger, e.g., $\Delta^{81}Br_{CH3Br(gas)\text{-}NaBr(aq)}$ is 0.95 ± 0.041‰ at 20 °C. For minerals, the fractionation between pure NaBr and $NaBr_{(aq)}$ is close to zero, but for Br-halite$_{(1/124)}$ ($Na_{125}Cl_{124}Br_1$) and $NaBr_{(aq)}$, the fractionation is 0.206 ± 0.041‰. Similar results were found for fractionation between Br-sylvite, Br-bischofite and Br-bearing solutions, and this indicates that Br isotope variations are closely related to Cl/Br ratios in minerals. By using the calculated isotope fractionation factors, Gao and Liu (2021) modelled bromine isotope variations during salt precipitation processes. At 20 °C, the total Br isotope variations were found to be between −0.45‰ and + 0.21‰. These variations are much smaller than bromine isotope variations observed in natural samples. Therefore, Gao and Liu (2021) suggest that other processes must be responsible to fractionate bromine isotopes to the extent as observed in nature. Because larger isotope fractionation is expected when equilibrium between species with different oxidation states takes place, the larger bromine isotope variations in nature can possibly be explained by a change of oxidation state. However, Gao and Liu (2021) state that fractionation factors increase with increasing Cl/Br ratio, and the highest ratio they simulated was 124. The Cl/Br ratio in the first halite that precipitates from a seawater brine is however approximately 600, so that it may be possible that if higher Cl/Br ratios could be simulated the fractionation factors that are obtained are much closer to the bromine isotope fractionation found in nature, suggesting that it may potentially not be necessary to explain the variations with changes in oxidation state, but just with an increase of the ratio between chlorine and bromine in the salt that precipitates.

Very recently Zhang et al. (2024) estimated the chlorine isotope fractionation using density functional theory for a large array of chloride salts and complex ions at temperature ranges between 100 and 900 °C. The obtained values for the fractionation of salt minerals agreed with the experimental observations that progressive precipitation of halite from brine lowers the $\delta^{37}Cl$ value of the residual fluid until the formation of potassium and magnesium chlorides. The results of this study were used to better understand ore-forming processes, both at lower and at higher temperatures.

8.4.3 Is Chlorine Isotope Fractionation Between Halite and Brine Possibly Dependent on the Seawater Chemistry?

In a recent publication, Eastoe and Meng (2024) commented on their observations from various salt deposits where the observed $\delta^{37}Cl$ range fell outside that range that was to be expected from published fractionation factors between NaCl, KCl and $MgCl_2$ and their saturated brines. They were able to explain these anomalous $\delta^{37}Cl$ values with the suggestion that NaCl that precipitates from very potassium rich solutions does so with an isotope fractionation that might be as low as 0.9996 ($\varepsilon^{37}Cl = -0.4‰$). Eastoe and Meng (2024) suggest that this large variation in potential fractionation factors is essential in order to explain $\delta^{37}Cl$ variations in salt deposits, when it is accepted that no secular changes in the $\delta^{37}Cl$ of the oceans existed during at least the Phanerozoic. Eastoe and Meng (2024) conclude with a statement that the dependence of the halite precipitation fractionation on the composition of the brine is suggested as an explanation, but one that conflicts with experimental data obtained from evaporation of synthetic Cambrian-Silurian seawater by Eggenkamp et al. (2019c). The cycles of halite precipitation fractionation may in part be analogous to cycles observed in the distribution coefficient of bromide between seawater and basal evaporite halite (Siemann and Schramm 2000, 2002; Siemann (2003). Variations in the halite precipitation fractionation may affect the $\delta^{37}Cl$ values in lacustrine evaporites and marine settings in which surface halite dissolves into inflowing seawater. It is hoped that future experimental fractionation determinations will be able to prove or disprove this hypothesis, and in the case it is disproven, that an acceptable explanation is found to explain the $\delta^{37}Cl$ values that fall outside the range expected for the determined fractionation factors.

8.5 Experimental Determination of Chlorine Isotope Fractionation During the Formation of Atacamite

Chlorine isotope data on atacamite [$Cu_2Cl(OH)_3$] are sparse, but those available span a large range from −3 to +6‰ (Eggenkamp and Schuiling 1995; Arcuri and Brimhall 2003; Reich et al. 2019). This large range indicates that supergene oxidation processes can result in significant fractionation of chlorine isotopes. In order to get a better constraint on the chlorine isotope variation in supergene oxidation zones Reich et al. (2019) measured the $\delta^{37}Cl$ of a large set of atacamite samples and reported also the first experimental determination of chlorine isotope fractionation between atacamite and water. To approximate the isotope fractionation factor between atacamite and water, Reich et al. (2019) synthetized atacamite following the procedures by Sharkey and Lewin (1971) and Le Roux et al. (2016). They prepared synthetic atacamite crystals at room temperature (23 °C) using the calcite replacement method, which involved adding 1 g of powdered calcite to 1 L of 0.1 M $CuCl_2$ solution, which was left to stir for 24 h to complete the replacement reaction. The formed atacamite crystals were washed with 0.5 M $Mg(NO_3)_2$ to remove excess Cu^{2+} and Cl^- from the mineral surface. $AgNO_3$ was added to the decanted wash water to check for the presence of Cl^- by precipitation of AgCl. The atacamite was rinsed with $Mg(NO_3)_2$ until it was chloride free. Salts were removed by rinsing with ultrapure deionized water, the crystals were dried overnight (at 50 °C) and gently crushed into a powder. The powder was analysed and showed to be roughly a 50:50 mixture of atacamite and its polymorph bollatackite. Both the chlorine isotope composition of the powder and the excess $CuCl_2$ solution were measured to determine the isotope fractionation factor. Reich et al. (2019) observed that the $\delta^{37}Cl$ of the atacamite powder was +0.25 ± 0.07‰ and of the excess $CuCl_2$ solution it was −0.50 ± 0.14‰. Overall this led to a $10^3\ln\alpha_{(atacamite-solution)}$ ($\varepsilon^{37}Cl$) of +0.75 ± 0.13‰.

8.6 Volcanic Systems

It was found that, under certain circumstances extreme $\delta^{37}Cl$ values were found in samples taken from volcanic systems. Eggenkamp (1994) for example recorded a $\delta^{37}Cl$ value as high as +9.5‰ in Giggenbach bottles with fumaroles collected from Lewotolo Volcano on Lomblen Island, Indonesia. At the main fumarole of Poás Volcano, Costa Rica, Sharp et al. (2010) also found very high $\delta^{37}Cl$ which even could be as high as +12‰. On the other hand, most chlorine isotope observations related to volcanic systems seem to have rather modest $\delta^{37}Cl$ values that are within 1‰ from SMOC (e.g. Eggenkamp 1994; Barnes et al. 2008, 2009).

8.6.1 Chlorine Isotope Fractionation Between $HCl_{(g)}$ and $Cl^-_{(aq)}$

Normally equilibrium fractionation between $HCl_{(g)}$ and $Cl^-_{(aq)}$ is expected to be the main fractionation process in volcanic systems. The fractionation factor for this system has been calculated as +1.7‰ at about 60 °C and is expected to be decreasing both at lower and higher temperatures by Schauble et al. (2003) while Czarnacki and Hałas (2012), taking into account the hydration of the chloride ion, even predict that the fractionation is at most 1.55‰ at about 60 °C while further showing the same characteristics. Sharp et al. (2010) conducted experiments in which they equilibrated $HCl_{(g)}$ with a 1 M HCl solution. They analysed the $\delta^{37}Cl$ of the gaseous hydrochloric acid as well as the solution and found that the difference in $\delta^{37}Cl$ between the two phases at 50 °C was 1.6‰, which is very close to the theoretically obtained value for this system. These fractionation data can explain the generally small variations in $\delta^{37}Cl$ values, but not directly the extreme data that were observed by Eggenkamp (1994) and Sharp et al. (2010) in some volcanic samples.

8.6.2 Chloride Isotope Fractionation During Ammonium Chloride Formation

Hoering and Parker (1961) conducted various experiments to determine and understand potential chlorine isotope fractionation in a period that no natural variations had ever been detected. One of these experiments consisted of the determination of the fractionation between chlorine isotopes in the process of the formation of ammonium chloride in equilibrium with hydrogen chloride in volcanic systems. They measured chlorine isotope data from the reaction:

$$NH_4{}^{37}Cl_{(s)} + H{}^{35}Cl_{(g)} \Leftrightarrow NH_4{}^{35}Cl_{(s)} + H{}^{37}Cl_{(g)} \qquad (8.9)$$

at 200 °C and found a fractionation factor α of 1.0009 ± 0.0003 which indicates that solid ammonium chloride is about 0.9‰ lighter than gaseous hydrogen chloride when it is precipitated in equilibrium with gaseous hydrogen chloride. Eggenkamp (1994) reported a $\delta^{37}Cl$ value of ammonium chloride sampled from the Vesuvius volcano in Italy of −4.9‰. Assuming an initial $\delta^{37}Cl$ of the hydrogen chloride of 0‰ such a low value could only be reached after 99.85% of all hydrogen chloride has precipitated as ammonium chloride when we assume that the salt is formed according to a simple Rayleigh type of fractionation. Considering this very negative value a different method of formation can not be excluded.

8.6.3 Kinetic Fractionation Due to Hydrogen Chloride Escaping from an Acid Solution

Sharp et al. (2010) conducted experiments to obtain information on the origin of the extreme $\delta^{37}Cl$ values they observed in samples from the Poás volcano fumarole in Costa Rica. They bubbled dry air through an Erlenmeyer flask filled with hydrochloric acid in concentrations between 0.6 and 2.9 M at temperatures of at most 95 °C for 10 to 20 min, and collected the hydrogen chloride that escaped from this solution in a second Erlenmeyer flask. They observed that the $\delta^{37}Cl$ of the hydrogen chloride in this second flask had highly increased values (+7.9 to +9.6‰, compared to +0.8‰ for the original hydrochloric acid solution). When this experiment was conducted with fuming hydrogen chloride (11.6 M) hardly any measurable fractionation was observed. Equilibrium in this system was not explained as equilibrium between liquid and gas phases with the same chemical bonds. The gas phase consists of diatomic hydrogen chloride molecules, whereas the chlorine in solution consists of the solvated chloride ion. Because the hydrogen-chloride bond in the diatomic molecule is much stronger than the loosely bound solvated chloride ion in aqueous solution, the heavy isotope ^{37}Cl tends to be preferentially partitioned into the hydrogen chloride gas (i.e., the stronger bond). Sharp et al. (2010) were able to explain the large isotope variations in this system through the occurrence of a Rayleigh type fractionation, where the very first fraction of hydrogen chloride that is transferred to the second flask has the most extreme $\delta^{37}Cl$ which was found in their fumarole experiment. Very important in this system is the fact that the evaporating hydrochloric acid solution is continuously condensing on the walls of the system, so that most hydrogen chloride returns to the original solution and only a small part is transferred to the fumarole outlet. They explained the observation that no fractionation occurred when the experiment was conducted with fuming hydrogen chloride solution with the fact that in this solution hydrogen chloride in solution consists overwhelmingly of non-dissociated HCl molecules in solution so that the same type of chemical bond is present in both the solution and the vapour. To understand the fractionation during simple evaporation they conducted different experiment where 1 litre of a 1 M hydrochloric acid solution was left to evaporate until only 0.1 litre of solution was left. The $\delta^{37}Cl$ of the hydrochloric acid solution increased during this evaporation process to +10.3‰ vs. SMOC, indicating that in this case the *light* isotope ^{35}Cl was escaping preferentially with a fractionation ($10^3 \ln\alpha$, $\varepsilon^{37}Cl$) of about −4‰.

The effect of evaporation from a hyperacid brine on the chlorine isotope composition of the residue has been studied by Rodriguez et al. (2018). They studied evaporation from a hyperacid (pH 0.1) sulphate-chloride brine water from the active lake of Kawah Ijen volcano (Indonesia). Absence of other sulphates and halides despite the high load of dissolved elements (initial TDS ca. 100 g/kg) can be attributed to increased solubility of metals, promoted by extensive formation of complexes between the variety of cations and the major anions (HSO_4^-, Cl^-, F^-) present. Chlorine deviations from a conservative behaviour point to losses of gaseous hydrogen chloride ($HCl_{(g)}$) during evaporation and consequently an increase in Br/Cl ratios. Chlorine isotope fractionation that accompanied the escape of $HCl_{(g)}$ showed a marked change in sign and magnitude in the course of progressive evaporation of the brine. The calculated factor of fractionation between $HCl_{(g)}$ and dissolved Cl for the initial interval (before 500 h) is positive ($1000\ln\alpha_{HCl(g)\text{-}Cldiss}$ = +1.55 ± 0.49‰ to +3.37 ± 1.11‰), indicating that, at first, the escaping chloride was isotopically heavier than the dissolved Cl remaining in the brine. Conversely, fractionation shifted to the opposite direction in the subsequent interval ($1000\ln\alpha_{HCl(g)\text{-}Cldiss}$ = −3.96 ± 0.08‰ to −5.62 ± 0.17‰), values that are in agreement with values reported previously (see above). It is proposed that Cl isotopic fractionation in highly acidic brines is controlled by the distribution of dissolved chlorine species, which changes from Cl^- to HCl^o dominance with progressive pH decrease.

8.7 High Temperature Equilibrium Fractionation in Rocks and Minerals

Sharp et al. (2010) determined the equilibrium isotope fractionation for the system:

$$nepheline + NaCl \Rightarrow sodalite \quad (8.10)$$

They reacted nepheline with a large excess of sodium chloride at a temperature of 825 °C. All nepheline reacted to sodalite in this reaction and the chlorine isotope composition of both the sodalite and the excess sodium chloride was determined. The $10^3\ln\alpha$ for this reaction (sodalite-NaCl) was observed to be −0.3 ± 0.1.

8.7.1 First-Principles Estimates of High Temperature Chlorine Isotope Fractionation

The first calculations on chlorine isotope fractionation at high temperatures were presented by Schauble and Sharp (2011). They reported a −0.7‰ fractionation between sodalite and HCl at 950 K and a −0.02‰ fractionation between sodalite and $NaCl_{(c)}$ at 1098 K. Small, negative values that agreed reasonably well with the experimental data obtained by Sharp et al. (2010). Later, Balan et al. (2019) presented first-principle models for a larger range of minerals, including rock-forming minerals. They presented β-factors for some rock forming minerals, forsterite, lizardite, anthrophyllite, sodalite and chlorapatite. Further they also presented β-factors for a large number of salt minerals, chlorine gas and hydrogen chloride. With the use of these β-factors it is possible to calculate, in principle, the isotope fractionation at various temperatures. A new set of first-principles calculations were published later by Liu et al. (2021). They presented β-factors for apatite group minerals, muscovite, phlogopite, tremolite and marialite. Liu et al. (2021) also took pressure effects into account in their calculations since they wanted to estimate chlorine isotope variations in the mantle too. Based on their calculations an average $\delta^{37}Cl$ of the mantle could be estimated. The $\delta^{37}Cl$ value for the bulk mantle was estimated to be in the range −0.69‰ to −0.37‰ with a mean value of −0.53 ± 0.16‰.

8.7.2 Chlorine Isotope Fractionation During Crystallisation of Sodalite and Eudialyte

Based on a large set of $\delta^{37}Cl$ measurements of the chlorine rich minerals sodalite ($Na_8Al_6Si_6O_{24}Cl_2$), tugtupite ($Na_8Be_2Al_2Si_8O_{24}Cl_2$), and eudialyte ($Na_{15}Ca_6Fe_3Zr_3Si_{26}O72(O,OH,H_2O)_4(Cl,F,OH)_2$) from a single, well understood, magmatic intrusion (Ilímaussaq, South Greenland), Eggenkamp et al. (2022) attempted to estimate chlorine isotope fractionation during crystallisation of these minerals. The total $\delta^{37}Cl$ variations in their dataset were between +0.1 and + 0.4‰ for sodalite (N = 16), 0 and + 0.3‰ for tugtupite (N = 6) and + 0.2 and + 1.7‰ for eudialyte (N = 31). Based on these data Cl isotope fractionation ($10^3\ln\alpha$ values) between minerals, melts and fluids in this well-defined natural system could be estimated. Cl isotope fractionation between magmatic sodalite and melt is slightly negative, at about −0.05‰, and decreases to about −0.1‰ at hydrothermal conditions. Cl isotope fractionation between tugtupite and hydrothermal fluid is slightly larger, at about −0.2‰. Magmatic eudialyte has $\delta^{37}Cl$ values that are approximately 0.35‰ higher than magmatic sodalite has, suggesting a Cl isotope fractionation between eudialyte and melt of about +0.3‰.

8.7.3 Bromine Isotope Fractionation During Crystallisation of Sodalite and Tugtupite

Based on a small set of sodalite ($Na_8Al_6Si_6O_{24}Cl_2$) and tugtupite ($Na_8Be_2Al_2Si_8O_{24}Cl_2$) samples Eggenkamp et al. (2021) were able give a first rough estimate of the bromine isotope fractionation during crystallisation of these minerals. Sodalite and tugtupite are chloride rich zeolite type tectosilicates. Sodalite is a magmatic mineral, but can also form at low temperatures under hydrothermal conditions. Tugtupite, a beryllium containing sodalite group mineral only forms under hydrothermal conditions when the beryllium content in the hydrothermal fluid is high enough (Markl 2001; Markl et al. 2001). δ^{81}Br values for magmatic samples are lower than for hydrothermal samples, suggesting a fractionation effect during the evolution of the bromide in the process from magmatic to hydrothermal crystallisation. Increasing δ^{81}Br values suggest negative fractionation between sodalite and melt in equilibrium. δ^{81}Br of magmatic sodalite is approximately +0.2‰. Because mineral-melt equilibrium can be compared to salt-brine equilibrium (see Sect. 8.4.1) it can be estimated that the isotope fractionation effect between a saturated solution and salt can be approximated to sodalite in equilibrium with melt too. In this way, bromine isotope fractionation ($10^3\ln\alpha_{\text{sodalite-melt}}$) between sodalite and melt can be estimated as −0.3‰, which indicates that δ^{81}Br of the melt is about +0.5 to +0.6‰. This would, if the melt is representative for the lithospheric mantle, also represent a possible lithospheric mantle value.

8.8 Fractionation Due to Acid Displacement in Aerosols

8.8.1 Observations of Chlorine Isotope Variations in Aerosols

When aerosols consisting of primary sodium chloride from sea spray react with atmospheric acid particles chlorine may escape from the aerosols as hydrogen chloride molecules. This process was already recognised a considerable time ago (Eriksson 1959). During this process the chloride ions in the aerosols are displaced with sulphate and nitrate ions (acid displacement). When they studied bulk aerosols Volpe and Spivack (1994) discovered that the δ^{37}Cl values of the aerosols show a positive correlation with the so called non-sea-salt sulphate concentrations and the chloride deficiency. Chloride deficiency is defined as the Na/Cl ratio in the aerosol as compared to the Na/Cl ratio in seawater. This effect could be explained through a kinetic isotope fractionation which was associated with the volatilisation of hydrogen chloride. From experiments that Volpe and Spivack (1994) conducted and which they compared with the data measured in natural aerosols it was possible to calculate an experimental fractionation factor for this effect of 0.9969. In a second paper on chlorine isotope variations from marine aerosols Volpe et al. (1998) sampled aerosols with different grain sizes and found differences between aerosol groups of different sizes. They observed that in smaller aerosols δ^{37}Cl increases with an increasing chloride deficiency, while in larger aerosols it decreases with increasing chloride deficiency. The effect that is seen in the smaller aerosols is the same effect as Volpe and Spivack (1994) observed in their study on bulk aerosols from Bermuda. In the larger aerosols Volpe et al. (1998) were able to calculate a relationship between the δ^{37}Cl and the chloride deficiency. They concluded that when during acid replacement reactions chloride in the particles is replaced with sulphate and nitrate the chloride is released in the form of isotopically heavy hydrogen chloride from the particles. The related isotope fractionation with this second process was either 1.00303 ± 0.0005 if a first order mechanism was assumed or 1.00280 ± 0.0005 when a zeroth order mechanism was assumed. Considering the scatter in the data it was not possible to decide which mechanism actually occurred.

8.8.2 Experimental Determination on Hydrogen Chloride Escape in the Reaction of Sodium Chloride with Sulphuric Acid

To test the hypothesis that the δ^{37}Cl of sodium chloride decreases as a result of escape of hydrogen chloride from a solution containing sodium chloride and sulphuric acid, Eggenkamp (1994) conducted the following experiment: A solution containing 0.2 M sodium chloride and 0.1 M sulphuric acid was prepared. This solution was evaporated to dryness at 50 °C and the residue was examined for its chloride isotope composition. Unlike the original sodium chloride, the dry residue was birefringent, and the chloride content had decreased from about 60 to 15%. Evidently, part of the chloride in the solution was evaporated due to the reaction:

$$2NaCl + H_2SO_4 \Rightarrow Na_2SO_4 + 2HCl \uparrow \quad (8.11)$$

The $\delta^{37}Cl$ of the remaining salt was 0.52 ± 0.05‰ higher than that of the original sodium chloride. This showed that $\delta^{37}Cl$ increases when part of the chloride escapes as hydrogen chloride gas. The result was used to make a rough estimate of the fractionation factor. A Rayleigh fractionation curve was drawn through the point where three quarters of chloride is removed with a total fractionation of +0.52‰. The fractionation factor ($\alpha_{gas\text{-}solution}$) based on this single measurement was about 0.99963 ($\varepsilon^{37}Cl_{(gas\text{-}solution)}$ = −0.37‰). Although considerably smaller than the value obtained by Volpe et al. (1998) it was clear that the same effect had taken place, and the different value obtained may be caused by the fact that in this case only a single, fairly rough experiment was conducted. The process that occurred in this simple experiment is comparable to the processes described by Sharp et al. 2010 and Rodríguez et al. (2018). In these studies it was suggested that during the first stage $\delta^{37}Cl$ of the residue increased and during the last stage decreased. This effect was the result of a change in the chloride species that evaporated, being dominated of chloride ions in the first stage and neutral HCl molecules in the last stage. The relatively modest overall fractionation obtained in this experiment can thus possibly be explained as being the sum of both processes, exactly as observed later in the evaporation experiment conducted by Rodríguez et al. (2018).

8.9 Chlorine Isotope Fractionation During Sublimation under Lunar Conditions

While the $\delta^{37}Cl$ variation on Earth is very modest, with the overwhelming majority of data between −2.0 and + 2.0‰, and only very rarely values beyond −10 and + 10‰, the variability in lunar samples is much larger. Sharp et al. (2010) showed that the $\delta^{37}Cl$ variation in lunar basalt and glass was at least 25‰ (from −1 to +24‰), while in a later study Wang et al. (2019) showed that the $\delta^{37}Cl$ of apatite from lunar meteorites could be as high as +81‰. These highly positive $\delta^{37}Cl$ values were explained from the volatilization of metal chlorides (such as NaCl, KCl and $FeCl_2$) under anhydrous circumstances (Sharp et al. 2010), where light chloride evaporated from the salt minerals. In order to understand the fractionation that took place during this process, Han et al. (2023) conducted NaCl and KCl sublimation experiments at various temperatures and pressures in an N_2 atmosphere to investigate the chlorine isotope fractionation behaviour. Han et al. (2023) conducted sublimation experiments on NaCl and KCl at pressures of $7*10^{-5}$ bar, 10^{-2} bar and 1 bar at temperatures (943 to 1061 K, 670 to 788 °C for NaCl and 945 to 1020 K, 672 to 747 °C for KCl) below the respective melting points of NaCl (1074 K, 801 °C) and KCl (1043 K, 770 °C). They observed that for NaCl the chlorine isotope fractionation factor ($\alpha_{gas\text{-}solid}$) was 0.9985 ± 0.0002 ($\varepsilon^{37}Cl$ = −1.5 ± 0.2‰) at 1 bar, 0.9958 ± 0.0004 ($\varepsilon^{37}Cl$ = −4.2 ± 0.4‰) at 10^{-2} bar and 0.99807 ± 0.00004 ($\varepsilon^{37}Cl$ = −1.93 ± 0.04‰) at $7*10^{-5}$ bar pressure. For KCl they observed chlorine isotope fractionation factors ($\alpha_{gas\text{-}solid}$) of 0.9988 ± 0.0003

($\varepsilon^{37}Cl = -1.2 \pm 0.3‰$) at 1 bar, 0.9972 ± 0.0003 ($\varepsilon^{37}Cl = -2.8 \pm 0.3‰$) at 10^{-2} bar and 0.9989 ± 0.0001 ($\varepsilon^{37}Cl = -1.1 \pm 0.1‰$) at $7*10^{-5}$ bar pressure. These results showed that chlorine isotope fractionation during sublimation of NaCl is larger than during sublimation of KCl, and surprisingly that the chlorine isotope fractionation under very low and very high pressure conditions is less than under intermediate pressure conditions.

8.10 Equilibrium Fractionation in Systems with Lesser Geologic Relevance

8.10.1 Equilibrium Fractionation Between Chloride and Hexchloroplatinate

Although not of geologic relevance Hoering and Parker (1961) also conducted an experiment determining the chlorine isotope fractionation in the reaction between platinum and chloride in which hexachloroplatinate ions are formed:

$$\left(Pt^{35}Cl_6\right)^{2-} + 6^{37}Cl^- \leftrightarrow \left(Pt^{37}Cl_6\right)^{2-} + 6^{35}Cl^- \qquad (8.12)$$

This reaction was studied to check their experimental procedures and to show that relatively small variations in the ratio of chlorine isotopes could indeed be detected by their analytical method. This experiment obtained an equilibrium fractionation factor α of 1.0009 ± 0.0003 ($\varepsilon^{37}Cl = 0.9 \pm 0.3‰$) at 23 °C, indicating that ^{37}Cl concentrates in the hexachloroplatinate formed in the reaction.

8.10.2 Equilibrium Between Hydrogen Chloride Gas and Chlorides Dissolved in Glacial Acetic Acid

Howald (1960) studied chlorine isotope fractionation between hydrogen chloride gas in equilibrium with chloride solutions in glacial acetic acid (anhydrous CH_3COOH). The purpose of this study was to understand the behaviour of ion-pairs in solutions of a low dielectric constant. The dielectric constant of glacial acetic acid is about 6.15 as compared to 80.37 for water at 20 °C (Maryott and Smith 1951). The experiments were done between hydrogen chloride in the gas phase and HCl, H_3OCl, $HgCl_2$, LiCl and $SrCl_2$ dissolved in glacial acetic acid. From the experiments it appeared that at higher salt concentrations in the acetic acid the isotope effect is negative, i.e. the $^{37}Cl/^{35}Cl$ ratio in the solution is lower than in the hydrogen chloride gas above, while at low concentrations the hydrogen chloride gas is heavier than the salt in solution. From the experimental data that were obtained Howald (1960) calculated the following isotope fractionation factors α, LiCl; 1.0040 ± 0.0003 ($\varepsilon^{37}Cl = 4.0 \pm 0.3‰$), $SrCl_2$; 1.0042 ± 0.0003 ($\varepsilon^{37}Cl = 4.2 \pm 0.3‰$), H_3OCl; 1.0039 ± 0.0003 ($\varepsilon^{37}Cl = 3.9 \pm 0.3‰$), $HgCl_2$; 1.0061 ± 0.0009 ($\varepsilon^{37}Cl = 6.1 \pm 0.9‰$) and HCl; 1.0059 ± 0.0002 ($\varepsilon^{37}Cl = 5.9 \pm 0.2‰$). The results indicate that ionic species, which exist as free chloride ions in glacial acetic acid (LiCl, $SrCl_2$, H_3OCl) have lower fractionation factors than more covalent species which are present as molecules in glacial acetic acid ($HgCl_2$, HCl).

8.11 Extreme Chlorine Equilibrium Isotope Fractionation at the Nano-Scale

While at macro scales chlorine isotope fractionation normally is relatively modest, Pokora et al. (2022) observed that at nano scales, especially at sub-van der Waals distances, extremely large chlorine isotopic fractionation can exist. They studied the influence of confinement on the chlorine isotope fractionation in a number of nanotubes with different lengths and radius and in fullerenes. Graphene was used as reference without confinement and consequently little isotope fractionation. Pokora et al. (2022) found that the chlorine isotope equilibrium fractionation of chloride trapped in different nanostructures show surprisingly large values which are related to the distance to the surrounding environment. The chlorine isotope fractionation depends on the distance of the chloride ion to the nearest atom of the nanostructure. The isotope fractionation factor α can be as high as 1.045 ($\varepsilon^{37}Cl$ = 45‰) at small distances (<1.6 Å). At larger distances the chlorine isotope fractionation ε decreases to values close to 0 for distances larger than 3 Å.

References

Agrinier P, Destrigneville C, Giunta T, Bonfacie M, Bardoux G, Andre J, Lucazeau F (2019) Strong impact of ion filtration on the isotopic composition of chlorine in young clay-rich oceanic sediment pore fluids. Geochim Cosmochim Acta 245:525–541

Agrinier P, Bonifacie M, Bardoux G, Lucazeau F, Giunta T, Ader M (2021) Chlorine isotope data of chlorides challenge the pore fluid paradigm. Geochim Cosmochim Acta 300:258–278

Alder BJ, Alley WE, Dymond JH (1974) Studies in molecular dynamics. XIV. Mass and size` dependence of the binary diffusion coefficient. J Chem Phys 61:1415–1420

Appelo CAJ, Postma D (2005) Geochemistry, groundwater and pollution.2nd edn. Balkema, Leiden

Arcuri T, Brimhall GH (2003) The chloride source for atacamite mineralization at the Radomiro Tomic porphyry copper deposit, Northern Chile. Econ Geol 98:1667–1681

Aston FW (1919) The constitution of the elements. Nature 104:393

Aston FW (1920a) Comment to "the separation of the element chlorine into normal chlorine and meta-chlorine, and the positive electron" by Harkins. Nature 105:231

Aston FW (1920b) The separation of the element chlorine into normal chlorine and meta-chlorine, and the positive electron. Withdrawal of comment. Nature 106:375

Atteberry RW, Boyd GE (1950) Separation of seventh group anions by ion-exchange chromatography. J Am Chem Soc 72:4805–4806

Balan E, Créon L, Sanloup C, Aléon J, Blanchard M, Paulatto L, Bureau H (2019) First-principles modeling of chlorine isotope fractionation between chloride-bearing molecules and minerals. Chem Geol 525:424–434

Barnes JD, Sharp ZD, Fischer TP (2008) Chlorine isotope variations across the Izu-Bonin-Mariana arc. Geology 36:883–886

Barnes JD, Sharp ZD, Fisher TP, Hilton DR, Carr MJ (2009) Chlorine isotope variations along the Central American volcanic front and back arc. Geochem Geophys Geosys 10:Q11S17

Beekman HE, Eggenkamp HGM, Appelo CAJ (2011) An integrated modelling approach to reconstruct complex solute transport mechanisms – Cl and $\delta^{37}Cl$ in pore water of sediments from a former brackish lagoon in The Netherlands. Appl Geochem 26:257–268

Bird RB, Steward WE, Lightfoot EN (2002) Transport phenomena.2nd edn. John Wiley & Sons

Bourg IC, Sposito G (2007) Molecular dynamics simulations of kinetic isotope fractionation during the diffusion of ionic species in liquid water. Geochim Cosmochim Acta 71:5583–5589

Bourg IC, Richter FM, Christensen JM, Sposito G (2010) Isotopic mass dependence of metal cation diffusion coefficients in liquid water. Geochim Cosmochim Acta 74:2249–2256

Brewer AK, Madorsky SL, Taylor JK, Dibeler VH, Bradt P, Parham OL, Britten RJ, Reid JG (1947) Concentration of isotopes of potassium by the counter-current electromigration method. J Res Natl Bur Stand 38:137–168

Brönsted JN, Hevesy G (1921) The separation of the isotopes of chlorine. Nature 107:619

Cameron AE, Herr W, Herzog W, Lundén A (1956) Isotopen-Anreicherung beim Brom durch electrolytische Überführung in geschmolzenem Bleibromid. Z Naturforschng 11a:203–205

Campbell DJ (1985) Fractionation of stable chlorine isotopes during transport through semipermeable membranes. MS. thesis. University of Arizona

Carslaw HS, Jaeger JC (1959) Conduction of heat in solids.2nd edn. Oxford University Press, London

Chapman DL (1920) The separation of the isotopes of chlorine. Nature 106:9

Columbia University (1942) Special report for meetings of executive Committee of Section S-1, Oct 24, 1942. MW Kellogg Co, Industrial Application of the Diffusion Method, OSRD

Crank J (1956) The mathematics of diffusion. Oxford University Press, p 347

Czarnacki M, Hałas S (2012) Isotope fractionation in aqua-gas systems: Cl2-HCl-Cl-, Br2-HBr-Br- and H2S-S2-. Isot Environ Health Stud 48:55–64

Desaulniers DE, Cherry JA, Fritz P (1981) Origin, age and movement of pore water in argillaceous quaternary deposits at four sites in southwestern Ontario. J Hydrol 50:217–231

Desaulniers DE, Kaufmann RS, Cherry JA, Bentley HW (1986) ^{37}Cl-^{35}Cl variations in a diffusion controlled groundwater system. Geochim Cosmochim Acta 50:1757–1764

Deyhle A, Kopf A, Frape S, Hesse R (2003) Evidence for fluid flow in the Japan trench forearc using isotope geochemistry (Cl, Sr, B): Results from ODP site 1150. Island Arc 13:258–270

Duursma EK, Hoede C (1967) Theoretical, experimental and field studies concerning molecular diffusion of radioisotopes in sediments and suspended solid particles of the sea. Part A: theories and mathematical calculations. Neth J Sea Res 3:423–457

Eastoe CJ, Meng FW (2024) Is Cl isotope fractionation in the seawater-evaporite system dependent on seawater chemistry? Chem Geol 644:121832

Eastoe CJ, Long A, Knauth LP (1999) Stable chlorine isotopes in the Palo Duro Basin, Texas: evidence for preservation of Permian evaporite brines. Geochim Cosmochim Acta 63:1375–1382

Eggenkamp HGM (1994) δ^{37}Cl; the geochemistry of chlorine isotopes. Geol Ultraiect 116:1–150. Thesis Utrecht University

Eggenkamp HGM (2014) The geochemistry of stable chlorine and bromine isotopes. Springer

Eggenkamp HGM (2015a) Comment on "Stable isotope fractionation of chlorine during the precipitation of single chloride minerals" by Luo, C.-G., Xiao, Y.-K., Wen, H.-J., Ma, H.-Z., Ma, Y.-Q., Zhang, Y.-L., Zhang, Y.-X. and He, M.-Y. [Applied Geochemistry 47 (2014) 141–149]. Appl Geochem 54:111–116

Eggenkamp HGM (2015b) Why are variations in bromine isotope compositions in the Earth's history larger than chlorine isotope compositions? Ann UMCS Section AAA 70:187–196

Eggenkamp HGM, Coleman ML (1997) Comparison of chlorine and bromine isotope fractionation in diffusion: geochemical consequences. 7th Ann Goldschmidt Conf, Tucson, AZ, p 213

Eggenkamp HGM, Coleman ML (2009) The effect of aqueous diffusion on the fractionation of chlorine and bromine stable isotopes. Geochim Cosmochim Acta 73:3539–3548

Eggenkamp HGM, Schuiling RD (1995) δ^{37}Cl variations in selected minerals: a possible tool for exploration. J Geochem Explor 55:249–255

Eggenkamp HGM, Middelburg JJ, Kreulen R (1994) Preferential diffusion of ^{35}Cl relative to ^{37}Cl in sediments of Kau Bay, Halmahera, Indonesia. Chem Geol (Isot Geosci Sect) 116:317–325

Eggenkamp HGM, Kreulen R, Koster van Groos AF (1995) Chlorine stable isotope fractionation in evaporites. Geochim Cosmochim Acta 59:5169–5175

Eggenkamp HGM, Bonifacie M, Ader M, Agrinier P (2016) Experimental determination of stable chlorine and bromine isotope fractionation during precipitation of salt from a saturated solution. Chem Geol 433:46–56

Eggenkamp HGM, Louvat P, Griffioen J, Agrinier P (2019a) Chlorine and bromine isotope evolution within a fully developed upper Permian natural salt sequence. Geochim Cosmochim Acta 245:316–326

Eggenkamp HGM, Louvat P, Agrinier P, Bonifacie M, Bekker A, Krupenik V, Griffioen J, Horita J, Brocks JJ, Bagheri R (2019b) The bromine and chlorine isotope composition of primary halite deposits and their significance for the secular isotope composition of seawater. Geochim Cosmochim Acta 264:13–29

Eggenkamp HGM, Louvat P, Agrinier A (2019c) Halogen stable isotope evolution during experimental evaporation of seawater-type brines. ESIR Isotope Workshop XV, June 23–27, 2019, Lublin, Poland. pp 79–81

Eggenkamp HGM, Marks MAW, Louvat P, Markl G (2021) Bromine isotope variations in magmatic and hydrothermal sodalite and Tugtupite and the estimation of Br isotope fractionation between melt and sodalite. Minerals 11:370

Eggenkamp HGM, Marks MAW, Bonifacie M, Bardoux G, Agrinier P, Markl G (2022) Cl isotope fractionation in magmatic and hydrothermal eudialyte, sodalite and tugtupite (Ilímaussaq intrusion, South Greenland). Chem Geol 604:120932

Eriksson E (1959) The yearly circulation of chlorines and sulfur in nature: Meteorological, geochemical and pedological implications. Tellus 1:375–403

Fick A (1855) Über Diffusion. Ann Phys 170:59–86

Fink CG, Urey HC, Lake DB (1934) The diffusion of hydrogen through metals: fractionation of hydrogen isotopes. J Chem Phys 2:105–106

Fortin MA, Watson EB, Stern R (2017) The isotope mass effect on chlorine diffusion in dacite melt, with implications for fractionation during bubble growth. Earth Planet Sci Lett 480:15–24

Gao CH, Liu Y (2021) First-principles calculations of equilibrium bromine isotope fractionations. Geochim Cosmochim Acta 297:65–81

Giunta T, Devauchelle O, Ader M, Locke R, Louvat P, Bonifacie M, Métivier F, Agrinier P (2017) The gravitas of gravitational isotope fractionation revealed in an isolated aquifer. Geochem Perspect Lett 4:53–58

Godon A, Jendrzejewski N, Castrec-Rouelle M, Dia A, Pineau F, Boulègue J, Javoy M (2004) Origin and evolution of fluids from mud volcanoes in the Barbados accretionary complex. Geochim Cosmochim Acta 68:2153–2165

Groen J, Velstra J, Meesters AGCA (2000) Salinization processes in paleowaters in coastal sediments from Suriname: evidence from δ^{37}Cl analysis and diffusion modelling. J Hydrol 234:1–20

Han ZY, Hui HJ, Wei HZ, Li WQ (2023) Isotopic fractionation of chlorine and potassium during chloride sublimation under lunar conditions. Geochim Cosmochim Acta

Harkins WD (1920a) The separation of the element chlorine into normal chlorine and meta-chlorine, and the positive electron. Nature 105:230–231

Harkins WD (1920b) The separation of the element chlorine into chlorine and meta-chlorine. Science 51:289–291

Harkins WD (1920c) The nuclei of atoms and the new periodic system. Phys Rev 15:73–94

Harkins WD, Hall RE (1916) The periodic system and the properties of the elements (Fifth paper on atomic structure). J Am Chem Soc 38:169–221

Harkins WD, Hayes A (1921) The separation of the element chlorine into isotopes (isotopic elements). The heavy fraction from diffusion. J Am Chem Soc 43:1803–1825

Harkins WD, Hayes A (1922) The separation of the element chlorine into isotopes (isotopic atomic species). Phys Rev 19:403–404

Harkins WD, Jenkins FA (1926) The separation of the element chlorine into isotopes. The light fraction. J Am Chem Soc 48:58–69

Harkins WD, Ligget TH (1924) The discovery and separation of the isotopes of chlorine and the whole number rule. J Phys Chem 28:74–82

Harkins WD, Wilson ED (1915) The changes of mass and weight involved in the formation of complex atoms. J Am Chem Soc 37:1367–1383

Hasegawa T, Nakata K (2018) A measurement method for isotope fractionation of ^{35}Cl and ^{37}Cl by a conventional through-diffusion experiment. Chem Geol 483:247–253

Herzog W, Klemm A (1958) Die Temperaturabhängigkeit des Isotopie-Effekts bei der elektrolytischen Wanderungen der Chlorionen in geschmolzenem Thallium(I)-chlorid. Z Naturforschg 13a:7–16

Heumann KG, Hoffmann R (1976) Chlorine isotope effects in ion exchange chromatography. Angew Chem Int Ed Eng 15:55

Hewlett RG, Anderson OE (1962) A history of the United States Atomic Energy Commission. Volume I. the new World, 1939/1946, Chapter 4 Commitment p 98

Hoering TC, Parker PL (1961) The geochemistry of the stable isotopes of chlorine. Geochim Cosmochim Acta 23:186–199

Howald RA (1960) Ion pairs. I. Isotope effects shown by chloride solutions in glacial acetic acid. J Am Chem Soc 82:20–24

Impey RW, Madden PA, McDonald IR (1983) Hydration and mobility of ions in solution. J Phys Chem 87:5071–5083

Jähne B, Heinz G, Dietrich W (1987) Measurement of the diffusion coefficients of sparingly soluble gases in water. J Geophys Res 92:10767–10776

Kaufmann RS (1984) Chlorine in groundwater: stable isotope distribution. Ph.D. thesis. University of Arizona

Klemm A (1943) Isotopenanreicherung bei Diffusion von Kupfer in Silbersulfid. Z Physik Chem 193:29–39

Klemm A, Lundén A (1955) Isotopenanreicherung beim Chlor durch electrolytische Überfürung in geschmolzenem Bleichlorid. Z Naturforschg 10a:282–284

Konstantinov BP, Bakulin EA (1965) Separation of chloride isotopes in aqueous solutions of lithium chloride, sodium chloride and hydrochloric acid. Russ J Phys Chem 39:315–318

Langvad T (1954) Separation of chlorine isotopes by ion-exchange chromatography. Acta Chem Scand 8:526–527

Le Roux SG, Miller JA, Dunford AJ, Clarke CE (2016) The dissolution kinetics of atacamite in the acid range and the stability of atacamite containing soils from Namaqualand, South Africa. Appl Geochem 64:22–29

Lindemann FA (1921) Discussion on isotopes. Proc R Soc London Series A 99:102–104

Liu X, Wei HZ, Li YC, Williams-Jones AE, Lu JJ, Jiang SY, Dong G, Ma J, Eastoe CJ (2021) Chlorine isotope mantle heterogeneity: constraints from theoretical first-principles calculations. Chem Geol 572:120193

Lundén A, Herzog W (1956) Isotopenanreicherung bei Chlor durch electrolytische Überführung in geschmolzenem Zinkchlorid. Z Naturforschg 11a:520

Lundén A, Lodding A (1960) Isotopenanreicherung bei Brom durch electrolytische Überführung in geschmolzenem Zinkbromid. Z Naturforschg 15a:320–322

Luo CG, Xiao YK, Ma HZ, Ma YQ, Zhang YL, He MY (2012) Stable isotope fractionation of chlorine during evaporation of brine from a saline lake. Chin Sci Bull 57:1833–1843

Luo CG, Xiao YK, Wen HJ, Ma HZ, Ma YQ, Zhang YL, Zhang YX, He MY (2014) Stable isotope fractionation of chlorine during the precipitation of single chloride minerals. Appl Geochem 47:141–149

Luo CG, Xiao YK, Wen HJ, Ma HZ, Ma YQ, Zhang YL, He MY (2015) Reply to the comment on the paper "stable isotope fractionation of chlorine during the precipitation of single chloride minerals". Appl Geochem 54:117–118

Madorsky SL, Strauss S (1948) Concentration of isotopes of chlorine by the counter-current electromigration method. J Res Natl Bur Stand 38:185–189

Marchese FT, Beveridge DL (1984) Pattern recognition approach to the analysis of geometrical features of solvation: application to the aqueous hydration of Li^+, Na^+, K^+, F^-, and Cl^-. J Am Chem Soc 106:3713–3720

Markl G (2001) Stability of Na-Be minerals in late-magmatic fluids of the Ilímaussaq alkaline complex, South Greenland. Geol Greenl Surv Bull 190:145–158

Markl G, Marks M, Schwinn G, Sommer H (2001) Phase equilibrium constraints on intensive crystallization parameters of the Ilímaussaq complex, South Greenland. J Petrol 42:2231–2258

Maryott AA, Smith ER (1951) Table of dielectric constants of pure liquids. Natl Bur Stand Circ 514:44

Middelburg JJ, De Lange GJ (1989) The isolation of Kau Bay during the last glaciation: direct evidence from interstitial water chlorinity. Neth J Sea Res 24:615–622

Musashi M, Oi T, Eggenkamp HGM (2004) Experimental determination of chlorine isotope separation factor by anion-exchange chromatography. Anal Chim Acta 508:37–40

Musashi M, Oi T, Eggenkamp HGM, Yato Y, Matsuo M (2007) Anion-exchange chromatographic study of the chlorine isotope effect accompanying hydration. J Chromatogr A 1140:121–125

Nuevo MJ, Morales JJ, Heyes DM (1995) Mass dependence of isotope self diffusion by molecular dynamics. Phys Rev E 51:2026–2032

Phillips FM, Bentley HW (1987) Isotopic fractionation during ion filtration: I. Theory. Geochim Cosmochim Acta 51:683–695

Pokora M, Paneth A, Paneth P (2022) Unprecedently large $^{37}Cl/^{35}Cl$ equilibrium isotopic fractionation on nanoconfinement of chloride anion. Sci Rep 12:1768

Powell DH, Barnes AC, Enderby JE, Neilson GW, Salmon PS (1988) The hydration structure around chloride ions in aqueous solution. Faraday Discuss Chem Soc 85:137–146

Ransom B, Spivack AJ, Kastner M (1995) Stable Cl isotopes in subduction-zone pore waters: implications for fluid-rock reactions and the cycling of chlorine. Geology 23:715–718

Reich M, Barnes JD, Breecker DO, Barra F, Milojevic C, Drew DL (2019) Chlorine isotope fractionation recorded in atacamite during supergene copper oxidation. Chem Geol 525:168–176

Richter FM, Liang Y, Davis AM (1999) Isotope fractionation by diffusion in molten oxides. Geochim Cosmochim Acta 63:2853–2861

Richter FM, Davis AM, DePaolo DJ, Watson EB (2003) Isotope fractionation by chemical diffusion between molten basalt and rhyolite. Geochim Cosmochim Acta 67:3905–3923

Richter FM, Mendybaev RA, Christensen JN, Hutcheon ID, Williams RW, Sturchio NC, Beloso Jr AD (2006) Kinetic isotopic fractionation during diffusion of ionic species in water. Geochim Cosmochim Acta 70:277–289

Richter FM, Watson EB, Mendybaev RA, Teng FZ, Janney PE (2008) Magnesium isotope fractionation in silicate melts by chemical and thermal diffusion. Geochim Cosmochim Acta 72:206–220

Riemann W, Lindenbaum S (1952) Analysis of mixtures of chloride and bromide by ion-exchange chromatography. Anal Chem 24:1199–1200

Rodríguez A, van Bergen MJ, Eggenkamp HGM (2018) Experimental evaporation of hyperacid brines: effects on chemical composition and chlorine isotope fractionation. Geochim Cosmochim Acta 222:467–484

Samoilov OY (1957) A new approach to the study of hydration of ions in aqueous solutions. Disc Faraday Soc 24:141–146

Schauble EA, Rossman GR, Taylor HP (2003) Theoretical estimates of equilibrium chlorineisotope fractionations. Geochim Cosmochim Acta 67:3267–3281

Schauble EA, Sharp ZD (2011) Modeling isotopic signatures of nebular chlorine condensation. In: 21th Annual V.M. Goldschmidt Conference, vol 75. Mineral Mag, Prague, Czech Republic, p 1810

Schreiner F, Fried S, Friedman AM (1982) Measurement of radionuclide diffusion in ocean floor sediment and clay. Nucl Technol 59:429–438

Sharkey JB, Lewin SZ (1971) Conditions governing the formation of atacamite and paratacamite. Am Mineral 56:179–191

Sharp ZD, Shearer CK, McKeagan KD, Barnes JD, Wang YQ (2010) The chlorine isotope composition of the Moon and implications for an anhydrous mantle. Science 329:1050–1053

Siemann MG (2003) Extensive and rapid changes in seawater chemistry during the Phanerozoic: evidence from Br contents in basal halite. Terra Nova 15:243–248

Siemann MG, Schramm M (2000) Thermodynamic modelling of the Br partition between aqueous solutions and halite. Geochim Cosmochim Acta 64:1681–1693

Siemann MG, Schramm M (2002) Henry's and non-Henry's law behavior of Br in simple marine systems. Geochim Cosmochim Acta 66:1387–1399

Soddy F (1920) The separation of the isotopes of chlorine. Nature 105:516

Spivack AJ, Kastner M, Ransom B (2002) Elemental and isotopic chloride geochemistry and fluid flow in the Nankai trough. Geophys Res Lett 29:1661–1665

Strydom J, Sterpenich J, Grgic D, Richard A, Eggenkamp HGM, Agrinier P, Louvat P, Mosser-ruck R, Gaire P, Gaucher E (2022) Experimental study of chemical evolution and isotope fractionation of Cl and Br in pore water expelled during strong clay compaction. Appl Geochem 140:105274

Tabor D (1991) Gases, liquids and solids.3rd edn. Cambridge University Press

Tan HB, Ma HZ, Xiao YK, Wei HZ, Zhang XY, Ma WD (2005) Characteristics of chlorine isotope distribution and analysis on sylvinite deposit formation based on ancient salt rock in western Tarim basin. Sci China (ser. D) 48:1913–1920

Tan HB, Ma HZ, Wei HZ, Xu JX, Li TW (2006) Chlorine, sulfur and oxygen isotopic constraints on ancient evaporite deposit in the Western Tarim Basin, China. Geochem J 40:569–577

Tan HB, Ma HZ, Zhang XY, Xu JX, Xiao YK (2009) Fractionation of chlorine isotope in salt mineral sequences and application: research on sedimentary stage of ancient salt rock deposit in Tarim Basin and western Qaidam Basin. Acta Petrol Sinica 25:955–962. (in Chinese with English abstract)

Thomson JJ (1913) Positive rays of electricity. Longmans, Green and Co, London, pp 114–115

Volpe C, Spivack AJ (1994) Stable chlorine isotopic composition of marine aerosol-particles in the Western Atlantic Ocean. Geophys Res Lett 21:1161–1164

Volpe C, Wahlen M, Pszenny AAP, Spivack AJ (1998) Chlorine isotopic composition of marine aerosols: implications for the release of reactive chlorine and HCl cycling rates. Geophys Res Lett 25:3831–3834

Wang Y, Hsu WB, Guan YB (2019) An extremely heavy chlorine reservoir in the Moon: insights from the apatite in lunar meteorites. Sci Rep 9:5727

Weaver TR (2000) Stable isotopes of chlorine. In: Cook P, Herczeg AL (eds) Environmental tracers in subsurface hydrology. Kluwer Academic Publishers, pp 492–498

Wei W, Kastner M, Spivack AJ (2008) Chlorine stable isotopes and halogen concentrations in convergent margins with implications for the Cl isotopes cycle in the ocean. Earth Planet Sci Lett 266:90–104

Willeke M (2003) Limits of the validity of the mass ratio independence of the Stokes–Einstein relation: molecular dynamics calculations and comparisons with the Enskog theory. Mol Phys 11:1123–1130

Xiao YK, Zhang CG (1992) High precision isotopic measurement of chlorine by thermal ionization mass spectrometry of the Cs_2Cl^+ ion. Int J Mass Spectrom Ion Proceses 116:183–192

Xiao YK, Liu WG, Zhou YM, Wang YH, Shirodkar PV (2000) Variations in isotopic compositions of chlorine in evaporation controlled salt lake brines of Qaidam Basin, China. Chin J Oceanol Limn 18:169–177

Zhang MM, Wei HZ, Lu JJ, Williams-Jones AE, Eastoe CJ, Li WQ, Xia ZG, Hohi SV, Ma J, Jiang SY, Wang JL (2024) Estimates of chlorine isotope fractionation factors using density functional theory: applications to ore-forming systems. Geochim Cosmochim Acta 367:1–15

9 Theoretical and Experimental Fractionation Between Species with Different Oxidation States

9.1 Theoretical Isotope Fractionation Between Chlorine and Bromine Species with Different Oxidation States

After it had become clear that equilibrium fractionation between chemical phases containing different hydrogen isotopes could be both theoretically and experimentally determined with perfect agreement (Urey and Rittenberg 1933; Rittenberg and Urey 1934; Rittenberg et al. 1934) interest in the equilibria between isotopes of the other light elements also grew.

Urey and Greiff (1935) published the first comprehensive theoretical study on isotope equilibrium fractionation between molecules of different oxidation state from most light elements, including chlorine and bromine. They calculated that for chlorine the equilibrium fractionation factor (enrichment factor) at 298.1 K between hydrogen chloride (HCl) and chlorine gas (Cl_2) was 1.003 ($\varepsilon^{37}Cl = +3‰$), while between hydrogen bromide (HBr) and bromine gas (Br_2) this factor was only 1.0004 ($\varepsilon^{81}Br = +0.4‰$), with the heavier isotope concentrating in the chlorine and bromine gas. The fractionation factors were also calculated at 273.1 K (1.004 [$\varepsilon^{37}Cl = +4‰$] for chlorine and 1.0005 [$\varepsilon^{81}Br = +0.5‰$] for bromine) and at 600 K (1.00015 [$\varepsilon^{37}Cl = +0.15‰$] for chlorine and 0.99997 [$\varepsilon^{81}Br = -0.03‰$] bromine). Interestingly, at the highest temperature the heavier bromine isotope is (only just) concentrated in hydrogen bromide. Several more equilibrium fractionation factors for chlorine and bromine were published later by Urey (1947) in his classic paper on equilibrium isotope fractionation. Fractionation factors between HCl, Cl_2, ClO_2, ClO_3^- and ClO_4^- and all their combinations were computed for five different temperatures (273.1, 298.1, 400, 500 and 600 K). Following these calculations chlorine isotope fractionation between species with different oxidation states can be very large. For example equilibrium fractionation between HCl and ClO_4^- is 1.080 [$\varepsilon^{37}Cl = +80‰$] at 298.1 K (25 °C). A more limited set of equilibrium fractionation factors was computed for bromine. The fractionation factors between HBr, Br_2 and BrO_3^- were determined for the same temperatures as for chlorine. Equilibrium fractionation for bromine is considerably less than for chlorine. For example, the largest value for a fractionation between bromine species in the table is the equilibrium between HBr and BrO_3^-, 1.008 [$\varepsilon^{81}Br = 8‰$] while the equivalent equilibrium fractionation for chlorine (for equilibrium between HCl and ClO_3^-) is 1.050 [$\varepsilon^{37}Cl = +50‰$] (both at 273.1 K). Newer data for fractionation factors on the equilibrium between HCl and Cl_2 were determined by Richet et al. (1977), Schauble et al. (2003), Czarnacki and Hałas (2012) and Balan et al. (2019). In these studies equilibrium fractionation for different temperature ranges with values between −120

H. Eggenkamp, *The Geochemistry of Stable Chlorine and Bromine Isotopes*, Advances in Isotope Geochemistry, https://doi.org/10.1007/978-3-031-75633-7_9

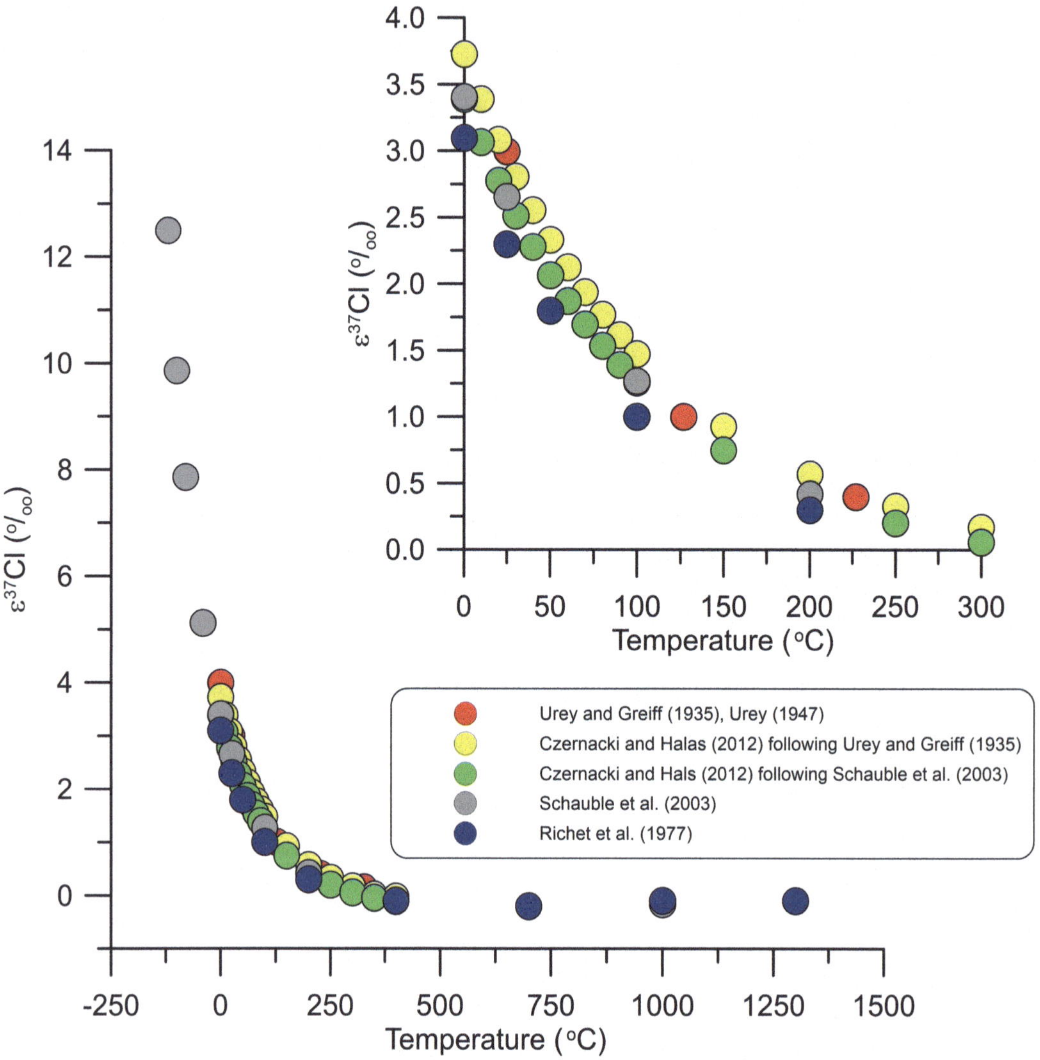

Fig. 9.1 Equilibrium isotope fractionation for the Cl_2-HCl equilibrium as published in the various papers in which this equilibrium was studied

and 1300 °C (153.15 and 1573.15 K) were calculated. Figure 9.1 shows the calculated equilibrium fractionation for the equilibrium between HCl and Cl_2 in a temperature versus fractionation factor diagram for these studies. In this figure it is clearly visible that the different calculations led to slightly different isotope fractionation factors. At low temperatures (0 °C, 273.15 K) these differences add up to almost 1‰, which is decreasing to about 0.3‰ at temperatures above 120 °C. As is also observed that the fractionation factor deceases below 1 for temperatures above approximately 340 °C. The calculations made by Richet et al. (1977) show that a minimum equilibrium fractionation factor of 0.9998 [$\varepsilon^{37}Cl = -0.8‰$] is reached at a temperature of about 1000 K (726.85 °C) and that it increases again (to ultimately 1.0000) above that temperature.

9.2 Experimental Determination of Isotope Fractionation Between Species with Different Oxidation States

9.2.1 Experimental Determination of Isotope Fractionation Between HX and X_2 (X = Cl, Br)

The first published experimental determination of any chemical equilibrium between chlorine species with different oxidation states is the determination by Hoering and Parker (1961) of the exchange reaction between hydrogen chloride and chlorine gas. The chemical and mass spectrometric procedures were tested by measuring the small equilibrium isotope effect in the exchange between gaseous hydrogen chloride and chlorine:

$$H^{37}Cl + \frac{1}{2}\,{}^{35}Cl_2 \Leftrightarrow H^{35}Cl + \frac{1}{2}\,{}^{37}Cl_2 \quad (9.1)$$

The rate to reach isotopic equilibrium in this system has been studied by Dodgen and Libby (1949) and Johnston and Libby (1951) using radioisotopes of chlorine. They found that the reaction was heterogeneously catalysed by the walls of Pyrex vessels with an exchange half time of 3 min. Hoering and Parker (1961) repeated the experiment published by Dodgen and Libby (1949) and Johnston and Libby (1951). The reactants were allowed to exchange for 24 h or longer. After separation, the two reactants were converted to silver chloride and subsequently to hydrogen chloride for isotope analysis. The magnitude of the isotope fractionation in this exchange reaction had been calculated by Urey (1947). Since both molecules are gases with a simple structure and for which good values of the vibrational frequencies were known, there was good reason to assume that the calculated value of the equilibrium fractionation factor is close to the true value. Hoering and Parker (1961) conducted five such exchange experiments, with obtained an average equilibrium fractionation factor of 1.0038 ± 0.0004 [$\varepsilon^{37}Cl = +3.8 \pm 0.4‰$]. The calculated values for this reaction are between 1.0031 and 1.0040 at 0 °C and 1.00266 and 1.0030 at 25 °C. The good agreement between the observed and the theoretical values gives confidence in the experimental procedures. Giunta et al. (2017) reported the results of experiments to determine chlorine isotope fractionation in the reaction to oxidise hydrochloric acid to chlorine gas with the use of hydrogen peroxide as oxidising agent at 25 °C and at 0 °C. Their apparatus consisted of a sealed glass reactor where liquid HCl (12 N) was mixed with liquid 30% H_2O_2. This mixture results in the oxidation of Cl^- and the formation of Cl_2 gas. As long as it is degassing, Cl_2 gas is flushed out of the solution using N_2 as a vector-gas from the reaction flask to a potassium hydroxide (KOH) solution (pH 14) where Cl_2 disproportionates into the soluble species Cl^- and ClO^-. After each experiment, the chlorine isotope composition was measured in the recovered KOH-trap solution, as well as in the residual HCl solution. Consistent with theory, the produced Cl_2 gas is always enriched in the heavier ^{37}Cl as compared to the initial HCl reservoir. The following isotope fractionation factors were obtained; At 0 °C $1000\ln\alpha_{(Cl2\text{-}Cl\text{-})} = +4.51(^{+1.65}/_{-0.49})‰$, and at 25 °C $1000\ln\alpha_{(Cl2\text{-}Cl\text{-})}$ is $+3.94(^{+0.69}/_{-0.18})‰$. These values are close to values obtained by Hoering and Parker (1961), suggesting that both experimental designs reflect the actual fractionation between HCl and Cl_2 very well. The experimental values are however higher than the calculated values, suggesting that the calculations led to slightly low values, and perhaps adjustments may be needed to the assumptions where these calculations were based upon.

The (fast) exchange reaction between bromine and hydrogen bromide containing a radioactive bromine isotope:

$$H^{@}Br + \frac{1}{2}Br_2 \Leftrightarrow HBr + \frac{1}{2}\,{}^{@}Br_2 \quad (9.2)$$

where $^{@}Br$ indicates a radioactive bromine isotope, has been experimentally (qualitative) determined already at an early date (Brejneva et al. 1936; Liberatore and Wiig 1940; Libby 1940), but has unfortunately never been repeated to establish experimentally the stable isotope exchange fractionation between the two gases.

9.2.2 Experimental Determination of Equilibrium Chlorine Isotope Fractionation Between Hypochlorous Acid ($ClOH_{aq}$) and Chloride ($Cl^-{}_{aq}$)

When chlorine gas is trapped in an alkaline solution it disproportionates into chloride (Cl^-) and hypochlorous acid (ClOH). Giunta et al. (2013) presented data on the equilibrium fractionation between Cl^- and ClOH. Unfortunately the exact method they applied in their experiment was not described in this abstract. $10^3 \ln \alpha(ClOH_{aq}\text{-}Cl^-{}_{aq})$ was determined as +3.04 ± 0.20‰ at ~20 °C, which is about 1‰ lower than was predicted by Schauble et al. (2003).

9.3 Chlorine Isotope Fractionation During Reduction from and Oxidation to Perchlorate

At the end of the twentieth century it became clear that perchlorates actually pose a health threat when they are present in drinking water (Urbansky 1998). Perchlorates were previously not known to exist in the environment except for their occurrence in the Atacama desert in Chile (Ericksen 1981), but it was realised that, especially in the South-Western United States, its presence in drinking water supplies could be high enough to possess health risks to the general population. Due to their stability in the environment it was also realised that remediation of perchlorates would be a difficult task. In the same period it was discovered that several specialised microbes were evolved that can couple growth to anaerobic reduction of perchlorate and completely reduce it to chloride (Romanenko et al. 1976; Stepanyuk et al. 1992; Malmqvist et al. 1994; Rikken et al. 1996; Bruce et al. 1999; Coates et al. 1999; Michaelidou et al. 2000). It was considered that the microbial reduction could be a good way to remediate perchlorates from the environment. In the same period it became possible to analyse the chlorine isotope composition of perchlorates (Ader et al. 2001; Sturchio et al. 2003).

9.3.1 Microbial Reduction of Perchlorates

Considering the fact that the calculated equilibrium fractionation between the +VII (perchlorate) and −I (chloride) oxidation states of chlorine is very large (80‰ at 25 °C; Urey 1947), it was realised that during microbial reduction a relatively large chlorine isotope fractionation could be expected, and it was also expected that it could be a perfect tracer for actual perchlorate reduction (Coleman et al. 2003). Coleman et al. (2003) tested the possible chlorine isotope fractionation with the help of *Azospira suillum*, a microbe that reduces perchlorate in an anoxic environment using acetate as proton donor. They found that in a medium with only acetate as electron donor and perchlorate as electron acceptor the chloride concentration increased considerably, while the perchlorate concentration decreased. During this process a considerable chlorine isotope fractionation was observed. During the process a Rayleigh type fractionation was observed with a fractionation between the original perchlorate and the produced chloride of −14.8 ± 0.7‰. Experiments by Sturchio et al. (2003) of the same process yielded comparable isotope fractionation data for this process between −16.6 and −12.9‰. Further experiments were done later by Sturchio et al. (2007) using both *Dechlorospirillum* sp. and *A. suillum* at 22 °C and 10 °C. They observed isotope fractionation values between perchlorate and chloride which were between −14.5 and −11.5‰. Ader et al. (2008) published the full experimental data and methodology used to calculate the results reported by Coleman et al. (2003) along with a re-evaluation of their results using an improved statistical treatment. Using this treatment a much more constraint experimental isotope fractionation of −14.94 ± 0.15‰ was concluded upon. On the basis of these results a hypothetical model was proposed for the biochemical control of the chlorine isotope fractionation induced by microbial perchlorate reduction.

As the calculated equilibrium fractionation between perchlorate and chloride is much larger than the observed fractionation during microbial reduction it is concluded that the microbial reaction is not occurring at isotope equilibrium between chloride and perchlorate. Ader et al. (2008) concluded that the observed fractionation is about equal to the expected fractionation between perchlorate and a hypothetical chlorine molecule or radical with oxidation state +VI. They hypothesised that this step is in thermodynamic equilibrium and is the rate determining step in the reduction reaction, while the following steps (until the formation of chloride) are not in thermodynamic equilibrium with the result that they do not show isotope fractionation.

9.3.2 Chlorine Isotope Fractionation During Oxidative Formation of Chlorate and Perchlorate

Urey (1947) predicted very large chlorine isotope fractionation factors in the equilibria between the oxidised species of chlorine. At 25 °C (298.1 K) the following $10^3 \ln\alpha$ values were predicted: ClO_4^--HCl: +80‰, ClO_4^--ClO_2^0: +52‰, ClO_4^--ClO_3^-: +35‰, ClO_3^--HCl: +43‰, ClO_3^--ClO_2^0: 16‰ and ClO_2^0-HCl: 27‰. Comparably large fractionation factors between chloride and oxidised species were calculated by Schauble et al. (2003). They indicate that under equilibrium conditions, $\delta^{37}Cl$ values of coexisting ClO_x species are positively correlated with the Cl oxidation state, with the following estimated fractionation factors with respect to Cl^- at 25 °C (298.1 K): +2.5‰ to +6‰ for Cl^0,+7‰ for Cl^{+1}, +7‰ to +9‰ for Cl^{+2}, +10‰ for Cl^{+3}, + 27‰ for Cl^{+4}, and + 75‰ for Cl^{+7} respectively.

Based on these theoretical fractionation factors extreme positive $\delta^{37}Cl$ values were expected in natural perchlorate and chlorate samples. In reality $\delta^{37}Cl$ values of perchlorates vary from negative values (−15 to −10‰) in Atacama Desert samples to positive values (+3 to +6‰) in New Mexico and Great Lakes samples and intermediate values (−4 to 0‰) in the Death Valley area (Bao and Gu 2004; Böhlke et al. 2005; Jackson et al. 2010, 2016; Sturchio et al. 2006, 2011; Poghosyan et al. 2014; Hatzinger et al. 2018). Although a larger range than for chloride chlorine isotope ratios, still a much more modest range than expected based on theoretical fractionation factors was observed in these studies. Only few analyses have been made of chlorine isotopes of natural chlorate occurrences (Jackson et al. 2021) and the results from these observations varied from −1.4 to +1.3‰ in Atacama Desert samples and from +0.8 to +1.0‰ in Death Valley samples, even a much smaller provisional range than for perchlorate samples. Estrada et al. (2017) already observed that during uptake of perchlorate through plants it seemed that little or no chlorine isotope fractionation is taking place.

To understand chlorine isotope fractionation during oxidation of chloride to chlorate and perchlorate, including the possible reaction pathways, Estrada et al. (2021) investigated the isotope effects during oxidation of dissolved and dry chloride by oxidation by ozone (O_3) and by UV light. They prepared solutions containing chloride, hypochlorite and chlorine dioxide and oxidised these via ozone or UV light and obtained solution containing chlorate and perchlorate. Dry chloride (in the form of NaCl) was only oxidised through ozone, and only (little) perchlorate was found as reaction product. In all other cases chlorate was the main reaction product and (also) little perchlorate was formed. Estrada et al. (2021) describe the complex reaction pathways that exist between chloride and chlorate/perchlorate under oxidising conditions and suggest that chlorate and perchlorate can be formed after various intermediate steps, each of course with their own fractionation factors and because various steps can be fast, it is not known if these equilibria are actually reached.

In order to study the actual chlorine isotope fractionation of the various reaction steps Estrada et al. (2021) set up oxidation experiments between chloride, hypochlorite and chlorite on one side and chlorate and perchlorate as reactants with the use of ozone and UV light as oxidative agent. The following chlorine isotope fractionation factors ($10^3 \ln\alpha$) were (by approximation) found: $NaCl \rightarrow NaClO_4$ (O_3), −3.6‰; $Cl^- \rightarrow ClO_4^-$ (O_3), +7.2‰; $ClO^- \rightarrow ClO_3^-$ (O_3), +10.6‰; $ClO^- \rightarrow ClO_3^-$ (UV), +1.0‰; $ClO^- \rightarrow ClO_4^-$ (O_3), +12.0‰; $ClO^- \rightarrow ClO_4^-$ (UV), +14.8‰; $ClO_2^- \rightarrow ClO_3^-$ (O_3), −0.1‰; $ClO_2^- \rightarrow ClO_3^-$ (UV), +2.4‰; $ClO_2^- \rightarrow ClO_4^-$ (O_3), +3.0‰; $ClO_2^- \rightarrow ClO_4^-$ (UV), +20.7‰. Due to low concentrations and/or fast reaction rates it was not possible to measure further $\delta^{37}Cl$ values and thus fractionation factors. The published results show however that measured fractionation factors vary widely depending on the oxidising agent (O_3 or UV light) and the starting ion. This indicates that the reaction pathway indeed is an important determining factor in the observed chlorine isotope fractionation. Interesting however is that in most cases during oxidation the $\delta^{37}Cl$ increases, as was expected. The obtained values however do not agree perfectly with the calculated values which may be due to the complex pathways and intermediate steps that can be, for example, too fast to reach equilibrium. Also, the obtained fractionation factors suggest that larger chlorine isotope variations should be found in natural samples than are actually observed. This indicates that the processes that rule the chlorine isotope variations of natural chlorate and perchlorate samples are even more complex than predicted in these laboratory experiments. As Estrada et al. (2021) state, "more experimental and theoretical assessments of the kinetics and mechanisms of ClO_4^- and ClO_3^- producing reactions, as well as the magnitude of the associated kinetic, equilibrium, and photoinduced isotope effects, are needed for the entire range of conditions for which ClO_4^- and ClO_3^- production may feasibly occur, from the stratosphere to the Earth surface".

References

Ader M, Coleman ML, Doyle SP, Stroud M, Wakelin D (2001) Methods for the stable isotopic analysis of chlorine in chlorate and perchlorate compounds. Anal Chem 73:4946–4950

Ader M, Chaudhuri S, Coates JD, Coleman M (2008) Microbial perchlorate reduction: a precise laboratory determination of the chlorine isotope fractionation and its possible biochemical basis. Earth Planet Sci Lett 269:604–612

Balan E, Créon L, Sanloup C, Aléon J, Blanchard M, Paulatto L, Bureau H (2019) First-principles modeling of chlorine isotope fractionation between chloride-bearing molecules and minerals. Chem Geol 525:424–434

Bao H, Gu B (2004) Natural perchlorate has its unique oxygen isotope signature. Environ Sci Technol 38:5073–5077

Böhlke JK, Sturchio NC, Gu B, Horita J, Brown GM, Jackson WA, Batista JR, Hatzinger PB (2005) Perchlorate isotope forensics. Anal Chem 77:7838–7842

Brejneva N, Roginsky S, Schilinsky A (1936) Methodology of the introduction of radioactive halogens to organic molecules. Acta Physicochim URSS 7:549–574

Bruce RA, Achenbach LA, Coates JD (1999) Reduction of (per)chlorate by a novel organism isolated from a paper mill waste. Environ Microbiol 1:319–331

Coates JD, Michaelidou U, Bruce RA, O'Connor SM, Crespi JN, Achenbach LA (1999) The ubiquity and diversity of dissimilatory (per)chlorate-reducing bacteria. Appl Environ Microbiol 65:5234–5241

Coleman ML, Ader M, Chaudhuri S, Coates JD (2003) Microbial isotopic fractionation of perchlorate chlorine. Appl Environ Microbiol 69:4997–5000

Czarnacki M, Hałas S (2012) Isotope fractionation in aqua-gas systems: Cl_2-HCl-Cl^-, Br_2-HBr-Br^- and H_2S-S^{2-}. Isot Environ Health Stud 48:55–64

Dodgen HW, Libby WF (1949) The exchange reaction between the hydrogen halides and the halogens in the gaseous state. J Chem Phys 17:951–957

Ericksen GE (1981) Geology and origin of the Chilean nitrate deposits. USGS Prof. Pap. 1188:37 pp

Estrada NL, Böhlke JK, Sturchio NC, Gu B, Harvey G, Burkey KO, Grantz DA, MacGrath MT, Anderson TA, Rao B, Sevanthi R, Hatzinger PB, Jackson WA (2017) Stable isotopic composition of perchlorate and nitrate accumulated in plants: hydroponic experiments and field data. Sci Total Environ 595:556–566

Estrada NL, Anderson TA, Böhlke JK, Guc BH, Hatzinger PB, Mroczkowski SJ, Rao B, Sturchio NC, Jackson WA (2021) Origin of the isotopic composition of natural perchlorate: experimental results for the impact of reaction pathway and initial ClO_x reactant. Geochim Cosmochim Acta 311:292–315

Giunta T, Labadi J, Eggenkamp HGM (2013) Experimental determination of chlorine isotope fractionation in Cl_2-Cl^-_{aq} and $ClOH_{aq}$-Cl^-_{aq}. 23th Annual V.M. Goldschmidt Conference. Firenze, Italy

Giunta T, Labidi J, Eggenkamp HGM (2017) Chlorine isotope fractionation between chloride (Cl^-) and dichlorine (Cl_2). Geochim Cosmochim Acta 213:375–382

Hatzinger PB, Bohlke JK, Sturchio NC, Izbicki J, Teague N (2018) Four-dimensional isotopic approach to identify perchlorate sources in groundwater: application to the Rialto-Colton and Chino Subbasins, Southern California (USA). Appl Geochem 97:213–225

Hoering TC, Parker PL (1961) The geochemistry of the stable isotopes of chlorine. Geochim Cosmochim Acta 23:186–199

Jackson WA, Bohlke JK, Gu B, Hatzinger PB, Sturchio NC (2010) Isotopic composition and origin of indigenous perchlorate and co-occurring nitrate in the southwestern United States. Environ Sci Technol 44:4869–4876

Jackson WA, Davila AF, Böhlke JK, Sturchio NC, Sevanthi R, Estrada N, Brundrett M, Lacelle D, McKay CP, Poghosyan A, Pollard W, Zacny K (2016) Deposition, accumulation, and alteration of Cl^-, NO_3^-, ClO_4^- and ClO_3^- salts in a hyper-arid polar environment: mass balance and isotopic constraints. Geochim Cosmochim Acta 182:197–215

Jackson WA, Brundrett M, Böhlke JK, Hatzinger PB, Mroczkowski SJ, Sturchio NC (2021) Isotopic composition of natural and synthetic chlorate ($\delta^{18}O$, $\Delta^{17}O$, $\delta^{37}Cl$, $^{36}Cl/Cl$): methods and initial results. Chemosphere 274:129586

Johnston WH, Libby WF (1951) The exchange reaction between hydrogen chloride and chlorine in the gaseous state. J Am Chem Soc 73:854–855

Libby WF (1940) Mechanism of the exchange reaction between gaseous bromine and hydrogen bromide. J Chem Phys 8:348

Liberatore LC, Wiig EO (1940) Exchange reaction of gaseous bromine and hydrogen bromide. J Chem Phys 8:165–170

Malmqvist A, Welander T, Moore E, Ternstrom A, Molin G, Stenstrom IM (1994) *Ideonella dechloratans* gen. nov., sp. nov., a new bacterium capable of growing anaerobically with chlorate as an electron acceptor. Syst Appl Microbiol 17:58–64

Michaelidou U, Achenbach LA, Coates JD (2000) Isolation and characterization of two novel (per) chlorate-reducing bacteria from swine waste lagoons. In: Urbansky ED (ed) Perchlorate in the environment. Kluwer Academic/Plenum, New York, NY, pp 271–283

Poghosyan A, Sturchio NC, Morrison CG, Beloso Jr AD, Guan Y, Eiler JM, Jackson WA, Hatzinger PB (2014) Perchlorate in the Great Lakes: isotopic composition and origin. Environ Sci Technol 48:11146–11153

Richet P, Bottinga Y, Javoy M (1977) A review of hydrogen, carbon, nitrogen, oxygen, Sulphur, and chlorine stable isotope fractionation among gaseous molecules. Annu Rev Earth Planet Sci 5:65–110

Rikken G, Kroon A, Van Ginkel C (1996) Transformation of (per)chlorate into chloride by a newly isolated bacterium: reduction and dismutation. Appl Microbiol Biotechnol 45:420–426

Rittenberg D, Urey HC (1934) The thermal decomposition of deuterium iodide. J Am Chem Soc 56:1885–1889

Rittenberg D, Bleakney W, Urey HC (1934) The equilibrium between the three hydrogens. J Chem Phys 2:48–49

Romanenko VI, Korenkov VN, Kuznetsov SI (1976) Bacterial decomposition of ammonium perchlorate. Mikrobiologiya 45:204–209

Schauble EA, Rossman GR, Taylor HP (2003) Theoretical estimates of equilibrium chlorine-isotope fractionations. Geochim Cosmochim Acta 67:3267–3281

Stepanyuk V, Smirnova G, Klyushnikova T, Kanyuk N, Panchenko L, Nogina T, Prima V (1992) New species of the *Acinetobacter* genus *Acinetobacter thermotoleranticus* sp. nov. Mikrobiologiya 61:347–356

Sturchio NC, Hatzinger PB, Arkins MD, Suh C, Heraty LJ (2003) Chlorine isotope fractionation during microbial reduction of perchlorate. Environ Sci Technol 37:3859–3863

Sturchio NC, Böhlke JK, Gu B, Horita J, Brown GM, Beloso Jr AD, Hatzinger PB, Jackson WA, Batista JR (2006) Stable isotopic compositions of chlorine and oxygen in synthetic and natural perchlorates. In: Gu B, Coates JD (eds) Perchlorate environmental occurrences, interactions, and treatment. Springer, New York, pp 93–109

Sturchio NC, Böhlke JK, Beloso Jr AD, Streger SH, Heraty LJ, Hatzinger PB (2007) Oxygen and chlorine isotopic fractionation during perchlorate biodegradation: laboratory results and implications for forensics and natural attenuation studies. Environ Sci Technol 41:2796–2802

Sturchio NC, Böhlke JK, Gu BH, Hatzinger PB, Jackson WA (2011) Isotopic tracing of perchlorate in the environment. In: Baskaran M (ed) Handbook of environmental isotope geochemistry, advances in isotope geochemistry. Springer, Berlin Heidelberg

Urbansky ET (1998) Perchlorate chemistry: implications for analysis and remediation. Biorem J 2:81–95

Urey HC (1947) The thermodynamic properties of isotopic substances. J Chem Soc 1947:562–581

Urey HC, Greiff LJ (1935) Isotope exchange equilibria. J Am Chem Soc 57:321–327

Urey HC, Rittenberg D (1933) Some thermodynamic properties of the H^1H^2, H^2H^2 molecules and compounds containing the H^2 atom. J Chem Phys 1:137–143

10 Isotope Fractionation Related to Organochlorine and Organobromine Compounds

10.1 Early Experimental Determination of Chlorine Isotope Fractionation

The first studies in which chlorine isotope variations in relationship with organic compounds were experimentally determined were applied during the 1950s. Probably the earliest of these studies were done by Bartholomew et al. (1954a, 1954b) who studied the chlorine isotope effect in reactions of 2-chloro-2-methyl-propane (*tert*-butyl chloride). In these studies essentially the isotope effect of the reaction

$$C_4H_9Cl \leftrightarrow C_4H_9\bullet + Cl\bullet \quad (10.1)$$

is studied. Two of these reactions were studied. The first of these experiments was done in an ethanol-silver nitrate solution. When this solution is used the back reaction can not take place due to the precipitation of chloride as silver chloride. As a result in the following step the $C_4H_9\bullet$ radical reacts with the solvent to form 2-ethoxy-2-methyl-propane (ethyl-*tert*-butyl-ether).

$$C_4H_9\bullet + C_2H_5OH \rightarrow C_4H_9OC_2H_5 + H\bullet \quad (10.2)$$

The other reaction they studied was the experiment in which 2-chloro-2-methyl-propane was allowed to react with an aqueous alcoholic sodium hydroxide solution. In this solution two possible reactions take place, either the $C_4H_9\bullet$ radical reacts with the solvent to form 2-ethoxy-2-methyl-propane (ethyl-*tert*-butyl-ether) or it reacts with the hydroxide ions so that 2-methyl-2-propanol (*tert*-butanol) is formed. In this experiment the back reaction is prevented by the alkaline nature of the solution as both 2-ethoxy-2-methyl-propane and 2-methyl-2-propanol do not react with chloride in alkaline solutions. Reactions were performed in such a manner that different fractions of the reaction products could be sampled. Chloride was precipitated as silver chloride and converted to chlorine gas for isotope analyses. The observations made on these experiments indicated that the reaction rate k for ^{35}Cl was slightly larger than for ^{37}Cl. The isotope effect for the reaction was determined as $k^{35}/k^{37} = 1.008 \pm 0.002$. This result is in the expected direction as the heavier isotope is bound more firmly to the carbon atom than the lighter isotope and consequently undergoes bond breaking reactions less rapidly.

A second study, unfortunately described in detail only in the PhD dissertation of Nair (1956), and briefly in Fry (1964), was done by Nair and Fry and studied the isotope effect of the reaction between chloride and cyanide ions in aqueous ethanol. They observed a comparable chlorine isotope effect of 1.0074 ± 0.0005. The next, published, experimental study was presented by Hill and Fry (1962) who studied isotope effects in the reactions of benzyl- and substituted benzyl-chlorides with various nucleophiles. In these experiments they studied the isotope effects of the substitution of the chlorine atom in aromatic

H. Eggenkamp, *The Geochemistry of Stable Chlorine and Bromine Isotopes*, Advances in Isotope Geochemistry, https://doi.org/10.1007/978-3-031-75633-7_10

compounds. They determined the chlorine isotope effects for unsubstituted, *p*-methoxy-, *p*-methyl-, *p*-chloro- and *p*-nitrobenzyl chlorides with each of three nucleophiles: water (plain hydrolysis), cyanide ion and thiosulfate ion. The reactions were done in an aqueous dioxane solution containing 80% dioxane by weight. The effect of these reactions was that the chloride in the benzenes was substituted with the nucleophiles. The isotope effect of these reactions was determined after precipitating the liberated chloride as silver chloride and subsequent reaction to chloromethane for isotope analysis. During these reactions plots were made that showed the progress of the reaction. Two types of plots were produced. First order plots, showing the decrease of the original chloride bearing aromatic compound, and second order plots that were made to understand the reactions with cyanide and thiosulphate ions and these plots show the increase of the substituted aromatic compounds. A clear difference exists between reactions where first order plots show a straight line, and reactions where they show a clear curvature. Secondly, plots were recognised which have curved first order plots and straight second order plots. The differences between the two types of plots were explained as follows: The first type describes first order kinetics, and the second type describes second order kinetics. Very interesting results were obtained for the isotope effects, which seemed to be related to the type of kinetics that was found. In the experiments where the first order plots were linear the isotope effect k^{35}/k^{37} is close to 1.0078, while in those experiments where the second order plots were linear it is close to 1.0058. Because of these differences it was concluded that there is a relationship between the kinetic behaviour of the reaction and the chlorine isotope effect. From these results it was also concluded that measuring the kinetic isotope effect in these reactions can be a valuable tool in mechanism studies and the understanding of these reactions.

During the 1970s James W. Taylor, who was also responsible for considerable improvements in the analytical methods for both chlorine and bromine isotopes (Taylor and Grimsrud 1969; Willey and Taylor 1978), presented several more studies in which chlorine (and even bromine) isotope effects were studied (Grimsrud and Taylor 1970; Turnquist et al. 1973; Williams and Taylor 1973, 1974; Graczyk and Taylor 1974; Julian and Taylor 1976; Graczyk et al. 1978; Willey and Taylor 1980). These chlorine isotope studies confirmed the rate of the chlorine isotope effects that were observed in the earlier studies. Depending on the reaction the chlorine kinetic isotope effects are in general somewhere between 1.005 and 1.010. Turnquist et al. (1973) also determined the temperature dependence of the chlorine isotope effects and discovered that they were close to linearly dependent with the reciprocal temperature, and decreased towards higher temperatures. Willey and Taylor (1980) presented the only study in which also bromine isotope effects are examined. They presented the bromine isotope effects for the solvolysis reaction of 2-bromo-2methyl-propane (*tert*-butyl bromide) in methanol and the reaction between bromo-butane (*n*-butyl bromide) and thiophenoxide anion in 0.25 M lithium methoxide at various temperatures. These reactions were chosen so that they were able to compare the results obtained from experiments to determine the bromine isotope effects to the results obtained by Turnquist et al. (1973) who did the chlorine equivalent experiments of these reactions. The results showed that there were considerably smaller bromine isotope effects than chlorine isotope effects. While Turnquist et al. (1973) found chlorine isotope effects in the solvolysis reaction of 2-chloro-2methyl-propane of 1.0109 at 10 °C to 1.0095 at 60 °C, the solvolysis reaction of 2-bromo-2methyl-propane varied from 1.00325 at 0 °C to 1.00303 at 40 °C (Willey and Taylor 1980). The reaction of chloro-butane with the thiophenoxide anion yield chlorine isotope effects between 1.0096 at 0 °C and 1.0079 at 60 °C, while the equivalent bromine reactions isotope effects were only 1.00191 at −10 °C to 1.00158 at 40 °C (see Fig. 10.1).

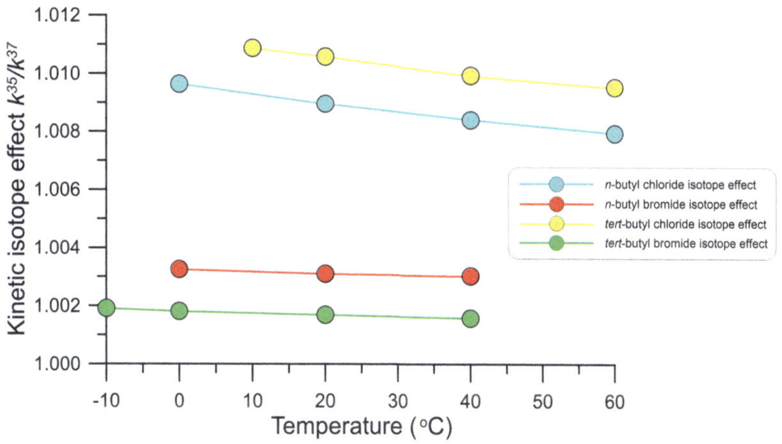

Fig. 10.1 Comparison of the results of experimentally determined kinetic isotope effects for reactions involving n-butyl-chloride and -bromide and tert-butyl chloride and -bromide

10.2 Theoretical Estimates on Chlorine Isotope Fractionation Between Organochlorine Molecules and Chloride

Schauble et al. (2003) calculated the reduced partition function ratios for a significant number of organochlorine compounds that can be used to calculate the theoretical isotope fractionation between two molecules in equilibrium. These data were all for species which are in principle in the gaseous state but it gives a good impression what size of fractionation can be expected. Czarnacki and Hałas (2012) also calculated the reduced partition function ratios for the aqueous chloride ion at various temperatures. These data can be combined to estimate equilibrium fractionation between organochlorine compounds and chloride in solution. The data calculated by Schauble et al. (2003) are of considerable interest as they indicate very clearly that isotope fractionation between organochlorine compounds and chloride (either as aqueous, dissolved chloride or as hydrogen chloride gas) can be relatively large (several ‰), but not as large as the equilibrium fractionation between chloride and perchlorates. When data calculated by Schauble et al. (2003) for organic molecules and Czarnacki and Hałas (2012) for aqueous chloride are combined an equilibrium fractionation of about 8‰ at 0 °C, 7‰ at 25 °C and 4.5‰ at 100 °C between these two phases (organochlorine and aqueous chloride) is expected. The calculations from Schauble et al. (2003) indicate that larger fractionation factors will be observed for molecules that contain more chloride atoms in them such as perchloroethylene as compared to trichloroethylene and tetrachloromethane as compared to chloromethane.

10.3 Modern Studies to Determine Chlorine Isotope Fractionation in Organic Compounds

After the first attempts to analyse the stable chlorine isotope composition of organochlorine compounds of environmental and geological interest (Van Warmerdam et al. 1995; Holt et al. 1997; Jendrzejewski et al. 1997) studies were started to extract organochlorine compounds from environmental samples (Sturchio et al. 1998; Holt et al. 2001). In the same period more studies on chemical fractionation were published while also the first studies on microbiological reduction of chlorine from organochlorine compounds were published.

10.3.1 Chemical Fractionation Studies

In their study on a more convenient (albeit much less precise) method to analyse the chlorine leaving group kinetic isotope effect for the S_N2 reduction of benzyl chloride to toluene by sodium borohydride in dimethyl suphoxide (DMSO) at 30 °C, equivalent to the studies by Hill and Fry (1962) and the group around JW Taylor, Westaway et al. (1998) determined the fractionation both by analysing the chlorine isotope composition with the chloromethane method and by fast atom bombardment mass spectrometry. They observed that the results of the isotope effect measurements from this reaction was within error equal at about 1.008 for both methods, a value equivalent to the value obtained by Hill and Fry (1962), but with large differences in precision between the two methods (see also Sect. 5.2.9).

More recently Audí-Miró et al. (2013) studied the effectiveness of the degradation of chlorinated ethenes by zero-valent iron. Zero-valent iron degradation of chlorinated ethenes is an efficient way to remove these compounds from contaminated sites through reductive dechlorination. The advantage to microbial degradation is twofold. It is more efficient so that toxic intermediates such as vinyl chloride do not accumulate in the environment while it also operates a second reductive pathway (vicinal β-dichloroelimination) during which two chlorine atoms are split off the chlorinated ethenes rapidly leading to harmless products such as ethene and ethane (Arnold and Roberts 2000; Liu et al. 2006; Elsner et al. 2008). Audí-Miró et al. (2013) determined both the chlorine and the carbon isotope fractionation during this process. During this process the apparent kinetic isotope effect for the chlorine isotopes was 1.008 ± 0.001 for trichloroethene and 1.013 ± 0.002 for *cis*-dichloroethene. Measuring both chlorine and carbon isotopes showed interesting results especially when the data were compared with the data previously obtained by Abe et al. (2009) for the reduction of *cis*-dichloroethene. While the slope in the δ^{37}Cl vs. δ^{13}C plot for the abiotic reduction with zero-valent iron obtained by Audí-Miró et al. (2013) was 3.1 ± 0.2 Abe et al. (2009) found a slope of 11.4 ± 0.6 when they reduced *cis*-dichloroethene through microbial reduction with the mixed culture KB-1, a slope that is almost four times as large. This observation was a clear indication that fundamentally different chemical processes are in effect between the two types of dechlorination.

10.3.2 Thermal Decomposition of 1,2-Dichloroethane in Air

One of the first studies in which attempts were made to use chlorine isotope variations to solve environmental problems concerned the thermal decomposition of 1,2-dichloroethane in air. 1,2-dichloroethane was used in the past as an additive to leaded car fuel and at temperatures above about 300 °C it starts to lose its chlorine atoms. Baumann and Heumann (1988) set up experiments to estimate the chlorine isotope fractionation during this decomposition. In their experiments they heated 1,2-dichloroethane in air to a temperature of about 300 °C en collected the first chloride that escaped from it. They analysed the chlorine isotope composition of the liberated chloride and found that its isotope composition was 2 to 3‰ lighter than that of the original 1,2-dichloroethane. At this temperature the decomposition only just started and only between 0.5 and 5% of the 1,2-dichloroethane was decomposed. Baumann and Heumann (1988) explained their results by the fact that the heavier isotopes is bound more strongly than the lighter molecule so that the lighter isotope escapes more easily from these molecules during thermal decomposition.

10.3.3 Chlorine Isotope Fractionation Induced by Microbes

Several microbes are able to either produce or decompose organochlorine compounds. Especially the formation of halomethanes from fungi (Harper 1985, 1994; Harper et al. 1988) has been studied well and currently something like 3700 halogen containing chemicals are known to be produced in nature (Gribble 2010). The organic cycle of chlorine is very well reviewed by Öberg (2002) who described the many paradoxes when organochlorine compounds in nature are considered.

10.3.3.1 Chlorine Isotope Effects During Microbially Induced Organochlorine Compound Production

Several organochlorine compounds are known to be produced in nature. To examine the isotope fractionation during these processes Reddy et al. (2002) determined the chlorine isotope effects during the enzyme catalysed chlorination of 1,3,5-trimothoxybenzene and 3,5-dimethylphenol. They actually extracted the catalyst Fe(III)-heme-chloroperoxidase from the fungus *Caldariomyces fumago* and treated this with either 1,3,5-trimothoxybenzene or 3,5-dimethylphenol, potassium chloride and hydrogen peroxide in a citrate/phosphate buffer (McIlvaine 1921) at pH 3. The potassium chloride used in these reactions had a known δ^{37}Cl of −0.76 ± 0.07‰. It was found that the 1,3,5-trimothoxybenzene was almost fully dichlorinated and its δ^{37}Cl was observed to be −12.06 ± 0.18‰. The 3,5-dimethylphenol was either trichlorinated (about 75% of it) or dichlorinated (about 25% of it) and the bulk δ^{37}Cl of this mixture was −11.08 ± 0.08‰. These results indicated that ^{35}Cl was preferentially incorporated in the substrate, and that the isotope fractionation of this process is rather large. As Reddy et al. (2002) indicated that anthropogenic chlorinated organic compounds have δ^{37}Cl values which are mostly between −5.1 and + 1.2‰ (Reddy et al. 2000; Reddy et al. 2001; Drenzek et al. 2002) natural and anthropogenic chlorinated organic compounds should be easily distinguishable in nature due to the much lower δ^{37}Cl values of natural chlorinated organic compounds that were expected to have values between −14 and −10‰ as compared to anthropogenic compounds which would have values close to 0‰.

The results of this study were partly confirmed by Aeppli et al. (2013a) who studied the chlorine isotope fractionation during two important pathways for enzymatic natural halogenation: chlorination by chloroperoxidase and flavin-dependent halogenases. Of these chloroperoxidase chlorination showed a very large fractionation of −12.6 ± 0.9‰ which is comparable to the results obtained by Reddy et al. (2002). On the other hand flavin-dependent halogenases showed no significant fractionation with a value of −0.3 ± 0.6‰. This large variation is attributed to the largely different reaction mechanisms employed by these enzymes. Aeppli et al. (2013a) also showed that the δ^{37}Cl in bulk organochlorine compound mixtures extracted from boreal forest soils were only slightly depleted in ^{37}Cl relative to inorganic chlorine. This showed to be in contrast to previous suggestions (Reddy et al. 2002) that chloroperoxidase chlorination plays a key role in production of soil chlorinated organic compounds. Aeppli et al. (2013a) conclude that this observation points to the additional involvement of either other chlorination pathways, or that dechlorination of naturally produced chlorinated organic compounds can neutralize δ^{37}Cl shifts caused by chloroperoxidase chlorination.

10.3.3.2 Fractionation of Chlorine During Microbial Decomposition of Organochlorine Compounds

During bacterial dehalogenation of organochlorine compounds bacteria obtain energy through reduction of organically bound chlorine to chloride ions which are then released into the environment. Potentially this process can cause isotope fractionation and this may be used to understand the processes that have taken place since a contamination has appeared. Due to the importance of these processes in remediation of organochlorine contamination in the environment much more attention has been applied to these processes than to the natural chlorination of organic compounds.

One of the first studies in which both chlorine and carbon isotope fractionation as a result of microbial degradation was determined was the experimental study by Heraty et al. (1999) on dichloromethane. They determined both the chlorine and carbon isotope composition in the remaining dichloromethane, and compared it to the original isotope compositions, after it was aerobically degraded by MC8b, a gram-negative methylotrophic bacterium related to *Methylobacterium* or *Ochrobactrum*. The isotope fractionation factor α that was determined from this reaction was 0.9962 ± 0.0003 ($\varepsilon^{37}Cl$ = −3.8 ± 0.3‰) for chlorine and 0.9576 ± 0.0015 ($\varepsilon^{13}C$ = −42.4 ± 1.5‰) for carbon, indicating that both $\delta^{37}Cl$ and $\delta^{13}C$ of the remaining dichloromethane decreased. These fractionation factors correspond to kinetic isotope effects of 1.0038 and 1.0424 respectively for chlorine and carbon.

Chlorine isotope fractionation during bacterial dehalogenation of dichloromethane was also shown by Zyakun et al. (2007). They determined the isotope fractionation of dichloromethane during microbial dehalogenation by *Methylobacterium dichloromethanicum* and *Albibacter methylovorans*. Contrary to the study by Heraty et al. (1999) they determined the isotope composition of the reacted dichloromethane directly on the dichloromethane molecule that was introduced into their mass spectrometer. To determine isotope fractionation masses 84 ($^{35}Cl_2CH_2$), 86 ($^{35}Cl^{37}ClCH_2$) and 88 ($^{37}Cl_2CH_2$) were compared. The dichloromethane that remained after the reaction had more positive isotope ratios than the original dichloromethane, which is quite the opposite when compared to the experimental observations from Heraty et al. (1999). The increased isotope ratio was explained by the process of diffusion of isotopically lighter dichloromethane through the cell walls of the bacteria so that the residual dichloromethane ends up being more positive. Considering this strong effect it is not clear if any microbial isotope effect did actually take place during their experiments.

Drenzek et al. (2004) studied the change in chlorine isotope composition during microbial decomposition of 2,3,4,5-tetrachlorobifenyl. During the reaction to 2,3,5-trichlorobifenyl the isotope composition of the 2,3,4,5-tetrachlorobifenyl did not change, even after 90% of the original amount had reacted. Assuming that other processes would fractionate the chlorine isotope composition it was concluded that biological activity could be discriminated against other isotope transformation processes.

Abe et al. (2009) determined chlorine and carbon isotope fractionation during both aerobic and reductive dechlorination of vinyl chloride and *cis*-dichloroethene. Their results showed that in the case of aerobic oxidation there were very different isotope enrichment factors for carbon and chlorine isotopes. In the case of carbon isotopes they found enrichment factors of −7.2‰ for vinyl chloride and −8.5‰ for *cis*-dichloroethene, while both compounds showed only an isotope enrichment factor of −0.3‰ for chlorine isotopes. In the case of reductive dechlorination the carbon isotope enrichment factors were much larger with values of −25.2‰ for vinyl chloride and −18.5‰ for *cis*-dichloroethene. The chlorine isotope enrichment factors were also larger than in the case of aerobic oxidation with values of −1.8‰ for vinyl chloride and −1.5‰ for *cis*-dichloroethene. Abe et al. (2009) concluded that these values were less than expected for a primary chlorine isotope effect and these data suggested that no cleavage of C-Cl bonds takes place during the initial rate-limiting step. Interestingly these data show a very different slope in the $\delta^{37}Cl$ vs. $\delta^{13}C$ plot for *cis*-dichloroethene compared to the slope measured by Audí-Miró et al. (2013) in their abiotic reduction experiments with zero valent iron, which made the latter conclude that during biotic and abiotic dechlorination very different dehalogenation pathways take place. Cretnik et al. (2014) determined both chlorine and carbon isotope effects of the dehalogenation of tetrachloroethene to trichloroethene and trichloroethene to *cis*-1,2-dichloroethene. They observed chlorine isotope effects of −5.0 ± 0.1‰ during dehalogenation of tetrachloroethene and −3.6 ± 0.2‰ during dehalogenation of trichloroethene. For carbon isotopes the effects were −19.0 ± 0.9‰ and −12.2 ± 1.0‰ respectively.

In a study on the potential microbial (fungal) degradation of 2,4,6-trichlorophenol Aeppli et al. (2013b) determined chlorine and carbon isotope fractionation that is the result of this reaction. During this reaction 2,4,6-trichlorophenol is oxidised, using the enzyme chloroperoxidase, extracted from the fungus *Caldariomyces fumago*, to 2,6-dichloro-*para*-benzoquinone and a chloride ion. It was observed that during the reaction the remaining, unreacted 2,4,6-trichlorophenol increased in $\delta^{37}Cl$ and that it followed a very good Rayleigh fractionation curve. From that curve it could be calculated that the $\delta^{37}Cl$ values of the liberated chloride, which were unfortunately not determined in this experiment, would be 2.1 ± 0.4‰ lower than that of the 2,4,6-trichlorophenol. This value would agree to the value of the apparent kinetic isotope effect of 1.0021 ± 0.0004. Interestingly it was concluded that the $\delta^{37}Cl$ of the reaction product 2,6-dichloro-*para*-benzoquinone would be constant and equal to the original $\delta^{37}Cl$ at the beginning of the reaction. The reason for this is that the two chlorine atoms at the unreactive 2,4,6-trichlorophenol position preserve the original $\delta^{37}Cl$, similar to the trichloroethene-to-cis-dichloroethene hydrogenolysis as determined by Hunkeler et al. (2009). Unfortunately Aeppli et al. (2013b) were unable to determine the $\delta^{37}Cl$ of the 2,6-dichloro-*para*-benzoquinone due to the instability of this compound, so that this behaviour, the constancy of its $\delta^{37}Cl$, could not be verified.

Trichloromethane ($CHCl_3$; chloroform) can be degraded to dichloromethane (CH_2Cl_2) by only a few bacterial strains. Heckel et al. (2017, 2019) observed that $\varepsilon^{37}Cl$ and $\varepsilon^{13}C$ for this process can either show identical isotope effects, thus having the same sign, or opposing isotope effects. They found that two closely related *Dehalobacter* sp. transform trichlororomethane to dichloromethane with opposing dual element isotope slopes (Heckel et al. 2019). One strain dechlorinates trichloromethane by the reductive dehalogenase CfrA and exhibited identical isotope effects to those measured with the corrinoid cofactor vitamin B12 (with a dual C-Cl slope, $\Lambda_{C/Cl} = 6.6 \pm 0.1$ and $\Lambda_{C/Cl} = 6.5 \pm 0.2$, respectively, where $\Lambda_{C/Cl}$ is defined as the slope in a $\delta^{13}C/\delta^{37}Cl$ plot), indicating that they likely share the same reaction mechanism (Heckel et al. 2017, 2019). Another strain uses the trichloromethane reductive dehalogenase TmrA and this strain exhibited an inverse chlorine isotope effect leading to an opposing dual element isotope slope ($\Lambda_{C/Cl} = -1.2 \pm 0.2$) (Heckel et al. 2019). Soder-Walz et al. (2022) determined isotope fractionation of both chlorine and carbon isotopes during degradation of trichloromethane to dichloromethane by another *Dehalobacter* strain. They studied chlorine and carbon isotope fractionation during trichloromethane transformation in cultured cells and in enzymatic assays. Trichloromethane transformation in three studied systems resulted in small but significant chlorine ($\varepsilon^{37}Cl = -0.9 \pm 0.1$‰ to -1.2 ± 0.2‰) and carbon ($\varepsilon^{13}C = -2.7 \pm 0.1$‰ to -4.1 ± 0.5‰) isotope fractionation. A characteristic and consistent dual C-Cl isotope fractionation pattern was observed for all studied systems, with a combined $\Lambda_{C/Cl} = 2.8 \pm 0.3$ (Soder-Walz et al. 2022). This $\Lambda_{C/Cl}$ differed significantly from values reported by Heckel et al. (2017, 2019). These findings indicate that isotope fractionation during microbial degradation of trichloromethane to dichloromethane differs significantly depending on the reaction pathways used by different bacterial strains and measurement of the isotope composition of trichloromethane and dichloromethane in a system is a useful tool to understand which degradation process has taken place in the system (Soder-Walz et al. 2022).

Due to its importance at contaminated sites and because it is very persistent in the environment studies on the degradation of hexachlorocyclohexane (HCH) have been initiated during the last 10 years. Liu et al. (2020a, 2020b) studied the isotope effects of the uptake of HCH isomers by wheat. They showed that $\delta^{37}Cl$ values of β-HCH remained stable in soil and wheat, indicating that no transformation takes place. In contrast, an increase in $\delta^{37}Cl$ values of α-HCH indicated its transformation in soil and wheat. Identification of 1,3,4,5,6-pentachlorocyclohexe ne indicated that this is the major metabolite of α-HCH transformation. This increase is different at different stages of wheat growth. This suggests that the preferential transformation of enantiomers varied at different growth stages.

The maximum values that could be expected for the chlorine kinetic isotope effects during dechlorination of organic compounds were evaluated by Świderek and Paneth (2012). According to experiments done by Streitwieser et al. (1958) on hydrogen isotopes in cyclopentyl tosylates it could be calculated that the maximum chlorine kinetic isotope effects would be 1.013. These values appeared too low when it was shown experimentally that during reduction of perchlorate (although this is no organic ion) the chlorine kinetic isotope effect is in the range of 1.015 to 1.017 (Coleman et al. 2003; Sturchio et al. 2003). Earlier Paneth (1992) calculated that that the maximum kinetic isotope effect could theoretically be as large as 1.019 on the basis of BEBOVIB calculations (Sims and Lewis 1985). Using this BEBOVIB methodology Burton et al. (1977) calculated that the chlorine kinetic isotope effect in the ionisation reaction of *tert*-butyl chloride could be as large as 1.0256, which was in serious disagreement with the experimentally determined value of 1.011 (Turnquist et al. 1973). Świderek and Paneth (2012) made new calculations which indicated that chlorine kinetic isotope effects in dehalogenation reactions from heavy compounds containing at least three chlorine atoms could show that these isotope effects would be as large as 1.025. They conclude that these large kinetic isotope effects can be associated with reactions in which chlorine is passed between two heavy atoms and is shielded from interactions with solvent molecules.

The experimental results that were obtained in these studies suggest that chloride isotope fractionation during microbial degradation of organochlorine compounds is very much dependant of the reaction path. The data obtained so far show that a reaction can either induce isotope fractionation, or show no fractionation et al. The processes that take place during degradation of chlorinated organic compounds are slowly better understood and it is important to realise that the chlorine isotope effects are dependent on the type and the pathway of the reactions that take place.

10.3.4 Chlorine Isotope Fractionation Associated with the Synthesis of Hexachlorocyclohexane

Hexachlorocyclohexane (HCH) is one of the most prominent persistent organic pollutants in the environment. While several isomers exist, only the isomer γ-HCH (known under the trade name Lindane), has insecticidal properties and has been used as pesticide in the past extensively. Nowadays HCH is banned according to the Stockholm convention from 2009. Due to its persistent character HCH isomers (during synthesis of Lindane also various other isomers are produced) are still widely found in the environment, especially since the unwanted isomers were considered waste and as such were discarded. Isotope studies on environmental and originally manufactured HCHs (Ivdra et al. 2017; Wu et al. 2019) showed large variations in all three isotope systems that are present in HCH isomers (chlorine, carbon and hydrogen). Large variations in isotope compositions between HCHs from various producers (Ivdra et al. 2017) also suggest that different producers use different production pathways and as such that isotope compositions can potentially be used to differentiate between producers.

Liu et al. (2022) conducted experiments to synthesise HCH through photochemical reactions between chlorine gas and benzene. HCHs are formed in this process in three steps, where in each step two chlorine atoms are added to the benzene (multistep radical dichlorination reaction). The rate limiting step is the first where benzene reacts with the chlorine to the hypothetical dichlorocyclohexadiene. The following steps ultimately lead to HCH isomers. The chlorine isotope fractionation ($\alpha Cl_{HCH/Cl2}$) between Cl_2 and HCH was determined in these experiments as similar, only ranging from 0.9744 ± 0.0002 ($\varepsilon^{37}Cl$ = −25.6 ± 0.2‰) to 0.9747 ± 0.0002 ($\varepsilon^{37}Cl$ = −25.3 ± 0.2‰), for α-, β- and γ-HCH, while a statistical smaller $\alpha_{ClHCH/Cl2}$ fractionation was observed for δ-HCH (0.9754 ± 0.0002, $\varepsilon^{37}Cl$ = −24.6 ± 0.2‰). In experiments that were applied with ^{13}C-labeled benzene and ^{2}H-labeled benzene, the $\alpha_{ClHCH/Cl2}$ values were lower than those in the other experiments. Liu et al. (2022) hypothesized that the addition of a chlorine atom to the carbon atom of the benzene ring requires higher activation energy when the reacting carbon is ^{13}C or substituted with ^{2}H, leading to a larger chlorine isotopic fractionation. The isotope fractionation shows that, starting from Cl_2 with a $\delta^{37}Cl$ of +6.2 ± 0.2‰, the earliest formed HCH isomers have the lowest $\delta^{37}Cl$ values (approximately between −20 and −14‰. While more benzene is reacted with chlorine, the $\delta^{37}Cl$ of later formed HCHs increases due to Rayleigh fractionation. Because $\delta^{37}Cl$ of HCH is lower than that of Cl_2, the $\delta^{37}Cl$ of the remaining Cl_2 slowly increases, resulting in also increasing $\delta^{37}Cl$ values in later formed HCHs. At the end of the experiments, when only 10 to 20% of the benzene was left $\delta^{37}Cl$ of the produced HCHs is between −12 and −8‰ in most experiments.

10.3.5 Kinetic Fractionation of Organochlorine Compounds in the Environment

Chlorinated organic compounds behave rather complex when they are present in the environment. Apart from fractionation that is caused by chemical or microbiological decomposition they can also fractionate kinetically in circumstances such as evaporation from or diffusion through a system. Although normally considered minor causes for fractionation compared to the decomposition of chlorinated organic compounds in certain circumstances the variations it causes can so be remarkable that the effects are described here.

10.3.5.1 Evaporation of Organochlorine Compounds

Already very early isotope studies on the distillation process of chloroform ($CHCl_3$) and carbon terachloride (CCl_4) in which both carbon and chlorine isotope compositions were studied indicated that while ^{13}C is enriched in the distillate, ^{37}Cl is enriched in the residue (Baertschi et al. 1953; Baertschi and Kuhn 1957). Interestingly it was concluded that there already were indications as early as in 1945 that ^{35}Cl was enriched in the distillate using density measurements, spectroscopic methods and nuclear reactions (Wieland 1943; Alder [unpublished], in Baertschi et al. 1953). Even earlier attempts to measure this fractionation (Grimm 1929; Rank and Kargarise 1951) were unable to indicate isotope fractionation due to the (density) methods used. Baertschi et al. (1953) discovered that the same characteristics were also found for oxygen isotopes and suggested that the origin of this effect is that perhaps a molecule with a heavier central atom exhibits a somewhat higher vapour pressure than the molecule with the lighter isotope as the central atom. These results appear to show the same effect as comparable experiments with boron trichloride that were applied by Green and Martin (1952). In this last study only the isotope fractionation of the central, boron, atom was determined and not that of the chlorine atoms.

The results obtained with chloroform and carbon tetrachloride by Baertschi et al. (1953) were confirmed by the experiments Huang et al. (1999) applied on chlorinated ethanes. They determined the chlorine isotope fractionation during evaporation of chlorinated aliphatic hydrocarbons. They evaporated two different solvents (trichloroethene [TCE] and dichloromethane [DCM]) at room temperature (24 ± 1 °C) and measured both the carbon and chlorine isotope composition of the fraction remaining. Both solvents showed the same characteristics during evaporation. Carbon and chlorine isotopes changed during evaporation according to a Rayleigh fractionation. Interestingly the $\delta^{37}Cl$ of the residue was more positive than the original solvent while the $\delta^{13}C$ in the residue decreased during evaporation. Fractionation factors α for the chlorine isotopes are 0.99818 ± 0.00022 ($\varepsilon^{37}Cl$ = −1.82 ± 0.22‰) in TCE and 0.99952 ± 0.00006 in DCE ($\varepsilon^{37}Cl$ = −0.48 ± 0.06‰) and for the carbon isotopes 1.00031 ± 0.00004 ($\varepsilon^{13}C$ = 0.31 ± 0.04‰) in TCE and 1.00065 ± 0.00002 ($\varepsilon^{13}C$ = 0.65 ± 0.02‰) in DCE. The mechanism for this inverse isotope fractionation was not well understood. The results of the chlorine and carbon isotope fractionation found in this study could be compared to the results obtained by Heraty et al. (1999) in experiments of microbial degradation of dichloromethane. In this study both $\delta^{37}Cl$ and $\delta^{13}C$ of the remaining solvent (dichloromethane) decreased and Huang et al.

(1999) concluded that the relationship between $\delta^{37}Cl$ and $\delta^{13}C$ in chlorinated aliphatic hydrocarbons can be used to distinguish evaporation from biodegradation in contaminated systems.

10.3.5.2 Fractionation Due to Diffusion of Chlorinated Organic Compounds in the Environment

Although it would be expected that organochlorine compounds could fractionate due to their diffusion in the environment this effect, according to Schmidt et al. (2004), did not significantly influence the isotopic composition. This effect may be limited as the mass difference between isotopologues of many organo-chlorine compounds is relatively small. In these cases isotope effects due to fractionation of the compounds may easily be masked by chlorine isotope effects of dehalogenation of the compounds (e.g. Lewandowicz et al. 2001).

Jeannottat and Hunkeler (2012) studied the chlorine and carbon isotope fractionation during volatilisation and diffusive transport of trichloroethene in the unsaturated zone. They found that the different processes produced very distinct isotope effects for carbon and chlorine isotopes which potentially made it possible to distinguish between younger and older organo-chlorine contaminated zones and between the processes that were responsible for the fractionation that occurred during these processes. Unlike the studies cited above Jeannottat and Hunkeler (2012) found a considerable chlorine isotope fractionation during diffusion controlled evaporation of trichloroethene from a sand column of −1.39 ± 0.06‰. This is considerable larger than the carbon isotope fractionation in the same process which was only +0.10 ± 0.05‰, effectively insignificant. This difference was explained by the different effects that apply to chlorine and carbon isotopes during the processes of vaporising the non-aqueous phase liquid and normal diffusion. In the case of carbon the isotope effect for the two processes are opposite, thus showing a heavier vapour than the liquid, cancelling out the negative isotope effect caused by diffusion, while in the case of chlorine the vapour is lighter in the case of non-aqueous phase liquid-vapour equilibrium so that the effect of the normal diffusion is even cumulated.

10.4 Overall Chlorine Isotope Effects Resulting from Environmental Processes

Zimmermann et al. (2020) summarised our current knowledge on methods to analyse chlorine isotope compositions of chlorinated organic compounds, isotope fractionation during various processes that occur in the environment and field applications.

With regard to physical processes that fractionate chlorine isotopes Zimmermann et al. (2020) discuss various possible processes. First, liquid-vapour fractionation that has been studied already at an early time (e.g. Baertschi et al. 1953; Bradley 1954), and can lead to, in principle, large chlorine isotope variations in the environment (Jeannottat and Hunkeler 2012). Diffusion was discussed as an important process to fractionate chlorine isotopes in organic compounds. It is mostly considered as diffusion in aqueous solutions and studied and recently reviewed by Wanner and Hunkeler (2015, 2019). Sorption can also lead to isotope fractionation as shown by Vakili (2017) who observed little to no carbon isotope effects during sorption, but rather significant ($\varepsilon^{37}Cl$ = 0.2‰ for *cis*-1,2-dichloroethene and $\varepsilon^{37}Cl$ = −0.85‰ for tetrachloroethene) chlorine isotope fractionation during sorption, where the lighter isotope was enriched in the sorpted phase. Zimmermann et al. (2020) state that chlorine isotope fractionation due to physical processes is mostly caused by various processes that take place at the same time, but has not been studied in much detail in field studies. They indicate that the reason for this limited interest is caused by the believe that physical processes have less influence on chlorine isotope variations than (bio)chemical degradation processes, and may be less pronounced once they have reached a steady state as discussed by Schmidt and Jochmann (2012).

Chlorine isotope fractionation due to biotic (mostly through bacteria) and abiotic (through chemical oxidation or reduction) degradation has been studied regularly and these studies showed large variations in both chlorine and carbon isotope fractionation. Zimmermann et al. (2020) summarised biotic studies by Heraty et al. (1999), Palau et al. (2014a, 2017), Rosell et al. (2019), Abe et al. (2009), Kuder et al. (2013), Cretnik et al. (2013, 2014), Wiegert et al. (2013), Renpenning et al. (2014) and Badin et al. (2014). Depending on the compound, proposed mechanism and culture the bulk chlorine isotope fractionation ($\varepsilon^{37}Cl$) varies mostly between −5.1 and −0.3‰, with one very negative fractionation at −12‰, while the bulk carbon isotope fractionation ($\varepsilon^{13}C$) varies between −42.4 and −0.7‰. Large variations are observed in both carbon and chlorine isotope fractionation, but they are all negative, indicating that the isotope values of both chlorine and carbon will decrease during biotic degradation of selected organic compounds (see Fig. 10.2). Zimmermann et al. (2020) also summarised abiotic studies by Torrentó et al. (2017), Rodriguez-Fernández et al. (2018), Palau et al. (2014b), Audi-Miró et al. (2013) and Liu et al. (2014). Here also large variations in both chlorine (from −6.2 to −0.44‰) and carbon (from −57 to −1.6‰) isotope fractionation were observed (see also Fig. 10.2). Just like in the case of biotic fractionation in all cases fractionation is negative, indicating decreasing isotope values during abiotic degradation too. Although the spread in the fractionation is rather large it appears there are no clear systematics, either between type of fractionation or compound that is degrading. Also for individual compounds, isotope fractionation can be very different depending on the culture or the reactant that causes the isotope fractionation.

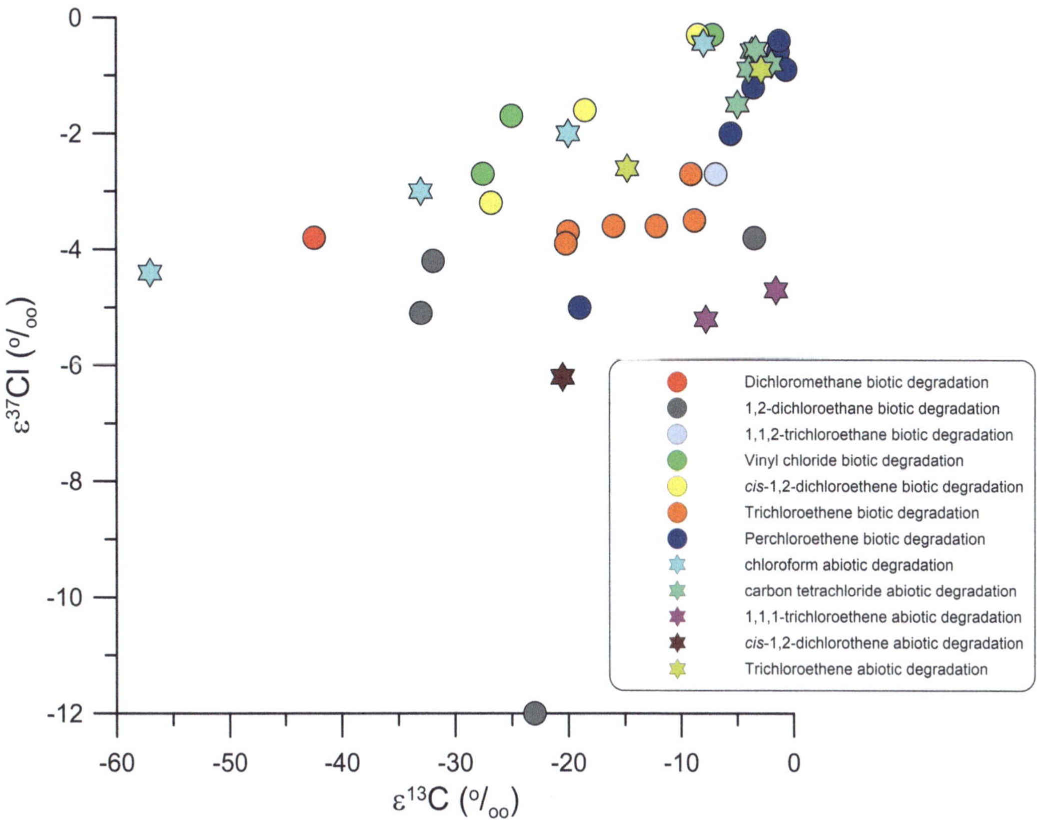

Fig. 10.2 Summary of the various $\varepsilon^{13}C$ and $\varepsilon^{37}Cl$ isotope fractionation data reviewed by Zimmermann et al. (2020)

The results of these laboratory studies indicate that application to field studies is, due to the large number of variables, very complex. In most cases multiple sources and degradation pathways have to be identified. Zimmermann et al. (2020) conclude that more systematic studies are necessary to fully understand effects of degradation of organochlorine compounds in the environment. What they consider especially important is the need for a reliable method of producing isotopically different, compound-specific sets of reference standards. Overall, in their review Zimmermann et al. (2020) have shown that there is great potential in the combination of chlorine and carbon isotopes from organic contaminants in the environment, but that several additional efforts are needed to overcome the many challenges that still exist, not only on the analytical part but also because of the lack of well understood fractionation factors and processes.

10.5 Isotope Fractionation of Bromine in Organic Compounds

Of all organobromine compounds bromomethane (CH_3Br) is by far the most important. This compound has the highest concentration of all long living organobromine compounds in the atmosphere (Schauffler et al. 1993) and is known to originate from both natural and anthropogenic sources. As bromomethane is about 50 times more efficient as chloromethane to break down atmospheric ozone even the small concentrations that are present in the stratosphere could be responsible for as much as 5% of the global loss of stratospheric ozone (WMO 1995; UNEP 1992). For that reason production of bromomethane was largely prohibited under the Montreal Protocol (Cincinelli et al. 2012). Considering however the fact that it is also produced naturally it is very important to understand the origin of bromomethane in the stratosphere and several isotope studies have been conducted to help understand its origin. These studies unfortunately concerned only the carbon and not the bromine isotope variations of bromomethane (Bill et al. 2002a, 2002b, 2004). Considering the fact that only few methods have been developed to analyse bromine isotopes in organic compounds (Sylva et al. 2007; Gelman and Halicz 2011; Holmstrand et al. 2010; Hitzfeld et al. 2011; Thornton et al. 2013; Zakon et al. 2016) it is perhaps not surprising that only a relatively small number of studies have been conducted to determine bromine isotope fractionation during production or decomposition of these organic compounds since the early studies by Willey and Taylor (1980).

The bromine isotope composition of atmospheric bromomethane has been determined for the first time by Horst et al. (2013). Their data indicated more negative $\delta^{81}Br$ values in subarctic samples from the north of Sweden as compared to more positive values in the Stockholm area. The bromomethane concentration in the subarctic samples were two to three times lower than the Stockholm samples, and it was concluded that the Stockholm samples were contaminated with industrial bromomethane.

The first attempt to estimate the isotope difference between natural and anthropogenic organobromine compounds was done by Carrizo et al. (2011). Although this study did not measure the actual fractionation during formation of a compound from its source material it compared industrially and naturally produced 2,4-bromophenol. They found that the $\delta^{81}Br$ value for an industrially produced 2,4-dibromophenol was $-1.1 \pm 0.9‰$ (1 SD) while a naturally produced 2,4-dibromophenol had a $\delta^{81}Br$ of $+0.2 \pm 1.6‰$ (1 SD). This difference indicated a statistically significant difference between the two samples of about 1.3‰. They concluded that different pathways of formation of a these compounds lead to different isotope compositions, implicating that significant isotope fractionation during their formation exist.

Horst (2013) and Horst et al. (2014) attempted to determine isotope fractionation of bromomethane during its release from plant matter. They investigated bromomethane formed from pectin (fortified with KBr) and from *Salicornia fruticosa*, a halophytic plant with a high salt content. They found that the bromomethane released from both the pectine and *Salicornia fruticosa* showed δ^{81}Br values that were about 2‰ lower than the original KBr suggesting a fractionation of this order during the formation of bromomethane.

During photolysis of brominated phenols both bromine and carbon isotope fractionate. Zakon et al. (2013) observed that bromine isotope fractionation, with values up to +5.1‰ was opposite in sign to carbon isotope fractionation with values between −12.6 to −23.4‰. These results may be attributed to the coexistence of both mass-dependent and mass-independent isotope fractionation of the carbon-bromine bond cleavage. These effects were found for different bromophenols that were dissolved in ethanol, but not in water. In water this effect was only found for 4-bromophenol, while in the case of 2-bromophenol no fractionation was observed and in the case of 3-bromophenol only fractionation of bromine was observed. It was suggested that these different results might be explained in the case of 2-bromophenol with the fact that carbon-bromine bond cleavage is not a rate-limiting step in the reaction, and in the case for 3-bromophenol that a mass independent fractionation effect takes place.

Kuntze et al. (2016) determined dual carbon-bromine isotope fractionation during different chemical and biological transformations of ethylene dibromide. Different correlations for carbon and bromine isotope fractionation ($\Lambda_{C/Br}$) was observed not only for processes following different transformation pathways, but also for abiotic and biotic processes with presumed same formal chemical degradation mechanisms. The studied processes resulted in a wide range of $\Lambda_{C/Br}$ values: $\Lambda_{C/Br} = 30.1$ was observed for hydrolysis of EDB in alkaline solution; $\Lambda_{C/Br}$ values between 4.2 and 5.3 were determined for dibromoelimination pathway with reduced corrinoids and Zn(0) particles; EDB biodegradation by *Ancylobacter aquaticus* and *Sulfurospirillum multivorans* resulted in $\Lambda_{C/Br} = 10.7$ and 2.4, respectively; Fenton-like degradation resulted in carbon isotope fractionation only, leading to infinite (∞) $\Lambda_{C/Br}$. Relatively large bromine isotope effects with ^{81}Br apparent kinetic isotope effects of 1.0012–1.002 and 1.0021–1.004 were observed for nucleophilic substitution and dibromoelimination, and reveal strong bromine isotope effects, that were not recognised before the study by Kuntze et al. (2016).

Recent developments in the application of compound-specific isotope analyses of bromine and the resulting knowledge on bromine isotope effects during degradation of brominated organic compounds in the environment were reviewed by Xiong et al. (2022). This review discusses the various analytical methods developed and discusses the application for identifying sources, tracing transformation pathways and biological metabolisms.

References

Abe Y, Aravena R, Zopfi J, Shouakar-Stash O, Cox E, Roberts JD, Hunkeler D (2009) Carbon and chlorine isotope fractioantion during aerobic oxidation and reductive dechlorination of vinyl chloride and cis-1,2-dichloroethane. Environ Sci Technol 43:101–107

Aeppli C, Bastviken D, Andersson P, Gustafsson Ö (2013a) Chlorine isotope effects and composition of naturally produced organochlorines from chloroperoxidases, flavin-dependent halogenases, and in forest soil. Environ Sci Technol 47:790–797

Aeppli C, Tysklind M, Holmstrand H, Gustafsson Ö (2013b) Use of Cl and C isotopic fractionation to identify degradation and sources of polychlorinated phenols: mechanistic study and field application. Environ Sci Technol 47:6864–6871

Arnold WA, Roberts AL (2000) Pathways and kinetics of chlorinated ethylene and chlorinated acetylene reaction with Fe(0) particles. Environ Sci Technol 34:1794–1805

Audí-Miró C, Cretnik S, Otero N, Palau J, Shouakar-Stash O, Soler A, Elsner M (2013) Cl and C isotope analysis to assess the effectiveness of chlorinated ethene degradation by zero-valent iron: evidence from dual element and product isotope values. Appl Geochem 32:175–173

Badin A, Buttet G, Maillard J, Holliger C, Hunkeler D (2014) Multiple dual C-Cl isotope patterns associated with reductive dechlorination of tetrachloroethene. Environ Sci Technol 48:9179–9186

Baertschi P, Kuhn W (1957) Dampfdruckunterschiede isotoper Verbindungen (Infrarot-Anteil der Dispersionswechselwirkung als Ursache für grössere Flüchtigkeit der schweren Molekelspezies). Helv Chim Acta 40:1084–1103

Baertschi P, Kuhn W, Kuhn H (1953) Fractionation of isotopes by distillation of some organic substances. Nature 171:1018–1020

Bartholomew RM, Brown F, Lounsbury M (1954a) Chlorine isotope effect. Nature 174:133

Bartholomew RM, Brown F, Lounsbury M (1954b) Chlorine isotope effect in reactions of *tert*-butyl chloride. Can J Chem 32:979–983

Baumann H, Heumann KG (1988) Chlorisotopenseparation als möglicher Indikator für Emissionsquellen chlorhaltiger Verbindungen. Fresenius Z Anal Chem 330:637–638

Bill M, Rhew RC, Weiss RF, Goldstein AH (2002a) Carbon isotope ratios of methyl bromide and methyl chloride emitted from a coastal salt marsh. Geophys Res Lett 29:4. https://doi.org/10.1029/2001GL012946

Bill M, Miller LG, Goldstein AH (2002b) Carbon isotope fractionation of methyl bromide during agricultural soil fumigations. Biogeochem 60:181e190

Bill M, Conrad ME, Goldstein AH (2004) Stable carbon isotope composition of atmospheric methylbromide. Geophys Res Lett 31:L04109

Bradley DC (1954) Fractionation of isotopes by distillation of some organic substances. Nature 173:260–261

Burton GW, Sims LB, Wilson JC, Fry A (1977) Calculation of ^{14}C, ^{37}Cl, and deuterium kinetic isotope effects in solvolysis of tert-butyl chloride. J Am Chem Soc 99:3371–3379

Carrizo D, Unger M, Holmstrand H, Andersson P, Gustafsson Ö, Sylva SP, Reddy CM (2011) Compound-specific bromine isotope compositions of one natural and six industrially synthesised organobromine substances. Environ Chem 8:127–132

Cincinelli A, Pieri F, Zhang Y, Seed M, Jone KC (2012) Compound specific isotope analysis (CSIA) for chlorine and bromine: a review of techniques and applications to elucidate environmental sources and processes. Environ Poll 169:112–127

Coleman ML, Ader M, Chaudhuri S, Coates JD (2003) Microbial isotopic fractionation of perchlorate chlorine. Appl Environ Microbiol 69:4997–5000

Cretnik S, Thoreson KA, Bernstein A, Ebert K, Buchner D, Laskov C, Haderlein S, Shouakar-Stash O, Kliegman S, McNeill K, Elsner M (2013) Reductive Dechlorination of TCE by chemical model Systems in Comparison to Dehalogenating bacteria: insights from dual element isotope analysis ($^{13}C/^{12}C$, $^{37}Cl/^{35}Cl$). Environ Sci Technol 47:6855–6863

Cretnik S, Bernstein A, Shouakar-Stash O, Löffler F, Elsner M (2014) Chlorine isotope effects from isotope ratio mass spectrometry suggest intramolecular C-Cl bond competition in Trichloroethene (TCE) reductive dehalogenation. Molecules 19:6450–6473

Czarnacki M, Hałas S (2012) Isotope fractionation in aqua-gas systems: Cl_2-HCl-Cl^-, Br_2-HBr-Br^- and H_2S-S^{2-}. Isot Environ Health Stud 48:55–64

Drenzek NJ, Tarr C, Eglinton TI, Heraty L, Sturchio NC (2002) Stable chlorine and carbon isotopic compositions of selected semi-volatile organochlorine compounds. Org Chem 33:437–444

Drenzek NJ, Eglinton TI, Wirsen CO, Sturchio NC, Heraty LJ, Sowers KR, Wu QZ, May HD, Reddy CM (2004) Invariant chlorine isotopic signatures during microbial PCB reductive dechlorination. Environ Pollut 128:445–448

Elsner M, Chartrand M, Vanstone M, Lacrampe-Couloume G, Sherwood Lollar B (2008) Identifying abiotic chlorinated ethene degradation: characteristic isotope patterns in reaction products with nanoscale zero-valent iron. Environ Sci Technol 42:5963–5970

Fry A (1964) Application of the successive labelling technique to some carbon, nitrogen and chlorine isotope effect studies of organic reaction mechanisms. Pure Appl Chem 8:409–420

Gelman F, Halicz L (2011) High-precision isotope ratio analysis of inorganic bromide by continuous flow MC-ICPMS. Int J Mass Spectrom 307:211–213

Graczyk DG, Taylor JW (1974) Chlorine kinetic isotope-effects in nucleophilic-substitution reactions – support for ion-pairs mechanism in reactions of p-methoxybenzyl chloride in 70 percent aqueous acetone. J Am Chem Soc 96:3255–3261

Graczyk DG, Taylor JW, Turnquist CR (1978) Use of chlorine kinetic isotope effects for evaluating ion-pairing in nucleophilic displacements at saturated carbon. J Am Chem Soc 100:7333–7339

Green M, Martin GR (1952) The vapour pressure of isotopic substances. 1. Boron trichloride. Trans Faraday Soc 48:416–421

Gribble GW (2010) Naturally occurring organohalogen compounds—a comprehensive update. Progress in the chemistry of organic natural products 91. Springer

Grimm HG (1929) Tests on the determination of vapour differences of isotopes through fractional distillation. I. Distallation testing on CCl_4. Z Phys Chem 2:181–199

Grimsrud EP, Taylor JW (1970) Chlorine kinetic isotope effects in nucleophilic displacements at a saturated carbon. J Am Chem Soc 92:739–741

Harper DB (1985) Halomethane from halide ion – a highly efficient fungal conversion of environmental significance. Nature 315:55–57

Harper DB (1994) Biosynthesis of halogenated methanes. Biochem Soc Trans 22:1007–1011

Harper DB, Kennedy J, Hamilton JTG (1988) Chloromethane biosynthesis in poroid fungi. Phytochemistry 27:3147–3153

Heckel B, Cretnik S, Kliegman S, Shouakar-Stash O, McNeill K, Elsner M (2017) Reductive outer-sphere single electron transfer is an exception rather than the rule in natural and engineered chlorinated ethene dehalogenation. Environ Sci Technol 51:9663–9673

Heckel B, Phillips E, Edwards E, Sherwood Lollar B, Elsner M, Manefield MJ, Lee M (2019) Reductive dehalogenation of trichloromethane by two different dehalobacter restrictus strains reveal opposing dual element isotope effects. Environ Sci Technol 53:2332–2343

Heraty LJ, Fuller ME, Huang L, Abrajano Jr T, Sturchio NC (1999) Isotope fractionation of carbon and chlorine by microbial degradation of dichloromethane. Org Geochem 30:793–799

Hill JW, Fry A (1962) Chlorine isotope effects in the reactions of benzyl and substituted benzyl chlorides with various nucleophiles. J Am Chem Soc 84:2763–2769

Hitzfeld KL, Gehre M, Richnow HH (2011) A novel online approach to the determination of isotopic ratios for organically bound chlorine, bromine and Sulphur. Rapid Commun Mass Spectrom 25:3114–3122

Holmstrand H, Unger M, Carrizo D, Andersson P, Gustafsson Ö (2010) Compound-specific bromine isotope analysis of brominated diphenyl ethers using GC-ICP-multicollector-MS. Rapid Commun Mass Spectrom 24:2135–2142

Holt BD, Sturchio NC, Abrajano TA, Heraty LJ (1997) Conversion of chlorinated volatile organic compounds to carbon dioxide and methyl chloride for isotopic analysis of carbon and chlorine. Anal Chem 69:2727–2733

Holt BD, Heraty LJ, Sturchio NC (2001) Extraction of chlorinated aliphatic hydrocarbons from groundwater at micromolar concentrations for isotopic analysis of chlorine. Environ Pollut 113:263–269

Horst A (2013) Stable bromine isotopic composition of methyl bromide. Method development and applications. PhD thesis. Stockholm University

Horst A, Thornton BJ, Holmstrand H, Andersson P, Crill PM, Gustafsson Ö (2013) Stable bromine isotopic composition of atmospheric CH_3Br. Tellus Ser B – Chem Phys Meteorol 65:21040

Horst A, Andersson P, Thornton BJ, Holmstrand H, Wishkerman A, Keppler F, Gustafsson Ö (2014) Stable bromine isotopic composition of methyl bromide released from plant matter. Geochim Cosmochim Acta 125:186–195

Huang L, Sturchio NC, Abrajano T, Heraty LJ, Holt BD (1999) Carbon and chlorine isotope fractionation of chlorinated aliphatic hydrocarbons by evaporation. Org Geochem 30:777–785

Hunkeler D, Van Breukelen BM, Elsner M (2009) Modelling chlorine isotope trends during sequential transformation of chlorinated ethenes. Environ Sci Technol 43:6750–6756

Ivdra N, Fischer A, Herrero-Martin S, Giunta T, Bonifacie M, Richnow HH (2017) Carbon, hydrogen and chlorine stable isotope fingerprinting for forensic investigations of hexachlorocyclohexanes. Environ Sci Technol 51:446–454

Jeannottat S, Hunkeler D (2012) Chlorine and carbon isotopes fractionation during volatilization and diffusive transport of trichloroethane in the unsaturated zone. Environ Sci Technol 46:3169–3176

Jendrzejewski N, Eggenkamp HGM, Coleman ML (1997) Sequential determination of chlorine and carbon isotopic composition in single microliter samples of chlorinated solvent. Anal Chem 69:4259–4266

Julian RL, Taylor JW (1976) Evaluation of possible S_N2 transition-state models for reaction of n-butyl chloride using chlorine kinetic isotope effects. J Am Chem Soc 98:5238–5248

Kuder T, van Breukelen BM, Vanderford M, Philp P (2013) 3d-CSIA: carbon, chlorine, and hydrogen isotope fractionation in transformation of TCE to ethene by a *Dehalococcoides* culture. Environ Sci Technol 47:9668–9677

Kuntze K, Kozell A, Richnow HH, Halicz L, Nijenhuis I, Gelman F (2016) Dual carbon-bromine stable isotope analysis allows distinguishing transformation pathways of ethylene dibromide. Environ Sci Technol 50:9855–9863

Lewandowicz A, Rudzinski J, Tronstad L, Widersten M, Ryberg P, Matsson O, Paneth P (2001) Chlorine kinetic isotope effects on the haloalkane dehalogenase reaction. J Am Chem Soc 123:4550–4555

Liu CC, Tseng DH, Wang CY (2006) Effects of ferrous iron on the reductive dechlorination of trichloroethylene by zero-valent iron. J Hazard Mater B136:706–713

Liu Y, Gan Y, Zhou A, Liu C, Li X, Yu T (2014) Carbon and chlorine isotope fractionation during Fenton-like degradation of trichloroethene. Chemosphere 107:94–100

Liu X, Wu L, Kümmel S, Merbach I, Lal R, Richnow HH (2020a) Compound specific isotope analysis and enantiomer fractionation to characterize the transformation of hexachlorocyclohexane isomers in a soil-wheat pot system. Environ Sci Technol 54:8690–8698

Liu X, Wu L, Kümmel S, Richnow HH (2020b) Characterizing the biotransformation of hexachlorocyclohexanes in wheat using compound-specific stable isotope analysis and enantiomer fraction analysis. J Hazard Mater 406:105420

Liu X, Wu LP, Kümmel S, Richnow HH (2022) Stable isotope fractionation associated with the synthesis of hexachlorocyclohexane isomers for characterizing sources. Chemosphere 296:133938

McIlvaine TC (1921) A buffer solution for colorimetric comparison. J Biol Chem 49:183–186

Nair PM (1956) PhD dissertation. University of Arkansas, Fayetteville, Arkansas

Öberg G (2002) The natural chlorine cycle – fitting the scattered pieces. Appl Microbiol Biotechnol 58:565–581

Palau J, Cretnik S, Shouakar-Stash O, Höche M, Elsner M, Hunkeler D (2014a) C and Cl isotopc fractionation of 1,2-dichloroethane displays unique $\delta^{13}C/\delta^{37}Cl$ patterns for pathway identification and reveals surprising C-Cl bond involvement in microbial oxidation. Environ Sci Technol 48:9430–9437

Palau J, Shouakar-Stash O, Hunkeler D (2014b) Carbon and chlorine isotope analysis to identify abiotic degradation pathways of 1,1,1-trichloroethane. Environ Sci Technol 48:14400–14408

Palau J, Yu R, Hatijah Mortan S, Shouakar-Stash O, Rosell M, Freedman DL, Sbarbati C, Fiorenza S, Aravena R, Marco-Urrea E, Elsner M, Soler A, Hunkeler D (2017) Distinct dual C-Cl isotope fractionation patterns during anaerobic biodegradation of 1,2-dichloroethane: potential to characterize microbial degradation in the field. Environ Sci Technol 51:2685–2694

Paneth P (1992) In isotopes in organic chemistry (eds: Buncel E, Saunders WH). Elsevier

Rank DH, Kargarise RE (1951) Vapour pressure of isotopic carbon tetrachlorides. Nature 168:740–741
Reddy CM, Heraty LJ, Holt BD, Sturchio NC, Eglinton T, Drenzek NJ, Xu L, Lake JL, Maruya KA (2000) Stable chlorine isotopic compositions of aroclors and aroclor-contaminated sediments. Environ Sci Technol 34:2866–2870
Reddy CM, Drenzek NJ, Eglinton TI, Heraty LJ, Sturchio NC, Shiner VJ (2001) Environ Sci Pollut Res 9:183
Reddy CM, Xu L, Drenzek NJ, Sturchio NC, Heraty LJ, Kimblin C, Butler A (2002) A chlorine isotope effect for enzyme-catalyzed chlorination. J Am Chem Soc 124:14526–14527
Renpenning J, Keller S, Cretnik S, Shouakar-Stash O, Elsner M, Schubert T, Nijenhuis I (2014) Combined C and Cl isotope effects indicate differences between corrinoids and enzyme (sulfurospirillum multivorans PceA) in reductive dehalogenation of tetrachloroethene, but not trichloroethene. Environ Sci Technol 48:11837–11845
Rodríguez-Fernández D, Heckel B, Torrentó C, Meyer A, Elsner M, Hunkeler D, Soler A, Rosell M, Domènech C (2018) Dual element (CCl) isotope approach to distinguish abiotic reactions of chlorinated methanes by Fe(0) and by Fe(II) on iron minerals at neutral and alkaline pH. Chemosphere 206:447–456
Rosell M, Palau J, Mortan SH, Caminal G, Soler A, Shouakar-Stash O, Marco-Urrea E (2019) Dual carbon – chlorine isotope fractionation during dichloroelimination of 1,1,2-trichloroethane by an enrichment culture containing *Dehalogenimonas sp.* Sci Total Environ 648:422–429
Schauble EA, Rossman GR, Taylor HP (2003) Theoretical estimates of equilibrium chlorine-isotope fractionations. Geochim Cosmochim Acta 67:3267–3281
Schauffler SM, Heidt LE, Pollock WH, Gilpin TM, Vedder JF, Solomon S, Lueb RA, Atlas EL (1993) Measurements of halogenated organic compounds near the tropical tropopause. Geophys Res Lett 20:2567–2570
Schmidt TC, Jochmann MA (2012) Origin and fate of organic compounds in water: characterization by compound-specific stable isotope analysis. Annu Rev Anal Chem 5:133–155
Schmidt TC, Zwank L, Elsner M, Berg M, Meckenstock RU, Haderlein SB (2004) Compound-specific stable isotope analysis of organic contaminants in natural environments: a critical review of the state of the art, prospects, and future challenges. Anal Bioanal Chem 378:283–300
Sims LB, Lewis ED (1985) In isotopes in organic chemistry (eds: Buncel E, Saunders WH). Elsevier
Soder-Walz JM, Torrentó C, Algora C, Wasmund K, Cortés P, Soler A, Vicent T, Rosell M, Marco-Urrea E (2022) Trichloromethane dechlorination by a novel Dehalobacter sp. strain 8M reveals a third contrasting C and Cl isotope fractionation pattern within this genus. Sci Tot Environ 813:152659
Streitwieser A, Jagow RH, Fahey RC, Suzuki S (1958) Kinetic isotope effects in the acetolyses of deuterated cyclopentyl tosylates. J Am Chem Soc 80:2326–2332
Sturchio NC, Clausen JL, Heraty LJ, Huang L, Holt BD, Abrajano TA (1998) Chlorine isotope investigation of natural attenuation of trichloroethene in an aerobic aquifer. Environ Sci Technol 32:3037–3042
Sturchio NC, Hatzinger PB, Arkins MD, Suh C, Heraty LJ (2003) Chlorine isotope fractionation during microbial reduction of perchlorate. Environ Sci Technol 37:3859–3863
Świderek K, Paneth P (2012) Extending limits of chlorine kinetic isotope effects. J Org Chem 77:5120–5124
Sylva SP, Ball L, Nelson RK, Reddy CM (2007) Compound-specific $^{81}Br/^{79}Br$ analysis by capillary gas chromatography/multicollector inductively coupled plasma mass spectrometry. Rapid Commun Mass Spectrom 21:3301–3305
Taylor JW, Grimsrud EP (1969) Chlorine isotopic ratios by negative ion mass spectrometry. Anal Chem 41:805–810
Thornton B, Horst A, Carrizo D, Holmstrand H, Andersson P, Crill PM, Gustafsson Ö (2013) A high-volume cryosampler and sample purification system for bromine isotope studies of methyl bromide. J Atmos Ocean Technol 30:2095–2107
Torrentó C, Palau J, Rodríguez-Fernández D, Heckel B, Meyer A, Domènech C, Rosell M, Soler A, Elsner M, Hunkeler D (2017) Carbon and chlorine isotope fractionation patterns associated with different engineered chloroform transformation reactions. Environ Sci Technol 51:6174–6184
Turnquist CR, Taylor JW, Grimsrud EP, Williams RC (1973) Temperature dependence of chlorine kinetic isotope effects for aliphatic chlorides. J Am Chem Soc 95:4133–4138
UNEP [United Nations Environment Programme] (1992) Methyl bromide: its atmospheric science, technology, and economics (Montreal Protocol Assessment Supplement)
Vakili F (2017) Stable isotope fractionations of chlorinated ethenes associated with physical processes. PhD Thesis. University of Waterloo, Ontario, Canada
Van Warmerdam EM, Frape SK, Aravena R, Drimmie RJ, Flatt H, Cherry JA (1995) Stable chlorine and carbon isotope measurements of selected chlorinated organic solvents. Appl Geochem 10:547–552
Wanner P, Hunkeler D (2015) Carbon and chlorine isotopologue fractionation of chlorinated hydrocarbons during diffusion in water and low permeability sediments. Geochem Cosmochim Acta 157:198–212
Wanner P, Hunkeler D (2019) Isotope fractionation due to aqueous phase diffusion – what do diffusion models and experiments tell? – a review. Chemosphere 219:1032–1043
Westaway KC, Koerner T, Fang YR, Rudziński J, Paneth P (1998) A new method of determining chlorine kinetic isotope effects. Anal Chem 70:3548–3552
Wiegert C, Mandalakis M, Knowles T, Polymenakou PN, Aeppli C, Macháčková J, Holmstrand H, Evershed RP,

Pancost RD, Gustafsson O (2013) Carbon and chlorine isotope fractionation during microbial degradation of tetra- and Trichloroethene. Environ Sci Technol 47:6449–6456
Wieland K (1943) Spektroskopische Bestimmung der Anreicherung eines Chlorisotopen mit minimalen Mengen von Quecksilber(II)-chlorid (Sublimat). Helv Chim Acta 26:1939–1944
Willey JF, Taylor JW (1978) Capacitive integration to produce high precision isotope ratio measurements on methyl chloride and bromide. Anal Chem 50:1930–1933
Willey JF, Taylor JW (1980) Temperature-dependence of bromine kinetic isotope effects for reactions of normal-butyl and *tert*-butyl bromides. J Am Chem Soc 102:2387–2391
Williams RC, Taylor JW (1973) Chlorine kinetic isotope-effect models. 1. Isotopic dependence in nominal C-Cl stretching vibrations of aliphatic chlorides and vibrational analysis of *tert*-butyl chloride ground-state. J Am Chem Soc 95:1710–1714
Williams RC, Taylor JW (1974) Chlorine kinetic isotope-effect models.2. Vibrational analysis and kie calculations of *tert*-butyl chloride transition-state models. J Am Chem Soc 96:3721–3726
WMO [World Meteorological Organization] (1995) Scientific Assessment of Ozone Depletion: 1994. Global Ozone Research and Monitoring Project Report No. 37, Geneva
Wu L, Liu Y, Liu X, Bajaj A, Sharma M, Lal R, Richnow HH (2019) Isotope fractionation approach to characterize the reactive transport processes governing the fate of hexachlorocyclohexanes at a contaminated site in India. Environ Int 132:105036
Xiong JK, Li GY, An TC (2022) New advance in the application of compound-specific isotope analysis (CSIA) in identifying sources, transformation mechanisms and metabolism of brominated organic compounds. Crit Rev Environ Sci Technol 52:3973–3996
Zakon Y, Halicz L, Gelman F (2013) Bromine and carbon isotope effects during photolysis of brominated phenols. Environ Sci Technol 47:14147–14153
Zakon Y, Halicz L, Lev O, Gelman F (2016) Compound-specific bromine isotope ratio analysis using gas chromatography/quadrupole mass spectrometry. Rapid Commun Mass Spectrom 30:1951–1956
Zimmermann J, Halloran LJS, Hunkeler D (2020) Tracking chlorinated contaminants in the subsurface using compound-specific chlorine isotope analysis: a review of principles, current challenges and applications. Chemosphere 244:125476
Zyakun AM, Firsova YE, Torgonskaya ML, Doronina NV, Trotsenko YA (2007) Changes of chlorine isotope composition characterize bacterial dehalogenation of dichloromethane. Appl Biochem Microbiol 43:593–597

Part IV

Natural Chlorine and Bromine Isotope Variations on Earth

11 Variations of Stable Chlorine and Bromine Isotopes in the Hydrosphere and in Evaporites

On Earth, chlorine and bromine are distributed quite systematically in the large geological reservoirs. Due to their chemical character, both elements are primarily concentrated in the fluid phase of any system and as a consequence, they are highly concentrated in the oceans on earth. The concentrations in rocks and sediments are relatively low, except for evaporites where a considerable part of the chloride and to a lesser extend the bromide that was once stored in the oceans is concentrated (Schilling et al. 1978; Knauth 1998; Hay et al. 2006). The main geological reservoirs that can be recognised concerning the distribution of chlorine and bromine are, as already discussed in Sect. 3.2.1, the oceans, the mantle, evaporites, sediment (pore water), the continental crust and the oceanic crust. Chlorine and bromine isotope variations in the hydrosphere and evaporites will be discussed in this chapter, the isotope variations in the other reservoirs will be discussed in the following chapters. The atmosphere is only a small reservoir. Still, the chlorine and bromine isotope measurements of chlorine and bromine present in the atmosphere are described in this book, as atmospheric chlorine and bromine have considerable influences on environmental processes. In the following chapters an overview will be given on our knowledge of chlorine and bromine isotope compositions of the terrestrial (and extraterrestrial) reservoirs, as well as their variations within these reservoirs. Chlorine isotopes have been studied, at least globally, in most of the reservoirs. Knowledge on bromine isotopes is still rudimentary, but sufficient information is available to present basic information on its distribution on earth.

As average $\delta^{37}Cl$ values of the major reservoirs appear to be relatively close to zero most variations within the reservoirs are the result of fractionation within the reservoirs. Fractionation during transfer between the reservoirs will take place, but is generally not large enough to change the average isotope composition of the reservoirs involved. The main processes that cause fractionation of chlorine and bromine isotopes have been described in Chap. 8. In this chapter a number of studies are presented in which isotope variations in the hydrosphere and the evaporites (which are the major chlorine and bromine sinks of geochemical relevance) are observed. I will evaluate how these natural variations are interpreted, either in the sense of fractionation in a certain environment or to use the chlorine and bromine isotope compositions as a tracer for the origin of chlorine or bromine in a system.

In the following sections the focus will be on chlorine isotope variations in fairly large detail. Bromine isotope variations will be discussed for as far as they have been studied. Since most studies still describe chlorine isotope variations only, the number of bromine isotope studies is significantly fewer than those on chlorine isotopes. For

H. Eggenkamp, *The Geochemistry of Stable Chlorine and Bromine Isotopes*, Advances in Isotope Geochemistry, https://doi.org/10.1007/978-3-031-75633-7_11

that reason, bromine isotope variations will only be discussed if information on them is available.

11.1 The Hydrosphere

Due to the high solubility of chloride and bromide both ions are heavily concentrated in the hydrosphere. This includes not only the oceans but also sediment formation waters where according to Land (1995) and Knauth (1998) even more chloride and bromide are present than in the oceans (see Chap. 3). As indicated in Table 3.2 most chloride and bromide in the surface reservoirs (crust, sediments and oceans) is present in aqueous solution, either in the oceans or in formation waters in sediments.

11.1.1 Oceans

11.1.1.1 Chloride

Considering the long residence time of chloride in the oceans (87,000,000 years, Berner and Berner 1987) it was already proposed in early studies (Hoering and Parker 1961) that ocean water might be used as reference standard for chlorine isotope studies. As no studies were done on natural samples which resulted in significant chlorine isotope variations before Kaufmann (1984) this standard was not defined, and formally it still is not defined. Although an isotope reference material for chlorine (NIST SRM 975) was available, and that was used to define the absolute atomic weight of chlorine (Shields et al. 1962), it was never really used as reference standard for chlorine stable isotope measurements and has been exhausted for some time. It has been replaced with SRM 975a (NIST 2001) which is depleted in δ^{37}Cl relative to SRM 975 by 0.23‰. Kaufmann (1984) measured the isotope composition of eight seawater samples from different locations and depths. The standard deviation of 12 measurements of these 8 samples was only 0.06‰. As it was also considered that ocean water is probably the ultimate source of most groundwater chloride, ocean water chloride was selected as the standard for stable chlorine isotope measurements. It has to be realised that, in this early research period at The University of Arizona, the main target was to study the chloride cycle in groundwater. Kaufmann (1984) suggested to call this standard SMOC (Standard Mean Ocean Chloride). This name was called a misnomer by Xiao et al. (2002) because in reality no formally accepted chlorine isotope ocean water standard exists. Godon et al. (2004a) collected 24 ocean water samples from all major oceans. Most of these waters were collected from the surface of the oceans and relatively close to the coast. Also samples from three depth profiles were analysed. These samples were analysed and all were within 0.1‰ of the laboratory reference ocean water standard. From this study it was concluded that there is no need to define a specific international standard for chlorine stable isotope measurements. This conclusion was confirmed later by Liu et al. (2013) who analysed ten seawater samples taken from different locations along the Chinese coast and found that all samples were within 0.01‰ (with a standard deviation of 0.006‰) of their seawater standard. As a result of these observations the name SMOC is used in most studies on chlorine stable isotopes and refers in general to the internal laboratory chloride standard of each of the laboratories studying chlorine isotope variations.

A few studies exist where apparently variations in chlorine isotope compositions in ocean water samples were observed. Shirodkar et al. (2003) measured chlorine isotope profiles at three locations on the Indian Ocean Mid-Oceanic Ridge. At all three locations significant variations in the δ^{37}Cl were found. Surprisingly at all three locations the δ^{37}Cl of surface water was highly positive relative to ISL354, a standard sodium chloride produced from ocean water collected from the Pacific Ocean. Samples from greater depth are more negative, at one station even as negative as −0.53‰. This decrease with depth correlates with the salinity of the water (Fig. 11.1) and it is explained that chlorine isotope variations are caused by evaporation of ocean water as described by Xiao et al. (2000). They suggested that δ^{37}Cl values of seawater are regulated by the evaporation to precipitation ratio. They suggest

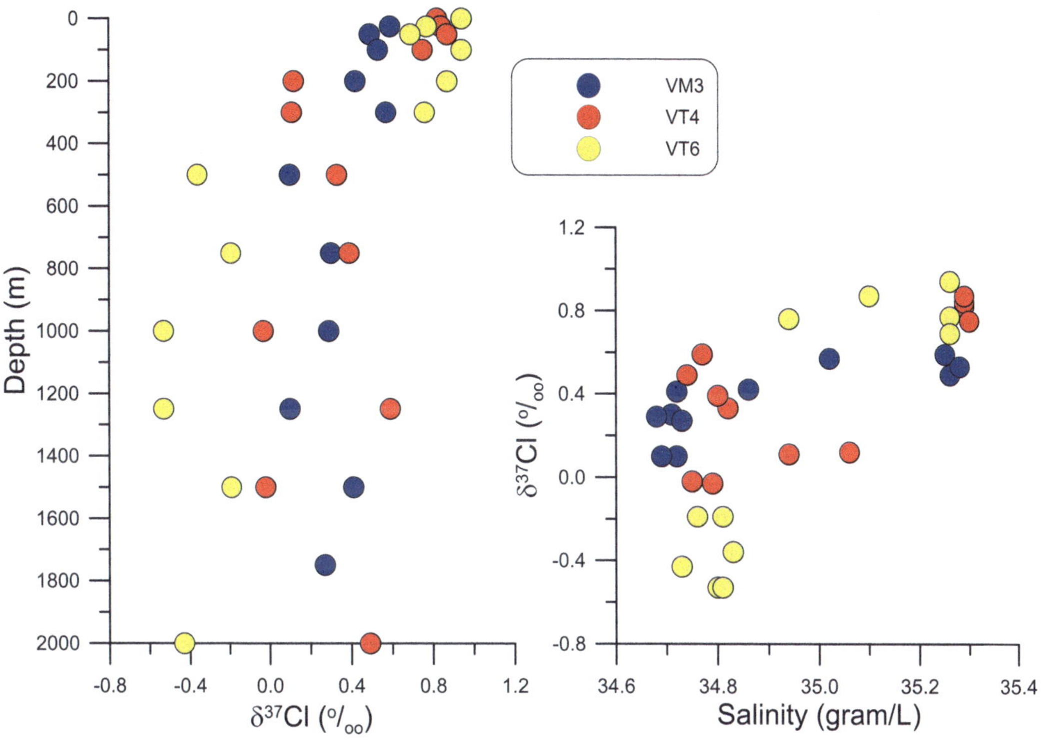

Fig. 11.1 δ^{37}Cl-depth and δ^{37}Cl-salinity diagrams from de data obtained by Shirodkar et al. (2003). In these figures it appears that δ^{37}Cl decreases with depth in the Indian Ocean. The relationship between δ^{37}Cl and salinity is related to the fact that salinity in these profiles decreases with depth, just like the δ^{37}Cl in these profiles. The large variations in δ^{37}Cl values of seawater samples observed in this study by Shirodkar et al. (2003) can be questioned, especially considering the fact that such large variations were never observed by other laboratories, especially considering the observations on a large set of different seawater samples from different locations measured recently at the Institut de physique du globe de Paris (P Agrinier, pers. comm)

that during evaporation the chloride of the seawater is volatilized as hydrogen chloride, causing enrichment of ^{37}Cl in surface seawater, explaining higher δ^{37}Cl values at the surface.

This effect appears to be confirmed by a study of ocean surface waters that were collected from the upper meter of four major world oceans (Shirodkar et al. 2006). In this study with samples from 51 stations it was observed that ocean surface water samples show δ^{37}Cl values between −0.76 and + 0.74‰ relative to ISL354. These variations were related to different air-sea fluxes. A statistical approach using Q-mode factor analysis determined four distinct groups of samples in this set. Although slightly different average δ^{37}Cl values were found in each of these four groups they were not significantly different from each other. The conclusion of this study was that δ^{37}Cl values of individual ocean water samples are constant, but a defined seawater sample has to be used when it is taken as a reference standard for isotopic analyses. The results from these studies indicate that, although δ^{37}Cl values of ocean water are between +1 and −1‰, it is not constant enough to use a general ocean water as standard. This conclusion is opposite to the conclusions reached by Kaufmann (1984), Godon et al. (2004a) and Liu et al. (2013). The results however, at least suggest that there might be some uncertainty about the constancy of the δ^{37}Cl of the oceans and it might be recommended to compare the ocean water in use as standard in the various laboratories with ISL954 which can be considered homogeneous and can act as fixed zero on the chlorine isotope scale. In spite of this recommendation most laboratories working on chlorine isotopes still work with their own laboratory ocean water standard, and no problems with this policy have been encountered yet. Following this, researchers in most laboratories are still convinced that the $\delta^{37}Cl$ of the oceans are invariable. Unfortunately the samples reported by Shirodkar et al. (2003, 2006) were all measured using the P-TIMS method at the Xining Laboratory, while most other laboratories still work with IRMS techniques. Considering the conflicting conclusions between the results obtained by P-TIMS from just one single laboratory and IRMS analyses from a variety of laboratories it would be very valuable if more interlaboratory comparison would be organised, and it would be especially useful if samples presented in the studies from Shirodkar (2003, 2006) would have been reanalysed in one or more of the laboratories using IRMS techniques.

11.1.1.2 Bromine

As the residence time in the oceans for bromide is even longer than for chloride (130,000,000 years; Broeker and Peng 1982) it is expected, just as for chloride that the bromine isotope composition of the oceans is constant. Unfortunately no larger or interlaboratory comparison study on the bromine isotope composition of the oceans has been done yet, but Du et al. (2013) analysed five different seawater samples which were taken from different locations along the coast of China. They observed that the variation between these samples was less than 0.1‰, and their preliminary conclusion was that the bromine isotope composition of the oceans would be, as it is for chlorine, constant so that it could indeed be used as international standard. This standard is still, as an analogue to chlorine, normally called SMOB or Standard Mean Ocean Bromide (Eggenkamp and Coleman 2000).

11.1.2 Sea Ice

The chlorine and bromine concentrations and isotope compositions in Antarctic sea-ice have been determined by Vallelonga et al. (2021). In Antarctic sea ice the bromine concentrations range from 710 ± 20 μg.L^{-1} to 31,440 ± 570 μg.L^{-1} and the chlorine concentrations range from 218 ± 12 mg.L^{-1} to 7945 ± 351 mg.L^{-1}. The concentrations of bromide and chloride are well correlated with a slope of 0.0038, which is almost the same as the ratio in seawater. The range in isotope compositions is rather large in this dataset with $\delta^{37}Cl$ ranging from −1.10 to +0.43‰ with a mean of −0.26‰ and standard deviation of 0.10‰ and $\delta^{81}Br$ ranging from −0.64 to +0.10‰ with a mean of −0.16‰ and standard deviation of 0.04‰. No correlation between the chlorine and bromine isotope compositions and the respective concentrations are found. However the range in measured isotope values is larger (for both Cl and Br) in samples with lower concentrations and diverge to values close to 0‰ in samples with the highest concentrations. The samples with the largest range in isotope compositions tend to be the deepest samples in the sea ice cores. Vallelonga et al. (2021) suggest multiple parameters that may influence isotope fractionation processes in sea ice, which include ice texture, age, salt content, and algal biomass content (represented by chlorophyll-a concentrations), but they did not found a single factor to consistently correlate with the isotope signatures. Vallelonga et al. (2021) consider that isotope fractionation results most possibly from brine rejection and desalinization processes.

11.1.3 Rain Water

Chlorine isotope compositions of rain water have been studied in detail for a coastal and an interior region of eastern Canada (Koehler and Wassenaar 2010). At the interior Bonner Lake, Ontario, site the $\delta^{37}Cl$ values of dissolved Cl^- in precipitation ranged from −3.5 to −1.2‰ (vs. SMOC) with an amount-weighted annual average of −2.3‰. At a coastal site, Bay D'Espoir, Newfoundland, $\delta^{37}Cl$ values of dissolved Cl^- values ranged from −3.1 to 0.0‰ with an amount-weighted annual average of −1.3‰. These negative $\delta^{37}Cl$ values provide evidence that atmospheric HCl is ^{37}Cl depleted, presumably from acidification of sea-salt aerosols, as was observed by Volpe and

Spivack (1994) and Volpe et al. (1998) in actual aerosols. Koehler and Wassenaar (2010) also measured the chlorine isotope composition of headwaters in two mountain rivers in Western Canada and these were also depleted in ^{35}Cl, with δ^{37}Cl values down to −3.1‰ in the samples with the lowest chloride concentrations. Samples with the lowest chloride concentration represent mostly rainwater indicating that also in the region where these rivers spring rainwater also has strongly negative δ^{37}Cl values.

11.1.4 Sediment Pore Water

Pore water is present in virtually all sedimentary basins. This pore water is, especially at greater depths, mostly saline, indicating that a large amount of both chloride and bromide is present in this water. Recent estimates are that the amount of chloride in these pore waters may be even larger than the total amount of chloride in the oceans and more than double the amount of chloride in evaporites at about $330*10^{20}$ g (Land 1995). The origin of all pore water ultimately is from the oceans as it was suggested by Knauth (1998) that before the development of the continental crust, all salt and brine currently found on the continents would have been entirely in the oceans. The composition of the pore waters is changed by chemical processes such as water-rock interaction, salt dissolution, albitization as well as physical processes such as diffusion, ion exchange and ion-filtration and mixing of water bodies with different compositions. During these processes not only the chemical but also the chlorine and bromine isotopic composition of the water can alter considerably.

11.1.4.1 Chlorine and Bromine Isotope Variations in Sediment Pore Waters

Chlorine isotopes in porewaters have been studied from environments such as shield brines (Canada (Desaulniers et al. 1986; Kaufmann et al. 1987; Stotler et al. 2010), Sweden (Wirt 1988; Sie and Frape 2002), Finland (Stotler et al. 2010) and Siberia (Shouakar-Stash et al. 2007; Alexeeva et al. 2015a, 2015b)), oil- and coal-field formation waters (Kaufmann et al. 1988, 1993; Eggenkamp 1994; Ziegler et al. 2001; Bagheri 2013; Louvat et al. 2016; Stotler et al. 2021) and sedimentary aquifer waters (Martin 1993; Li et al. 2012; Wang and Frape 2015; Giunta et al. 2017; Yu et al. 2019).

In general, most samples have negative δ^{37}Cl values, with a median value of about −0.2‰. That indicates that for chlorine negative δ^{37}Cl data are more common than positive data. What is well known regarding chlorine isotopes is the narrow range of the data, with the majority of measured values between −0.6 and + 0.1‰ vs. SMOC. And except for a few extreme δ^{37}Cl measurements virtually all samples have δ^{37}Cl values that are not more than 2‰ from SMOC.

Because the bromide concentrations in sedimentary pore waters are in many cases relatively high the bromine isotope composition has been studied regularly in this type of samples already from the early 2000s. One of the first observations that can be made is the striking dissimilarity between chlorine and bromine isotope compositions of pore waters. It was expected, before bromine isotope studies had been done, that, as chlorine and bromine have an apparently comparable chemistry, their isotope geochemistries would be reasonably similar. Already in the first study reporting natural bromine isotope measurements (Eggenkamp and Coleman 2000) it was shown that this was definitely not the case. The earliest papers in which bromine isotope measurements were published are Eggenkamp and Coleman (2000), Shouakar-Stash et al. (2007), Stotler et al. (2010) and Boschetti et al. (2011). During the last years the number of bromine isotope measurements has increased considerably, and they are now regularly measured together with chlorine isotopes in the same samples (e.g. Alexeeva et al. 2015a, 2015b; Stotler et al. 2021; Wang and Frape 2015; Giunta et al. 2017; Yu et al. 2019). Interestingly, and opposite to the observations on chlorine isotopes, δ^{81}Br values tend to be mostly positive in pore waters. Their median value for example is about +0.4‰ relative to SMOB, as compared to about −0.2‰ relative to SMOC for chlorine.

11.1.4.2 Diffusion of Chloride in Pore Waters

Diffusion is perhaps the most thoroughly evaluated fractionation process in natural aqueous systems. The reason is that due to the high diffusion coefficient of chloride in water, as well as the relatively large fractionation between the two chlorine isotopes (when compared to other possible fractionation effects) natural variations can be significant. Several studies have been published in which the isotope fractionation due to diffusion has been determined from the chlorine isotope variations in ground and pore water profiles. Among these is one of the earliest chlorine isotope studies on natural samples (Desaulniers et al. 1986). In this study diffusion of chloride from a saline bedrock was examined. In more shallow samples the chloride concentration was lower and the isotope ratio of these samples was indeed lower than in the deeper samples confirming that ^{35}Cl diffuses faster than ^{37}Cl. In this study it was also observed that the δ^{37}Cl of the deepest samples, taken from the bedrock, was very positive with values between 1.5 and 2.0‰. When these values were measured they were the most positive δ^{37}Cl ever observed. They could not be explained at the moment of publication, and these very positive values are still difficult to explain even considering the progress of our knowledge (e.g. Shouakar-Stash 2008). It is however possible that this type of relatively positive values are caused by ion-filtration processes that took place in the bedrock after Phillips and Bentley (1987) showed that ion-filtration could be responsible for very positive δ^{37}Cl values. The results obtained in the study by Desaulniers et al. (1986) facilitated the calculation of the isotope fractionation due to diffusion. The fractionation factor they obtained during this study was however lower than fractionation factors found in both earlier and later experimental studies. The reason for this discrepancy may be due to the fact that natural systems are always more difficult to interpret than experimental systems.

Eggenkamp et al. (1994) determined the diffusion coefficient in a relatively simple natural system. During the last glacial period Kau Bay on the Indonesian island of Halmahera was separated from the ocean and became a freshwater lake due to the low sea level. When the sea-level increased again after the glacial period ended the bay water became saline again and salt water started to diffuse into the freshwater sediment of the former lake. As this was a simple system with a more or less constant sedimentation rate and salinity of the overlying water it was possible to determine the diffusion and its fractionation using a simple analytical model. In more complex systems it was only possible to approach the actual diffusion and fractionation using relatively simple analytical solutions such as the case of seawater intrusion along the Surinam coast (Groen et al. 2000). In this system there may have been alternation of more and less fresh water periods and as a result of this the diffusion and fractionation could only be determined approximately. To interpret the diffusion model properly for this kind of complex diffusion systems a different approach is needed. Beekman et al. (2011) managed to model both the chloride concentration as well as the chlorine isotope composition in a rather complex system by using a numerical technique. In their model the system was separated into thin layers each representing 50 cm of sediment and diffusion over time within and between these layers was determined. This way they managed to model both the chloride concentration and the chlorine isotope composition very satisfactorily. Numerical modelling has also been used to estimate the diffusion of chloride together with its isotope fractionation in the Paris Basin (Le Gal La Salle et al. 2013; Rebeix et al. 2014), while earlier studies used a more analytical approach to determine the fractionation in this complex basin that contains very thick layers of sediment (Lavastre et al. 2005). The chlorine isotope composition in porewaters taken from a 1900 meter deep drill core in the Japanese Mobara Gas field (Mahara et al. 2013) also proved useful to determine the pore water mobility. Chloride concentration and chlorine isotope composition could be used together to estimate the porewater advection rate as 3 mm/year during the past 220,000 years using a fractionation factor of 1.002, approximately halfway fractionation factors published by Desaulniers et al. (1986; 1.0012) and Eggenkamp et al. (1994; 1.003) and close to the experimentally determined value of 1.0017 as determined by Eggenkamp and Coleman (2009). The analyses of chlorine isotope ratios in porewater samples from cores taken in Lake Kinneret (Stiller et al. 2009) which show clearly a diffusive isotope profile also proved that chlorine isotopes could be used to distinguish between different sources. While several 5 m cores taken from the bottom of this lake all showed an increase in chloride with depth in some of the cores the chlorine isotopes of the porewater increased with depth while others decreased with depth. While these profiles indicated diffusion of chloride between a chloride source at depth and the lake water, it was suggested that the isotopic differences observed were indicative of spatial variability in the porewater brine sources. In the case of Lake Kinneret it could be that long term diffusion may have altered the original, negative $\delta^{37}Cl$ of the Dear Sea Rift Valley brine (Stiller et al. 1998) from beneath the floor of the central part. If the hypothesis described by Stiller et al. (2009) is correct, then the apparent contradiction between the positive $\delta^{37}Cl$ values that are attributed to a pelagic brine source and the negative $\delta^{37}Cl$ values of three groups of Dear Sea Rift Valley type saline springs could be reconciled because each of the latter represents a deep, isolated brine reservoir (Kolodny et al. 1999), in which the original $\delta^{37}Cl$ signature of residual brines could have been preserved.

11.1.5 Chlorine Isotope Variations in Deep Sea Pore Waters

Ransom et al. (1995) discovered that porewaters taken from subduction zones showed very negative $\delta^{37}Cl$ values that could be as low as −8‰. Ransom et al. (1995) analysed the chlorine isotope compositions of porewaters from two active and one passive oceanic margin and found that porewaters from the passive margin, of the coast of West Africa, have $\delta^{37}Cl$ values very close to 0‰, while active margins showed (very) negative values. In the Barbados subduction zone samples were taken less than 1 m below the seafloor and even at these shallow depths $\delta^{37}Cl$ values as low as −4.7‰ were observed, indicating very active advective fluid flow. At the Nankai subduction zone samples show these negative values at about 1000 meter below the seafloor. In this region fluids move via a very complex network of microfaults as described by Kastner et al. (1993). Ransom et al. (1995) explained these negative values as the result of diagenetic and metamorphic dehydration and transformation reactions. They assumed that the low values were the result of dehydration of hydrous silica minerals which were found to have very positive $\delta^{37}Cl$ when measured by P-TIMS (Magenheim et al. 1994, 1995). Godon et al. (2004b) also studied the porewater at the Barbados accretionary complex and concluded that the very low $\delta^{37}Cl$ endmember in their systems were the result of ionfiltration. It is clear however that not in all deep ocean porewaters very low $\delta^{37}Cl$ values are found. In their study on the chlorine isotope composition of porewaters from the Brazos-Trinity Basin IV in the Gulf of Mexico Li and Jiang (2013) found $\delta^{37}Cl$ in the range between −0.81 and + 0.71‰, averaging "close to seawater" at −0.18‰ without being able to explain the observed variation properly.

In their study on porewaters from the Blake ridge gas-hydrate field in the west Atlantic Hesse et al. (2006) also encountered very negative $\delta^{37}Cl$ values at great depth. In their samples they also found a decrease of the chloride concentration with depth, which, according to Egeberg and Dickens (1999), was caused by the melting process of gas-hydrates. It was concluded however by Hesse et al. (2006) that over 50% of the chlorinity reduction actually did was occur before these hydrates melted. They concluded that the lower chlorinity was caused by advection of a strongly ^{37}Cl depleted low chlorinity fluid. The origin of this low chlorinity fluid remained however unknown.

11.1.5.1 Possible Relationship Between Very Low $\delta^{37}Cl$ in Deep Sea Porewaters and Serpentinites

The $\delta^{37}Cl$ values in these deep marine porewaters may be related to the chlorine isotope compositions of massive seafloor serpentinites. Sharp and Barnes (2004) indicated that the water-soluble fraction from these serpentinites may have rather negative $\delta^{37}Cl$ values of −1.5‰ in older and values close to zero in more recent serpentinites. The insoluble fraction has $\delta^{37}Cl$ values which are about 0.33‰ more positive than the soluble fraction. In a much larger study Barnes and Sharp (2006) showed that there is a considerable variation in the chlorine isotope characteristics between different cores where the chlorine isotope composition is measured on serpentinites. They also observed that the chemical composition of cores with different chlorine isotope characteristics is different. Barnes and Sharp (2006) indicated that non-zero bulk $\delta^{37}Cl$ values indicate that chloride in serpentinites was incorporated during the original hydration and is not an artefact of seawater infiltration during drilling. Cores with positive $\delta^{37}Cl$ values are most likely explained by open system fractionation during hydrothermal alteration, with preferential incorporation of ^{37}Cl from seawater into the serpentinite and loss of residual isotopically light chloride back to the ocean. Fluid/rock ratios were probably low as evidenced by the presence of water-soluble salts. The cores with isotopically negative chloride were characterized by a thick overlying sedimentary package that was in place prior to serpentinisation. Barnes and Sharp (2006) believed that the low $\delta^{37}Cl$ values of these cores were the result of hydration of ultramafic rock by infiltrating aqueous pore fluids from the overlying sediments. The resulting serpentinites inherit the characteristic negative $\delta^{37}Cl$ values of the pore waters. Chlorine stable isotopes can be used to identify the source of the serpentinising fluid and ultimately discern chemical and tectonic processes involved in serpentinisation. These studies thus indicated that the low $\delta^{37}Cl$ values found in some of the serpentinite cores may be the result and not the origin of the low $\delta^{37}Cl$ values in subduction porewaters.

Bonifacie et al. (2008) found a range of negative $\delta^{37}Cl$ values in oceanic serpentinites. They concluded that these negative values were likely the result from water-rock interaction with fluids with negative $\delta^{37}Cl$ values. The homogeneity of $\delta^{37}Cl$ values from a high-pressure olivine-bearing vein and metaperidotite samples that they observed implied that progressive loss of chlorine inherited from oceanic alteration throughout subduction did not significantly fractionate chlorine isotopes. Bonifacie et al. (2008) concluded that chlorine that recycled in subduction zones via metaperidotites should thus show a range of $\delta^{37}Cl$ values similar to the range found in oceanic serpentinised peridotites.

The chlorine isotope composition, as well as the halogen chemistry of serpentinites, during the progressive dehydration of serpentinites during subduction was studied by John et al. (2011). They hoped to understand how relevant serpentinites would be for the recycling of halogens during subduction and deep mantle recharge. They observed that the halogen compositions of the analysed samples indicate input from seawater and sedimentary sources during initial serpentinisation of either subducting lithospheric mantle during slab bending or forearc mantle by uprising slab fluids. During dehydratation of the serpentinites the fluids that were released were enriched in bromide and iodide relative to chloride, so that the relative chloride contents of the serpentinite increased. On the other hand they observed that the $\delta^{37}Cl$ range of the deserpentinisation samples showed only a narrow range with overlap on $\delta^{37}Cl$ values from seafloor serpentinites. Based on these observations they concluded that while no significant chlorine isotope fractionation occurred during the subduction dehydration of serpentinites, subduction of serpentinites and their dehydration strongly affects the halogen budget of the mantle.

The various studies on serpentinites indicate that stable isotopes, among which chlorine stable isotopes, are effective tracers of serpentinising fluids. Barnes et al. (2013) state that the final isotopic composition of serpentinites will be determined by the isotopic composition of the serpentinising fluid, the temperature of interaction, and the water/rock ratio. For example, high $\delta^{18}O$ values may be the result of low-temperature serpentinisation by seawater (large $\Delta^{18}O_{serp-water}$) or interaction with an ^{18}O-enriched slab derived fluid. In their study on serpentinites from the Franciscan Complex, western California, USA Barnes et al. (2013) observed in their samples $\delta^{37}Cl$ values that ranged from −0.1 to +1.7‰ vs. SMOC with 12 of the 15 samples falling between +0.2 and + 0.9‰ (with an average value for all samples of +0.6 ± 0.5‰). These data support serpentinisation via seawater on the seafloor. Extreme negative or positive $\delta^{37}Cl$ values reflecting the influence of a non-seawater fluid are not present in the data set, except for one sample with a value of +1.7‰.

In their discussion Barnes et al. (2013) summarise that chlorine isotope measurements of obducted serpentinites and metaperidotites show no chlorine isotope fractionation during prograde subduction metamorphism and that the chlorine isotope composition of obducted serpentinites is thought to preserve a record of seafloor tectonic processes (Barnes et al. 2006; Bonifacie et al. 2008; John et al. 2011). Serpentinites formed via interaction with seawater have $\delta^{37}Cl$ values of about +0.2 to +0.5‰ (Barnes and Sharp 2006). Serpentinites formed from interaction with sedimentary pore fluids and/or sediment-derived fluid have $\delta^{37}Cl$ values of about −2.0 to −0.5‰ (Barnes and Sharp 2006; Barnes et al. 2006). Rare serpentinites with $\delta^{37}Cl$ values above +1.0‰ (n = 9) are restricted to serpentine seamounts (n = 2) and obducted serpentinites from Elba Island, Italy (n = 1) and HP/UHP localities in the western Alps (n = 6) (Barnes et al. 2006, 2008; John et al. 2011; Selverstone and Sharp 2011). Serpentinites with high $\delta^{37}Cl$ values are postulated to reflect a fluid component from the subducted slab, either high pressurised metasediments or altered oceanic crust (John et al. 2010; Selverstone and Sharp 2011). Post-serpentinization interaction with meteoric water does not alter the chlorine isotopic composition of serpentinites (Barnes et al. 2006).

11.1.5.2 Possible Chlorine Isotope Fractionation in Deep Sea Porewaters Due to Ion Filtration

Although the potential of ion filtration as a process to fractionate chlorine isotopes has been studied both experimentally and theoretically already since the early 1980s (Campbell 1985; Phillips and Bentley 1987), studies in which it is recognised as one of the major processes to explain chlorine isotope variations in natural environments are rare. In several studies it is suggested as a potential process for the observed large isotope variation (e.g. Kaufmann et al. 1988; Coleman et al. 1994; Ziegler et al. 2001; Woulé-Ebongué et al. 2005; Lavastre et al. 2005; Bonifacie et al. 2007b; Li et al. 2012) but it was in many cases doubtful if ion-filtration was an important process and it was also indicated in these studies that the results were either inconclusive or that ion-filtration as a major process could even be rejected as the process that was responsible for the observed chlorine isotope variations. Ion filtration is also mentioned as the potential process that can explain young low-temperature, high-porosity clay-rich oceanic sediments where strongly ^{37}Cl-depleted pore fluids have been sampled in convergent margins (Ransom et al. 1995; Spivack et al. 2002; Deyle et al. 2003; Godon et al. 2004b; Wei et al. 2008), in passive margins (Hesse et al. 2000) and ridge flanks (Bonifacie 2005; Bonifacie et al. 2007b). In these studies, it was observed that $\delta^{37}Cl$ depletions starting from 0‰ at the interface with seawater, down to very negative values as low as −8‰ at some sites, at sampling points buried deep in the sedimentary pile where they are the most isolated from seawater. However, only Godon et al. (2004b), Bonifacie (2005), and Bonifacie et al. (2007a, 2007b) provided arguments to support ion filtration in the production of the strongly ^{37}Cl-depleted pore fluids and concluded that the very negative $\delta^{37}Cl$ values that they encountered at the Manon site, at the outer edge of the Barbados accretionary complex were the result of ion filtration. In the system they studied it was found that the porewaters were on mixing lines between seawater with a $\delta^{37}Cl$ value of 0‰ and endmember water samples that were characterised with very negative $\delta^{37}Cl$ values of −4.51, −5.19 and −5.24‰ at three different locations. As the chloride concentrations of these low $\delta^{37}Cl$ endmembers are relatively high diffusion was excluded as a potential source for these samples. Godon et al. (2004b) discussed that following the theoretical studies by Phillips and Bentley 1987) a ^{37}Cl depletion of 5‰ agrees with a 20-fold increase of chloride behind the membrane. Godon et al. (2004b) proposed a model for the generation of these low $\delta^{37}Cl$ fluids as a residual negative $\delta^{37}Cl$ fluid reservoir that is created at the bottom of a prism or a sediment pile. Further compaction and/or fracturing and/or dewatering of the slab may flush out these fluids and focus them towards the décollement zone. Mixing between the fluids and ultimately with seawater and water that has been released during gas hydrate destabilizations may explain the data set within the individual cores and between the different structures. Godon et al. (2004b) suggested that low $\delta^{37}Cl$ values that were found in a few other early chlorine isotope studies and that were not explained in these original studies (Eggenkamp and Coleman 1997; Hesse et al. 2000; Hesse 2003) could also be caused by ion-filtration. In a later study Hesse et al. (2006) could however not confirm this suggestion, and they were unable to identify the source of their low $\delta^{37}Cl$ fluid. Also most other studies excluded ion filtration as the process to generate the observed extremely low $\delta^{37}Cl$ values. The results of these studies indicate that it is still unsure as to how chlorine isotope variations observed in several natural environments can be related to the process of ion-filtration. The general conclusion of these early studies was that potentially ion-filtration is a process that is able to fractionate chloride in solutions very efficiently it appears to be very difficult to detect its signature in natural environments.

Agrinier et al. (2019) studied the chlorine isotope composition in the pore fluids from two IODP sites (C001 and C0012) in the Shikoku basin on the Philippine plate at the front of the Nankai subduction zone. At both sites, $\delta^{37}Cl$ values show a regular decrease with depth from about 0‰ (seawater value) at the seafloor to minimum values of −8.5 and −6.5‰ respectively in deep, hydrogeologically isolated, low permeability, sediment levels. At some depths, local increases of $\delta^{37}Cl$ with values up to 6‰ interrupt the overall decreasing trends. These increases in the $\delta^{37}Cl$ can be interpreted as if external fluids are injected into the sedimentary pile at locations of sedimentary strata with higher permeability. Agrinier et al. (2019) use conservation equations to demonstrate that these $\delta^{37}Cl$ depth profiles can be modelled using the effect of the compaction of growing clay-rich sedimentary piles on the water and chloride (including the isotopes ^{35}Cl and ^{37}Cl). Their model shows that a small chlorine isotope fractionation ($\varepsilon^{37}Cl_{\text{expulsed fluid-residual fluid}}$ = 1 to 6‰) may explain the observed negative $\delta^{37}Cl$ values at the bottom of the sedimentary pile. The complementary positive $\delta^{37}Cl$ must accumulate at the top but are most likely erased by dilution with seawater chlorides across the seafloor interface. The chlorine isotope fractionation range obtained by Agrinier et al. (2019) is in agreement with the theory of ion filtration of Phillips and Bentley (1987) in which the mobility of chlorides through semi-permeable clay membrane is determined by ion repulsion. Interestingly Agrinier et al. (2019) conclude that the very negative $\delta^{37}Cl$ pore waters do not seem to contribute largely to subduction zone magmatic, tectonic, metamorphic products where this ^{37}Cl-depleted chlorine is not documented. They consider this a strong argument to consider that most of these clay-rich sediment pore waters are released back into the ocean rather than subducted. In order to check the results obtained from the two sediment cores drilled near the Nankai subduction zone in a wider variety of tectonic contexts Agrinier et al. (2021) evaluated the $\delta^{37}Cl$ and chemical compositions of a further 11 different sediment cores drilled at many different ODP-IODP sites (Srivastava et al. 1987a, 1987b, 1987c, 1987d; Peirce et al. 1989a, 1989b, 1989c; Davis et al. 1997b, 1997c; Hesse et al. 2000; Deyle et al. 2003; Paull et al. 2005a, 2005b; Sacks et al. 2005a, 2005b; Richter et al. 2005; Bonifacie

2005; Bonifacie et al. 2007b; Saito et al. 2010a, 2010b; Henry et al. 2012a, 2012b; Agrinier et al. 2019, 2021). They evaluated sedimentary processes from a wide range of sedimentation types and conditions ranging from high to very low rates of sedimentation (between 0.137 m/ka and 0.003 m/ka) and high to low rates of compaction (gradient of porosity between −0.16%/m and −0.018%/m). All studied cores show relatively constant chloride contents but with a systematic decrease in $\delta^{37}Cl$ with depth from 0‰ down to −8.5‰ (depending on the core). Interestingly, $\delta^{37}Cl$ data are highly correlated with $\delta^{18}O$. $\delta^{37}Cl$ in porefluids decreased in all studied sediment cores regardless of tectonic or sedimentary setting. The trends could not be explained by water-rock interaction reactions because low temperature minerals have chloride contents that are too low to compensate for $\delta^{37}Cl$ decrease that is observed in pore fluids. For that reason, Agrinier et al. (2021) hypothesized that fluid-specific processes are responsible for the $\delta^{37}Cl$-depletions of the fluids and that $\delta^{37}Cl$-enriched chlorides were expelled out of the sediments into the ocean. After reviewing the fluid-specific processes that are known to change chlorine isotope ratios in chlorides, Agrinier et al. (2021) ruled out that diffusion and gravitational isotope fractionation could generate the observed isotope pattern. The flow of a $\delta^{37}Cl$ -depleted fluid from the underlying basaltic basement into the sediments could explain the $\delta^{37}Cl$ data. But the mechanism that produces depletion in $\delta^{37}Cl$ of the fluid remains unknown. It cannot be chloride exchanges between fluids and rocks. Agrinier et al. (2021) show that compaction-induced ion filtration of chlorides through clay-rich membranes can produce the observed pore fluid $\delta^{37}Cl$-depletions, with isotope fractionation ($\varepsilon^{37}Cl$) ranging from 0 to +8‰ between the chloride of the expelled fluid (the permeate) and the chloride of the residual fluid (the retentate). Agrinier et al. (2021) observed that smectite-rich sediments are associated with higher isotopic fractionation, while illite/chlorite-rich sediments are associated with intermediate values and clay-poor sediments are associated with the lowest isotope fractionation. This suggests that chlorine isotope fractionation may be controlled by the surface charge that is associated with specific clay minerals. Calculations by Agrinier et al. (2021) showed that compaction-induced filtration has the capacity to produce ^{18}O-depletion for oxygen isotope fractionation ($\varepsilon^{18}O$) between the expelled fluid and the retentate ranging from 0 to 5‰. ^{18}O-enrichment in the expelled fluid is in agreement with the experimental data of Haydon and Graf (1986). The results of the studies by Agrinier et al. (2019, 2021) indicate that ion-filtration must certainly be considered as a potential important mechanism for fractionating (chlorine) isotope species in sediment pore waters.

11.2 Evaporites

Evaporites are sediments, and are all the result of evaporation of water masses (either seawater or terrestrial lake water). Thus, their origin is in all cases the hydrosphere. The resulting evaporite sediments as such are able to give geochemical information on the original brine from which they have precipitated and for that reason the isotope geochemistry of the evaporites will be described in this chapter.

11.2.1 The Chlorine Isotope Composition of Marine Evaporite Deposits

Due to the high chloride contents of evaporites these deposits have been studied in considerable detail and as a result the chlorine isotope compositions of evaporites are fairly well understood. Most evaporites origin from ocean water and reflect processes that take place during the precipitation of salt from ocean water in enclosed basins and from processes taking place with the evaporites under influence of changing pressure and temperature. Evaporites can also be terrestrial in origin. In this case the evaporite deposits are precipitated from originally fresh water that flows into a closed basin and that is evaporated to dryness. The isotope composition of these deposits is potentially different from marine deposits due to the different mode of formation. Terrestrial deposits form from river water which has a different chemical composition than ocean water. As the chemical composition of river water is also much more variable than the composition of ocean water the composition of the salt deposits and its chlorine isotope composition is also expected to be different. Although rivers often contain relatively high calcium and bicarbonate concentrations calcium carbonate normally precipitates at an early stage during the evaporation process and the remaining river brine is, just like ocean water, dominated by sodium and chloride.

Marine evaporites are formed from the evaporation of seas and oceans. The chemical composition of this type of deposits is studied very well both theoretically and experimentally for more than a century, starting with a very large numbers of papers written by Van't Hoff, who presented over 50 papers on experimental studies to determine the different salts that precipitate from seawater and the order in which they form as well as the changes due to environmental processes (pressure, temperature) after deposition, of which Van't Hoff and Mayerhoffer (1897) was the first study, and which were later published in book form (Van't Hoff 1905, 1909, 1912). Still probably the best monograph describing the origin and composition of salt deposits is written by Braitsch (1962, 1971). The most common salt that precipitates from ocean water is halite (NaCl) and this mineral dominates most marine salt deposits, and thus determines the chlorine isotope composition of most deposits.

The chlorine isotope composition of Phanerozoic salt deposits have been studied in considerable detail (Eggenkamp et al. 1995, 2019a, 2019b; Eastoe and Peryt 1999; Eastoe et al. 1999, 2001, 2007; Meng et al. 2018; Tang et al. 2018; Hussain et al. 2020, 2021; Zhao et al. 2021; Yu et al. 2023). During precipitation of halite from brine a small chlorine isotope fractionation of about 0.3‰ takes place, while during precipitation of potassium and magnesium salts the chlorine isotope fractionation is negligible or slightly negative. Eggenkamp et al. (1995) calculated the consequence of this small fractionation for the isotope composition of salt deposits and concluded that starting from ocean water the first salt to be deposited must have more positive δ^{37}Cl values, while later salt must be more negative δ^{37}Cl values than ocean water. Overall the variation was expected to be relatively small, with values between −0.5 and + 0.3‰. This was confirmed in all studies on marine salt deposits. The summary compiled by Eastoe et al. (2007), containing over 230 chlorine isotope analyses from Phanerozoic marine salt deposits showed that the more than 90% of the samples had δ^{37}Cl values between +0.5 and −0.5‰, and no samples were outside the range of −1 to +1‰ relative to SMOC, the current ocean composition. These results confirmed the relatively small fractionation of chlorine from seawater during precipitation, and are also a very strong indication that the chlorine isotope composition of the oceans did not vary significantly, taking into account the errors of the measurements and secondary fractionation effects in the salts after deposition, during the whole of the Phanerozoic.

Eggenkamp et al. (2019b) discussed the δ^{37}Cl in relation with the Br contents of the halite. They observed a clear negative relationship between the δ^{37}Cl and the Br/Cl ratio in the salt. Their observations are discussed in more detail below in Sect. 11.2.5 because they have been analysed in relationship to the bromine isotope composition. Their observations were confirmed by Meng et al. (2018). Yu et al. (2023) studied the δ^{37}Cl of halite in relationship to prospecting for potash deposits and observed high Br/Cl ratios to be related to low δ^{37}Cl values, also agreeing with the conclusions from Eggenkamp et al. (2019b). In other presumably primary halite deposits however, Tang et al. (2018) observed however δ^{37}Cl values that are mostly between −0.6 and −0.4‰ at Br/Cl ratios of 0.00031 to 0.00045, indicating significantly lower δ^{37}Cl values than expected based in the observations made by Eggenkamp et al. (2019b). Further, δ^{37}Cl values reported by Hussain et al. (2020, 2021) where mostly between −0.4 and 0.0‰, but with very low Br/Cl ratios (0.000003 to 0.000005). Du et al. (2022) found little to no correlation between the Br/Cl ratio and the δ^{37}Cl in halite from the Mesozoic-Cenozoic Lanping–Simao Basin, Southeastern

Tibetan Plateau, China. These recent studies suggest that the relationship between the $\delta^{37}Cl$ and the Br/Cl ratio in halite is not as straightforward as discussed in Eggenkamp et al. (2019b) and it is suggested to analyse in detail the $\delta^{37}Cl$ versus the Br/Cl ratio relationship in future studies on primary halite deposits. Zhao et al. (2021) determined both the Cl and the B isotope composition of a series of Badenian salt samples from the Carpatia Foredeep basin. These salts are presumably marine, although some very high (+0.6 to +0.8‰) $\delta^{37}Cl$ values were measured that can not normally be explained by precipitation form normal seawater. No Br concentrations were measured here, but based on the $\delta^{11}B$ values it was concluded that the salts were precipitated not from pure ocean water, but from mixtures between ocean water and three different brines with distinct origins, an evaporite brine with low $\delta^{37}Cl$ and $\delta^{11}B$ comparable to seawater, a sedimentary basin brine with high $\delta^{11}B$ and slightly negative $\delta^{37}Cl$, and a dissolved diapiric halite with high $\delta^{37}Cl$ and $\delta^{11}B$ comparable to seawater. It was concluded that the combination of Cl and B isotopes can also help to distinguish multiple sources for the origin of a salt sample, which appears in many cases to be more complex than originally assumed.

11.2.2 The Chlorine Isotope Composition of Terrestrial Evaporite Deposits

There is one region from which Eastoe et al. (2007) did not take into account the analysed chlorine isotope results in their study on Phanerozoic evaporites because these evaporites are clearly not marine but of terrestrial origin. This region is the very large inland Tarim Basin in the Xingjang Uygur Autonomous Region in China which has an area of approximately 560,000 km^2 and that has been studied very extensively for its chlorine isotope geochemistry as it was discovered that the chlorine isotope composition of the salt deposits in this basin could be used as a proxy for the exploitation of potassium deposits.

As described by Tan et al. (2005) this basin has been formed from two huge subbasins, the Kuba subbasin and the Shaqa subbasin during the Triassic period. During late Cretaceous and early Paleogene periods Paleo-Mediterranean seawater invaded regularly. Due to tectonic activities the basins were systematically depressed or upheaved and broken up in a series of sub-depressions. As a result of this, seawater intrusions were separated from the ocean in bays, lagoons and salt lakes from which salt was later deposited.

In these studies (Xiao et al. 1997, 2000; Liu et al. 1997; Tan et al. 2005, 2006, 2009; Luo et al. 2012) it was found that the chlorine isotope variations in the basins are large. In these studies a good correlation was found between the chloride contents and the $\delta^{37}Cl$ of lake brines, where in the most concentrated brines very low $\delta^{37}Cl$ values of less than −2‰ were found. The $\delta^{37}Cl$ of the rock salt deposits have a sequence of values from >0‰ for early halite, <0‰ for late halite, <−0.5‰ for sylvite and < −1‰ for carnallite (Tan et al. 2009). The very negative $\delta^{37}Cl$ values in the carnallite deposits are not in agreement with data for carnallite deposited from seawater as predicted by Eggenkamp et al. (1995), according to which the value could not be lower than about −0.7‰ and it may be that the low observed values in the Tarim Basin are due to the fact that these deposits show an important terrestrial component and thus a possibly different chlorine isotope evolution.

A summary of the behaviour of chlorine isotopes in arid non-marine basins is given by Eastoe (2016). Eastoe (2016) observed that the total range in chlorine isotope compositions in non-marine evaporites is far greater (−3 to +3‰) than the range in marine evaporites (−1 to +1‰). Eastoe (2016) found that basins could be characterised by highly positive, near-zero of negative $\delta^{37}Cl$ values. Values near 0‰ were mostly observed in basins in western and central China. The $\delta^{37}Cl$ values in these basins are close to SMOC and the small chlorine isotope variability can be explained with small scale isotope fractionation during precipitation and dissolution of the salt. Positive $\delta^{37}Cl$ values were observed in western China and western USA. $\delta^{37}Cl$ values up to +5.5‰ were seen in salty weathered crust samples in the Stafford Basin in Arizona. These high values perhaps are caused from small incremental changes in δ^{37}Cl that resulted from outward diffusion of chloride depleted in ^{37}Cl, leaving an enriched salt reservoir at depth within the crust. Diffusion at a spatial scale of 1 to 10 cm, and a time scale of days, can be sufficient if repeated over a sufficient number of cycles of wetting and drying. The δ^{37}Cl in the groundwater can also be significantly increased. Values above +1‰ in the Deep River basin (North Carolina) and China Lake (California), can be representative of halite-bearing mudstone or fossil evaporite brine. The last basin type, characterised by negative δ^{37}Cl below −1‰ could at the moment of writing only be explained by the input of chloride from rainwater that can be very negative. Unfortunately chloride concentrations in rainwater are normally very low, and huge amounts of rain, and a very long timespan, are necessary to lead to large evaporite deposits with negative δ^{37}Cl values.

11.2.3 Indirect Indications for the Bromine Isotope Composition of Evaporites Before They Could Be Measured Directly

Because the bromine contents in evaporites, especially in halite deposits, is very low the isotope composition of this bromide could not be measured directly before the development of specialised analytical techniques (Eggenkamp and Louvat 2018). For that reason, no direct bromine isotope measurements were known from evaporite deposits before 2019 (Eggenkamp et al. 2019a, 2019b). When evaporites are formed the bromide in the solution is concentrated in the brine resulting in very low bromide contents in the precipitated halite. This effect explains why evaporites contain 18% of all chlorine present in the bulk silicate earth, but only 1.2% of all bromine. When these evaporites dissolve they result in porewaters with Br/Cl ratios which are considerably lower than those in seawater. As described above, this type of porewater had been observed at several locations and they showed surprisingly similar bromine isotope values. Shouakar-Stash et al. (2007) studied the chlorine and bromine isotope composition of Siberian shield brines and reported four samples in this category with δ^{81}Br values of +0.73, +0.65, +0.71 and + 0.20‰ respectively. Boschetti et al. (2011) determined the δ^{81}Br of three porewaters from northern Italy with very low Br/Cl ratios and observed values of +0.66, +0.88 and + 0.12‰ respectively. In a study from the Iranian Kangan Gasfield aquifer (Bagheri 2013; Bagheri et al. 2014) a δ^{81}Br of +0.87‰ was found in a water sample with a low Br/Cl ratio which was identified as dissolved halite from the Kuh-e-Namak Salt Diapir. Although this set of data was only of a limited size before 2019, the data suggested that the average δ^{81}Br in halite deposits is about 0.60 ± 0.29‰ (1 SD). This value is much more positive than would be expected from the fractionation factor between sodium bromide and a saturated sodium bromide solution (ε^{37}Cl = −0.16 to −0.07‰; Eggenkamp 1995; Eggenkamp et al. 2011, 2016). If it were to assume that bromide precipitates from ocean brine with a δ^{81}Br of 0‰ we would expect that bromide in halite deposits has a δ^{81}Br much closer to 0‰ than is actually observed in halite and halite solutions.

11.2.4 Indications for the Existence of Large Br Isotope Variations in Phanerozoic Evaporites

Eastoe et al. (2007) concluded, based on the chlorine isotope compositions of Phanerozoic evaporites, that the chlorine isotope composition of the oceans did not vary significantly during Phanerozoic times. Very interesting in relation herewith are data presented by Shouakar-Stash et al. (2006) at the 16th Goldschmidt Conference in Melbourne. In their presentation they showed that in pre-Mississippian formation waters from the Williston Basin (North Dakota, these samples were 355 to 500 Ma old) the bromine isotope data showed a variation that was much larger than the variation in the chlorine isotope data.

While the chlorine isotope data varied between approximately −0.74 and + 0.47‰, the bromine isotope data varied from approximately −1.50 to +2.83‰. There also appeared to be a relation between the chlorine and bromine isotope data in this set as the Upper Ordovician, Yeoman, formation waters showed the most depleted values for both δ^{81}Br and δ^{37}Cl, while the Upper Devonian, Bakken, formation waters showed the most enriched values for both isotopes. When the results of the bromine and the chlorine isotope measurements were plotted versus time (Fig. 11.2) this variation and the similar trend for chlorine and bromine isotopes became clearly visible.

Shouakar-Stash et al. (2006) suggested that these data indicate that there is a systematic temporal variation of seawater δ^{81}Br with time. This observation may be related to the fact that the bromide concentration in seawater in fact varied considerably over Phanerozoic times. Siemann (2003) showed that the bromide content in basal halite (which was defined as the first halite that precipitates from seawater when this evaporates) varied from about 28 μg/g basal halite to more than 100 μg/g basal halite. As the bromide content in halite is related to the bromide content in the brine this indicates that the variation in the bromide concentration during Phanerozoic times was much larger than the variation in the chloride concentration and this may be related to the sug-

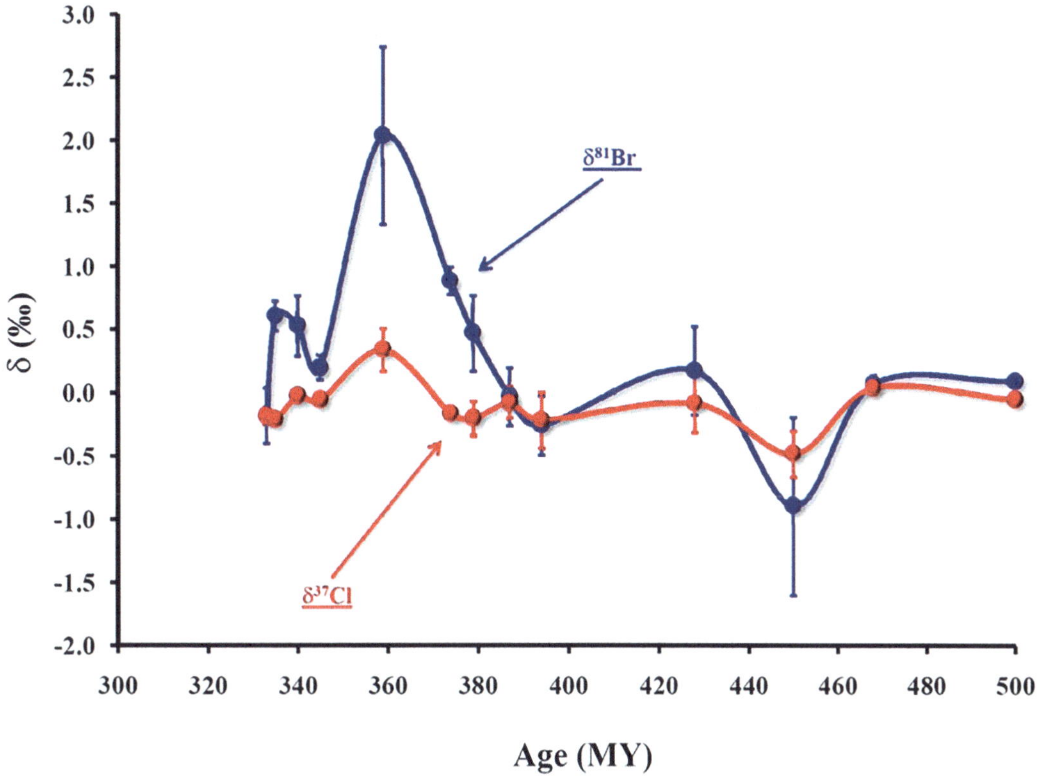

Fig. 11.2 $\delta^{81}Br_{SMOB}$ (‰) and $\delta^{37}Cl_{SMOC}$ (‰) versus Age (MY) of the Williston Basin formation waters (Mississippian – Cambrian) after eliminating samples with TDS vales less than 200,000 mg/L, samples that suffered dilution or halite dissolution and samples that were collected outside the area from 102.50° W to 104.00° W longitudes and 46.50° N to 49.00° N latitudes. The bars represent the isotopic ranges in each specific formation, and the dots represent the average isotopic values of these stratigraphic units. Figure courtesy of Orfan Shouakar-Stash (2008)

gestion of a relatively large bromine isotope variation in ocean water during Phanerozoic times.

Eggenkamp (2015) summarised the data that suggested a larger isotope variability on earth for bromine as compared to chlorine. Although the data available at that time were only limited, they clearly indicated this effect. Chlorine and bromine are present in nature mostly in the form of the halide ion, and are considered very conservative ions, i.e. they have little chemical interactions with other substances. For that reason it was contra intuitive that isotope variations for bromine are larger than for chlorine as it has a much larger atomic weight and as a result the mass ratio between the two stable isotopes is smaller, normally leading to less isotope fractionation and consequently smaller isotope variations. It was suggested by Eggenkamp (2015) that this discrepancy could be caused because the bromide ion is easy to oxidise to bromine. Although not realised often in geochemical studies, the difference in oxidation-reduction behaviour between chlorine and bromine made it possible that bromine was discovered at all (Balard 1826; Löwich 1827), and it makes it possible that the two elements were separated for chemical analyses in the nineteenth century and isotope analysis in the twenty-first century (Dechan 1886, Friedheim and Meyer 1892, Eggenkamp and Coleman 2000). For that reason Eggenkamp (2015) suggested that the larger than expected bromine isotope variations were caused by the difference in redox behaviour. It was shown that oxidation of bromide to bromomethane through biological processes is about 100 times as efficient as the oxidation of chloride in Scottish salt marshes (Blei et al. 2010) and based upon this Eggenkamp (2015) estimated that overall the reaction/production rate of CH_3Br is 26.9 times larger than the reaction/production rate of CH_3Cl. Contrary, isotope fractionation between chloride and chloromethane is larger than between bromide and bromomethane, and they are approximated as $\varepsilon^{37}Cl = -4.09‰$ and $\varepsilon^{81}Br = 0.7‰$, estimated from calculations from Schauble et al. (2003) and measurements from Horst et al. (2014). Integrated over the Phanerozoic, assuming release of methyl halides at the above rate and fractionation, Eggenkamp (2015) estimated a decrease in $\delta^{37}Cl$ in the oceans of 0.2‰ and of $\delta^{81}Br$ of 0.8‰. Eggenkamp (2015) concluded that this effect could be responsible for the difference in total variation of $\delta^{81}Br$ and $\delta^{37}Cl$ on Earth. Later studies however revealed that a different process was responsible for the differences in isotope variation between the two isotopes (Eggenkamp et al. 2019a, 2019b; Gao and Liu 2021).

11.2.5 Systematic Chlorine and Bromine Isotope Variation in Evaporites

In two studies Eggenkamp et al. (2019a, 2019b) examined both the chlorine and bromine isotope variation in evaporites. Eggenkamp et al. (2019a) studied the variation of both isotopes in a fully developed salt deposit in the northern Netherlands. In this deposit, where all major salt minerals, from early stage halite to last stage carnallite are found both the chlorine and bromine isotope evolution can be traced over the full evaporation sequence of ocean water. Eggenkamp et al. (2019a) observed that the behaviour of chlorine and bromine isotopes in these deposits differed significantly. They observed that the chlorine isotope variations follow previously predicted characteristics, showing slightly positive $\delta^{37}Cl$ (relative to seawater) in halite (NaCl) dominated layers (up to +0.05‰), decreasing to moderately negative values in carnallite ($KMgCl_3.6H_2O$) and bischofite ($MgCl_2.6H_2O$) dominated layers (down to −0.55‰). Eggenkamp et al. (2019a) were the first to also measure bromine isotope compositions in the same salt samples as where chlorine isotope compositions were measured. They observed that $\delta^{81}Br$ values decreased quickly in layers dominated by halite (from +0.2 to −0.5‰) and, contrary to the $\delta^{37}Cl$ values in the same samples, increased again in layers that were dominated by carnallite and bischofite (to just above SMOB). Their preliminary observations suggested that the Br isotope fractionation ($\varepsilon^{81}Br$) during precipitation of halite can be as high as +1.3, 3.8 times as large as Cl isotope fractionation.

In their second study, Eggenkamp et al. (2019b) investigating only primary halite samples from 18 evaporite deposits spanning the whole of the Phanerozoic, as well as the Ediacaran, Tonian and Orosirian. They observed that a fairly good relationship existed between the Br/Cl ratio and the δ^{37}Cl of most primary halite deposits from ages between the Tonian and the Messinian. In these samples δ^{37}Cl decreased systematically with increasing Br/Cl ratios, indicating that both the Br/Cl ratio and the δ^{37}Cl are good indications for the evolutionary stage of primary halite. Differences from this observation were seen in recent samples taken from salt pans, that showed much higher Br/Cl ratios and also higher δ^{37}Cl values. This different behaviour is probably caused by the enforced evaporation of ocean water in salt pans to increase precipitation of halite. Also the extremely old Orosirian show different behaviour. These samples have very low Br/Cl ratios, probably caused by repeated dissolution and precipitation of the salt. Interestingly these samples have δ^{37}Cl values that are within a very narrow range (close to −0.1‰), also suggesting strong reworking (mixing) of the salt, that also suggests a similar process that caused the low Br/Cl ratios. All other samples suggested a negative relationship between δ^{37}Cl (decreasing from about +0.3 to −0.2‰) and the Br/Cl ratio (increasing from about 0.00004 to 0.0006), indicating decreasing δ^{37}Cl in line with increasing the Br in a salt sample. This effect confirms that the chlorine isotope fractionation between halite and saturated brine indeed is positive with an ε^{37}Cl of about 0.3‰. These observations were used to estimate the secular variation of δ^{37}Cl in the oceans over the past billion years and they led to the conclusion that the δ^{37}Cl of the oceans has been very constant over that long period with a value of about −0.10 ± 0.25‰. Recently Agrinier et al. (2023) determined the δ^{37}Cl range in the oceans based on δ^{37}Cl variations in pore fluids. It was concluded that δ^{37}Cl of the ocean water should vary according to the isotope composition of detrital inputs. Agrinier et al. (2023) calculated that the expected secular range of δ^{37}Cl variation in the oceans would be between −0.3 and + 0.3‰. This is consistent with the range of δ^{37}Cl in seawater predicted by the study from Eggenkamp et al. (2019b) on evaporites deposited during the last 2 Ga. In the case of bromine this effect is not very visible. Eggenkamp et al. (2019b) could only determine a significant negative relationship between the δ^{81}Br and the Br/Cl ratio in halite samples three ages (Permian, Mid-Paleozoic and Ediacaran-Cambrian). These correlations suggested an ε^{81}Br between halite and the saturated brine between +0.5 and 1.0‰. Interestingly however was the fact that no samples were found in the region with low Br/Cl ratios and low δ^{81}Br values. This is interpreted as confirmation of a high ε^{81}Br as no low δ^{81}Br are observed in the earliest stages of halite precipitation. Unfortunately the scatter in δ^{81}Br values were so large that it could only be concluded that the δ^{81}Br values in the oceans during the last billion years could have varied anywhere between −0.6 and + 1.0‰.

Ercan et al. (2019) determined the chemical and isotopic compositions of Miocene salt in the Tuz Gölü basin in central Anatolia (Turkey). They observed in the halite facies low δ^{37}Cl (from −0.06 to 0.06‰) and low δ^{81}Br (from 0.03 to 0.95‰) suggesting that the brine in this basin was mostly derived from seawater. However, the chemical compositions as well as the boron isotope compositions that they encountered suggested that other effects must also have taken place and they state that except for ocean water other watertypes, such as saline pore water and fresh groundwaters must have been involved in the formation of these halite deposits. Overall, with the combination of a full suite of chemical and isotope analyses, Ercan et al. (2019) showed that even when the chlorine and bromine isotope compositions indicate a major marine influence in the composition of halite, other characteristics can indicate that other (minor) influences or admixtures must also be present. Thus, this way it is possible to give a better overview on full evolution of the origin of, in this case, salt deposits.

11.3 Relationships Between Bromine and Chlorine Isotope Ratios in Water and Salt Samples

Bromine isotopes in inorganic systems have mostly been studied thoroughly in formation water settings. Sediment formation waters, which contain generally very high chloride and bromide concentrations are the type of samples where it is most easily possible to study bromine isotopes in detail, as in most other settings bromide concentrations are too low to study its isotope geochemistry in detail. Because in recent years methods have become available to measure bromine isotopes in samples with (much) lower quantities of bromide (e.g. Louvat et al. 2016; Eggenkamp and Louvat 2018) some studies were published where also the bromine isotope composition in salt samples was investigated. The first small set of data in which bromine isotope variations in natural samples were shown was published by Eggenkamp and Coleman (2000). Although it was already shown by Willey and Taylor (1978) and Xiao et al. (1993) that it is possible to measure bromine isotopes using methods comparable to the measurement of chlorine isotopes, by analysing CH_3Br molecules in IRMS or Cs_2Br^+ ions in P-TIMS, no studies on the natural variation of bromine isotopes were made in these early days, perhaps due to the expected difficulty to separate bromine from a very large excess of chlorine. Another reason that no attempts to measure bromine isotope compositions were made may be the fact that it was assumed, due to the comparability of the chemistry of chlorine and bromine, that the bromine isotope composition of a sample would mirror its chlorine isotope composition. This effect was shown for example in the simple diffusion experiments presented by Eggenkamp and Coleman (1997). In these experiments it was shown that the fractionation factor of bromine due to diffusion was just about half the fractionation factor of chlorine due to diffusion, and further that both diffusion systems (chlorine and bromine) showed the same characteristics. To the surprise of the researchers it was found already in the first study on natural bromine isotope variations that the bromine isotope variation in a set of formation waters from the Norwegian Oseberg field was completely opposite to the chlorine isotope variation in the same samples. The relation between chlorine and bromine isotopes in this dataset was negative and the correlation coefficient was significantly below 1 which indicated that different fractionation processes for chlorine and bromine isotopes must have taken place. No interpretation was presented by Eggenkamp and Coleman (2000) for their results as the chlorine isotope data from this set were only published on a later date by Ziegler et al. (2001). After Shouakar-Stash et al. (2005) published their method to analyse bromine isotopes using an online technique they analysed the bromine isotope composition in several studies on formation waters from Precambrian shields in Siberia (Shouakar-Stash et al. 2007; Alexeeva et al. 2015a, 2015b) and Canada and Fennoscandia (Stotler et al. 2010). The combination of chlorine and bromine isotopes was also determined in a set of Italian formation waters from the Northern Apennine Foredeep Basin in northern Italy (Boschetti et al. 2011), various sedimentary basins in the US and Canada (Shouakar-Stash et al. 2006; Shouakar-Stash 2008; Giunta et al. 2017; Wang and Frape 2015), the Kangan gas field in Iran (Bagheri 2013; Bagheri et al. 2014), and from groundwater from the Great Artesian Basin in Australia (Gwynne et al. 2013). The first combination of chlorine and bromine isotope values from China were measured in Neogene formation water samples from the North China Plain by Chen et al. (2014), followed by formation waters from the northern sedimentary basin of the South China Sea (Yu et al. 2019). Eissa et al. (2016) measured the combination of chlorine and bromine isotopes in relatively shallow (hand dug) wells in Egypt, while Moussaoui et al. (2023) studied the chlorine and bromine isotope characteristics in comparable samples from Tunisia. Surface (sabka) samples from the UAE were studied by Saeed et al. (2021). Louvat et al. (2016), in their study to develop a new method to analyse bromine isotopes by ICP-MS, measured the bromine isotope compositions of 21 samples that were previously described (and their chlorine isotope compositions measured) in Eggenkamp (1994). Chlorine and bromine isotopes from very low concentration samples were measured by Henkemans et al. (2018) in water samples from a freshwater lake in Greenland. Very deep geothermal waters were studied by Pinti et al. (2019, 2020). Coal bed methane fluids finally were studied by Stotler et al. (2021). Chlorine and bromine isotopes in salt samples were studied in Ercan et al. (2019) and Eggenkamp et al. (2019a, 2019b).

Chloride and bromide concentrations in formation waters and salt deposits have been studied extensively as proxies for their origin (e.g. Valyashko 1956; Rittenhouse 1967; Kühn 1968; Matray et al. 1989; Davies et al. 1998). Looking at the chloride and bromide concentrations Rittenhouse (1967) was able to determine five groups of formation waters, as discussed in detail in Sect. 3.1.1.2. These five groups were defined by him as follows: I. Formation waters in which the bromine and total mineralisation approximate that what would be expected from simple evaporation of seawater ("Evaporation of seawater brine" on Fig. 3.1) or dilution of seawater with water of low mineralisation and bromide content ("Seawater dilution" on that figure). II. formation waters which are similar to formation waters of group I except for having about twice as much bromine for a specified total mineralisation. The additional bromine may have been added during early diagenesis. III. Formation waters in which the total-dissolved-solids content is greater than that of seawater but the bromine content is less than would be expected from simple concentration of seawater. These waters appear to have dissolved some halite. IV. Formation waters in which the total dissolved solids content is less than that of seawater but the bromine content is less than would be expected from simple dilution of seawater with water of low total mineralisation and bromine content. These waters may have been diluted with water having fairly high dissolved solids concentrations but low bromine content, or may have been group III waters diluted with water of low total mineralisation and bromine content. V. These formation waters have a high salinity and higher bromide contents than waters from group II. Some of these waters are associated with thick salt deposits and appear to be the residual brine which is left after salt deposition.

Only six studies were published from which both chloride and bromide concentrations and chlorine and bromine isotopes were measured when the first edition of this book was written (Eggenkamp and Coleman (2000), Ziegler et al. (2001), Shouakar-Stash et al. (2007), Stotler et al. (2010), Boschetti et al. (2011) and Bagheri et al. (2014).

The most extreme samples with regard to chlorine isotopes are still the samples from the Norwegian Oseberg field with $\delta^{37}Cl$ values as low as −4.7‰. These samples have chloride and bromide concentrations which are very close to seawater. The low $\delta^{37}Cl$ very clearly indicate that these waters are not simply seawater. Ziegler et al. (2001) explained the low $\delta^{37}Cl$ values as the result of the combination of strong diffusion and membrane-filtration effects. Unfortunately the bromine isotope composition in this study have not been interpreted together with the chlorine isotope compositions as they were not published together.

Samples from later studies were interpreted in tandem. Shouakar-Stash et al. (2007) analysed a large set of formation brines from Siberia. They were able to divide their dataset into four groups, one of which contained of fresh waters with bromide concentrations too low to be analysed for bromine isotopes. The other groups however could be distinguished based on chlorine and bromine concentrations and isotopes. The first group, Siberia A, consists of calcium-chloride brines that are believed to be the residual of an evaporated paleo-seawater, which is indicated by Br/Cl ratios in these samples which are much higher than the Br/Cl ratio in seawater, and isotope compositions which are fairly close to 0‰ for both isotopic systems. The second group, Siberia B consists of sodium chloride brines that are derived mainly from halite dissolution, as is shown by very low Br/Cl ratios. These samples are characterised by relatively positive $\delta^{81}Br$ data, and from this set it was concluded for the first time that evaporites might be characterised by generally positive $\delta^{81}Br$ data. The third grouping consisted of mostly very saline waters (although not as saline as group Siberia A) from which the origin could not be clearly determined, although the chemical and isotopic characteristics of this group suggested that they were the product of various complex scenarios such as metamorphism, water-rock interaction, permafrost freezing and mixing. This group showed very large variations in both chlorine and bromine isotope compositions, among which the most positive bromine isotope data from which also chlorine isotope date were measured. Because of the large variation in chlorine and bromine isotope compositions it was impossible to obtain a clear explanation of the origin of these samples and for that reason a complex origin was suggested.

Stotler et al. (2010) analysed the chlorine and bromine isotope compositions of a large group of 42 samples from the Canadian shield and 18 samples from the Fennoscandian shield. Based on the chloride and bromide concentrations these samples would mostly fall into group 2 of Rittenhouse's (1967) classification. The Br/Cl ratio of the Canadian samples averages around 0.009 and of the Fennoscandian samples around 0.007, which is between twice and three times the ratio in seawater. The $\delta^{37}Cl$ values of the Canadian samples vary between −0.78 and + 0.98‰ and the $\delta^{81}Br$ between +0.01 and + 1.29‰. For the Fennoscandian samples the chlorine isotopes vary between −0.54 and + 1.52‰ and the bromine between +0.26 and + 2.04‰. A weak positive correlation between chlorine and bromine isotopes (with a r^2 of 0.36) was observed in this set. At one site with serpentinite rocks present it was observed the $\delta^{37}Cl$ isotope values showed a large range while these samples showed only very small variations in $\delta^{81}Br$ values and this was attributed to ion filtration through serpentinite, which apparently affected the chloride but not the bromide ions. Comparisons with other isotope systems, such as $^{87}Sr/^{86}Sr$, indicated that water-rock interactions at some sites were likely to influence both halogen isotope systems. The large variation of $\delta^{37}Cl$ and $\delta^{81}Br$ values observed in the samples measured by Stotler et al. (2010) did not support a marine origin for these brines. This indicated that, if a seawater origin were to be considered for the fluids, a process or combination of processes significantly altered both the chlorine and the bromine isotope signatures. These unidentified processes probably explain the shift of the Br/Cl ratios from 0.035 to higher values showing that the samples in group II from Rittenhouse (1967) in reality do have a non seawater origin.

A large set of 23 formation waters from the Emilia-Romagna region of the Northern Apennine Foredeep in Northern Italy have been analysed for (amongst others) chloride and bromide as well as chlorine and bromine isotopes (Boschetti et al. 2011). Based on the total mineralisation of the samples the set is divided into three groups; brackish, saline and brine samples. The Br/Cl ratios indicate that the brackish samples are mostly diluted seawater (average Br/Cl is 0.0041), although a few of them have very low Br/Cl ratios indicating evaporite dissolution followed with subsequent dilution with fresher water. These samples appear to have the most positive $\delta^{81}Br$ values, showing a further independent indication for positive $\delta^{81}Br$ values in evaporite deposits. The saline samples seem to be altered samples (group II, Br/Cl is on average 0.0061) while the brine samples show Br/Cl ratios with are in between the brackish and saline samples (average Br/Cl is 0.0049). On the $\delta^{81}Br$ vs. $\delta^{37}Cl$ plot the other samples show a considerable scatter. This is explained in part as a possible contribution of marine organogenic bromine. Most samples however seem to describe a rough positive trend with a ~ 0.5 slope. This is explained by probable halogen (both chloride and bromide) diffusion from brines toward aquifers with lower salinities. This slope seems to agree with the fact that the isotope fraction factor for chlorine diffusion is about double the isotope fractionation for bromide diffusion as shown by Eggenkamp and Coleman (2009).

Three of the four samples analysed by Bagheri et al. (2014) show Br/Cl ratios that are between this ratio in seawater and double this amount and their isotopes indicate some moderate secondary processes. One of the samples has a low Br/Cl ratio indicating an evaporite origin, which is confirmed by a rather positive δ^{81}Br value.

The first formation water samples from China from which both chlorine and bromine isotopes were measured were published by Chen et al. (2014). The North China Plain, consisting of the Jizhong Depression and Huanghua Depression, has a complex history that was not well understood yet. For that reason Chen et al. (2014) considered to use both chlorine and bromine isotopes for continued discussion. In their samples δ^{37}Cl and δ^{81}Br range from −0.40‰ to +0.07‰, and + 0.28‰ to +1.22‰, respectively. Chen et al. (2014) show that the formation waters in both depressions are of meteoric (river or lake water) origin followed by evaporation and mixing with seawater. The difference between the two tectonic units is that the evolution of formation waters in the Jizhong Depression is similar to that of salt lakes in arid regions, showing clear evaporation, while formation waters in the Huanghua Depression indicate more interaction with seawater. This geochemical difference reflects the tectonic difference between two depressions. Chen et al. (2014) also concluded that δ^{81}Br behaves more conservatively than δ^{37}Cl in the evolution of groundwater under certain circumstances. In two northern sedimentary basins, the Beibuwan Basin and the Zhujiangkou Basin, of the South China Sea oil field δ^{37}Cl and δ^{81}Br were used to distinguish formation waters between these two basins (Yu et al. 2019). Comparing to the Beibuwan Basin, the formation waters in Zhujiangkou Basin were more enriched in ^{81}Br, and more depleted in ^{37}Cl. The formation waters in the Zhujiangkou Basin were stored in a relatively more closed environment and might have been affected more by evaporation, whereas those in the Beibuwan Basin might experience more influence from halite dissolution by ancient precipitation through open fracture structures.

Alexeeva et al. (2015a, 2015b) made a thorough survey of δ^{37}Cl and δ^{81}Br in saline ground waters and brines in sedimentary and intrusive rocks of the Siberian Platform. Preliminary results of the isotope compositions in ground brines of the Siberian Platform confirm the meteoric nature of the sodium chloride brines, as well as the formation of calcium chloride brines from evaporated paleo-seawater.

In very stagnant pore water in principle isotope fractionation can occur due to gravity. Giunta et al. (2017) studied this effect in an isolated aquifer in the Illinois Basin. In this basin the pore water must have been stagnant for a period of at least 20, and possibly up to 300 million years. They observed that both δ^{37}Cl and δ^{81}Br increased with depth in the isolated aquifers. Giunta et al. (2017) explained this increase by a gravitational effect, where the heavier isotope preferentially sinks to the base of the aquifer compared to the lighter isotope.

Shallow and deep groundwater, as sampled from both hand dug and drilled wells in the Wadi Watir delta of the Sinai peninsula of Egypt were studied by Eissa et al. (2016). The ground water has elevated concentrations, including chloride and bromide, from which the Br/Cl ratios are significantly different from seawater. The isotope characteristics of chloride and bromide in groundwater from the Wadi Watir delta deep groundwaters ($-0.01 < \delta^{37}$Cl $< +0.22$‰ and + 0.22‰ $< \delta^{81}$Br $< +0.41$‰) are different from the recharge water ($-0.48 < \delta^{37}$Cl < -0.06‰ and + 0.94‰ $< \delta^{81}$Br $< +1.28$‰). This indicates that the main groundwater recharge source comes from the catchment along the main stream channel entering the delta. Mass balance mixing models show that groundwater in the main well field contains 4 to 10% deep saline groundwater, but in some wells along the coast they even contain 2 to 6% seawater and 18 to 29% deep saline groundwater. Most groundwater recharge occurs near the outlet of the main channel. The most important factors that control the salinity of the groundwater are excessive pumping and the availability of recharge water. A study of shallow groundwater from Tunisia, also sampling both shallow hand dug wells and deeper boreholes was presented by Moussaoui et al. (2023). They studied the groundwater salinisation process in the Djebeniana coastal aquifer in south-east Tunisia. In this area two sample clusters were observed, the first more saline samples with δ^{37}Cl from −0.3‰ to +0.16‰ and δ^{81}Br from +0.21‰ to +0.39‰, and the second with δ^{37}Cl from 0.8‰ to

+0.09‰ and δ^{81}Br from +0.28‰ to +0.54‰. While there is much overlap in the isotope data from chlorine and bromine in this set the two groups could be better constrained on salinity (chloride and bromide concentrations) and δ^{18}O. based on the combination of all measured parameters Moussaoui et al. (2023) recognised four different processes that could be associated with the salinisation of the groundwater; recent recharge from salt dissolved by rainwater, leakage from deep mineralised water from the Miocene aquifer, seawater intrusion, and ion-exchange reactions.

That it is also possible to measure not only chlorine isotope variations but also bromine isotope variations in very dilute samples was shown by Henkemans et al. (2018), who determined both isotopes in a series of freshwater lakes from Greenland. δ^{81}Br values in this set were determined in samples with Br concentrations as low as 0.08 mg.L^{-1}. Henkemans et al. (2018) observed that evaporation was a key process that impacted both lake chemistry and isotopic signatures in the ice marginal area. The importance of this effect decreased towards the coast. Evaporative processes were especially apparent in the δ^{37}Cl and δ^{81}Br signatures of the dilute lake-waters.

The origin of sabkha brines present in the coastal aquifer from Abu Dhabi could be explained with the use of δ^{81}Br and δ^{37}Cl of these brines (Saeed et al. 2021). They showed that both the δ^{81}Br and the δ^{37}Cl of the sabkha aquifer (δ^{81}Br is +1.17‰, δ^{37}Cl is +0.25‰) were statistically indistinguishable from the δ^{81}Br and the δ^{37}Cl of the underlying brine (δ^{81}Br is +1.11‰, δ^{37}Cl is +0.21‰) and were statistically very distinct from the nearby seawater (where both δ^{81}Br and δ^{37}Cl are 0.0‰). Thus, the observations made by Saeed et al. (2021) provide primary direct evidence that the major source of the sabkha brines is from discharging underlying geologic brines and not from the adjacent sea water.

Coal bed methane fluids were studied by Stotler et al. (2021). They observed in these fluids from the Pennsylvanian Cherokee Group coals of the Cherokee Basin in south-east Kansas, USA in general very high δ^{81}Br values (above +2.0‰) while the δ^{37}Cl values are much more "normal", that is around 0‰. Based on the isotope composition and the hydrochemistry they were able to distinguish two samples groups. Lower salinity and higher temperature fluids from deeper units are leaking up into the Cherokee Group and mix with a higher salinity fluid that has higher δ^{81}Br values. Variations in δ^{37}Cl indicate unknown processes other than mixing affect the salinity. This process does not appear to be related to evaporation, evaporite dissolution or diffusion. Although not yet fully understood, interactions with organically bound bromine and soluble chloride within the coal could possibly have the strongest effect on the δ^{81}Br and the δ^{37}Cl values, resulting in the relatively high δ^{81}Br values and regular δ^{37}Cl values in these fluids.

Pinti et al. (2019, 2020) presented the chlorine and bromine isotope compositions of a large set of geothermal groundwaters. Because these samples are supposed to reflect mantle conditions they will be discussed in more detail in the next chapter.

In the papers discussed here 721 samples from which chlorine and bromine isotopes and concentrations were measured. Figure 11.3 shows the chloride and bromide concentrations of these 721 water and salt samples. In this figure, the salt samples are divided between halite and potash samples. These last group are defined by K + Mg + Ca > Na (molar) and are salts precipitated in later stage evaporites. Water samples are divided into four groups, samples with low Br/Cl ratios (Br/Cl < 0.001) that originate from the dissolution of halite, samples with high K + Mg + Ca concentrations that are either brines that have precipitated most sodium as halite or waters of this type that has been diluted with fresh water, and waters with high Na and Cl concentrations that is either diluted or evaporated seawater. Water samples with less than 1000 mg/L chloride are defined as fresh water in Fig. 11.3. Figure 11.3 contains a subset of the samples also shown in Fig. 3.1. While Fig. 3.1 shows a full overview of the distribution of chloride and bromide in different sample types, such as water, salts and other minerals, Fig. 11.3 only shows those water and salt samples from which both the chlorine and the bromine isotope composition has been published.

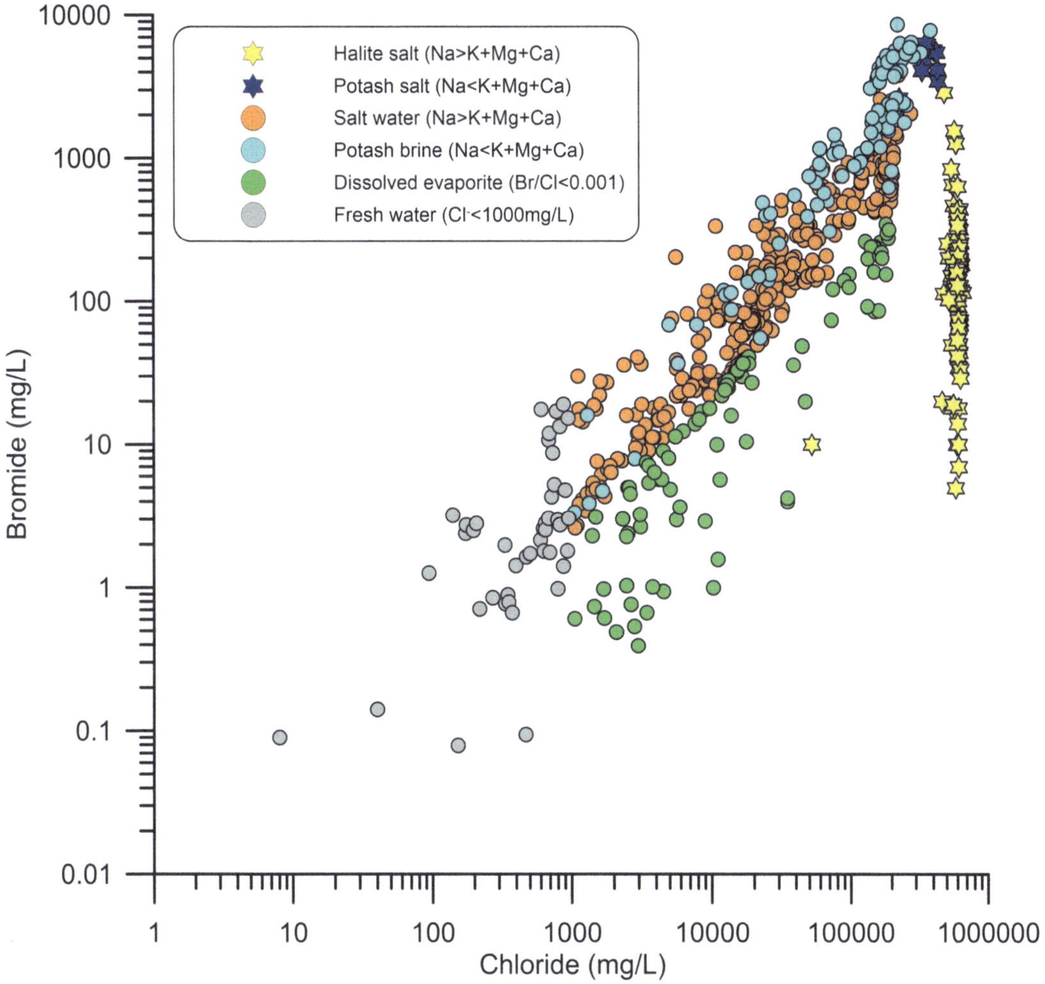

Fig. 11.3 Chloride and bromide concentrations from all known salt and water samples from which both chlorine and bromine isotope data are known. Data are split up in six groups based on the characterisation of the samples

Figure 11.4 shows the oxygen and hydrogen isotopes, $\delta^{18}O$ and $\delta^{2}H$, of the water samples in this set in order to determine if specific groups of samples with respect to the halogen isotopes show different behaviours on the water isotopes too.

Finally, in Fig. 11.5 the distribution of the chlorine and bromine isotopes of the different groups is shown.

Figure 11.3 shows that the six different samples types plot in different areas of the chloride-bromide landscape. As discussed before in more detail in Chap. 3, water samples dominated by sodium and chloride ions plot for the most part in the central part of the figure along the line through the seawater composition. Because halite that precipitates from seawater has little bromide incorporated, the Br/Cl ratio increases in the brine. When this brine is later diluted by freshwater the Br/Cl ratio does not change and for that reason these samples mostly plot above the seawater dilution/evaporation line. Also, salt that precipitates from these brines contains mostly potassium and magnesium salts, that incorporate significantly more bromide than halite does, and for that reason also plot at the top end of the figure. In precipitated halite only little bromide is incorporated. For that reason halite samples plot

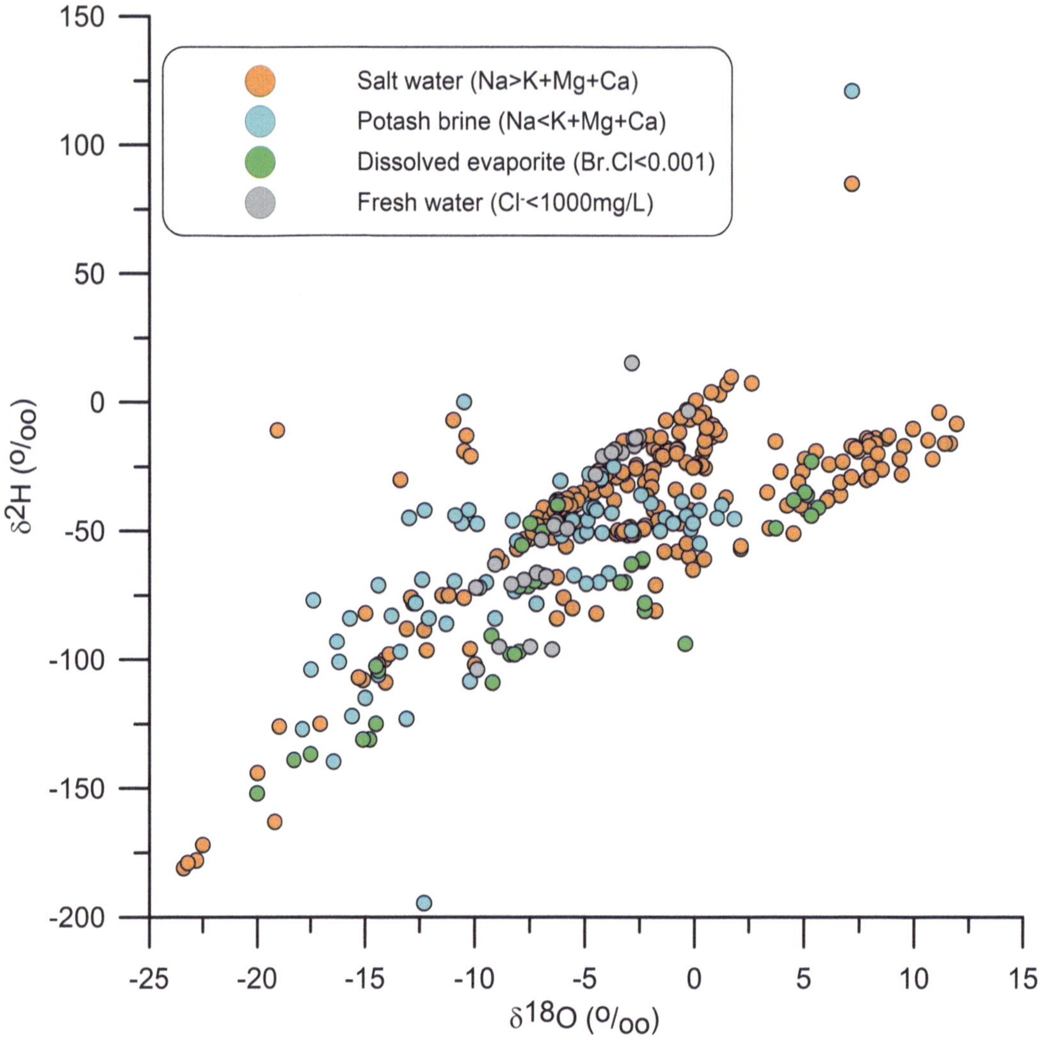

Fig. 11.4 Oxygen and hydrogen isotope composition of the water samples from which both chlorine and bromine isotope data are known. Groups are the same as in Fig. 11.3

at the far right of Fig. 11.3, showing that little bromide in co-precipitated with halite. When halite salt dissolves in fresh water the Br/Cl ratio will remain low, resulting in water samples with low Br/Cl ratios, that are found below the NaCl dominated waters in the figure. In Fig. 11.4 only water samples are represented since salt samples contain little or no water which is not normally analysed for its isotope composition. Water samples plot in a field that is bound by the meteoric waterline at the left and extends towards higher $\delta^{18}O$ values forming a kind of a triangle, but these are in fact water samples that deviate from the meteoric waterline through evaporation, an effect that increases the $\delta^{18}O$ significantly more than the $\delta^{2}H$ (Clayton et al. 1966). The triangle converges to a point where both $\delta^{18}O$ and $\delta^{2}H$ are strongly negative. The lower boundary in Fig. 11.4 represents samples close to the meteoric waterline or that consist of ocean-water mixed with meteoric water. The samples deviating from this line towards higher $\delta^{18}O$ values represent samples that are formed though intense evaporation, taking place under kinetic conditions at low relative humidities (Craig 1961; Clayton et al. 1966; Hoefs 1987). Although all four water-types are

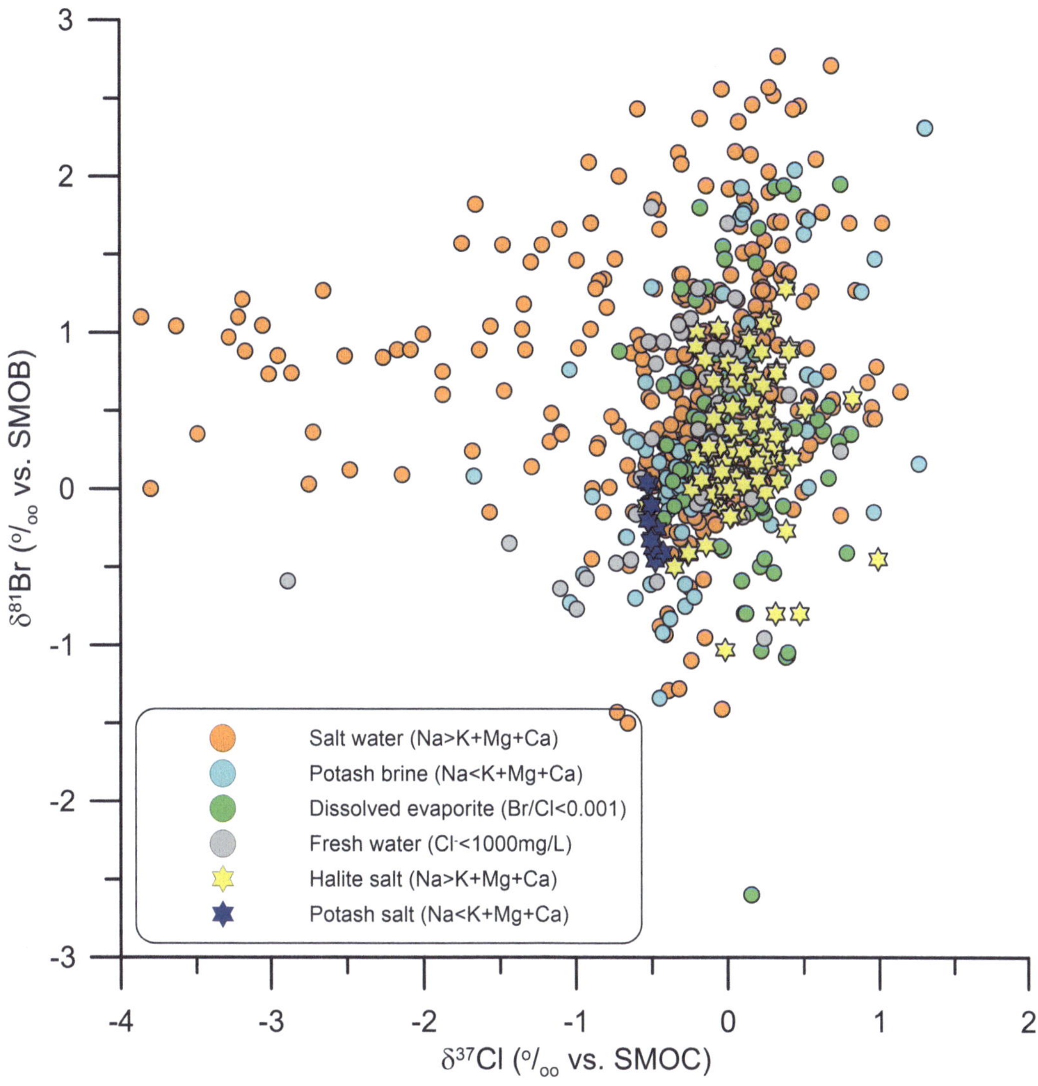

Fig. 11.5 Chlorine and bromine isotope compositions of water and salt samples. The sample groups are the same as defined in the text and in Fig. 11.3

represented in both the meteoric waters and the water representing strong evaporation it appears that water samples representing dissolved evaporite (with low Br/Cl ratios) are found more in the evaporated water types, while the evolved brines (K + Mg + Ca > Na) and their diluted equivalents tend to concentrate along the meteoric water line. NaCl dominated waters are mostly found in in the field with relatively high $\delta^{18}O$ values. These relative concentrations suggest that in many cases different water types originate from different climatic environments, although it definitely is not in all cases clear. Figure 11.5, that shows the $\delta^{81}Br$ and the $\delta^{37}Cl$ values from the samples represented in Fig. 11.3 suggests that there is a large overlap in halogen isotope composition within and between the different sample types. Still, different samples types dominate in different parts of the $\delta^{81}Br$ vs. $\delta^{37}Cl$ plot. Most samples do not plot far from the ocean water value, with $\delta^{37}Cl$ values (25th to 75th percentile) between −0.33 and + 0.16‰ and $\delta^{81}Br$ values between +0.06 and + 0.84‰. However, between the different sample types the variation is different from the overall values. Potash salt samples mostly have $\delta^{37}Cl$ values close to −0.50‰ and $\delta^{81}Br$ values between −0.36 and −0.10‰ and halite salt samples have most $\delta^{37}Cl$ values between −0.06 and + 0.24‰ and $\delta^{81}Br$ values between 0.00 and + 0.51‰. Water samples generally show different ranges, with the dissolved evaporite samples having most (25th to 75th percentile) $\delta^{37}Cl$ values between −0.03 and + 0.27‰ and $\delta^{81}Br$ values between +0.07 and + 0.65‰. Potash water samples show a different variability, with $\delta^{37}Cl$ values mostly more negative between −0.35 and + 0.04‰ and $\delta^{81}Br$ values slightly more positive between +0.05 and + 0.37‰. The largest range is observed in the NaCl dominated water-samples, which have a modest range for the most median samples ($\delta^{37}Cl$ values between −0.42 and + 0.15‰ and $\delta^{81}Br$ values between +0.03 and + 1.00‰), but they have the largest total range with $\delta^{37}Cl$ values that can be as low as −7‰ and $\delta^{81}Br$ values that can be as high as +3‰. Figure 11.5 does not show these extreme values, because otherwise most values would just plot at a small area close to the ocean water isotope composition, and for that reason the plot is clipped at −4 and + 2‰ for chlorine and at −3 and + 3 for bromine. Table 11.1 shows for all sample types as well as the total range of samples the main statistical parameters indicating the differences between the sample types but also the large overlap between the sample types. The overlap is the result of the small overall variability of chlorine and bromine isotope compositions found on Earth in combination with the relatively large isotope fractionation processes that take place between samples that are related and of the same sample type.

Finally in Fig. 11.6 two histograms are shown indicating the overall distribution of the chlorine and bromine isotope values.

In this figure clearly the differences between the chlorine and bromine isotope distributions can be seen. Most observations for chlorine are just below 0‰, while for bromine it is above the value for seawater. It is also visible that in the case of chlorine only few observations fall outside the range between −1 and + 1‰. In the case of bromine the histogram is less steep, with less observations in the most common bins and most observations actually fall within the range between −1 and + 2‰ relative to SMOB.

Table 11.1 Basic statistical characteristics of the water and salt samples from which both chlorine and bromine isotopes have been measured. Differences between the six sample types defined in the text are clearly shown in these data

	Average ± standard deviation (‰)	10% percentile	1st quartile	Median	3rd quartile	90% percentile
All samples, $\delta^{37}Cl$	−0.25 ± 1.08	−0.81	−0.33	−0.07	+0.16	+0.38
Halite salt, $\delta^{37}Cl$	+0.09 ± 0.23	−0.15	−0.06	+0.05	+0.24	+0.33
Potash salt, $\delta^{37}Cl$	−0.50 ± 0.03	−0.53	−0.51	−0.50	−0.49	−0.46
Salt water, $\delta^{37}Cl$	−0.46 ± 1.33	−1.56	−0.42	−0.15	+0.15	+0.35
Potash brine, $\delta^{37}Cl$	−0.14 ± 0.43	−0.54	−0.35	−0.20	+0.04	+0.46
Dissolved evaporite, $\delta^{37}Cl$	+0.22 ± 0.88	−0.26	−0.03	+0.10	+0.27	+0.61
Fresh water, $\delta^{37}Cl$	−0.29 ± 0.56	−0.83	−0.49	−0.16	0.00	+0.15
All samples, $\delta^{81}Br$	+0.46 ± 0.71	−0.21	+0.06	+0.32	+0.84	+1.37
Halite salt, $\delta^{81}Br$	+0.26 ± 0.44	−0.18	0.00	+0.22	+0.51	+0.88
Potash salt, $\delta^{81}Br$	−0.23 ± 0.17	−0.41	−0.36	−0.26	−0.10	+0.03
Salt water, $\delta^{81}Br$	+0.60 ± 0.75	−0.09	+0.13	+0.42	+1.00	+1.58
Potash brine, $\delta^{81}Br$	+0.28 ± 0.63	−0.29	−0.05	+0.19	+0.37	+1.25
Dissolved evaporite, $\delta^{81}Br$	+0.36 ± 0.72	−0.43	+0.07	+0.32	+0.65	+1.35
Fresh water, $\delta^{81}Br$	+0.37 ± 0.67	−0.58	−0.15	+0.45	+0.88	+1.07

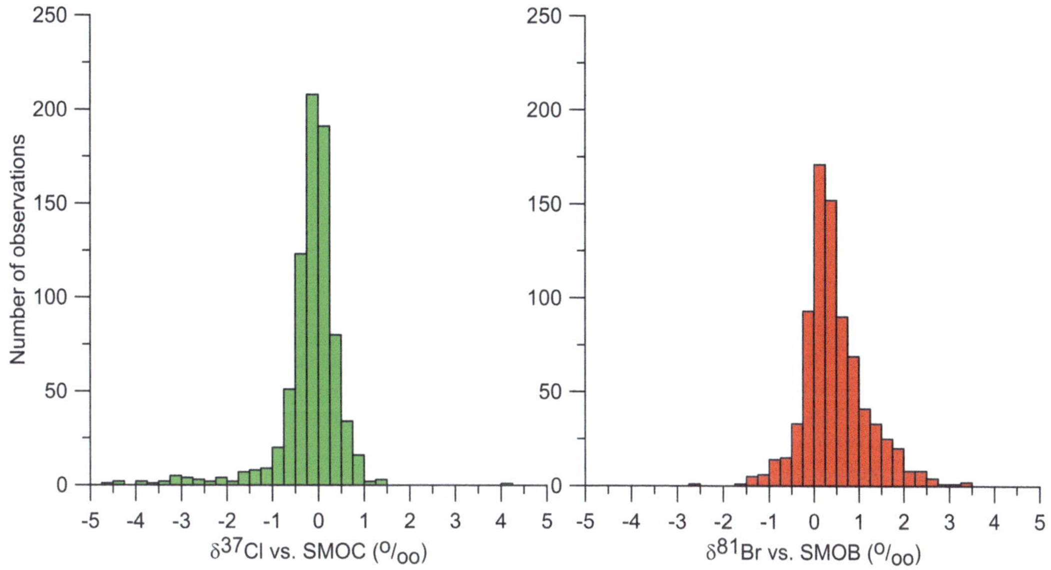

Fig. 11.6 Histograms showing the distribution of $\delta^{37}Cl$ and $\delta^{81}Br$ in the water and salt samples from which both isotopes have been measured

References

Agrinier P, Destrignevile C, Giunta T, Bonifacie M, Bardoux G, Andre J, Lucazeau F (2019) Strong impact of ion filtration on the isotopic composition of chlorine in young clay-rich oceanic sediment pore fluids. Geochim Cosmochim Acta 245:525–541

Agrinier P, Bonifacie M, Bardoux G, Lucazeau F, Giunta T, Ader M (2021) Chlorine isotope data of chlorides challenge the pore fluid paradigm. Geochim Cosmochim Acta 300:258–278

Agrinier P, Gieskes J, Subbarao G, Bardoux G, Bonifacie M (2023) Chloride exchanges between oceanic sediments and seawater: constraints from chlorine isotopes. Geochim Cosmochim Acta 361:10–23

Alexeeva LP, Alexeev SV, Kononov AM, Teng M, Liu YD (2015a) Chlorine and bromine isotopes in brines of the Western Yakutia. Isvestia 13:19–30. (in Russian with English abstract)

Alexeeva LP, Alexeev SV, Kononov AM, Teng M, Liu YD (2015b) Halogen isotopes (^{37}Cl and ^{81}Br) in brines of the Siberian platform. Proc Earth Planet Sci 13:47–51

Bagheri R (2013) Hydrochemistry and sources of connate water in the Zagros aquifers. PhD Thesis. Shiraz University, Iran

Bagheri R, Nadri A, Raeisi E, Eggenkamp HGM, Kazemi GA, Montaseri A (2014) Hydrochemical and isotopic ($\delta^{18}O$, $\delta^{2}H$, $^{87}Sr/^{86}Sr$, $\delta^{37}Cl$ and $\delta^{81}Br$) evidence for the origin of saline formation water in a gas reservoir. Chem Geol 384:62–75

Balard AJ (1826) Sur une substance particulière contenue dans l'eau de la mer. Annal Chim Phys 2me series 32:337–381

Barnes JD, Sharp ZD (2006) A chlorine isotope study of DSDP/ODP serpentinized ultramafic rocks: insights into the serpentinization process. Chem Geol 228:246–265

Barnes JD, Selverstone J, Sharp ZD (2006) Chlorine isotope chemistry of serpentinites from Elba, Italy, as an indicator of fluid source and subsequent tectonic history. Geochem Geophys Geosys 7:Q08015

Barnes JD, Sharp ZD, Fischer TP (2008) Chlorine isotope variations across the Izu-Bonin-Mariana arc. Geology 36:883–886

Barnes JD, Eldam R, Lee CTA, Errico JC, Loewy S, Cisneros M (2013) Petrogenesis of serpentinites from the Franciscan complex, western California, USA. Lithos 178:143–157

Beekman HE, Eggenkamp HGM, Appelo CAJ (2011) An integrated modelling approach to reconstruct complex solute transport mechanisms – Cl and $\delta^{37}Cl$ in pore water of sediments from a former brackish lagoon in The Netherlands. Appl Geochem 26:257–268

Berner EK, Berner RA (1987) The global water cycle: geochemistry and environment. Prentice-Hall, Inc, Englewood Cliffs, NJ

Blei E, Heal MR, Heal KV (2010) Long-term CH_3Br and CH_3Cl flux measurements in temperate salt marshes. Biogeosciences 7:3657–3668

Bonifacie M (2005) Le cycle du Chlore terrestre: les échanges manteau-océan. PhD thesis IPGP-Université Paris VII- Denis Diderot, 300p

Bonifacie M, Jendrzejewski N, Agrinier P, Coleman M, Pineau F, Javoy M (2007a) Pyrohydrolysis-IRMS determination of silicate chlorine stable isotope compositions. Application to oceanic crust and meteorite samples. Chem Geol 242:187–201

Bonifacie M, Monnin C, Jendrzejewski N, Agrinier P, Javoy M (2007b) Chlorine stable isotopic composition of basement fluids of the eastern flank of the Juan de Fuca ridge (ODP leg 168). Earth Planet Sci Lett 260:10–22

Bonifacie M, Busigny V, Mével C, Philippot P, Agrinier P, Jendrzejewski N, Scambelluri M, Javoy M (2008) Chlorine isotopic composition in seafloor serpentinites and high-pressure metaperidotites. Insights into oceanic serpentinization and subduction processes. Geochim Cosmochim Acta 72:126–139

Boschetti T, Toscani L, Shouakar-Stash O, Iacumin P, Venturelli G, Mucchino C, Frape SK (2011) Salt waters of the Northern Apennine Foredeep Basin (Italy):origin and evolution. Aquat Geochem 17:71–108

Braitsch O (1971) Salt deposits, their origin and composition. Springer-Verlag, p 288

Braitsch O (1962) Entstehung und Stoffbestand der Salzlagerstätten. Springer-Verlag, p 232

Broeker WS, Peng TH (1982) Tracers in the sea. Eldigio Press Lamont Doherty Geological Observatory

Campbell DJ (1985) Fractionation of stable chlorine isotopes during transport through semipermeable membranes. MS. thesis. University of Arizona

Chen LZ, Ma T, Du Y, Yang J, Liu L, Shan HM, Liu CF, Cai HS (2014) Origin and evolution of formation water in North China plain based on hydrochemistry and stable isotopes (^{2}H, ^{18}O, ^{37}Cl and ^{81}Br). J Geochem Explor 145:250–259

Clayton RN, Friedman I, Graf DL, Mayeda TK, Meents WF, Shimp NF (1966) The origin of saline formation waters. 1. Isotopic composition. J Geophys Res 71:3869–3882

Coleman ML, Eggenkamp HGM, Curtis CD (1994) Chlorine stable isotopes as a tracer of waters expelled from mud-rocks. ICOG 7. US Geol Survey Circular 1108:64

Craig H (1961) Isotopic variations in meteoric water. Science 133:1702–1703

Davies SN, Whittemore DO, Fabryka-Martin J (1998) Use of chloride/bromide ratios in studies of potable water. Ground Water 36:338–350

Davis EE, Fisher AT, Firth JV, et al (1997b) Rough basement transect (Sites 1026 and 1027). In: Proc ODP, Init Repts College Station, TX (Ocean Drilling Program), 168:101–160

Davis EE, Fisher AT, Firth JV, et al (1997c) Buried basement transect (Sites 1028, 1029, 1030, 1031, and 1032). In Proc ODP, Init Repts College Station TX (Ocean Drilling Program), 168:161–212

Dechan M (1886) Detection and estimation of iodine, bromine, and chlorine. J Chem Soc 49:682–685

Desaulniers DE, Kaufmann RS, Cherry JA, Bentley HW (1986) ^{37}Cl-^{35}Cl variations in a diffusion controlled groundwater system. Geochim Cosmochim Acta 50:1757–1764

Deyle A, Kopf A, Frape S, Hesse R (2003) Evidence for fluid flow in the Japan trench forearc using isotope geochemistry (Cl, Sr, B): Results from ODP Site 1150. Island Arc 13:258–270

Du Y, Ma T, Yang J, Liu L, Shan HM, Cai HS, Liu CF, Chen LZ (2013) A precise analytical method for bromine stable isotopes in natural waters by GasBench II-IRMS. Int J Mass Spectrom 338:50–56

Du SR, Zheng MP, Lü YY, Zhang YS, Zhang XF, Miao ZY, Li MM (2022) Origin of Halite in the Lanping–Simao Basin, Southeastern Tibetan Plateau, China: Evidence from Sr and Cl Isotopes. Acta Geol Sin (Engl Ed) 96:2028–2039

Eastoe CJ (2016) Stable chlorine isotopes in arid non-marine basins: instances and possible fractionation mechanisms. Appl Geochem 74:1–12

Eastoe CJ, Peryt T (1999) Multiple sources of chloride in Badenian evaporites, Carpathian mountains: stable chlorine isotope evidence. Terra Nova 11:118–123

Eastoe CJ, Long A, Knauth LP (1999) Stable chlorine isotopes in the Palo Duro Basin, Texas: evidence for preservation of Permian evaporite brines. Geochim Cosmochim Acta 63:1375–1382

Eastoe CJ, Long A, Land LS, Kyle JR (2001) Stable chlorine isotopes in halite and brine from the Gulf Coast Basin: brine genesis. Chem Geol 176:343–360

Eastoe CJ, Peryt TM, Petrychenko OY, Geisler-Cussey D (2007) Stable chlorine isotopes in Phanerozoic evaporites. Appl Geochem 22:575–588

Egeberg PK, Dickens G (1999) Thermodynamic and pore-water halogen constraints on gas hydrate distribution at ODP site 997 (Blake Ridge). Chem Geol 153:53–79

Eggenkamp HGM (1994) $\delta^{37}Cl$; the geochemistry of chlorine isotopes. Geol Ultraiect 116:1–150

Eggenkamp HGM (1995) Bromine stable isotope fractionation: Experimental determination on evaporites. European Union of Geosciences, 8th Congress. Terra Nova 7 Abst. Supp. 1:331

Eggenkamp HGM (2015) Why are variations in bromine isotope compositions in the Earth's history larger than chlorine isotope compositions? Ann UMCS Sectio AAA 70:185–196

Eggenkamp HGM, Coleman ML (1997) Comparison of chlorine and bromine isotope fractionation in diffusion: Geochemical consequences. 7th Annual Goldschmidt Conference, Tucson, AZ 213

Eggenkamp HGM, Coleman ML (2000) Rediscovery of classical methods and their application to the measurement of stable bromine isotopes in natural samples. Chem Geol 167:393–402

Eggenkamp HGM, Coleman ML (2009) The effect of aqueous diffusion on the fractionation of chlorine and bromine stable isotopes. Geochim Cosmochim Acta 73:3539–3548

Eggenkamp HGM, Louvat P (2018) A simple distillation method to extract bromine from natural water and salt samples for isotope analysis by multi-collector inductively coupled plasma mass spectrometry. Rapid Commun Mass Spectrom 32:612–618

Eggenkamp HGM, Middelburg JJ, Kreulen R (1994) Preferential diffusion of ^{35}Cl relative to ^{37}Cl in sediments of Kau Bay, Halmahera, Indonesia. Chem Geol (Isot Geosci Sect) 116:317–325

Eggenkamp HGM, Kreulen R, Koster van Groos AF (1995) Chlorine stable isotope fractionation in evaporites. Geochim Cosmochim Acta 59:5169–5175

Eggenkamp HGM, Bonifacie M, Ader M, Agrinier P (2011) Fractionation of Cl and Br isotopes during precipitation of salts from their saturated solutions. 21th Annual V.M. Goldschmidt Conference. Prague, Czech Republic. Mineral Mag 75:798

Eggenkamp HGM, Bonifacie M, Ader P, Agrinier P (2016) Experimental determination of stable chlorine and bromine isotope fractionation during precipitation of salt from a saturated solution. Chem Geol 433:46–56

Eggenkamp HGM, Louvat P, Griffioen J, Agrinier P (2019a) Chlorine and bromine isotope evolution within a fully developed upper Permian natural salt sequence. Geochim Cosmochim Acta 245:316–326

Eggenkamp HGM, Louvat P, Agrinier P, Bonifacie M, Bekker A, Krupenik V, Griffioen J, Horita J, Brocks JJ, Bagheri R (2019b) The bromine and chlorine isotope composition of primary halite deposits and their significance for the secular isotope composition of seawater. Geochim Cosmochim Acta 264:13–29

Eissa MA, Thomas JM, Pohll G, Shouakar-Stash O, Hershey RL, Dawoud M (2016) Groundwater recharge and salinization in the arid coastal plain aquifer of the Wadi Watir delta, Sinai, Egypt. Appl Geochem 71:48–62

Ercan HÜ, Karakaya MC, Bozdağ A, Karakaya N, Delican A (2019) Origin and evolution of halite based on stable isotopes (δ^{37}Cl, δ^{81}Br, δ^{11}B and δ^{7}Li) and trace elements in Tuz Gölü Basin, Turkey. Appl Geochem 105:17–30

Friedheim C, Meyer RJ (1892) Über die quantitative Trennung und Bestimmung von Chlor, Brom und Jod. Z Anorg Chem 1:407–422

Gao CH, Liu Y (2021) First-principles calculations of equilibrium bromine isotope fractionations. Geochim Cosmochim Acta 297:65–81

Giunta T, Devauchelle O, Ader M, Locke R, Louvat P, Bonifacie M, Métivier F, Agrinier P (2017) The *gravitas* of gravitational isotope fractionation revealed in an isolated aquifer. Geochem Perspect Lett 4:1736

Godon A, Jendrzejewski N, Eggenkamp HGM, Banks DA, Ader M, Coleman ML, Pineau F (2004a) A cross calibration of chlorine isotopic measurements and suitability of seawater as the international reference material. Chem Geol 207:1–12

Godon A, Jendrzejewski N, Castrec-Rouelle M, Dia A, Pineau F, Boulègue J, Javoy M (2004b) Origin and evolution of fluids from mud volcanoes in the Barbados accretionary complex. Geochim Cosmochim Acta 68:2153–2165

Groen J, Velstra J, Meesters AGCA (2000) Salinization processes in paleowaters in coastal sediments from Suriname: evidence from δ^{37}Cl analysis and diffusion modelling. J Hydrol 234:1–20

Gwynne R, Frape S, Shouakar-Stash O, Love A (2013) ^{81}Br, ^{37}Cl and ^{87}Sr studies to assess groundwater flow and solute sources in the southwestern great Artesian Basin, Australia. Prog Earth Planet Sci 7:330–333

Hay WW, Migdisov A, Balukhovsky AN, Wold CN, Flögel S, Söding E (2006) Evaporites and the salinity of the ocean during the Phanerozoic: implications for climate, ocean circulation and life. Palaeogeogr Palaeoclimatol Palaeoecol 240:3–46

Haydon PR, Graf DL (1986) Studies of smectite membrane behavior: temperature dependence, 20–180°C. Geochim Cosmochim Acta 50:115–121

Henkemans E, Frape SK, Ruskeeniemi T, Anderson NJ, Hobbs M (2018) A landscape-isotopic approach to the geochemical characterization of lakes in the Kangerlussuaq region, West Greenland. Arct Antarct Alp Res 50:e1420863

Henry P, Kanamatsu T, Moe K, The Expedition 333 Scientists (2012a) Site C0011, Proc. IODP, 333, vol 333. Integrated Ocean Drilling Program Management International, Inc, Tokyo, pp 1–96

Henry P, Kanamatsu T, Moe KT, Strasser M, The IODP Expedition 333 Scientific Party (2012b) IODP expedition 333: return to Nankai through subduction inputs sites and coring of mass transport deposits. Sci Drill 14:4–17

Hesse R (2003) Pore water anomalies of submarine gas-hydrate zones as tool to assess hydrate abundance and distribution in the subsurface: what have we learned in the past decade? Earth Sci Rev 61:149–179

Hesse R, Frape SK, Egeberg PK, Matsumoto R (2000) Stable isotope studies (Cl, O and H) of interstitial waters from site 997, Blake ridge gas hydrate field, West Atlantic. In: Paull CK, Matsumoto R, Wallace PJ, Dillon WD (eds) Proceedings of the Ocean Drilling Program, Scientific Results, vol 164. Ocean Drilling Program, College Station, TX, pp 129–137

Hesse R, Egeberg PK, Frape SK (2006) Chlorine stable isotope ratios as a tracer for pore-water advection rates in a submarine gas-hydrate field: implication for hydrate concentration. Geofluids 6:1–7

Hoefs J (1987) Stable isotope geochemistry.3rd edn. Springer-Verlag, Berlin Heidelberg

Hoering TC, Parker PL (1961) The geochemistry of the stable isotopes of chlorine. Geochim Cosmochim Acta 23:186–199

Horst A, Andersson P, Thornton BJ, Holmstrand H, Wishkerman A, Keppler F, Gustafsson Ö (2014) Stable bromine isotopic composition of methyl bromide released from plant matter. Geochim Cosmochim Acta 125:186–195

Hussain SA, Han FQ, Han J, Widory D (2020) Chlorine isotopes unravel conditions of formation of the Neoproterozoic rock salts from the salt range formation, Pakistan. Can J Earth Sci 57:698–708

Hussain SA, Han FQ, Ma Z, Hussain A, Mughal MS, Han J, Alhassan A, Widory D (2021) Origin, and evolution of Eocene rock salts in Pakistan and implications for paleoclimate studies: insights from chemistry and Cl stable isotopes. Front Earth Sci 9:644485

John T, Layne GD, Haase KM, Barnes JD (2010) Chlorine isotope evidence for crustal recycling into Earth's mantle. Earth Planet Sci Lett 298:175–182

John T, Scambelluri M, Frische M, Barnes JD, Bach W (2011) Dehydration of subducting serpentinite: implications for halogen mobility in subduction zones and the deep halogen cycle. Earth Planet Sci 308:65–76

Kastner M, Elderfield H, Jenkins WJ, Gieskes JM, Gamo T (1993) Geochemical and isotopic evidence for fluid flow in the western Nankai subduction zone, Japan. In: Hill IA, Taira A (eds) Proceedings of the Ocean Drilling Program, Scientific Results, Leg 131. ODP, College Station, Texas, pp 397–416

Kaufmann RS (1984) Chlorine in groundwater: stable isotope distribution. Ph.D. thesis. University of Arizona, Tucson, Ariz

Kaufmann RS, Frape S, Fritz P, Bentley H (1987) Chlorine stable isotope composition of Canadian shield brines. In: Fritz P, Frape SK (eds) Saline water and gases in crystalline rocks, vol 33. Geol Assoc Canada, Special Paper, pp 89–93

Kaufmann RS, Long A, Campbell DJ (1988) Chlorine isotope distribution in formation waters, Texas and Louisiana. AAPG Bull 72:839–844

Kaufmann RS, Frape SK, McNutt R, Eastoe CJ (1993) Chlorine stable isotope distribution of Michigan basin formation waters. Appl Geochem 8:403–407

Knauth LP (1998) Salinity history of the Earth's early ocean. Nature 395:554–555

Koehler G, Wassenaar LI (2010) The stable isotopic composition ($^{37}Cl/^{35}Cl$) of dissolved chloride in rainwater. Appl Geochem 25:91–96

Kolodny Y, Katz A, Starinsky A, Moise T, Simon E (1999) Chemical tracing of salinity sources in Lake Kinneret (Sea of Galilee), Israel. Limnol Oceanogr 44:1035–1044

Kühn R (1968) Geochemistry of the German potash deposits. Geol Soc Am Spec Pap 88:427–504

Land LS (1995) The role of saline formation water in crustal cycling. Aquat Geochem 1:137–145

Lavastre V, Jendrzejewski N, Agrinier P, Javoy M, Evrard M (2005) Chlorine transfer out of a very low permeability clay sequence (Paris Basin, France): ^{35}Cl and ^{37}Cl evidence. Geochim Cosmochim Acta 69:4949–4961

Le Gal La Salle C, Matray JM, Bensenouci F, Michelot JL, Dauzères A, Wittebroodt C, Frape S, Shouakar Stash O, Rebaix R, Lancelot J (2013) Modelling Cl-concentration and $\delta^{37}Cl$ profiles in porewater across a 250m-thick indurated clayrock at the tournemire URL (France). Prog Earth Planet Sci 7:471–474

Li YP, Jiang SY (2013) Major cation/chlorine ratio and stable chlorine isotopic compositions of sediment interstitial water in the Brazos-Trinity Basin IV from the Gulf of Mexico (IODP 308). J Asian Earth Sci 65:52–50

Li XQ, Zhou AG, Liu YD, Ma T, Liu CF, Liu L, Yang J (2012) Stable isotope geochemistry of dissolved chloride in relation to hydrogeology of the strongly exploited quaternary aquifers, North China plain. Appl Geochem 27:2031–2041

Liu WG, Xiao YK, Wang QZ, Qi HP, Wang YH, Zhou YM, Shirodkar PV (1997) Chlorine isotopic geochemistry of salt lakes in the Qaidam Basin, China. Chem Geol 136:271–279

Liu YD, Zhou AG, Gan YQ, Liu CF, Yu TT, Li XQ (2013) An online method to determine chlorine stable isotope composition by continuous flow isotope ratio mass spectrometry (CF-IRMS) coupled with a Gasbench II. J South Cent Univ 20:193–198

Louvat P, Bonifacie M, Giunta T, Michel A, Coleman M (2016) Determination of bromine stable isotope ratios from saline solutions by 'wet plasma' MC-ICPMS including a comparison between high and low-resolution modes, and three introduction systems. Anal Chem 88:3891–3898

Löwich C (1827) Über Brombereitung und eine auffallende Zersetzung des Aethers durch Chlor. Mag Pharm 21:31–36

Luo CG, Xiao YK, Ma HZ, Ma YQ, Zhang YL, He MY (2012) Stable isotope fractionation of chlorine during evaporation of brine from a saline lake. Chin Sci Bull 57:1833–1843

Magenheim AJ, Spivack AJ, Volpe C, Ransom B (1994) Precise determination of stable chlorine isotopic ratios in low-concentration natural samples. Geochim Cosmochim Acta 58:3117–3121

Magenheim AJ, Spivack AJ, Michael PJ, Gieskes JM (1995) Chlorine stable isotope composition of the oceanic crust: implications for Earth's distribution of chlorine. Earth Planet Sci Lett 131:427–432

Mahara Y, Ohta T, Tokunaga T, Matsuzaki H, Nagao K, Nakata E, Miyamoto Y, Kubota T (2013) Pore-water mobility: distribution of $\delta^{37}Cl$, $^{36}Cl/Cl$, $^{129}I/^{127}I$ and dissolved ^{4}He concentration in the core drilled in the Mobara gas field, Japan. Nucl Instrum Methods Phys Res B 294:597–601

Martin JG (1993) Stable chlorine isotopes in the Milk River aquifer, Alberta, Canada: implications for the geochemical evolution of the aquifer system. MSc thesis,. Univ Arizona

Matray JM, Meunier A, Thomas M, Fontes JC (1989) Les eaux de formation du Trias et du Dogger du Bassin Parisien: Histoire et effets diagénétiques sur les réservoirs. Bull Soc Nat Elf-Aquitaine (Prod) 13:484–504

Meng F, Zhang Y, Galamay AR, Bukowski K, Ni P, Xing E, Ji L (2018) Ordovician seawater composition. Evidence from fluid inclusions in halite. Geol Quart 62:344–352

Moussaoui I, Rosa E, Cloutier V, Neculite CM, Dassi L (2023) Chemical and isotopic evaluation of groundwater salinization processes in the Djebeniana coastal aquifer, Tunisia. Appl Geochem 149:105555

NIST (2001) Standard reference material® 975a. Isotopic Standard for Chlorine, Gaithersburg, MD

Paull CK, Matsumoto R, Wallace PJ, Shipboard Scientific Party (2005a) Geochemistry of interstitial water of ODP hole 164–997A. PANGAEA

Paull CK, Matsumoto R, Wallace PJ, Shipboard Scientific Party (2005b) Geochemistry of interstitial water of ODP hole 164–997B. PANGAEA

Peirce J, Weissel J, et al (1989a) Leg 121 background and objectives: ODP leg 121. In: Proc ODP Init Repts, 121:1–31

Peirce J, Weissel J, et al (1989b) Site 757: ODP Leg 121. In: Proc ODP Init Repts, 121:305–358

Peirce J, Weissel J, et al (1989c) Site 758: ODP Leg 121. In: Proc ODP Init Repts, 121:359–453

Phillips FM, Bentley HW (1987) Isotopic fractionation during ion filtration: I. Theory. Geochim Cosmochim Acta 51:683–695

Pinti DL, Castro MC, López-Hernández A, Hernández Hernández MA, Shouakar-Stash O, Richard L, Nuñez-Hernández S, Hall CM, Ramírez-Montes M (2019) Signature of ongoing brine reinjection on noble gas isotopes and fluid chemistry at Las Tres Vírgenes geothermal field, Mexico. J Volcanol Geotherm Res 377:33–42

Pinti DL, Shouakar-Stash O, Castro MC, López-Hernández A, Hall CM, Rocher O, Shibata T, Ramírez-Montes M (2020) The bromine and chlorine isotopic composition of the mantle as revealed by deep geothermal fluids. Geochem Cosmochim Acta 276:14–30

Ransom B, Spivack AJ, Kastner M (1995) Stable Cl isotopes in subduction-zone porewaters: implications for fluid rock reactions and the cycling of chlorine. Geology 23:715–718

Rebeix R, Le Gal La Salle C, Jean-Baptiste P, Lavastre V, Fourré E, Bensenouci F, Matray JM, Landrein P, Shouakar-Stash O, Frape SK, Michelot JL, Lancelot J (2014) Chlorine transport processes through a 2000 m aquifer/aquitard system. Mar Pet Geol 53:102–116

Richter C, Salisbury MH, Shinohara M et al (2005) Shipboard scientific Party, geochemistry of interstitial water of ODP hole 195–1201D. PANGAEA

Rittenhouse G (1967) Bromine in oil field waters and its use in determining possibilities of origin of these waters. AAPG Bull 17:1213–1228

Sacks IS, Suyehiro K, Acton GD, Shipboard Scientific Party (2005a) Geochemistry of interstitial water of ODP hole 186–1150A. PANGAEA

Sacks IS, Suyehiro K, Acton GD, Shipboard Scientific Party (2005b) Geochemistry of interstitial water of ODP hole 186–1150B. PANGAEA

Saeed W, Shouakar-Stash O, Wood WW, Frape S (2021) Isotopically enriched Geogenic δ^{81}Br and δ^{37}Cl: primary evidence for the ascending brine model. Groundwater 59:671–676. https://doi.org/10.1111/gwat.13096

Saito S, The Expedition 322 Scientists (2010a) Site C0011. In: Saito S, Underwood MB, Kubo Y (eds) Proc. IODP, 322, and the expedition 322 scientists, vol 322. Integrated Ocean Drilling Program Management International, Inc, Tokyo, pp 1–159

Saito S, The Expedition 322 Scientists (2010b) Site C0012. In: Saito S, Underwood MB, Kubo Y (eds) Proc. IODP, 322, and the expedition 322 scientists, vol 322. Integrated Ocean Drilling Program Management International, Inc, Tokyo, pp 1–121

Schauble EA, Rossman GR, Taylor HP (2003) Theoretical estimates of equilibrium chlorine-isotope fractionations. Geochim Cosmochim Acta 67:3267–3281

Schilling JG, Unni CK, Bender ML (1978) Origin of chlorine and bromine in the oceans. Nature 273:631–636

Selverstone J, Sharp ZD (2011) Chlorine isotope evidence for multicomponent mantle metasomatism in the Ivrea zone. Earth Planet Sci Lett 310:429–440

Sharp ZD, Barnes JD (2004) Water-soluble chlorides in massive seafloor serpentinites: a source of chloride in subduction zones. Earth Planet Sci Lett 226:243–254

Shields WR, Murphy TJ, Garner EL, Dibeler VH (1962) Absolute isotopic abundance ratios and the isotopic weight of chlorine. J Am Chem Soc 84:1519–1522

Shirodkar PV, Xiao YK, Hai L (2003) Boron and chlorine isotopic signatures of seawater in the central Indian ridge. Curr Sci 85:313–320

Shirodkar PV, Xiao YK, Sarkar A, Dalal SG, Chivas AR (2006) Influence of air-sea fluxes on chlorine isotopic composition of ocean water: implications for constancy in δ^{37}Cl – a statistical inference. Environ Int 32:235–239

Shouakar-Stash O (2008) Evaluation of stable chlorine and bromine isotopes in sedimentary formation fluids. PhD Thesis. University of Waterloo

Shouakar-Stash O, Frape SK, Drimmie RJ (2005) Determination of bromine stable isotopes using continuous-flow isotope ratio mass spectrometry. Anal Chem 77:4027–4033

Shouakar-Stash O, Frape SK, Rostron BJ, Drimmie RJ (2006) Variations of the δ^{81}Br and δ^{37}Cl stable isotope signature for pre-Mississippian formation waters of the Williston Basin. Abstract presented at the Goldschmidt Conference. Geochim Cosmochim Acta 70(suppl):A589

Shouakar-Stash O, Alexeev SV, Frape SK, Alexeeva LP, Drimmie RJ (2007) Geochemistry and stable isotope signatures, including chlorine and bromine isotopes, of the deep groundwaters of the Siberian platform, Russia. Appl Geochem 22:589–605

Sie PMJ, Frape SK (2002) Evaluation of the groundwaters from the Stripa mine using stable chlorine isotopes. Chem Geol 182:565–582

Siemann MG (2003) Extensive and rapid changes in seawater chemistry during the Phanerozoic: evidence from Br contents in basal halite. Terra Nova 15:243–248

Spivack AJ, Kastner M, Ransom B (2002) Elemental and isotopic chloride geochemistry and fluid flow in the Nankai trough. Geophys Res Lett 29:1661–1665

Srivastava SP, Arthur M, Clement B, et al (1987a) Introduction: ODP Leg 105, Proc. Init. Repts. (Part A). ODP 105:5–20

Srivastava SP, Arthur M, Clement B, et al (1987b) Site 645: ODP Leg 105, Proc. Init. Repts. (Part A). ODP 105:61–418

Srivastava SP, Arthur M, Clement B, et al (1987c) Site 646: ODP Leg 105, Proc. Init. Repts. (Part A). ODP 105:419–674

Srivastava SP, Arthur M, Clement B, et al (1987d) Site 647: ODP Leg 105, Proc. Init. Repts. (Part A). ODP 105:675–905

Stiller M, Nissenbaum A, Kaufmann RS, Long A (1998) Cl-37 in the Dead Sea system – preliminary results. Appl Geochem 13:953–960

Stiller M, Rosenbaum JM, Nishri A (2009) The origin of brines underlying Lake Kinneret. Chem Geol 262:293–309

Stotler RL, Frape SK, Shouakar-Stash O (2010) An isotopic survey of $\delta^{81}Br$ and $\delta^{37}Cl$ of dissolved halides in the Canadian and Fennoscandian shields. Chem Geol 274:38–55

Stotler RL, Kirk MF, Newell KD, Goldstein RH, Frape SK, Gwynne R (2021) Stable bromine isotopic composition of Coal Bed Methane (CBM) produced water, the occurrence of enriched ^{81}Br, and implications for fluid flow in the midcontinent, USA. Minerals 11:358

Tan HB, Ma HZ, Xiao YK, Wei HZ, Zhang XY, Ma WD (2005) Characteristics of chlorine isotope distribution and analysis on sylvinite deposit formation based on ancient salt rock in western Tarim basin. Sci China (Ser. D) 48:1913–1920

Tan HB, Ma HZ, Wei HZ, Xu JX, Li TW (2006) Chlorine, sulfur and oxygen isotopic constraints on ancient evaporite deposit in the Western Tarim Basin, China. Geochem J 40:569–577

Tan HB, Ma HZ, Zhang XY, Xu JX, Xiao YK (2009) Fractionation of chlorine isotope in salt mineral sequences and application: research on sedimentary stage of ancient salt rock deposit in Tarim Basin and western Qaidam Basin. Acta Petrol Sinica 25:955–962. (in Chinese with English abstract)

Tang QM, Zhang X, Zhu ZJ, Wei YY, Liu YW, Liu Z (2018) Chlorine isotopic composition and potash-forming indicator of halites of the Triassic Jialingjiang formation in the Dianjiang depression, eastern Sichuan Basin. Acta Geol Sin 92:1671–1679. (in Chinese with English abstract)

Vallelonga P, de Gois JS, Borges DLG, Costas-Rodríguez M, Gkinis V, Lannuzel D, Spolaor A, Vanhaecke F (2021) Concentration and isotope composition of bromine and chlorine in Antarctic Sea ice. Goechim Cosmochim Acta 293:18–27

Valyashko MG (1956) Geochemistry of bromide in the process of salt deposition and the use of the bromide content as a genetic and prospecting tool. Geochem Int 6:570–587

Van't Hoff JH (1905) Zur Bildung der ozeanischen Salzablagerungen. Teil 1. 85 pp. Fr. Vieweg

Van't Hoff JH (1909) Zur Bildung der ozeanischen Salzablagerungen. Teil 2. 90 pp. Fr. Vieweg

Van't Hoff JH (1912) Untersuchungen über die Bildungsverhältnisse der ozeanischen Salzablagerungen. 374 pp. Leipzig; Akad Verl Ges (Reprint of the original papers taken from Sitzungsb Kgl Preuß Akad, 1897–1908

Van't Hoff JH, Mayerhoffer W (1897) Untersuchungen über die Bildungsverhältnisse der oceanischen Salzablagerungen, insbesondere des Stassfurter Salzlagers. Sitzungsb Kgl Preuß Akad Wissenschaften 1897:69–75

Volpe C, Spivack AJ (1994) Stable chlorine isotopic composition of marine aerosol-particles in the Western Atlantic Ocean. Geophys Res Lett 21:1161–1164

Volpe C, Wahlen M, Pszenny AAP, Spivack AJ (1998) Chlorine isotopic composition of marine aerosols: implications for the release of reactive chlorine and HCl cycling rates. Geophys Res Lett 25:3831–3834

Wang YZ, Frape SK (2015) Chlorine and bromine isotopic analyses of groundwaters: DGR-3 and DGR-4. Rep NWMO-TR-2015-20, Dept Earth Environ Sci, Univ Waterloo

Wei W, Kastner M, Spivack AJ (2008) Chlorine stable isotopes and halogen concentrations in convergent margins with implications for the Cl isotopes cycle in the ocean. Earth Planet Sci Lett 266:90–104

Willey JF, Taylor JW (1978) Capacitive integration to produce high precision isotope ratio measurements on methyl chloride and bromide. Anal Chem 50:1930–1933

Wirt L (1988) The origin of chloride in groundwater in the Stripa granite, Sweden. MSc thesis. Univ Arizona

Woulé-Ebongué V, Jendrzejewski N, Walgenwitz F, Pineau F, Javoy M (2005) Chlorine isotope residual salt analysis: a new tool to investigate formation waters from core analyses. AAPG Bull 89:1005–1018

Xiao YK, Liu WG, Qi HP, Zhang CG (1993) A new method for the high precision isotopic measurement of bromine by thermal ionization mass spectrometry. Int J Mass Spectrom Ion Processes 123:117–123

Xiao YK, Liu WG, Zhou YM, Sun DP (1997) Isotope compositions of chlorine in brine and saline minerals. Chin Sci Bull 42:406–409

Xiao YK, Liu WG, Zhou YM, Wang YH, Shirodkar PV (2000) Variations in isotopic compositions of chlorine in evaporation controlled salt lake brines of Qaidam Basin, China. Chin J Oceanol Limn 18:169–177

Xiao YK, Zhou YM, Wei HZ WQZ, Liu WG, Eastoe CJ (2002) A secondary isotopic reference material of chlorine from selected seawater. Chem Geol 182:655–661

Yu HT, Ma T, Du Y, Chen LZ (2019) Genesis of formation water in the northern sedimentary basin of South China Sea: clues from hydrochemistry and stable isotopes (D, ^{18}O, ^{37}Cl and ^{81}Br). J Geochem Explor 196:57–65

Yu HD, Zou H, Wu JH, Bagas L, Gong DX, Cheng HF, Huang CC, Zhu H (2023) Prospects for Eocene potash deposition in the east Tethyan domain: new evidence from fluid inclusions and chlorine isotope of halite crystals in the Kuqa Basin, NW China. Ore Geol Rev 154:105313

Zhao Y, Wei HZ, Liu X, Wang YJ, Jiang SY, Eastoe CJ, Peryt TM (2021) Isotope evidence for multiple sources of B and Cl in Middle Miocene (Badenian) evaporites, Carpathian Mountains. Appl Geochem 124:104819

Ziegler K, Coleman ML, Howarth RJ (2001) Palaeohydrodynamics of fluids in the Brent Group (Oseberg field, Norwegian North Sea) from chemical and isotopic compositions of formation waters. Appl Geochem 16:609–632

12 The Chlorine and Bromine Isotope Compositions of the Mantle and the Core

Although chloride and bromide concentrations of the mantle are low, due to its vast size the mantle can be considered as a major reservoir for both elements. It contains approximately one fifth (depending on the source of the data) of all chlorine and bromine on Earth and the isotope composition of its chlorine and bromine is expected to be representative for the chlorine and bromine isotope composition of the Earth after the differentiation of the mantle form the core. For that reason a considerable effort has been made by various research groups to obtain proper, average, values for the chlorine isotope composition of the mantle. In spite of these efforts there still is some serious discussion of what the average chlorine isotope composition of the mantle should be due to the fact that the different research groups concluded towards different values for the average chlorine isotope composition of the mantle. No studies on the bromine isotope composition of mantle material have been made yet, although a few studies have been published that present bromine isotope values that may be related to mantle material and for that reason some suggestions on the possible bromine isotope composition of the mantle may be given.

Mantle material exchanges with the lithosphere at various locations on earth. At mid-ocean ridges it surfaces to form the oceanic lithosphere and in subduction zones it is returned to the mantle. Basalt from the mid-ocean ridges has been used to determine the isotope composition of the mantle, while the isotope composition of pore waters in oceanic margins has been studied to understand the processes that take place during subduction of the oceanic crust. These studies have been used to examine the global cycle of chlorine and the processes that take place in this cycle. More recently the chlorine (and bromine) isotope composition of the mantle has also been estimated from very deep geothermal fluids and "concentrated" chlorine and bromine from extremely evolved rocks. Whether the values obtained in these studies reflect a valid average mantle isotope value is not yet well understood.

The deepest region of the Earth is the core, consisting for more than 90% of iron and nickel (McDonough 2014). In Chap. 3 it was estimated that the core consists a significant part of the Earth's chlorine. Due to its distance form the surface, it is not possible to measure the chlorine isotope composition of the core directly. However, because the composition of the core agrees rather well with that of iron meteorites and based on the chlorine isotope composition of iron meteorites it may be possible to obtain an educated guess for a possible chlorine isotope composition of the core.

H. Eggenkamp, *The Geochemistry of Stable Chlorine and Bromine Isotopes*, Advances in Isotope Geochemistry, https://doi.org/10.1007/978-3-031-75633-7_12

12.1 The Chlorine (and Bromine) Isotope Composition of the Mantle

12.1.1 The Chlorine Isotope Composition of the Mantle Based on Rocks Derived from the Mantle

The chlorine isotope composition of the mantle has been a research target ever since the development of precise techniques for the measurement of very small amounts of chloride in rocks (Magenheim et al. 1994). The chlorine isotope composition of the mantle in these studies was determined from the values obtained from the freshest, least altered mid-ocean ridge basalts. In spite of the relatively large number of studies on the chlorine isotope composition of the mantle even today there is still no fixed agreement on the chlorine isotope composition of the mantle due to conflicting results of the measurements that were obtained in these different studies on the potential chlorine isotope composition of the mantle. Magenheim et al. (1995) were the first to publish a comprehensive set of chlorine isotope data from MORB glasses. Their data, analysed by P-TIMS (Magenheim et al. 1994) after extraction of chloride from rock samples by pyrohydrolysis, were all clearly enriched in ^{37}Cl. The results showed trends towards more positive δ^{37}Cl values for samples containing lower chloride concentrations and higher F/Cl ratios. From these results it was concluded that the δ^{37}Cl value of the mantle, which was assumed to be equal to the values for MORB glasses that were not affected by high-level assimilation of wallrock into magma chambers, is somewhere between +3.0 and + 7.2‰ vs. SMOC with an average value of +4.7‰. Interestingly the chlorine isotope composition of mantle carbonatites was later estimated to have values close to 0‰ (Eggenkamp and Koster van Groos 1997) which led to the suggestion that the mantle was not a homogeneous reservoir with respect to the chlorine isotope composition.

During the early years of the twenty-first century methods were developed to measure chlorine isotope compositions in rocks using IRMS techniques after pyrohydrolysis of rock samples (Barnes et al. 2006; Bonifacie et al. 2007). Using these techniques also the chlorine isotope compositions of MORB specimens were analysed. In the course of these studies the very positive δ^{37}Cl values that were observed by Magenheim et al. (1995) were not confirmed. It was concluded by both Sharp et al. (2007) and Bonifacie et al. (2008) that the very positive δ^{37}Cl values that were measured by Magenheim et al. (1994, 1995) were analytical artefacts due to the introduction of sulphates or, even more devastating, fluorides in the TIMS system, as was already discussed by Stewart (2000) in his PhD dissertation. Stewart (2000) observed that especially the introduction of fluoride ions into the TIMS ionization chamber caused suppression of Cs_2Cl^+ ionization, leading to inconsistencies of the isotope measurements. Interesting with this regard is a discussion based on the study by Wei et al. (2008a). In their comment Sharp and Barnes (2008) point to the problems of having fluorides in samples which are analysed by TIMS. In their reply Wei et al. (2008b) state that there cannot be a problem with fluoride in their samples as they prepare them in a similar way as Sharp and Barnes (2008) do, as silver chloride precipitate from a solution containing chloride ions. Considering the high solubility of silver fluoride the final silver chloride precipitate can only have very low fluoride contents. They state that due to this effect the results they obtained cannot be analytical artefacts resulting in high δ^{37}Cl values. Although the method used by Wei et al. (2008a) was different from the method developed by Magenheim et al. (1994), it is also clear from the description of their method that Magenheim et al. (1994) did remove both sulphates and fluorides from their samples using ion-exchange techniques. It is not exactly known however what the remaining fluoride concentration in their samples was before they were introduced into the mass spectrometer. The exact reason for the large difference in measured δ^{37}Cl values between the different analytical methods may not be perfectly understood yet.

Although the results obtained by Sharp et al. (2007) and Bonifacie et al. (2008) are closer to each other than to the values obtained by Magenheim et al. (1995) they are in clear disagreement. Sharp et al. (2007) observed that the MORB glasses they measured mostly had $\delta^{37}Cl$ values between +0.5 and −0.5‰, with only two samples showing data at about −1‰. The average value of their samples was −0.2‰ with a standard deviation of 0.5‰. They observed no significant correlation between $\delta^{37}Cl$ and chloride concentration although the two most negative samples have relatively high chloride concentrations. Bonifacie et al. (2008) observed different systematics. They observed a significant correlation between chloride concentration and $\delta^{37}Cl$. Samples with lower chloride concentrations showed lower $\delta^{37}Cl$ data and it was concluded that the value of the mantle would be ≤ −1.6‰ based on the $\delta^{37}Cl$ values of the samples with the lowest chloride concentrations. These very negative data were apparently confirmed by measurements from Layne et al. (2009) who, based on SIMS measurements from two very low chloride glassy MORB samples, containing as little as 12 ppm chloride, concluded that the mantle $\delta^{37}Cl$ range could even be extended to ≤ −3‰. At the Goldschmidt Conference in Prague (2011) both the New Mexico group (Sharp et al. 2011) as well as the Paris group (Agrinier et al. 2011) presented posters showing arguments in favour of their own respective views. Sharp et al. (2013) presented new data which confirmed their view that no significant difference between surface and mantle reservoirs exist while on the other hand Nakamura et al. (2011) earlier presented data that showed a good agreement between their samples measured by TIMS and those of Bonifacie et al. (2007). While Barnes and Sharp (2017) suggest that low (and high) average $\delta^{37}Cl$ values for the mantle may reflect analytical artefacts or are the result from interpretation of non-representative samples, most average chlorine isotope mantle values presented to this day suggest that no final agreement has been reached yet on what would be the real average chlorine isotope composition of the mantle.

Figure 12.1 shows the $\delta^{37}Cl$ values presented by Magenheim et al. (1995), Sharp et al. (2007), Bonifacie et al. (2008) and Layne et al. (2009), as well as those analysed by John et al. (2010) to show how each of these studies obtained their respective conclusion on the average mantle value. Considering the large discrepancies between the three studies it is striking how well the correlations between chloride concentration and $\delta^{37}Cl$ are for Magenheim et al. (1995) and Bonifacie et al. (2008), so that in both studies it is understandable that there was little space for drawing different conclusions.

It is however clear that the mantle is actually a very heterogeneous reservoir. This was already argued by Eggenkamp and Koster van Groos (1997), although their discussion was based on the later disqualified MORB data from Magenheim et al. (1995). More recent data presented by John et al. (2010) also clearly suggest a large variability in mantle chlorine isotope values. John et al. (2010) observed $\delta^{37}Cl$ values between −1.6 and + 2.9‰, measured by SIMS, for Ocean Island Basalts which are also regarded as representing a typical mantle reservoir. These values, many of which are positive (they are also incorporated in Fig. 12.1 for comparison), can not be explained by dilution of the chlorine isotope composition by seawater. Again however, four samples of Hawaiian ocean island basalts analysed by Sharp et al. (2013) all showed values of −0.2 ± 0.2‰. The combination of the different results obtained in the various studies indicate that, although still considerable research is needed, including intensive collaboration between the laboratories that work in this field to understand both the differences obtained between the laboratories and the real value of the chlorine isotope composition of mantle material, most data suggest a $\delta^{37}Cl$ that is, on average, relatively close to SMOC.

12.1.2 Mantle Metasomatism

Selverstone and Sharp (2011) studied the chlorine isotope composition in a large set of peridotite samples from the Finero peridotite to study the metasomatism that has appeared and to study the processes that are responsible for the

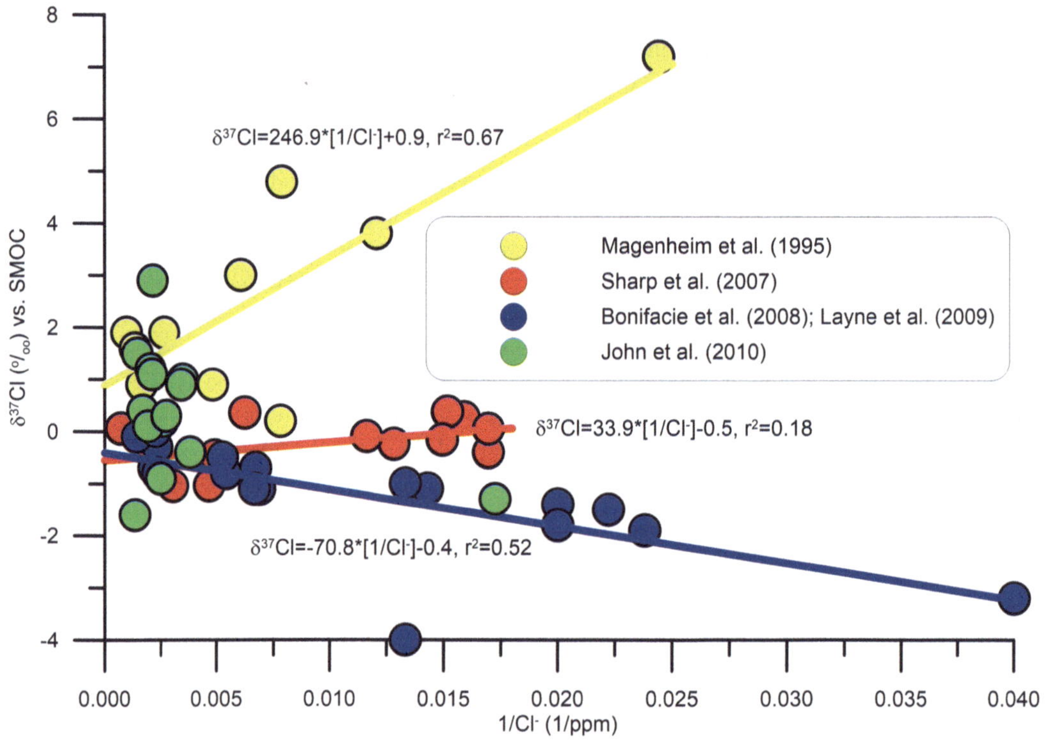

Fig. 12.1 δ^{37}Cl values and reciprocal chloride concentrations of MORB samples reported by Magenheim (1994), Sharp et al. (2007), Bonifacie et al. (2008), Layne et al. (2009) and John et al. (2010). The data show the estimates for the δ^{37}Cl value of the mantle were developed based on the measured δ^{37}Cl values of the samples presented in the respective studies

fractionation of chlorine isotopes in this region. They combined chlorine, hydrogen and oxygen stable isotope geochemistry with petrologic and major and trace element data to further constrain sources and settings of the metasomatic agents and to assess fluid infiltration mechanisms. They used relatively unmetasomatised samples from the Balmuccia peridotite as their control population (a total of 13 samples with a δ^{37}Cl range between −0.7 to +0.4‰). Four different rock types were analysed and the δ^{37}Cl in the samples ranged from −2.0 to +2.1‰. This variation could not be reconciled with metasomatism by a single homogeneous fluid. They found that effectively the variability in the dataset could be described with three end-members, one with a δ^{37}Cl value close to zero and with very negative δD values, one with negative and one with positive δ^{37}Cl values, these last two with less negative δD values. Although a large variation in chlorine isotope compositions of more than 4‰ was found in this dataset it was not yet possible to indicate the effects that actually fractionate the chlorine isotopes. Selverstone and Sharp (2011) stated that at this moment relatively little is known about the different chlorine isotope reservoirs within both the oceanic and continental crust and chlorine might fractionate during subduction-related dehydration. In spite of this they indicated that their study shows that chlorine isotope geochemistry adds clearly new information about the processes that are associated with fluid-rock interaction, even if it was not yet possible to name the exact processes that take place in these environments.

12.1.3 $\delta^{37}Cl$ and $\delta^{81}Br$ of Geothermal Fluids as Indicators for the Isotope Composition of the Mantle

Based on the chlorine and bromine isotope composition of geothermal water samples that are characterised by $^3He/^4He$ ratios between 7.7 and 8.2Ra (where Ra is the atmospheric $^3He/^4He$ ratio) Pinti et al. (2020) estimated that the average $\delta^{37}Cl$ of the mantle is +0.9‰ and the average $\delta^{81}Br$ is +0.7‰. High $^3He/^4He$ ratios between 7.7 and 8.2Ra are indicative of a mantle origin because the mantle contains relatively little α emitters that increase the ^4He content of the helium so that ^3He has relatively high concentrations. It was observed that samples with the highest $^3He/^4He$ ratios also showed the highest $\delta^{37}Cl$ and $\delta^{81}Br$ values. This led to the conclusion that the chlorine and bromine isotope composition of the primitive mantle was high, possible even higher than +3‰, and that its value decreased during the Earth's history to subduction of isotopically lighter chlorine and bromine. According to Pinti et al. (2020) the high isotope ratio of the mantle was inherited from the Moon forming impact. Because a high $\delta^{37}Cl$ of +3‰ had also been observed by John et al. (2010) in some OIBs they consider that this is a valid possibility for the original $\delta^{37}Cl$ of the mantle. Pinti et al. (2020) speculate that variations in isotopic values measured in mantle rocks and fluids could reflect heterogeneous regassing of subducted chlorine and bromine and the inefficient homogenization of recycled material in the MORB source, as was also suggested in other studies.

12.1.4 Estimation of the Isotope Composition of the Mantle from Chlorine Rich Minerals

Chlorine and bromine do not fit well in most minerals and for that reason they concentrate in the fluid (molten or aqueous) phase during the evolution of a melt or brine. Because chlorine and bromine continue to concentrate in the melt phase, in magmatic systems chlorine and bromine crystallise for the largest part in the last stages of rock formation. As a result, it could be hypothesized that the last crystallising, chloride and bromide rich, minerals contain a large (and perhaps the largest) part of the originally present chloride and bromide from the melt, and as such the isotope composition of this chloride or bromide might be representative for the average isotope composition of the original melt. With this in mind Eggenkamp et al. (2022) measured the chlorine isotope composition of a set of sodalite and eudialyte samples from the Ilímaussaq intrusion in South Greenland. It was found that both sodalite and eudialyte fractionate chlorine upon crystallisation from the melt. Based on the fractionation factors from these minerals it was concluded that the $\delta^{37}Cl$ of the melt was approximately +0.3‰. This value may be, according to Eggenkamp et al. (2022), representative for the lithospheric mantle below South Greenland. Whether this may also be representative for the whole mantle is of course not known. Eggenkamp et al. (2021) also studied the bromine isotope composition of a few bromide rich sodalite and tugtupite samples from the Ilímaussaq intrusion in South Greenland. In this study it is suggested that a relatively large isotope fractionation exists between the melt and crystallised sodalite. The data presented suggest that the melt below Ilímaussaq in Greenland had a $\delta^{81}Br$ of +0.5 to +0.6‰. This value is in agreement with the value of +0.7‰ as suggested by Pinti et al. (2020) based on the $\delta^{81}Br$ values of deep geothermal fluids. If this value indeed is representative for the mantle this indicates that the $\delta^{81}Br$ of the mantle is relative high compared to ocean water. Because there are only a relatively small number of bromine isotope data measured of only two (unrelated) geochemical systems, from which it is not understood well how these systems relate to the interaction between the mantle and the oceans it is not yet possible to conclude conclusively that the mantle has a positive $\delta^{81}Br$ and how this suggested value relates to the $\delta^{81}Br$ of the oceans.

12.1.5 Heterogeneity of Chlorine Isotopes in the Mantle Based on First-Principles Calculations

While it is obvious that the chlorine isotope composition of the mantle is heterogeneous, it is not yet clear what processes did cause this heterogeneity. Liu et al. (2021) use first-principles theoretical calculations on common chlorine rich minerals to estimate the chlorine isotope fractionation between these different chlorine rich minerals, and to use the values to understand mantle heterogeneity. The results obtained by Liu et al. (2021) show that, at ambient pressure and temperature conditions, the reduced isotope partition function ratios (the so called β-factors) are strongly correlated with the metal-chloride bond length or bond strength. That results in calculations that the chorine isotope fractionation between chlorine bearing minerals and o-existing aqueous fluids, at equilibrium, varies between −6 and + 3‰. Based on their calculations for the chlorine isotope fractionation factors and the estimations for the chlorine content and fluxes within and between different reservoirs as estimated by McDonough (2000); Hay et al. (2006); Barnes and Sharp (2017) and Eggenkamp (2014), Liu et al. (2021) estimate that the average δ^{37}Cl of the mantle lies somewhere in the range between −0.69 to −0.37‰ with a mean value of −0.53 ± 0.16‰.

12.1.6 Recapitulation of the Isotope Composition(S) of the Mantle

Based on the measurements of different types of mantle material and theoretical calculations it is clear that large chlorine isotope heterogeneities do exist in the mantle. These large heterogeneities mask the overall average, chlorine isotope composition of the mantle because it is not always clear if the measured samples are representative for the whole of the mantle. That may be the main reason that different studies present different average values. Data presented in these different studies suggest average values of +4.7 ± 1.8‰ (Magenheim et al. 1995), −0.1 ± 0.2‰ (Eggenkamp and Koster van Groos 1997), −0.1 ± 0.4‰ (Sharp et al. 2007), ≤ −1.6‰ (Bonifacie et al. 2008), ≤3‰ (Layne et al. 2009), −0.2 ± 0.3‰ (Sharp et al. 2013), +0.9‰ (Pinti et al. 2020), −0.53 ± 0.16‰ (Liu et al. 2021) and ~ +0.3‰ (Eggenkamp et al. 2022). From these data the high values from Magenheim et al. (1995) are by now disqualified as analytical artefacts, and the low values presented by Bonifacie et al. (2008) and Layne et al. (2009) are not yet clearly confirmed in other bulk studies, and need further confirmation in other studies. However, according to Sharp et al. (2013) these low values are due to analytical errors inherent in the analysis of very small samples in IRMS instruments. The positive δ^{37}Cl values obtained by Pinti et al. (2020) and Eggenkamp et al. (2022) were not determined on primary mantle material and the material they analysed may show some additional isotope fractionation during its extraction from this primary mantle material. Overall, it is not yet possible to estimate an exact average δ^{37}Cl of the mantle. Based on the available data the best suggestion may be that the average δ^{37}Cl of the mantle is relatively close to SMOC, with at most a few tenths (up to perhaps 0.3) of a per mill (either positive or negative) deviation from SMOC.

In the case of bromine, it even is more difficult to give an estimation for the average mantle isotope composition. Here, no direct measurements have been obtained yet on primary mantle material such as MORB or OIB. The only suggestions are given by δ^{81}Br measurements of deep geothermal fluids (Pinti et al. 2020) and the chlorine rich magmatic mineral sodalite (Eggenkamp et al. 2021). The data obtained in these studies suggest, and that is different as compared to the values obtained for chlorine, an average δ^{81}Br that is positive and possibly somewhere between +0.5 and + 1‰ relative to SMOB.

12.2 Possible Chlorine Isotope Composition of the Core

The Earth's core can not be reached due to its fast distance from the surface and unlike the mantle there is no exchange of material with surface reservoirs. For that reason it is very difficult to obtain any information on the possible chlorine isotope composition of the core, in spite of it being (possibly) one of the largest chlorine reservoirs of the Earth. The composition of the core is most likely comparable to the compositions of iron meteorites which can be considered as comparable to and possibly even representative for the composition of the Earth's core (Bottke et al. 2006; Hirschmann et al. 2021). Recent studies on the iron isotope composition of iron meteorites indeed confirmed the similarity between core formation and the origin of iron meteorites (Ni et al. 2020). For those reasons, it can be suggested that the chlorine isotope composition of iron meteorites may also be representative for the chlorine isotope composition of the Earth's core.

The chlorine isotope composition of a large set of iron meteorites was determined by Gargano and Sharp (2019). Within this set both the chlorine isotope composition as well as the chloride concentrations vary to a large extend. Interestingly, the δ^{37}Cl values from iron meteorites with chloride concentrations which are from the same order of magnitude as estimated for the Earth's core as estimated in Chap. 3 are all negative with δ^{37}Cl values roughly between −6 and −3‰. Gargano and Sharp (2019) suggest that these lower values are representative for the primordial δ^{37}Cl of the solar nebula, because they consider that there are no relevant mechanisms to decrease the chlorine isotope composition of the meteorite parent bodies, while, due to volatilization of light chloride the residual chloride can increase in δ^{37}Cl. Because of the suggestion that these iron meteorites may be representative for the Earth's core too this may suggest that the δ^{37}Cl of the core is relatively negative with a value that is lower than −5‰ and that can be as low as −7.2‰, the lowest value encountered in any iron meteorite. This interpretation also suggests a large difference between the chlorine isotope compositions of the core and the mantle. This difference may, according to Gargano and Sharp (2019), be due to a significant loss of volatile chlorine (approximately 60% of the originally present chlorine) during the formation of planet Earth.

References

Agrinier P, Coleman M, Bonifacie M (2011) A review of the analytical accuracy of Cl isotope measurements in rocks by various techniques: a way to explain inconsistency between results. In: 21th Annual V.M. Goldschmidt conference, vol 75. Mineral Mag, Prague, Czech Republic, p 413

Barnes JD, Sharp ZD (2017) Chlorine isotope geochemistry. Rev Mineral Geochem 82:345–378

Barnes JD, Selverstone J, Sharp ZD (2006) Chlorine isotope chemistry of serpentinites from Elba, Italy, as an indicator of fluid source and subsequent tectonic history. Geochem Geophys Geosyst 7:Q08015

Bonifacie M, Jendrzejewski N, Agrinier P, Coleman M, Pineau F, Javoy M (2007) Pyrohydrolysis-IRMS determination of silicate chlorine stable isotope compositions. Application to oceanic crust and meteorite samples. Chem Geol 242:187–201

Bonifacie M, Jendrzejewski N, Agrinier P, Humler E, Coleman M, Javoy M (2008) The chlorine isotope composition of earth's mantle. Science 319:1518–1520

Bottke WF, Nesvorný D, Grimm RE, Morbidelli A, O'Brien DP (2006) Iron meteorites as remnants of planetesimals formed in the terrestrial planet region. Nature 436:821–824

Eggenkamp HGM (2014) The geochemistry of stable chlorine and bromine isotopes. In: Hoefs J (ed) Advances in isotope geochemistry. Springer, Heidelberg

Eggenkamp HGM, Koster van Groos AF (1997) Chlorine stable isotopes in carbonatites: evidence for isotopic heterogeneity in the mantle. Chem Geol 140:137–143

Eggenkamp HGM, Marks MAW, Louvat P, Markl G (2021) Bromine isotope variations in magmatic and hydrothermal sodalite and Tugtupite and the estimation of Br isotope fractionation between melt and sodalite. Fortschr Mineral 11:370

Eggenkamp HGM, Marks MAW, Bonifacie M, Bardoux G, Agrinier A, Markl G (2022) Cl isotope fractionation in magmatic and hydrothermal eudialyte, sodalite and tugtupite (Ilímaussaq intrusion, South Greenland). Chem Geol 604:120932

Gargano A, Sharp Z (2019) The chlorine isotope composition of iron meteorites: evidence for the Cl isotope composition of the solar nebula and implications for extensive devolatilization during planet formation. Meteorit Planet Sci 54:1619–1631

Hay WW, Migdisov A, Balukhovsky AN, Wold CN, Flögel S, Söding E (2006) Evaporites and the salin-

ity of the ocean during the Phanerozoic: implications for climate, ocean circulation and life. Palaeogeogr Palaeoclimatol Palaeoecol 240:3–46
Hirschmann MM, Bergin EA, Blake GA, Ciesla FJ, Li J (2021) Early volatile depletion on planetesimals inferred from C–S systematics of iron meteorite parent bodies. Proc Natl Acad Sci 118:e2026779118
John T, Layne GD, Haase KM, Barnes JD (2010) Chlorine isotope evidence for crustal recycling into Earth's mantle. Earth Planet Sci Lett 298:175–182
Layne GD, Kent AJR, Bach W (2009) $\delta^{37}Cl$ systematics of a backarc spreading system: the Lau Basin. Geology 37:427–430
Liu X, Wei HZ, Li YC, Williams-Jones AE, Lu JJ, Jiang SY, Dong G, Ma J, Eastoe CJ (2021) Chlorine isotope mantle heterogeneity: constraints from theoretical first-principles calculations. Chem Geol 572:120193
Magenheim AJ, Spivack AJ, Volpe C, Ransom B (1994) Precise determination of stable chlorine isotopic ratios in low-concentration natural samples. Geochim Cosmochim Acta 58:3117–3121
Magenheim AJ, Spivack AJ, Michael PJ, Gieskes JM (1995) Chlorine stable isotope composition of the oceanic crust: implications for Earth's distribution of chlorine. Earth Planet Sci Lett 131:427–432
McDonough WF (2000) The composition of the earth. In: Teisseyre R, Majewski E (eds) Earthquake thermodynamics and phase transformations in the earth's interior, pp 3–23
McDonough WF (2014) Compositional model for the Earth's core. Treatise on Geochemistry 2nd Edition 3.16:559–577
Nakamura N, Nyquist LE, Shih CY, Fujitani T, Okano O (2011) Stable chlorine isotopes and elemental chlorine by thermal ionization mass spectrometry and ion chromatography: Martian meteorites, carbonaceous chondrites and standard rocks. 42nd Lunar and Planetary Science Conference, abstract 2513
Ni P, Chabot NL, Ryan CJ, Shahar A (2020) Heavy iron isotope composition of iron meteorites explained by core crystallization. Nat Geosci 13:611–615
Pinti DL, Shouakar-Stash O, Castro MC, Lopez-Hernández A, Hall CM, Rocher O, Shibata T, Ramírez-Montes M (2020) The bromine and chlorine isotopic composition of the mantle as revealed by deep geothermal fluids. Geochim Cosmochim Acta 276:14–30
Selverstone J, Sharp ZD (2011) Chlorine isotope evidence for multicomponent mantle metasomatism in the Ivrea zone. Earth Planet Sci Lett 310:429–440
Sharp ZD, Barnes JD (2008) Comment to "Chlorine stable isotopes and halogen concentrations in convergent margins with implications for the Cl isotopes cycle in the ocean" by Wei et al. A review of the Cl isotope composition of serpentinites and the global chlorine cycle. Earth Planet Sci Lett 274:531–534
Sharp ZD, Barnes JD, Brearly AJ, Chaussidon M, Fisher TP, Kamenetsky VS (2007) Chlorine isotope homogeneity of the mantle, crust and carbonaceous chondrites. Nature 446:1062–1065
Sharp ZD, Selverstone J, Mercer JA (2011) The Cl isotope composition of the mantle revisited. In: 21th Annual V.M. Goldschmidt conference, vol 75. Mineral Mag, Prague, Czech Republic, p 1848
Sharp ZD, Mercer JA, Jones RH, Brearley AJ, Selverstone J, Bekker A, Stachel T (2013) The chlorine isotope composition of chondrites and earth. Geochim Cosmochim Acta 107:189–204
Stewart MA (2000) Geochemistry of dikes and lavas from Hess deep: implication for crustal construction processes beneath mid-oceanic ridges and the stable chlorine isotope geochemistry of mid-ocean ridge basalts. PhD dissertation Thesis. Duke University, Durham, North Carolina
Wei W, Kastner M, Spivack A (2008a) Chlorine stable isotopes and halogen concentrations in convergent margins with implications for the cl isotopes cycle in the ocean. Earth Planet Sci Lett 266:90–104
Wei W, Kastner M, Spivack A (2008b) Reply to comment on: "chlorine stable isotopes and halogen concentrations in convergent margins with implications for the Cl isotopes cycle in the ocean". Earth Planet Sci Lett 274:535

13 Chlorine and Bromine Isotope Variations in the Earth's Lithosphere

13.1 Oceanic Crust

As mentioned in Chap. 12 analyses of the most pristine oceanic crust samples available were used to determine the chlorine isotope composition of the mantle. These studies attempted to determine the chlorine isotope composition of the mantle though the analysis of samples that show clear mantle characteristics. The oceanic crust however, is for the largest part contaminated by chloride that descends from assimilation with seawater, and it is generally agreed that the oceanic crust has an average $\delta^{37}Cl$ value which is close to 0‰. This follows not only from the data reported by Sharp et al. (2007, 2013), who measured chlorine isotope data close to 0‰ in most of their samples, but this is also inferred from the data reported by Bonifacie et al. (2008) and Magenheim et al. (1995) who both concluded that in samples with higher chlorine concentrations $\delta^{37}Cl$ values appear to converge to 0‰.

Overall however, when the individual measurements of oceanic crust samples are considered the figure is not as simple as it seems. If we were to consider results as obtained by Magenheim et al. (1995) or Bonifacie et al. (2008) it would be fine to conclude that MORB glasses fall on a line between basalt contaminated by seawater, with a $\delta^{37}Cl$ close to 0‰ and a (very) positive or negative mantle end-member in samples with the lowest chloride concentrations. Data obtained by Sharp et al. (2007, 2013) deserve a different approach. Although the average value for the oceanic crust is slightly, but hardly significantly, less than 0‰, thus more or less equal to SMOC, the variation is significant. With a standard deviation of the measurements of about 0.5‰ this indicates that the variation is much larger than the precision of the measurements, even considering that the precision of measurements for rock samples is not as good as for aqueous samples. This large variation clearly indicates that there must be processes that fractionate chlorine isotopes in the oceanic crust, during extraction of rock from the mantle in volcanic systems, and, obviously, that the mantle itself is a heterogeneous reservoir.

13.2 Continental Lithosphere

13.2.1 Chlorine and Bromine Isotopes in Rock Forming Minerals

The chlorine isotope composition of continental magmatic rocks has not been studied as extensively as that of oceanic rocks. Eggenkamp (1994) analysed six rocks from the Proterozoic (1168 ± 21 Ma; Blaxland et al. 1976) Ilímaussaq intrusion in Greenland and found $\delta^{37}Cl$ values mostly close to 0‰ (−0.02 to +0.11‰) with one exception of a sample with a value of +0.32‰. Markl et al. (1997) studied the chlorine isotope

H. Eggenkamp, *The Geochemistry of Stable Chlorine and Bromine Isotopes*, Advances in Isotope Geochemistry, https://doi.org/10.1007/978-3-031-75633-7_13

geochemistry of some lower crust samples from Lofoten (Norway) which indicated only minor involvement of mantle derived chlorine. $\delta^{37}Cl$ values of these rocks showed a large range of values between −1.12 and + 0.79‰. The only information that can be obtained from these few data is that the $\delta^{37}Cl$ of the continental crust is not very homogeneous but will generally have values between approximately +1 and −1‰ with an average value close to SMOC.

Eggenkamp et al. (2021a, 2022) studied the chlorine isotope composition from a large set of samples of chlorine rich minerals from the Ilímaussaq intrusion from South Greenland, as well as the bromine isotope compositions of a smaller set of sodalite and tugtupite samples from this intrusion. The total range in chlorine isotope compositions in their study was comparable to the range observed in the earlier studies. An attempt was made to relate the chlorine and bromine isotope compositions of these minerals to that of the mantle, as discussed in Chap. 12, although the rocks from which the samples were taken are normally considered lithospheric. The samples studied by Eggenkamp et al. (2022) suggested that during crystallisation of eudialyte and sodalite from the melt Rayleigh fractionation appeared due to chlorine isotope fractionation between the melt and the minerals. This can explain, at least in part, the isotopic heterogeneity that was observed in the earlier studies, when only a small number of non-related mineral and rock samples were measured. Eggenkamp et al. (2021b) also presented a suite of SIMS measurements on sodalite and eudialyte samples from the Ilímaussaq intrusion. They observed that in samples were both sodalite and eudialyte are present, $\delta^{37}Cl$ in eudialyte is higher than in adjacent sodalite. Between sodalite grains $\delta^{37}Cl$ can vary a few tenths of a per mill, while between eudialyte grains, variations can even be higher. In hydrothermal eudialyte $\delta^{37}Cl$ increases significantly at the grain boundary with values up to +3.5‰. These data show that SIMS can reveal significantly more information than bulk isotope measurements. Hoare et al. (2021) studied halogen (F, Cl, Br and I) and $\delta^{37}Cl$ systematics from kimberlite magmatism from the North Atlantic Craton (NAC) of West and South-west Greenland. It was observed that regional varieties exists in halogen systematics, with northern NAC kimberlite samples from Majuagaa that show mantle-like $\delta^{37}Cl$ values of −0.2 to −0.5‰ (based on the discussion in Sharp et al. 2013). While kimberlite dykes from Nigerdlikasik and Pyramidefjeld near the southern craton margin have positive $\delta^{37}Cl$ values of +0.4‰ to +1.3‰, in addition to a relative Cl and Br enrichment. This is consistent with the assimilation of recycled crust-derived halogens by the kimberlite magma. These samples are most probably not derived from mantle material, in spite of the fact that the $\delta^{37}Cl$ range agrees with data from Greenland that suggest that mantle related $\delta^{37}Cl$ values may be more positive than SMOC (Eggenkamp et al. 2022). The bromine isotopes measured in 13 sodalite and tugtupite samples from Ilímaussaq (Eggenkamp et al. 2021a) suggest that in general the $\delta^{81}Br$ in rocks are positive with respect to SMOB, with an average $\delta^{81}Br$ of +0.23 ± 0.07‰ in magmatic sodalite, and higher values of on average + 0.36 ± 0.08‰ in hydrothermal sodalite and + 0.71 ± 0.17‰ in hydrothermal tugtupite. These values suggest that the $\delta^{81}Br$ is in general positive in rocks, and that it increases, due to a relative large isotope fractionation, towards later stage facies.

13.2.2 Chlorine Isotopes in Fluid Inclusions

To assess whether chlorine stable isotopes could be useful to estimate the origin of chlorine in ore forming fluids a few studies are known in which the chlorine isotope composition was determined in fluid inclusions.

Eastoe et al. (1989) determined the chlorine isotope composition in several Mississippi valley type deposits from Tennessee and porphyry copper deposits from Arizona, Utah, Chile and Papua New Guinea. In their samples, which contained the leached chloride from 250 to 500 grams of rock, $\delta^{37}Cl$ values between −1.0 and + 0.8‰ were found. Although these were only preliminary results it was concluded that the high $\delta^{37}Cl$

values in the Mississippi valley type samples were related to formation waters, and that the low δ^{37}Cl waters had a different, then unknown origin. The samples from the porphyry copper deposits could not yet be explained but there seemed to be an isotopic difference between early and late chloride in some systems. It was stated however that the differences found in these samples contained indications that various processes took place, so that more research could prove useful.

Banks et al. (2000a, 2000b) presented δ^{37}Cl measurements on fluid inclusions from the Capitan pluton in New Mexico and the southwest England batholite. In their studies the chlorine isotope composition was determined by P-TIMS at the University of Leeds and as this method introduced the possibility to analyse rather small samples only 0.5 to 1 gram of rock was necessary to leach fluid from for isotope analysis. It was found that the fluid inclusions from the two different locations had clearly distinct characteristics. Those from New Mexico had δ^{37}Cl values clustering around 0‰ and Br/Cl ratios of about 0.0005, while the English samples had δ^{37}Cl values around +1.9‰ and Br/Cl values between 0.0010 to 0.0025. It was concluded that the chloride in the samples from New Mexico was derived from local evaporite sequences, while the samples from SW-England seemed to be more of a deep magmatic source. It has to be realised however that a deep origin for these samples was concluded because in those days the consensus was that the mantle had a strongly positive δ^{37}Cl following the conclusions by Magenheim et al. (1995). More recent studies have shown that these values should be either close to zero or negative (Sharp et al. 2007; Bonifacie et al. 2008). This actually indicates that if this study would have been done more recently a different explanation would be needed, and that the original conclusions can not be valid anymore.

Gleeson and Smith (2009), Nahnybida et al. (2009) and Richard et al. (2011) presented studies in which the δ^{37}Cl is measured on fluids extracted from Swedish, USA and Canadian ore systems. In these studies the chlorine isotope composition was determined on fluid inclusion leachates and were analysed at the Stable Isotope Hydrology and Ecology Laboratory of Environment Canada in Saskatoon (Saskatchewan, Canada) following the technique described by Wassenaar and Koehler (2004). Gleeson and Smith (2009) determined chlorine isotope compositions in leachates from three spatially associated iron oxide with copper and gold mineralising systems from Norbotten, Sweden. They found δ^{37}Cl values that were mostly negative. It was possible to determine three different populations of samples based on the Br/Cl ratios and the δ^{37}Cl values. Based on their results Gleeson and Smith (2009) concluded that the chloride in the two systems with mostly negative δ^{37}Cl values (Greenstone- and Porphyry hosted Cu-Au deposits with δ^{37}Cl values between −5.6 to −1.3‰ and late stage veins in Fe-oxide-apatite deposits with δ^{37}Cl values between −3.1 and −1.0‰) the chloride was of mantle origin, except for the most negative samples. This was based on the observation that in most recent mantle studies negative δ^{37}Cl values appeared the most common (Bonifacie et al. 2007, 2008; Layne et al. 2009) in mantle derived rocks although Sharp et al. (2007) indicated values much closer to zero. The values in the third system, the Nautanen Deformation Zone are mostly fairly close to zero so that these are concluded to be of a crustal, (meta-)evaporitic origin. Nahnybida et al. (2009) studied Br/Cl ratios and δ^{37}Cl values in fluid inclusions from porphyry copper deposits from Butte, Montana and Bingham Canyon, Utah (both in the USA). Br/Cl ratios of samples from the Butte copper deposit cluster below the value in seawater and overlap with mantle values. δ^{37}Cl values in these samples are between −2.3 and −0.8‰. Br/Cl ratios in fluid inclusions from the Bingham Canyon deposit have a larger range, up to $3.7*10^{-3}$. The δ^{37}Cl values in these samples are mostly lower with values between −4.1 and −0.9‰. Nahnybida et al. (2009) conclude that, in the absence of any process known at the moment that they conducted their study that can produce significant chlorine isotope fractionation at high temperatures, in the porphyry system at Bingham, and to a lesser extent at Butte, the negative chlorine isotopic signatures were inherited from a subducting slab and generated at low temperatures. Richard et al. (2011) studied the chlorine isotope composition and the Cl/Br ratios in fluid inclusions of uranium deposits from the Paleoproterozoic Athabasca Basin in Canada. In this study the $\delta^{37}Cl$ values were between −0.85 and + 0.36‰ and were interpreted with an evaporated seawater origin. Variations of these values were interpreted as secondary minor modifications of its seawater origin such as mixing with small amounts of evaporites, dilution with connate and/or formation waters, leaching of chloride from biotites or water loss by hydration reactions in alteration haloes linked to the uranium deposition.

The role of fluid mixing in the formation of zinc-lead deposits in south-west China was studied by Xu et al. (2022) with the use of fluid inclusions. These deposits are characterised by low-temperature metallogeny, no spatial relationship with plutons and strong water-rock interaction. This results in the difficulty of constraining the fluid origin and the process of ore formation by traditional geochemistry. For that reason, Xu et al. (2022) employed the chemical and chlorine isotope compositions of bulk fluid inclusions from ore-bearing and ore-barren quartz veins to constrain possible sources of ore-forming fluids and the process of metal precipitation in hydrothermal systems. The data obtained by Xu et al. (2022) indicated a two end-member mixing process between the ore-bearing veins with low Br/Cl ratios and high $\delta^{37}Cl$ values and the ore-barren veins with high Br/Cl ratios and $\delta^{37}Cl$ values that represent evaporated seawater. For the formation of vein-type zinc-lead deposits, Xu et al. (2022) were able to construct a genetic fluid mixing model that involved magmatic hydrothermal fluids related to activity from deep magma and evaporated seawater.

13.2.3 Chlorine Isotope Variations in Lithospheric Apatite

The development of SIMS techniques for the measurement of chlorine isotope ratios (Layne et al. 2004; Godon et al. 2004) made it possible to determine the isotope composition of individual chlorine rich mineral phases in rock. This is opposed to most earlier measurements where bulk measurements of the whole rock samples were made. A major disadvantage of bulk measurements is that average chloride concentrations are generally very low, which results in large, and difficult to handle sample quantities. The advantage of SIMS measurements is that individual measurements can be made of small spots (10 to 25 μm) in individual mineral grains so that not only individual minerals can be characterised, but even variations within individual grains.

Apatite is one of the most common chloride bearing minerals in the lithosphere and for that reason it is an ideal candidate to study the isotope composition of chlorine by SIMS. The normally high chloride content makes it possible to obtain good quality isotope compositions on the small spots. Also important and potentially interesting is of course that within individual apatite grains chlorine isotope profile can be determined, which can supply information on the processes that existed during the growth of individual grains.

Apatite was already used to determine the chlorine isotope ratio at an early date. In very early studies the target from these studies was to determine a precise atomic weight of the element chlorine, such as in the studies by Curie (1921); Gleditsch and Samdahl (1922); Dorenfeldt (1922) and Harkins and Stone (1926). In later studies attempts were made to investigate potential variations in the chlorine isotope ratio of different apatite samples, but these were not found yet (Morton and Catanzaro 1964). The first exact stable isotope measurements on apatite samples were made by Eggenkamp (1994) who determined the chlorine isotope compositions of just two apatite samples to gain insight in potential isotope variations in minerals.

After the development of SIMS methods to measure chlorine isotopes it became easier to measure the chlorine isotope composition in chlorine rich minerals and apatite was an obvious candidate for these studies. Kusebauch et al. (2015) were the first to apply such a study on apatites from the Bamble sector in Norway. The purpose of this study was to use apatite as a probe of metamorphic fluids that formed the rocks in this area. Chlorine isotope values of individual apatite grains were heterogeneous and ranged from −1.2 to +3.7‰. High $\delta^{37}Cl$ values were generally correlated with OH-rich zones of replaced apatite, whereas low $\delta^{37}Cl$ values were measured in F-rich zones of replaced apatite and in Cl-apatite of probably magmatic origin. Though apatite $\delta^{37}Cl$ values followed the general bulk trend, individual $\delta^{37}Cl$ signatures seemed to reflect the highly localized composition of interfacial fluid at the reaction front. Partitioning of chloride and chlorine stable isotopes between highly saline fluids and apatite in this system is a complex process and most likely controlled by unknown factors in addition to the Cl concentration. Andersson et al. (2019) used the chlorine isotope composition, in combination with a large suite of other parameters, of a set of fluor-apatites from the Olserum-Djupedal REE-phosphate mineralisation in south-east Sweden, as a tracer of the source, the chemistry and the evolution of ore-forming fluids. The range in $\delta^{37}Cl$ values is between −0.4 and + 1.6‰ in granite-hosted vein fluorapatite. The $\delta^{37}Cl$ values decrease in a sequence from granite-hosted fluorapatite to metasediment-hosted primary fluorapatite, to metasediment hosted primary fluorapatites, down to a value of −1‰. The $\delta^{37}Cl$ values of secondary fluorapatite slightly increase again, and range, depending on the area, from −0.7 and + 0.3‰, or even +1.0‰ in one exceptional region. In their sample set, Andersson et al. (2019) presume that only little chlorine isotope fractionation occurs during the formation of apatite from the ore forming fluid. Thus, they conclude that chlorine can be used as tracers of the ore-forming fluid. They conclude to the same result for the halogen ion ratios Br/Cl and I/Cl. Overall they conclude that apatite is sensitive to the composition of the ore-forming fluid and thus, that the chemical and isotopic composition of apatite can be used to trace the origin, chemistry, and evolution of fluids in hydrothermal ore-forming settings. Wudarska et al. (2020) studied the chlorine isotope compositions in one of the oldest surviving rock suites, the more than 3.7 billion years old Isua Supracrustal Belt in south-west Greenland. In this region $\delta^{37}Cl$ values between −0.8 and + 1.6‰ were observed. The highest vales are seen in banded iron formation samples. Contrary to Andersson et al. (2019) it was concluded that the $\delta^{37}Cl$ values in this set of samples cannot document uniquely the primary signatures of the apatite crystals. Nonetheless, they consider the chlorine isotope measurements as a helpful addition to assess secondary overprint in these apatite samples.

The chlorine isotope compositions of apatite samples are now measured regularly, not only on terrestrial samples but also on apatite samples from meteorites and space return sample material. These measurements are made at different laboratories and for that reason it is important that inter-laboratory studies are applied to assure that the values reported in the various studies can be compared. As a first attempt Wudarska et al. (2021) organised an inter-laboratory characterisation on the chlorine isotope composition of six apatite (two chlor-apatites and four fluor-apatites) reference materials. All SIMS analyses were done at the GFZ German Research Centre for Geosciences in Potsdam, Germany, while the IRMS analyses were done at three different laboratories, the Institut de physique du gobe de Paris (IPGP), France, the Department of Geological Sciences, University of Texas, Austin USA and the Department of Earth Sciences, University of Delaware, Newark, USA. The SIMS analyses of the six reference materials showed $\delta^{37}Cl$ values within a relatively narrow range of +0.09 to +0.42‰ with a standard deviation of the measurements that was in all cases less than (or equal) to 0.25‰. This made Wudarska et al. (2021) conclude that all reference materials have comparable chlorine isotope compositions. The IRMS measurements at the three laboratories showed relatively small systematic differences, where the measurements at the IPGP had the lowest values, the values determined in Delaware were about 0.2‰ higher and the values determined in Texas another about 0.2‰ higher. These differences may be caused by the different techniques to extract the chloride from the apatite crystals, because the ISL-354 standard that was measured at all three laboratories showed values that were much closer to the accepted value. These results indicate that for the measurement of chlorine isotope determinations on apatite samples (and probably also chlorine isotope determinations of other chlorine rich minerals) it would be best to internationally agree on a method to extract chloride from this type of samples. That way it may be guaranteed that the full extraction and measurement procedure is comparable between all laboratories and probably results may also be within a narrower range, so that mineral samples can be measured both by SIMS and IRMS with a better precision. It is hoped that such an effort can be made within the near future.

13.3 Chlorine Isotope Variations in Volcanic Systems

Due to their often extreme chemical compositions, chlorine isotopes in volcanic systems have been studied from early on. Eggenkamp (1994) measured the chlorine isotope compositions in a large set of volcanic gases, springs and lakes from Indonesia. From these data it appeared that samples close to the meteoric waterline (based on hydrogen and oxygen isotope compositions) had mostly negative δ^{37}Cl values, while samples with relatively high δ^{18}O values (samples that indicated geothermal alteration) mostly had positive δ^{37}Cl values. This difference was explained with different origins of the chloride in these samples, with meteoric waters containing chloride from seaspray and weathered rocks in the samples close to the meteoric waterline, while samples showing geothermal alteration were supposed to receive their chloride from hydrogen chloride degassing. It was suggested that this explanation is in agreement with experimental work done by Hoering and Parker (1961) who determined the isotope equilibrium between ammonium chloride and hydrogen chloride and found that the hydrogen chloride in this equilibrium concentrated the more positive chloride. Later experimental studies by Sharp et al. (2010) seemed to confirm these conclusions. At this moment no studies have been published yet that show bromine isotope data in volcanic systems.

Most chlorine isotope studies from volcanic related material were done as a proxy to understand the composition of the mantle, or to determine interaction between subduction slab lithospheric material and the mantle at subduction zones (e.g. Eggenkamp and Koster van Groos 1997; Barnes et al. 2008, 2009; Bonifacie et al. 2008; Layne et al. 2009; John et al. 2010; Sharp et al. 2013; Rizzo et al. 2013; Chiaradia et al. 2014; Agusto and Varekamp 2016; Rodriguez et al. 2016; Liotta et al. 2017). The reason for this is that due to the high temperature of volcanic rocks and fluids, as well as the incompatibility of chloride for rocks (and thus the concentration of chloride in the fluid phase) the isotope fractionation between these phases is negligible (Schauble et al. 2003; Liebscher et al. 2006). As such variations in volcanic material reflect variations in the origin of the magma that either may be mantle material, subducted lithosphere of most common mixtures of both sources.

Barnes et al. (2009) studied the variation on the chlorine isotope composition in different volcanoes from Central America (from Guatamala at about 14°N to Costa Rica at about 10°N). The samples analysed in their study, containing of ash, tephra and lava samples, showed δ^{37}Cl values between −2.6 and + 3.0‰. The purpose of this study was to use chlorine isotopes as a tracer for the fluids that sourced from the subducting Cocos slab. It was found that the δ^{37}Cl values could be related to the origin of the material analysed. The volcanoes from the south and the north (Costa Rica and El Salvador/Guatamala) suggest a mantle source as they show δ^{37}Cl values close to 0‰ or slightly negative, assuming that mantle values are close to 0‰ as determined by Sharp et al. (2007). The volcanoes which are located more in the centre of the region (Nicaragua) show either more positive and more negative δ^{37}Cl values which are explained as either sediment or hydrothermally altered rock (altered oceanic crust or possible serpentinites). This is for example confirmed by high Ba/La ratios in samples from these volcanoes which indicate a sediment component and high ^{87}Sr/^{86}Sr ratios which indicate altered oceanic crust. Finally negative $\delta^{37}Cl$ values that are found in Honduran volcanoes in the back-arc are possibly indicative for serpentinite derived fluids. In all these cases it was possible to explain the variations of the chlorine isotope compositions in the volcanic rocks from these volcanoes from the source, indicating that no isotope fractionation of chlorine in the volcanic system occurred. A comparable study was done on the Quaternary volcanic arc of Ecuador by Chiaradia et al. (2014). Chiaradia et al. (2014) investigated the stable chlorine isotope compositions and Cl concentrations of volcanic rocks from five Quaternary volcanoes in Ecuador that are situated on an across-arc transect encompassing the frontal arc (Pilavo, Pululahua, Pichincha volcanoes), main arc (Chacana caldera complex) and back-arc (Sumaco volcano). In general they observed that chlorine isotope compositions decrease in this region away from the trench. This result agrees with earlier models that magmas in this region are formed by a steadily decreasing melt fraction from the mantle induced by a steadily decreasing amount of fluid released by the subducted slab away from the trench. High δ^{37} Cl values of the frontal arc volcanoes (up to +3‰) imply that the chlorine carried by slab fluids derives, at least partly, from ^{37}Cl-rich subducted terrigenous sediments plus subordinate amounts of altered oceanic crust and serpentinite. The anomalously high chlorine contents (up to 0.2 wt.% Cl) and high $\delta^{37}Cl$ values (up to +1.5‰) from the back-arc volcano can be explained by preferential partial melting of mantle portions metasomatized by slab fluids during a Jurassic subduction event. Based on first order changes of $\delta^{37}Cl$ values across the arc, Chiaradia et al. (2014) observed that the $\delta^{37}Cl$ values at each volcanic centre are systematically lowered due to intracrustal evolution processes occurring in magmatic reservoirs at mid-crustal levels. Rizzo et al. (2013) investigated studied volcanic gases and rocks at the Etna volcano (Italy). While $\delta^{37}Cl$ values of both gases and rocks are within a relatively narrow range relative to SMOC (0.0 ± 0.7‰) Cl/K values of the rock samples suggest that the source of the chlorine is compatible with the depleted mantle, although contaminated with some altered oceanic crust. The conclusion reached by Rizzo et al. (2013) was in agreement with indications obtained from more commonly applied isotope tracers for mantle processes.

In a study to test the possibility to measure $\delta^{37}Cl$ values in melt inclusions by SIMS, Manzini et al. (2017) measured the chlorine isotope compositions in six glass standards. The $\delta^{37}Cl$ values in these reference samples ranged from −1.1 to +1.7‰. Melt intrusions were then measured in olivine samples from the Lesser Antilles, Vanuatu and the Aeolian islands arcs that displayed $\delta^{37}Cl$ ranges from −1.9 to +0.6‰, −1.7 to +0.4‰ and −3.4 to −1.4‰ respectively. The data (combined with chemical data such as Cl/K ratios) suggest that the chloride mainly originates from serpentinites (lithospheric serpentinites or serpentinized mantle wedge) in the Lesser Antilles and Vanuatu arcs. In contrast, chlorine added to the mantle wedge beneath the Aeolian Islands may record a higher proportion of fluids derived from subducting sediments. The observed differences of $\delta^{37}Cl$ values in melt inclusions from the Lesser Antilles and Vanuatu arcs and the Aeolian Islands may directly reflect the different subduction geometries present.

Ranta et al. (2021) determined the chlorine isotope composition of different types of silicic and intermediate Icelandic volcanic rocks. The results are compared to chlorine isotope compositions of Icelandic basalts from Halldórsson et al. (2016). Ranta et al. (2021) observed that $\delta^{37}Cl$ values of silicic rocks were shifted to more negative values compared to associated basalts and intermediate rocks by up to 2.9‰. Such large shifts cannot be explained by well understood processes inherent to silicic magma genesis. Therefore, Ranta et al. (2021) suggest that the low $\delta^{37}Cl$ values in silicic rocks are attributed to assimilation of magmatic brines that are formed and stored in long lived crustal magma mushes. The results obtained by Ranta et al. (2021) indicated that magmatic brine assimilation is a fundamental, but previously unrecognised part of rhyolite genesis.

13.3.1 Chlorine Isotope Effects Resulting from Volcanic Degassing

Barnes et al. (2014) determined the chlorine concentrations and isotope compositions resulting from eruptive degassing of obsidian clasts from Mono Craters volcanic field in California. While CO_2, H_2O and δ^2H values in these obsidian clast demonstrate a clear chemical and isotopic evidence for eruptive degassing this is not shown by chloride and $\delta^{37}Cl$ characteristics. It is expected that chlorine follows the degassing to a lesser extend that the other parameters as chlorine diffuses rather slowly in the cooling melt. Barnes et al. (2014) collected obsidian samples from two occurrences, tuff layers representing a single eruptive sequence and rhyolitic obsidian samples from high-silica flows forming the domes and coulées in the region. $\delta^{37}Cl$ values of the pyroclastic fall obsidians (−1.9‰ to −0.1‰) overlap with those of the dome samples (−1.2‰ to 0.0‰). The similar chloride concentrations between the pyroclastic fall and dome obsidians argue for lack of chlorine degassing, despite H_2O and CO_2 loss. The wide range in $\delta^{37}Cl$ values can be indicative of isotopic heterogeneities in the magma source due to assimilation of sedimentary material or fluxing of mantle-derived chlorine to a crustal melt.

Chlorine isotope ratios, in combination with boron and oxygen isotope ratios from olivine hosted melt intrusions (OHMIs) were studied by Bouvier et al. (2019) from various ocean arc settings (Lesser Antilles, Vanuatu, Aeolian, NE Japan and Izu-Bonin arcs) to determine the application of these isotope systems to study recycling of crust and sediment and its effects on the compositions of the mantle wedge. Bouvier et al. (2019) observed different characteristics in samples from the different arc settings. For example, the $\delta^{37}Cl$ values from the Aeolian and Izu-Bonin arcs were significantly lower (at about −2.5‰) than the $\delta^{37}Cl$ values from the other three arcs (at about −1‰). Also, differences in boron and oxygen isotopes were observed between the different arcs, that were not related to the chlorine isotope variations. The chlorine isotope differences in OHMIs were ultimately explained as the result of sources with different chlorine isotope compositions that were preserved in these melt inclusions, contrary to chlorine in the bulk rock where the isotope signature could have been lost due to diffusive fractionation during chlorine degassing. Bouvier et al. (2022a) studied the chlorine isotope composition of melt inclusions in a single hand-sample from the Stromboli volcano in Italy. The melt inclusions had very different chemical compositions and $\delta^{37}Cl$ values that ranged from −2.6 to +1.2‰. This range is larger than the range observed by Manzini et al. (2017) discussed in the previous section, and much larger than the range observed in bulk rocks from the Stromboli volcano as reported by Liotta et al. (2017). Like in the study by Barnes et al. (2014) the $\delta^{37}Cl$ variation is not related to degassing. In this study it is also not related to fractional crystallisation. From their dataset Bouvier et al. (2022a) concluded that the $\delta^{37}Cl$ is the result of mixing between two end-members, one with a high $\delta^{37}Cl$ (>1‰) that shows influence of a mantle source and a second with a low $\delta^{37}Cl$ (<2‰) that shows evidence of subducted sediments. Bouvier et al. (2022a) concluded that the $\delta^{37}Cl$ data thus help to refine the two sources that were initially identified from bulk rock studies. In the Central America Volcanic Arc Bouvier et al. (2022b) determined chlorine isotopes in melt inclusions to study element recycling during subduction in this arc. In different regions different chlorine isotope characteristics were observed with restricted ranges (less than 1‰ variations in single samples) in the northern part of the arc and wider ranges (larger than 1‰, and up to 3.8‰ in single samples) in more southern parts of the arc. Based on the data, including information on trace elements and existing knowledge on this region, it was concluded that three end-members contributed to this subduction zone. One originating from serpentinite with a $\delta^{37}Cl$ of about +0.6‰, a melt-like component with a $\delta^{37}Cl$ that might be as low as −2.5‰ and a possibly amphibole-bearing metasomatized mantle with a high $\delta^{37}Cl$ value up to +3.0‰. It was concluded in this case that the isotope differences in melt inclusions resulted from late stage processes such as mixing of different magma batches actually after the melt inclusions were entrapped in the magma.

The different studies of chlorine isotope variations in volcanic sources that have undergone degassing show that this degassing is not shown in the chlorine isotope composition in melt inclusions and that variations in $\delta^{37}Cl$ are the result of other processes that have taken place. As such it is a useful parameter to determine other processes than degassing that occur in this kind of situations.

13.3.2 Extremely High $\delta^{37}Cl$ Values in Fumaroles

Interestingly anomalously high $\delta^{37}Cl$ values are found in fumaroles in several volcanoes in Central America (Sharp et al. 2010). These values could be explained as the result of fractionation in the fumarole system (Sharp et al. 2010) as discussed in Sect. 8.6.3. This same kind of very highly anomalous $\delta^{37}Cl$ values were observed in early studies by Eggenkamp (1994) in Indonesian fumaroles and Wahrenberger et al. (1997) in fumaroles from the Russian Federation and Italy but could not be explained at that time. Very high $\delta^{37}Cl$ values in Poás Volcano (Costa Rica) were later confirmed to exist by Rodríguez et al. (2013) who also confirmed that the $\delta^{37}Cl$ values in the fumaroles are variable over time.

13.3.3 Changes of Chlorine Isotope Variations with Time in Volcanic Systems

Musashi et al. (2008) observed variations in the $\delta^{37}Cl$ values of a time series of volcanic spring-water samples at the Kusatsu-Bandaiko hot spring in Japan. Musashi et al. (2008) observed that the $\delta^{37}Cl$ values in this hot spring increased due to increased volcanic activity at the Kusatsu-Shirane volcanic region. Li et al. (2015) observed that magmatic chloride from La Soufrière in Guadeloupe and Mount Pelée in Martinique efficiently trapped in thermal springs and displays negative $\delta^{37}Cl$ values, which are −0.65‰, consistent with a slab-derived origin but distinct from the isotope compositions of chloride in surface reservoirs. Based on $\delta^{37}Cl$ measurements from a 30 year thermal springs archive from Guadeloupe (covering the period between 1976 and 2008), Li et al. (2015) observed that magmatic chloride can be flushed out of a hydrothermal system within 30 to 80 years, much faster than other tracers. They concluded that due to this effect $\delta^{37}Cl$ is potentially a unique proxy to monitor cessation and revival of arc volcanoes. From an extended time series of archived water samples taken from lake Poás Rodríguez et al. (2016) observed that the $\delta^{37}Cl$ values in lake Poás itself also vary over time. Relatively positive values mark samples taken during an interval of peak activity of the lake (1980–1986), when chloride concentrations (up to 73,100 ppm) and temperatures (up to 65 °C) were high, pH values were below 0, and its volume was strongly reduced. This elevated $\delta^{37}Cl$ is attributable to a preferential depletion of ^{35}Cl due to evaporation of $HCl_{(g)}$, or to a distinctive input flux into the lake. These observations show that chlorine isotope compositions can be used to monitor the volcanic activity in a region. A better understanding of the exact meaning of the variations is still needed, but once better understood these variations can be used in future volcanic monitoring and may hopefully be useful to forecast volcanic activity and eruptions.

References

Agusto M, Varekamp J (2016) The Copahue volcanic-hydrothermal system and applications for volcanic surveillance. In: Tassi F et al (eds) Copahue volcano, active volcanoes of the world. Springer, Berlin Heidelberg

Andersson SS, Wagner T, Jonsson E, Fusswinkel T, Whitehouse MJ (2019) Apatite as a tracer of the source, chemistry and evolution of ore-forming fluids: the case of the Olserum-Djupedal REE-phosphate mineralisation, SE Sweden. Geochim Cosmochim Acta 255:163–187

Banks DA, Green R, Cliff RA, Yardley BWD (2000a) Chlorine isotopes in fluid inclusions: determination of the origins of salinity in magmatic fluids. Geochim Cosmochim Acta 64:1785–1789

Banks DA, Gleeson SA, Green R (2000b) Determination of the origin of salinity in granite-related fluids:

evidence from chlorine isotopes in fluid inclusions. J Geochem Explor 69–70:309–312

Barnes JD, Sharp ZD, Fischer TP (2008) Chlorine isotope variations across the Izu-Bonin-Mariana arc. Geology 36:883–886

Barnes JD, Sharp ZD, Fisher TP, Hilton DR, Carr MJ (2009) Chlorine isotope variations along the central American volcanic front and back arc. Geochem Geophys Geosyst 10:Q11S17

Barnes JD, Prather TJ, Cisneros M, Befus K, Gardner JE, Larson TE (2014) Stable chlorine isotope behavior during volcanic degassing of H_2O and CO_2 at mono craters, CA. Bull Volcanol 76:805

Blaxland AB, Van Breemen O, Steenfelt A (1976) Age and origin of agpaitic magmatism at Ilímaussaq, South Greenland: Rb-Sr study. Lithos 9:31–38

Bonifacie M, Jendrzejewski N, Agrinier P, Coleman M, Pineau F, Javoy M (2007) Pyrohydrolysis-IRMS determination of silicate chlorine stable isotope compositions. Application to oceanic crust and meteorite samples. Chem Geol 242:187–201

Bonifacie M, Jendrzejewski N, Agrinier P, Humler E, Coleman M, Javoy M (2008) The chlorine isotope composition of earth's mantle. Science 319:1518–1520

Bouvier AS, Manzini M, Rose-Koga E, Nichols ARL, Baumgartner LP (2019) Tracing of Cl input into the sub-arc mantle through the combined analysis of B, O and Cl isotopes in melt inclusions. Earth Planet Sci Lett 507:30–39

Bouvier AS, Rose-Koga EF, Chapius A (2022a) Deciphering degassing and source effects in cl isotopes in melt inclusions: the possible role of amphibole in the magma source of Stromboli (Aeolian Island Arc). Front Earth Sci 9:793259

Bouvier AS, Portnyagin MV, Flemetakis S, Hoernle K, Klemme S, Berndt J, Mironov NL, Johm T (2022b) Chlorine isotope behavior in subduction zone settings revealed by olivine-hosted melt inclusions from the Central America Volcanic Arc. Earth Planet Sci Lett 581:117414

Chiaradia M, Barnes JD, Cadet-Voisin S (2014) Chlorine stable isotope variations across the quaternary volcanic arc of Ecuador. Earth Planet Sci Lett 396:22–33

Curie I (1921) Sur le poids atomique du chlore dans quelques mineraux. Compte Rend Sean 172:1025–1028

Dorenfeldt M (1922) Relative determination of the atomic weight of chlorine in Bamle apatite. J Am Chem Soc 45:1577–1579

Eastoe CJ, Guilbert JM, Kaufmann RS (1989) Preliminary evidence for fractionation of stable chlorine isotopes in ore forming hydrothermal systems. Geology 17:285–288

Eggenkamp HGM (1994) δ^{37}Cl; the geochemistry of chlorine isotopes. Geol Ultraiect 116:1–150. Thesis Utrecht University

Eggenkamp HGM, Koster van Groos AF (1997) Chlorine stable isotopes in carbonatites: evidence for isotopic heterogeneity in the mantle. Chem Geol 140:137–143

Eggenkamp HGM, Marks MAW, Louvat P, Markl G (2021a) Bromine isotope variations in magmatic and hydrothermal sodalite and tugtupite and the estimation of Br isotope fractionation between melt and sodalite. Fortschr Mineral 11:370

Eggenkamp HGM, Marks MAW, Ludwig T, Markl G (2021b) Intragranular halogen (F, Cl, Br), S and δ^{37}Cl variability as determined by SIMS in sodalite and eudialyte from the Ilímaussaq intrusion, South Greenland, GeoKarlsruhe 2021. https://doi.org/10.48380/dggv-hcj1-5702

Eggenkamp HGM, Marks MAW, Bonifacie M, Bardoux G, Agrinier P, Markl G (2022) Cl isotope fractionation in magmatic and hydrothermal eudialyte, sodalite and tugtupite (Ilímaussaq intrusion, South Greenland). Chem Geol 604:120932

Gleditsch E, Samdahl B (1922) Radioactivité sur le poids atomique de chlore dans un mineral ancien, l'apatede Balme. Compte Rend Sean 174:746–748

Gleeson SA, Smith MP (2009) The sources and evolution of mineralising fluids in iron oxide-copper-gold systems, Norrbotten, Sweden: constraints from Br/cl ratios and stable cl isotopes of fluid inclusion leachates. Geochim Cosmochim Acta 73:5658–5672

Godon A, Webster JD, Layne GD, Pineau F (2004) Secondary ion mass spectrometry for the determination of δ^{37}Cl. Part II. Intercalibration of SIMS and IRMS for aluminosilicate glasses. Chem Geol 207:291–303

Halldórsson SA, Barnes JD, Stefánsson A, Hilton DR, Hauri EH, Marshall EW (2016) Subducted lithosphere controls halogen enrichments in the Iceland mantle plume source. Geology 44:679–682

Harkins WD, Stone SB (1926) The isotopic composition and atomic weight of chlorine from meteorites and from minerals of non-marine origin. J Am Chem Soc 48:938–949

Hoare BC, Tomlinson EL, Barnes JD, Tappe S, Marks MAW, Epp T, Caulfield J, Riegler T (2021) Tracking halogen recycling and volatile loss in kimberlite magmatism from Greenland: evidence from combined F-cl-Br and δ^{37}Cl systematics. Lithos 384–385:105976

Hoering TC, Parker PL (1961) The geochemistry of the stable isotopes of chlorine. Geochim Cosmochim Acta 23:186–199

John T, Layne GD, Haase KM, Barnes JD (2010) Chlorine isotope evidence for crustal recycling into Earth's mantle. Earth Planet Sci Lett 298:175–182

Kusebauch C, John T, Whitehouse MJ, Engvik AK (2015) Apatite as probe for the halogen composition of metamorphic fluids (Bamble sector, SE Norway). Contrib Mineral Petrol 170:34

Layne GD, Godon A, Webster JD, Bach W (2004) Secondary ion mass spectrometry for the determination of δ^{37}Cl. Part I. Ion microprobe analysis of gasses and fluids. Chem Geol 207:277–289

Layne GD, Kent AJR, Bach W (2009) δ^{37}Cl systematics of a backarc spreading system: The Lau Basin. Geology 37:427–430

Li L, Bonifacie M, Aubaud C, Crispi O, Dessert C, Agrinier P (2015) Chlorine isotopes of thermal springs in arc volcanoes for tracing shallow magmatic activity. Earth Planet Sci Lett 413:101–110

Liebscher A, Barnes J, Sharp Z (2006) Chlorine isotope vapor-liquid fractionation during experimental fluid-phase separation at 400 °C/23 MPa to 450 °C/42 MPa. Chem Geol 234:340–345

Liotta M, Rizzo AL, Barnesw JD, D'Auria L, Mertelli M, Bobrowski N, Wittmer J (2017) Chlorine isotope composition of volcanic rocks and gases at Stromboli volcano (Aeolian Islands, Italy): inferences on magmatic degassing prior to 2014 eruption. J Volcanol Geotherm Res 336:168–178

Magenheim AJ, Spivack AJ, Michael PJ, Gieskes JM (1995) Chlorine stable isotope composition of the oceanic crust: implications for Earth's distribution of chlorine. Earth Planet Sci Lett 131:427–432

Manzini M, Bouvier AS, Barnes JD, Bonifacie M, Rose-Koga EF, Ulmer P, Métrich N, Bardoux G, Williams J, Layne GD, Straub S, Baumgartner LP, John T (2017) SIMS chlorine isotope analyses in melt inclusions from arc settings. Chem Geol 449:112–122

Markl G, Musashi M, Bucher K (1997) Chlorine stable isotope composition of granulites from Lofoten, Norway: implications for the cl isotope composition and for the source of cl enrichment in the lower crust. Earth Planet Sci Lett 150:95–102

Morton RD, Catanzaro EJ (1964) Stable chlorine isotope abundances in apatites from Odergardens Verk, Norway. Nor Geol Tidsskr 44:307

Musashi M, Oi T, Eggenkamp HGM, Matsuo M (2008) Chlorine isotope fractionation associated with volcanic activity at the Kusatsu-Bandaiko hot spring in Japan. Isot Environ Health Stud 44:305–313

Nahnybida T, Gleeson SA, Rust BG, Wassenaar LI (2009) Cl/Br ratios and stable chlorine isotope analysis of magmatic–hydrothermal fluid inclusions from Butte, Montana and Bingham Canyon, Utah. Miner Deposita 44:837–848

Ranta E, Halldórsson SA, Barnes JD, Jónasson SA (2021) Chlorine isotope ratios record magmatic brine assimilation during rhyolite genesis. Geochem Perspect Lett 16:35–39

Richard A, Banks DA, Mercadier J, Boiron MC, Cuney M, Cathelineau M (2011) An evaporated seawater origin for the ore-forming brines in unconformity-related uranium deposits (Athabasce Basin, Canada): cl/Br and δ^{37}Cl analysis of fluid inclusions. Geochim Cosmochim Acta 75:2792–2810

Rizzo AL, Caracausi A, Liotta M, Paonita A, Barnes JD, Corsaro RA, Martelli M (2013) Chlorine isotopic composition of volcanic gases and rocks at Mount Etna (Italy) and inferences on the local mantle source. Earth Planet Sci Lett 371:134–142

Rodríguez A, Martínez M, Eggenkamp HGM, van Bergen MJ (2013) Chlorine isotope variations in the hyper-acid lake system of Poás volcano, Costa Rica. IAVCEI 2013 Scientific Assembly, 58

Rodríguez A, Eggenkamp HGM, Martínez-Cruz M, van Bergen MJ (2016) Chlorine isotope and cl–Br fractionation in fluids of Poás volcano (Costa Rica): insight into an active volcanic–hydrothermal system. J Volcanol Geotherm Res 325:70–85

Schauble EA, Rossman GR, Taylor HP (2003) Theoretical estimates of equilibrium chlorine-isotope fractionations. Geochim Cosmochim Acta 67:3267–3281

Sharp ZD, Barnes JD, Brearly AJ, Chaussidon M, Fisher TP, Kamenetsky VS (2007) Chlorine isotope homogeneity of the mantle, crust and carbonaceous chondrites. Nature 446:1062–1065

Sharp ZD, Barnes JD, Fischer TP, Halick M (2010) An experimental determination of chlorine isotope fractionation in acid systems and applications to volcanic fumaroles. Geochim Cosmochim Acta 74:264–273

Sharp ZD, Mercer JA, Jones RH, Brearley AJ, Selverstone J, Bekker A, Stachel T (2013) The chlorine isotope composition of chondrites and earth. Geochim Cosmochim Acta 107:189–204

Wahrenberger C, Eastoe CJ, Seward TM, Dietrich V (1997) Stable chlorine isotope composition of volcanic gas condensates. In: 7th Annual Goldschmidt Conference, Tucson, AZ, p 213

Wassenaar LI, Koehler G (2004) On-line technique for the determination of the δ^{37}Cl of inorganic and total organic cl in environmental samples. Anal Chem 76:6384–6388

Wudarska A, Słaby E, Wiedenbeck M, Birski Ł, Wirth R, Götze J, Lepland A, Kusebauch C, Kocjan I (2020) Chlorine isotope composition of apatite from the >3.7 Ga Isua Supracrustal belt, SW Greenland. Fortschr Mineral 10:27

Wudarska A, Słaby E, Wiedenbeck M, Barnes JD, Bonifacie M, Sturchio NC, Bardoux G, Couffignal F, Glodny J, Heraty L, John T, Kusebauch C, Mayanna S, Wilke FDH, Deput E (2021) Inter-laboratory characterisation of apatite reference materials for chlorine isotope analysis. Geostand Geoanal Res 45:121–142

Xu L, Luo CG, Wen HJ, Zhou ZB, de Fourestier J (2022) The role of fluid mixing in the formation of vein-type Zn-Pb deposits in eastern Guizhou Province, SW China: insights from elemental compositions and chlorine isotopes of fluid inclusions. J Asian Earth Sci 239:105403

14 Isotope Variations of Chlorates and Perchlorates in the Environment

14.1 Introduction

It is only realised since about 2000 that perchlorate is a naturally occurring anion outside the Atacama desert in Chile (see e.g. DasGupta et al. 2005; Parker et al. 2008). Interestingly this knowledge developed in the same period when the first methods were developed to analyse the stable chlorine isotope compositions of perchlorates (Ader et al. 2001; Sturchio et al. 2003). The analytical methods to analyse the chlorine isotope compositions were quickly followed by methods to analyse the oxygen isotope composition of perchlorates as well (Bao and Gu 2004; Böhlke et al. 2005).

Perchlorate is a potentially hazardous ion, and the discovery of its widespread occurrence in drinking water and food resources (Gullick et al. 2001; Kirk et al. 2003; Jackson et al. 2005) sparked more interest in knowledge of its natural behaviour as well as studies on how to remove it from the human food chain.

Basically perchlorate found in the environment can be either naturally or artificially produced. Its largest natural source is the Atacama Desert in Chile (Ericksen 1981) where it is present in concentrations of up to 0.6% in nitrate ores. At the time of writing this report the Atacama Desert was the only known natural source of perchlorates. This perchlorate has however transported to many places in the world as the nitrate deposits in Chile were in the past the main source for fertilizer nitrate globally. More recently it has also been discovered that perchlorate is concentrated in other dry areas in the world such as in the Western United States (Rao et al. 2007). The processes by which it is produced in nature are not known for sure yet, but it appears that the most important mode of production is atmospheric oxidation of chlorine by ozone (e.g. Simonaitis and Heicklen 1975; Jaegle et al. 1996; Rao et al. 2010). Artificial perchlorate is produced in large quantities (about 10^7 kg.year^{-1}; DasGupta et al. 2006) by the electrolysis of sodium chloride brines (Schumacher 1960) for various military, aerospace and other industrial applications such as solid rocket fuel, explosives and fireworks, road flares and electroplating solutions.

Virtually all studies on the development of methods to concentrate perchlorates to measurable quantities from natural sources and to analyse the chlorine and oxygen isotope compositions of these samples, as well as the tracer studies in the same samples have been done by a single research group led by Neil C. Sturchio. The results of these studies have been discussed comprehensively in two chapters in books on perchlorate geochemistry and environmental isotope geochemistry (Sturchio et al. 2006; Sturchio et al. 2011), and a detailed report prepared under contract to the Department of Defence Strategic Environmental Research and Development Program (SERDP; Hatzinger et al. 2017). By

H. Eggenkamp, *The Geochemistry of Stable Chlorine and Bromine Isotopes*, Advances in Isotope Geochemistry, https://doi.org/10.1007/978-3-031-75633-7_14

necessity this chapter is no more than a very short summary of these publications.

14.2 Chlorine and Oxygen Stable Isotope Composition of Perchlorates

14.2.1 Natural Perchlorate

The perchlorate isotope composition of samples from two broad areas have been determined in various studies. It was found that samples originating from different regions had clearly distinguishable isotope characterisations. The chlorine isotope compositions of perchlorates from the Atacama Desert were analysed by Böhlke et al. (2005); Sturchio et al. (2006) and Jackson et al. (2010). They observed that the chlorine isotope composition of the Atacama perchlorates were in the range from −14.5 to −11.8‰. These data complemented earlier measured $\delta^{18}O$ and $\Delta^{17}O$ data reported by Bao and Gu (2004) who showed that $\delta^{18}O$ values of natural (Atacama) perchlorate are between −9.3 and −4.2‰ and $\Delta^{17}O$ values between +8.93 and + 9.57‰. These very positive $\Delta^{17}O$ values indicate an atmospheric origin of the Atacama perchlorates, presumably due to ozone oxidation.

The isotope composition has also been determined for a batch of samples from desert soils in the Southwestern United States (Jackson et al. 2010). Within these samples two different groups could be determined that were differentiated based on their different isotope compositions. Perchlorates from an area near Death Valley are recognisable by very positive $\Delta^{17}O$ data (up to +18.4‰). These studies indicated that all natural perchlorates measured today show positive $\Delta^{17}O$ values indicating mass independent formation (Thiemens 2006) of these perchlorates. Interestingly these samples have $\delta^{37}Cl$ and $\delta^{18}O$ values which are distinct from these values for Atacama desert perchlorate. Both the $\delta^{37}Cl$ and the $\delta^{18}O$ of these samples are much more positive with values between −4 and 0‰ and 0 and 26‰ respectively.

On the other hand samples extracted from groundwaters and the unsaturated zone soil leachates from the Southern High Plains and Middle Rio Grande Basin (Texas and New Mexico) showed more positive $\delta^{37}Cl$ values (+3 to +6‰) than the Death Valley samples, and $\delta^{18}O$ values which are equal to the lighter Death Valley samples. Interestingly $\Delta^{17}O$ values are close to 0‰ or slightly positive, up to +2.6‰. These lower $\Delta^{17}O$ values as compared to Atacama or Death Valley perchlorate are explained from either a different oxidation process (by UV radiation rather than by ozone), or by slow exchange with water oxygen in slightly wetter climatic conditions (Jackson et al. 2010).

14.2.2 Synthetic Perchlorate

The isotopic characteristics of synthetic perchlorate appears to be very different from the known natural isotope variations (Sturchio et al. 2006). A considerable set of synthetic perchlorates was shown to have $\delta^{37}Cl$ values relatively close to zero (−3.1 to +1.6‰) and rather negative $\delta^{18}O$ values (−24.8 to −12.5‰). The $\delta^{37}Cl$ values indicate the formation of synthetic perchlorate by oxidation from brines which is a relatively efficient process. The large range in oxygen isotope composition is caused by the meteoric water used in oxidation and mass dependent fractionation of the oxygen, as this process is rather inefficient (Sturchio et al. 2006). All $\Delta^{17}O$ values of the synthetic perchlorates are 0.0 ± 0.1‰ indicating that the fractionation during their formation is mass dependent.

14.3 Isotopic Compositions of Perchlorates as Natural Tracers for its Origin

As it was possible to determine that there are at least three (or four) isotopically distinct groupings of perchlorates (Bao and Gu 2004; Böhlke et al. 2005; Sturchio et al. 2006, Jackson et al. 2010) it was expected that the isotope composition of perchlorates in ground water could be used as a tracer for the origin of the perchlorate. It was argued that the isotope composition of perchlorates is retained in ground water due to the slow exchange with oxygen from water (Böhlke et al. 2009) and because biodegradation produces both systematic stable isotope enrichments, but that does not significantly affect $\Delta^{17}O$ or $^{36}Cl/Cl$ ratios (not discussed in this book). This effect makes it possible to detect biodegradation together with the initial sources of the perchlorate (Sturchio et al. 2007; Hatzinger et al. 2009). Sturchio et al. (2008) determined the isotope composition of perchlorates extracted from groundwaters from the Chino Basin in California. Isotope data showed that the origin of these waters was predominantly Atacama perchlorate, with minor mixture with Southern High Plains perchlorates. The results showed that extensive fertilisation has taken place in the past with nitrates mined from the Atacama Desert. Böhlke et al. (2009) analysed the isotope compositions of perchlorates from groundwater from Long Island, New York. Samples taken here showed either an Atacama or synthetic origin of the analysed perchlorates, with no indication of mixing. A third study by Slaten et al. (2010) on groundwater samples from Pasadena, California indicated that the origin of the perchlorate in this region was from at least three sources, two of which were synthetic. Also perchlorate contamination in groundwater from Pomona, California could be determined using this technique. In this region the results showed that the contamination was dominantly from Atacama perchlorate that was derived from the use of imported Chilean nitrate fertilizer for citrus cultivation in the past (Sturchio et al. 2014). Bennett (2015) did not agree with this conclusion, mainly because the proposed end-members were not properly characterised in this region, so that it would not be impossible that the source of the perchlorate was local instead of Chilean. According to Bennett (2015) ^{36}Cl were especially missing. This was rebutted by Sturchio (2015) who stated that although ^{36}Cl was not measured for this specific project ^{36}Cl data were known in the Pomona region and these data pointed to a Chilean origin of the perchlorate.

Sturchio et al. (2012) applied the above techniques to map perchlorate plumes on a regional scale in the Southeastern San Bernardino Basin, California. In this basin there are two very different sources of perchlorate. One source is a contaminated former rocket testing site. The perchlorate originating from this site is predominantly synthetic. The other perchlorate type is mostly agricultural and consists of Atacama type perchlorate. It is possible to follow both perchlorate types in the groundwater. Downstream the two perchlorate plumes do intersect. At several locations in the groundwater plumes the ground water was sampled and both types could be followed along the groundwater path. From the isotopic compositions of the samples it was suggested that a third, more diffuse source of perchlorate is also present. This source is proposed as Southern High Plains perchlorate and it may be present in minor quantities of up to 20% of the perchlorate in the samples. Interestingly the plumes intersect downstream of the originating sites and in several samples it is clear that the perchlorate is a mixture of both sources. The existence of isotopically distinct perchlorate plumes in the studied area is consistent with other data obtained from the samples and the area such as groundwater levels and flow paths as well as the historical land use and areal distribution of potential perchlorate sources.

Perchlorate concentrations in the Great Lakes are between 0.05 and 0.13 µg/L. Poghosyan et al. (2014) measured the isotope composition of this perchlorate ($\delta^{37}Cl$, $\delta^{18}O$ and $\Delta^{17}O$) in order to trace sources and behaviour of perchlorates in this largest freshwater system on Earth. Perchlorate $\delta^{37}Cl$ values from the Great Lakes range between +3.0‰ (Lake Ontario) and + 4.0‰ (Lake Superior), whereas $\delta^{18}O$ values range from −4.1‰ (Lake Superior) to +4.0‰ (Lake Erie). The perchlorate has mass-independent oxygen isotopic variations with positive $\Delta^{17}O$ values (+1.6‰ to +2.7‰) divided into two distinct groups: Lake Superior (+2.7‰) and the other four lakes (~ + 1.7‰). The stable isotope results indicate that perchlorate in the Great Lakes is predominantly of natural origin, having isotope compositions resembling that measured for indigenous perchlorate from pre-industrial groundwaters of the western USA.

Because perchlorate is potentially hazardous to humans it is important to understand the sources of the perchlorate that is ingested by the population. As a first approximation Poghosyan et al. (2016) analysed the isotope composition of urine from two populations, one in Atlanta, USA and one in Taltal, Chile (Atacama region). Urinary perchlorate from the Atlanta region resembles indigenous natural perchlorate from the western USA ($\delta^{37}Cl$ = +4.1 ± 1.0‰; $^{36}Cl/Cl$ = 1811(±136) × 10^{-15}), and urinary perchlorate from the Taltal, Chile region is similar to natural perchlorate in nitrate salt deposits from the Atacama Desert of northern Chile ($\delta^{37}Cl$ = −11.0 ± 1.0‰; $^{36}Cl/Cl$ = 254(±40) × 10^{-15}). Neither urinary perchlorate resembled the isotopic pattern found in synthetic perchlorate. These results indicate that natural perchlorate of regional provenance is the dominant exposure source for the two sample populations, and that chlorine isotope ratios provide a robust tool for elucidating perchlorate exposure pathways.

During World War I the north of France was for years on the front line between Germany and France. Enormous amounts of ammunition has been exploded in that region so that it is suspected that perchlorates used as ammunition actually contaminated the soil in this region. As the same regions is also an important agricultural area it is also suspected that former use of Chilean nitrate as fertiliser may have left some perchlorate in the soil. For that reason Cao et al. (2020) monitored the groundwater near the city of Reims in the north of France for a period of 2 years. Perchlorate concentrations are high in groundwater in this region (>4 μg/L). Based on the chlorine ($\delta^{37}Cl$ = ~0‰) and oxygen ($\delta^{18}O$ = −30 to −20‰, $\Delta^{17}O$ = ~0‰) isotope composition of the perchlorate in the groundwater the origin of the perchlorate in this region is mainly synthetic, proving the military source of perchlorate contamination in the study area. Considering the long persistence time of the explosive residues related to unexploded ammunitions and ammunition destruction activities, Cao et al. (2020) concluded that the perchlorate contamination in the groundwater of the study area is unlikely to decrease in the near future.

Based on the isotope compositions of chlorine ($\delta^{37}Cl$) and oxygen ($\delta^{18}O$, $\Delta^{17}O$) and the isotope abundance of radioactive ^{36}Cl in perchlorate samples can be discriminated between different sources for the perchlorate. Hatzinger et al. (2022) were able to characterise the origin of the perchlorate in eight representative groundwater wells in the Boardman-Umatilla area near the Columbia River in northeastern Oregon, USA. They found that natural perchlorate was present in all eight wells and was the main source in five of them. In the three other wells synthetic perchlorate was present and in one of the wells at least some Atacama Desert perchlorate could be detected. This study by Hatzinger et al. (2022) illustrates the role of irrigation recharge as a mechanism for producing high concentrations of indigenous natural perchlorate in groundwater by flushing accumulated salts from the unsaturated zone in semi-arid regions.

The results of these studies show the strength of the isotope studies on the transport of perchlorates. Considering the potential hazards of perchlorates in drinking waters it is obvious that it is important to know the origin of the perchlorate sources. With this knowledge it will be possible to develop better plans to remediate known sites and it also gives indispensable information on the most important sites that need to be cleaned. The discovery that the isotope composition of perchlorates can be used to trace its fate in the environment is probably one of the most powerful applications of the use of chlorine isotopes (in combination with the use of other isotopes) discovered to this date.

14.4 The Isotope Composition of Natural Chlorate

Jackson et al. (2021) were the first to analyse the isotope composition of natural chlorate samples. They analysed natural chlorate from the Atacama Desert in Chile, and Death Valley, USA. Chlorate extracted from Atacama samples had $\delta^{18}O$ from

+7.0 to +11.1‰, $\Delta^{17}O$ from +5.7 to +6.4‰, $\delta^{37}Cl$ from −1.4 to +1.3‰, and $^{36}Cl/Cl$ from $48*10^{-15}$ to $104*10^{-15}$. Chlorate from Death Valley samples had $\delta^{18}O$ from −6.9 to +1.6‰, $\Delta^{17}O$ from +0.4 to +2.6‰, $\delta^{37}Cl$ from +0.8 to +1.0‰, and $^{36}Cl/Cl$ from $14*10^{-15}$ to $44*10^{-15}$. Positive $\Delta^{17}O$ values of natural chlorate indicate that its production involved reaction with ozone, while its chlorine isotope composition is consistent with a tropospheric or near-surface source of chlorine. Additional isotope analyses of chlorate could provide stronger constraints on its production mechanisms and/or post-formational alterations, with applications for environmental forensics, global biogeochemical cycling of chlorine, and the origin of chlorate and perchlorate detected on Mars.

References

Ader M, Coleman ML, Doyle SP, Stroud M, Wakelin D (2001) Methods for the stable isotopic analysis of chlorine in chlorate and perchlorate compounds. Anal Chem 73:4946–4950

Bao H, Gu B (2004) Natural perchlorate has its unique oxygen isotope signature. Environ Sci Technol 38:5073–5077

Bennett P (2015) Comments on "Isotopic tracing of perchlorate sources in groundwater from Pomona, California" by Neil C. Sturchio, Abelardo Beloso Jr., Linnea J. Heraty, Stephen Wheatcraft, Rina Schumer. Appl Geochem 52:191–194

Böhlke JK, Sturchio NC, Gu B, Horita J, Brown GM, Jackson WA, Batista JR, Hatzinger PB (2005) Perchlorate isotope forensics. Anal Chem 77:7838–7842

Böhlke JK, Hatzinger PB, Sturchio NC, Gu B, Abbene I, Mroczkowski SJ (2009) Atacama perchlorate as an agricultural contaminant in groundwater: isotopic and chronologic evidence from Long Island, New York. Environ Sci Technol 43:5619–5625

Cao FF, Struchio NC, Ollivier P, Devau N, Heraty LJ, Jaunat J (2020) Sources and behavior of perchlorate in a shallow chalk aquifer under military (World War I) and agricultural influences. J Hazard Mat 398:123072

DasGupta PK, Martinelango PK, Jackson WA, Anderson TA, Tian K, Tock RW, Rajagopalan S (2005) The origin of naturally occurring perchlorate: the role of atmospheric processes. Environ Sci Technol 39:1569–1575

DasGupta PK, Dyke JV, Kirk AB, Jackson WA (2006) Perchlorate in the United States. Analysis of relative source contributions to the food chain. Environ Sci Technol 40:6608–6614

Ericksen GE (1981) Geology and origin of the Chilean nitrate deposits. USGS Prof Pap 1188, 37 pp

Gullick RW, Barhorst TS, LeChevallier MW (2001) Occurrence of perchlorate in drinking water sources. J Am Water Works Assoc 93:66–77

Hatzinger P, Bohlke JK, Sturchio NC, Gu B, Heraty LJ, Borden RC (2009) Fractionation of stable isotopes in perchlorate and nitrate during in situ biodegradation in a sandy aquifer. Environ Chem 6:44–52

Hatzinger PB, Harvey G, Jackson WA, Böhlke JK, Sturchio NC, Gu BH, Grantz D, Burkey K, McGrath M (2017) Identification and characterization of natural sources of perchlorate. Final Report of Phase II. SERDP Project ER-1435

Hatzinger PB, Böhlke JK, Jackson WA, Gu BH, Mroczkowski SJ, Struchio NC (2022) Isotopic discrimination of natural and anthropogenic perchlorate sources in groundwater in a semi-arid region of northeastern Oregon (USA). Appl Geochem 139:105232

Jackson WA, Preethi J, Laxman P, Tian K, Smith PN, Yu L, Anderson TA (2005) Perchlorate accumulation in forage and edible vegetation. J Agric Food Chem 53:369–373

Jackson WA, Bohlke JK, Gu B, Hatzinger PB, Sturchio NC (2010) Isotopic composition and origin of indigenous perchlorate and co-occurring nitrate in the southwestern United States. Environ Sci Technol 44:4869–4876

Jackson WA, Brundrett M, Böhlke JK, Hatzinger PB, Mroczkowski SJ, Sturchio NC (2021) Isotopic composition of natural and synthetic chlorate ($\delta^{18}O$, $\Delta^{17}O$, $\delta^{37}Cl$, $^{36}Cl/Cl$): methods and initial results. Chemosphere 274:129586

Jaegle L, Yung YL, Toon GC, Sen B, Blavier J (1996) Balloon observations of organic and inorganic chlorine in the stratosphere: the role of $HClO_4$ production on sulfate aerosols. Geophys Res Lett 23:1749–1752

Kirk AB, Smith EE, Tian K, Anderson TA, DasGupta PK (2003) Perchlorate in milk. Environ Sci Technol 37:4979–4981

Parker DR, Seyfferth AL, Reese BK (2008) Perchlorate in groundwater: a synoptic survey of "pristine" sites in the conterminous Unites States. Environ Sci Technol 42:1465–1471

Poghosyan A, Struchio NC, Morrison CG, Beloso AD, Guan YB, Eiler JM, Jackson WA, Hatzinger PB (2014) Perchlorate in the Great Lakes: isotopic composition and origin. Environ Sci Technol 48:11146–11153

Poghosyan A, Morel-Espinosa M, Valentin-Blasini L, Blount BC, Ferreccio C, Steinmaus CM, Struchio NC (2016) Chlorine isotopic composition of perchlorate in human urine as a means of distinguishing among exposure sources. J Expo Sci Environ Epidemiol 26:324–328

Rao B, Anderson TA, Orris GJ, Rainwater KA, Rajagopalan S, Sandvig RM, Scanlon BR, Stonestrom DA, Walvoord MA, Jackson WA (2007) Widespread natural perchlorate in unsaturated zones

of the Southwest United States. Environ Sci Technol 41:4522–4528
Rao B, Anderson TA, Redder A, Jackson WA (2010) Perchlorate formation by ozone oxidation of aqueous chlorine/oxychlorine species: role of Cl_xO_y radicals. Environ Sci Technol 44:2961–2967
Schumacher J (1960) Perchlorates: their properties, manufacture, and uses. American Chemical Society Monograph Series 146. Reinhold, New York
Simonaitis R, Heicklen J (1975) Perchloric acid: possible sink for stratospheric chlorine. Planet Space Sci 23:1567–1569
Slaten S, Fields KA, Santos S, Barton A, Rectanus HV, Bhargava M (2010) Integrated environmental forensics approach for evaluating the extent of dissolved perchlorate originating from multiple sources. Environ Forensic 11:72–93
Sturchio NC (2015) Reply to the comments by P. Bennett on "isotopic tracing of perchlorate sources in groundwater of Pomona, California" by N.C. Sturchio, A. Beloso Jr., L.J. Heraty, S. Wheatcraft, and R Schumer. Appl Geochem 52:195–196
Sturchio NC, Hatzinger PB, Arkins MD, Suh C, Heraty LJ (2003) Chlorine isotope fractionation during microbial reduction of perchlorate. Environ Sci Technol 37:3859–3863
Sturchio NC, Böhlke JK, Gu B, Horita J, Brown GM, Beloso Jr AD, Hatzinger PB, Jackson WA, Batista JR (2006) Stable isotopic compositions of chlorine and oxygen in synthetic and natural perchlorates. In: Gu B, Coates JD (eds) Perchlorate environmental occurrences, interactions, and treatment. Springer, New York, pp 93–109
Sturchio NC, Böhlke JK, Beloso Jr AD, Streger SH, Heraty LJ, Hatzinger PB (2007) Oxygen and chlorine isotopic fractionation during perchlorate biodegradation: laboratory results and implications for forensics and natural attenuation studies. Environ Sci Technol 41:2796–2802
Sturchio NC, Beloso Jr AD, Heraty LJ, LeClaire J, Rolfe T, Manning KR (2008) Isotopic evidence for agricultural perchlorate in groundwater of the western Chino Basin, California. In: Sixth international conference on remediation of chlorinated and recalcitrant compounds, Monterey, CA, 18–22 May 2008
Sturchio NC, Böhlke JK, Gu BH, Hatzinger PB, Jackson WA (2011) Isotopic tracing of perchlorate in the environment. In: Baskaran M (ed) Handbook of environmental isotope geochemistry, advances in isotope geochemistry. Springer-Verlag, Berlin Heidelberg
Sturchio NC, Hoaglund III JR, Marroquin RJ, Beloso Jr AD, Heraty LJ, Bortz SE, Patterson TL (2012) Isotopic mapping of groundwater perchlorate plumes. Ground Water 50:94–102
Sturchio NC, Beloso Jr A, Heraty LJ, Wheatcraft S, Schumer R (2014) Isotopic tracing of perchlorate sources in groundwater from Pomona, California. Appl Geochem 43:80–87
Thiemens MH (2006) History and applications of mass-independent isotope effects. Annu Rev Earth Planet Sci 34:217–262

15 Isotope Variations of Chlorinated and Brominated Organic Compounds in Natural Environments

15.1 Introduction

The first analytical techniques that were developed to analyse the isotope compositions of organochlorine compounds were dual-inlet techniques that required large amounts of sample for analysis. As the concentrations of organochlorine and organobromine compounds, even at contaminated sites, are relatively low, using dual-inlet would require the processing of large amounts of contaminated water or soil to obtain a sufficient amount of these compounds.

It was however realised early that with the use of Compound Specific Isotope Analysis (CSIA) it would be possible to analyse the isotope composition of different compounds in a mixture. Using CSIA the isotope composition is analysed by a continuous flow method, and as a consequence much lower amounts of sample are needed for the analysis of a single sample. Perhaps the first (Lojkasek-Lima et al. 2012) study in which the potential of CSIA for tracing chlorinated organic compounds such as tetrachloroethene and trichloroethene from contaminated sites was applied by Hunkeler et al. (2004) in sandy aquifers in Canada that were contaminated with organochlorine compounds. In this study the tracing was done solely using carbon isotopes, as methods to measure chlorine isotopes by CSIA were not developed yet.

After the development of CSIA methods for chlorine isotopes (Shouakar-Stash et al. 2006; Sakaguchi-Söder et al. 2007) it became possible to study not only the carbon, but also the chlorine isotope compositions of chlorinated organic compounds during transport from contaminated sites. It was already realised from the onset of the analyses of chlorine isotopes in chlorinated organic compounds (Holt et al. 1997; Jendrzejewski et al. 1997, 2001) that the combination of the two isotope systems ($\delta^{13}C$ and $\delta^{37}Cl$) showed a much more detailed characterisation on the origin and evolution of these compounds. Chlorine isotope compositions of organic compounds from environmental samples could be measured from the moment that methods were developed to extract them in sufficient amounts from natural samples (Holt et al. 2001). In some modern studies the chlorinated organic compounds are extracted from water and vapour using Multi Layer Samplers, a technique that can take weeks to obtain sufficient sample (Ronen et al. 1986, 2010), which need not necessarily be a problem for monitoring purposes.

15.2 Chlorine

15.2.1 Studies on the Dispersion of Chlorinated Solvents in Ground Waters

Chlorinated ethenes, especially perchlorethene and trichloroethene are commonly used chemicals

H. Eggenkamp, *The Geochemistry of Stable Chlorine and Bromine Isotopes*, Advances in Isotope Geochemistry, https://doi.org/10.1007/978-3-031-75633-7_15

in the dry-cleaning industry. Also in other industries, such as metal-working, paint industries, organic synthesis and the military are in various cases a source of contamination by chlorinated organic compounds (Jendrzejewski et al. 2001). As a result of their common use, they can be a very significant source of pollution in the atmosphere, soils and aquifers (e.g. Stiber et al. 1998; Sturchio et al. 1998). Characterising these pollutants and the different processes affecting them is thus an important issue. In this respect, isotopic geochemistry of carbon and chlorine can give additional information to investigate these compounds and their by-products. In recent years several studies have been published in which these organochlorine compounds have been extracted from contaminated sites to establish their origin. A few of these studies will be discussed in more detail as examples of the possibilities that the combined analysis of carbon and chlorine isotopes can deliver to understand how these compounds behave in contaminated aquifers.

Hunkeler et al. (2011) described the behaviour of chlorinated ethenes near a former industrial facility in the vicinity of Tel Aviv, Israel. On this site, that once was occupied with mostly metal-working and plating facilities, the aquifer was contaminated with several chlorinated ethenes. This contamination occurred as the former factories disposed of their used chlorinated ethenes at the factory grounds. At this location, consisting of sandstones the water table is located some 18 meters below the ground surface. Samples were taken at various depths, both in the unsaturated zone and below the water table. Samples were taken using a Multi Layer Sampler as described by Ronen et al. 1986, 2010). From these samples the organic compounds were determined by gas chromatography and the carbon isotope compositions were determined according to the techniques described by Hunkeler and Aravena (2000) and the chlorine isotope compositions using the methods described by Shouakar-Stash et al. (2006). They found that the concentrations of the chlorinated organic compounds were highest at the boundary between the saturated and the unsaturated zone and decreased in both zones away from the water table. Both carbon and chlorine isotopes, for the samples in which both isotopes were measured, showed no variation in the unsaturated zone, except, perhaps, as the results were non-conclusive in those samples taken closest to the land surface. In the saturated zone it appeared that the carbon isotopes behaved inconclusive, at some locations they increased, at others they decreased or were invariant. At the only site were chlorine isotope compositions were analysed the ratios decreased in value. These variations however appeared also to be very limited. The transport of the chlorinated ethenes was modelled numerically and the results of this modelling confirmed the observed relationship between the distances from the water table, the decreasing concentration and invariable carbon isotope composition for situations that are close to steady state. Before the steady state was reached the carbon isotope composition was expected to be lower away from the water interface due to processes such as diffusion. In the saturated zone the numerical modelling showed a steep decrease of the concentration with increasing distance from the water table combined with a steep decrease in the isotope composition for both $\delta^{13}C$ and $\delta^{37}Cl$. The different results for the unsaturated and the saturated zones appeared to be the result from the difference in speed in which equilibrium is reached in the two zones. While in the unsaturated zone the equilibrium can be reached within a year in the saturated zone this can take 1000s of years. The results obtained in this study (Hunkeler et al. 2011) indicate that carbon and chlorine isotopes can be used to link vapour phase contamination to their source even if extensive migration of the vapours occurs. The results from numerical modelling revealed that constant isotope ratios as observed in this study are only expected for systems that are close to steady state.

In a second case study Lojkasek-Lima et al. (2012) described the transport behaviour of trichloroethene trough a bedrock aquifer consisting of dolostone. In this study the transport was described fully within the saturated zone. The reason for this study was to understand the source of trichoroethene contamination that caused the closure of a municipal water-well. In the vicinity of this well there were several industries from which it was known that they had caused trichoroethene contamination, of which one had created a monitoring network. Samples from the municipal well, as well as from the monitoring wells were analysed for the concentration of chemical species as well as the chlorine and carbon isotope compositions of trichoroethene and *cis*-1,2-dichlorothene. Large variations in both chlorine and carbon isotope compositions were found between the different wells. For example, for trichloroethylene the carbon isotope varied between −35.6 and −21.8‰ and the chlorine isotope composition between +1.6 and +3.2‰. In the case of *cis*-1,2-dichlorothene the carbon isotope variation was between −36.3 and −18.9‰ and the chlorine isotope varies between 2.4 and 4.7‰. These data were plotted on a map and this indicated that the combined chlorine and carbon isotope composition of the municipal well samples were clearly different from the samples taken from the monitoring wells. From these results it was concluded that the contamination of the municipal well could not be caused by the industrial property with the monitoring network. The study by Lojkasek-Lima et al. (2012) demonstrated that compound specific isotope analysis of both chlorine and carbon would have been useful if it were applied much earlier to examine the sources of trichloroethylene that was found in the municipal well and if bedrock monitoring wells had been located at several depths beneath each of the potential trichloroethylene contributing properties. The results obtained using compound-specific isotope analysis also show that microbial reductive dechlorination of trichloroethylene occurs in some parts of the bedrock aquifer. At this site, the use of compound specific isotope analysis for chlorine and carbon in combination with analyses of volatile chlorinated organic compounds and redox parameters proved to be important due to the complexity introduced by biodegradation in the complex fractured rock aquifer. In this case it was highly recommended that this study was redone as soon as the municipal well is back into full operation, and make sure it is properly monitored.

In a study on isotope variations during degradation of chlorinated ethenes Wiegert et al. (2013) investigated a method to efficiently monitor this in situ degradation. They reported a field application of combined stable chlorine and carbon isotope analysis of tetrachloroethene and trichloroethene to investigate their biodegradation in a heavily contaminated aquifer. To analyse the chlorine isotope composition of the contaminants they applied the gas chromatography-quadrupole mass spectrometry (GCqMS) method developed by Aeppli et al. (2010). Both chlorine and carbon isotopes showed evidence of ongoing tetrachloroethene transformation. Applying published carbon isotope enrichment factors they were able to evaluate the extent of in situ tetrachloroethene degradation which was between 11 and 78%. Modelling both chlorine and carbon isotopes through a reactive transport model indicated that combined PCE and TCE mass load was dechlorinated by less than 10%, and that cis-dichloroethene was not further dechlorinated. The combined use of chlorine and carbon isotopes in this study made it possible to estimate chlorine isotope enrichment factors, which were between −7.8 and −0.8‰. Interestingly they found characteristic values for the ratio of chlorine and carbon isotope enrichment factors for the reduction of tetrachloroethene at their field site. This study also demonstrated the benefit of compound specific isotope analyses of both chlorine and carbon and the beneficent use of chlorine isotope variations to evaluate tetrachloroethene and trichloroethene dechlorination.

In a study in the industrial complex in Wonju, South Korea (Kaown et al. 2014) it was possible to assess the contamination of groundwater with chlorinated solvents such as trichloroethene and carbon tetrachloride. At the study site it was observed that trichloroethylene concentrations ranged between zero and 10,066 μg/L while carbon tetrachloride concentrations ranged between zero and 985 μg/L. Close to the contaminated site a location were asphalt was tested using chlorinated solvents which were disposed of without any appropriate treatment was located. Residual solvents were found both downstream along the flowline of groundwater as well as upstream due to ground water pumping to facilitate water for a laundry facility. In the upgradient area towards the laundry facility both trichloroethylene and carbon tetrachloride metabolites were detected while only trichloroethylene metabolites were detected in the region at lower altitudes along the flowline of the groundwater. These differences were apparently caused by the fact that downgradient anaerobic dechlorination took place while upgradient aerobic dechlorination took place. As the study region was an industrial area it was discovered that several small pockets of chlorinated solvents originated from several small contaminant sources. This complicated the study towards the origin of the solvent plumes. Although it was rather complex to differentiate among the contaminant sources in this complicated industrial area with several small and one large source it was possible to identify the main sources in this region using $\delta^{37}Cl$ and $\delta^{13}C$ analyses in combination with hydrogeochemical data. Freshly recharged water with depleted $\delta^{37}Cl$ and $\delta^{13}C$ values of trichloroethylene from the upgradient recharge zone of the main sources affected the plumes and changed the $\delta^{37}Cl$ and $\delta^{13}C$ values of trichloroethylene in the local source zones. Despite the complicated states of contamination, dual isotope analysis data could usefully be applied to evaluate the fate and transport of the contaminants as well as to differentiate among five different contaminant sources. Kaown et al. (2014) concluded that the results they obtained supported the usefulness of compound specific isotope analysis as a method for differentiating multiple plume sources. The results obtained from this study could be used to assign remediation responsibilities to the responsible parties. Dispersion of chlorinated ethenes, trichloroethene and tetrachloroethene) from different sources was determined by Palau et al. (2014) from a site in Spain. At this site large carbon isotope variations were observed, but little chlorine isotope variations. $\delta^{37}Cl$ only ranged from +0.53 to +0.66‰. Still, the data were used to investigate the role of biodegradation in groundwater, and the multi isotope approach was considered a valuable tool for characterising the contamination in complex sites.

When the sources of chlorinated hydrocarbon contaminations are removed from a site this will have effects on the long term chlorine and carbon isotope patters at the site. Rodríguez-Fernández et al. (2018a) studied these effects on the concentrations and isotope compositions of chlorinated methanes from a site in Spain where the sources of the contamination (such as barrels, tank, pit and wastewater pipe sources) were removed from the site. Tetrachloromethane in general showed for both carbon and chlorine isotope values an increasing trend in wells from the contaminated site, confirming the tertrachloromethane degradation processes. During this process $\delta^{37}Cl$ values could increase by 4 to 6‰, for chloroform from −4 to +2‰ and for tetrachloromethane from +4 to +8‰. Processes that took place at the field sites were determined from dual C-Cl isotope plots, and for chloroform related to the effects as described by Torrentó et al. (2017; chloroform reference slopes of oxidation, alkaline hydrolysis and hydrogenolysis + reductive elimination), Heckel et al. (2017; outer-sphere single electron transfer), Rodríguez-Fernández et al. (2018b; chloroform inferred pathways from biodegradation with vitamin B12), and for tetrachloromethane as described by Rodríguez-Fernández et al. (2018b; hydrogenolysis (Fe(0)), Rodríguez-Fernández et al. (2018b; inferred pathways from biodegradation with vitamin B12 from microcosms) and Rodríguez-Fernández et al., (2018c; degradation by pyrite and by magnetite at pH 12. The data obtained by Rodríguez-Fernández et al. (2018a) established a base for discerning the predominant or combined pathways for the natural attenuation of chlorinated methanes or for assessing the effectiveness of remediation strategies using recycled minerals or metallic iron.

Zimmermann et al. (2020) wrote an overview on principles, challenges and applications on the use of chlorine isotope analyses to track chlorinated contaminants in the subsurface. In their review they summarised the different methods that were developed to measure chlorine isotopes as well as the enrichment factors that were determined in various studies. They also summarised the many field studies that were published. The most recent studies at the time of publication of this review were, Palau et al. (2017); Torrentó et al. (2017); Wanner et al. (2017); Rodríguez-Fernández et al. (2018a), and Murray et al.

(2019). The case studies described in this section show that especially the use of the combination of both chlorine and carbon isotopes in the study of dispersion of chlorinated organic compounds in groundwaters is a very successful technique to understand the fate of these compounds in the environment. Due to the use of the dual isotope technique, in combination with the measurements of the concentrations of these compounds, it is possible to understand how they disperse on a site, how they degrade to compounds containing less chloride atoms per molecule and how solvents from different sources with different isotope compositions behave and mix with each other. Considering these successful case studies this is a very successful way to understand the fate of these contaminants in the environment. If it would become more common to analyse (also) the chlorine isotope compositions of these compounds, which are currently only measured in a limited number of laboratories, this will become a very useful application for chlorine isotopes in the future, as was clearly shown by Zimmermann et al. (2020).

15.2.2 Effects of Diffusion of Chlorinated Hydrocarbons in Water and Sediment

Wanner and Hunkeler (2015) studied the effects of diffusion of trichloroethene and dichloroethane. While the effects of diffusion are relatively small, significant chlorine isotope fractionation was observed in a clay core from a contaminated landfill. The $\delta^{37}Cl$ was +1.79‰ at the top of the clay core decreasing to +1.28‰ at 40 cm depth, indicating a $\delta^{37}Cl$ decrease of 0.51‰.

15.2.3 The Use of Chlorine Isotopes to Test the (Anthropogenic) Origin of a Chlorinated Compound

Holmstrand et al. (2010) determined the chlorine isotope composition of the persistent and bioaccumulating compound tris-(4-chlorophenyl) methane to elucidate whether its main source is natural or anthropogenic. In this study they extracted blubber from the Baltic grey seal (*Halichoerus grypus*) by continuous acetonitrile partitioning (Jensen et al. 1992), and the tris-(4-chlorophenyl)methane was isolated from the extract by preparative-capillary gas chromatography (Holmstrand et al. 2007). Chlorine isotope analysis was subsequently performed by sealed-tube combustion in conjunction with thermal-ionization mass spectrometry. The $\delta^{37}Cl$ of tris-(4-chlorophenyl)methane was -3.5 ± 0.5‰, similar to the previously reported $\delta^{37}Cl$ of technical grade p,p′-DDT (Holmstrand et al. 2006). The data obtained from the blubber is not consistent with a putative marine natural source of tris-(4-chlorophenyl)methane, as enzymatic (biotic) production is reported to give values of $\delta^{37}Cl < -10$‰ (Reddy et al. 2000, 2002). The $\delta^{37}Cl$-tris-(4-chlorophenyl)methane data obtained by Holmstrand et al. (2010) thus support the hypothesis that tris-(4-chlorophenyl)methane is produced as a byproduct during DDT synthesis and is released into the environment through the same pathways as DDT.

15.2.4 Isotope Effects During Environmental Dispersion of Hexachlorocyclohexanes

Hexachlorocyclohexanes (HCH, $C_6H_6Cl_6$) belong to the most relevant persistent organic pollutants (Voldner and Li 1995). HCH is synthesised trough the chlorination of benzene, and in that process a large number of hexachlorocyclohexane stereo-isomers are formed. The mixture consists mainly of five different HCH isomers (60–70% α-HCH, 10–15% γ-HCH, 5–12% β-HCH, 6–10% δ-HCH, 1–2% ε-HCH, and about 1–2% other isomers and impurities (Vijgen 2006; Xiao et al. 2004). HCHs were primarily produced because of the insecticide properties of γ-HCH (Lindane), which means that the other isomers that were produced were essentially just waste, and these were often (uncontrolled) dumped in the environment. HCHs have been used since 1942 as insecticides and were banned in the European Union in 2000, and are slowly phased out in the rest of the world as a result of the Stockholm Convention on Persistent Organic Pollutants. Because of the persistence of HCH in the environment many studies were applied to understand its fate in the environment. Because HCHs consist of three elements, carbon, hydrogen and chlorine, from which the isotope compositions can also be measured, isotope studies are an important element of these studies. In order to understand differences in isotope compositions of the main HCH stereo-isomers from different manufacturers Ivdra et al. (2017) measured the $\delta^{37}Cl$, $\delta^{13}C$ and $\delta^{2}H$ from a large set (77) of HCH samples, both mixtures and pure chemicals. The results showed a large isotope variation for all three isotope systems, $\delta^{37}Cl$ ranged from −6.69 to +0.54‰, $\delta^{13}C$ from −35.9 to −22.7‰ and $\delta^{2}H$ from −233 to +1‰. The chlorine isotope compositions could only be determined from (23) pure chemicals. Based on the combination of all three isotope systems Ivdra et al. (2017) were able to divide their sample set into five groups, but only one of these was characterised by a different, very negative, chlorine isotope composition compared to the other groups. The HCHs from this group are *α*- and *β*-HCH, isomers without insecticidal properties. The negative chlorine isotope ratios of these samples may result from specific synthetic and/or purification processes, applied to enrich non-active isomers from technical mixtures or probably even from chemical waste after the isolation of *γ*-HCH. In all other groups the $\delta^{37}Cl$ was more modest with values that were between −1.60 and + 0.54‰, reflecting a near SMOC source for the chlorine used in the HCH production.

Kannath et al. (2019) studied potential degradation of HCH with the dehydrochlorination reaction promoted by the hydroxyl ion in aqueous solution. They also studied chemical dehydrochlorination by the catalysed reaction via dehydrochlorinase. Kannath et al. (2019) concluded that based on the values of the isotope effects, both processes seem to occur via the same net mechanism. The behaviour of HCH during uptake in food was studied by Wu et al. (2019). Wu et al. (2019) observed that chlorine isotope values is of HCHs in milk, dung and seal blubber were much heavier (up to +3.79‰) compared to the upper limit of the reported source range. So, apparently metabolism of HCH in the digestive system of cows and seal causes strong isotope fractionation, while the concentration of HCH in the body decreased. As a result the residual fraction of HCHs in the organism does not represent the exposure of the organisms adequately. In a different system Liu et al. (2020a) studied how the chlorine isotope compositions of HCHs changed during uptake by wheat from the soil during its growth. In this study $\delta^{37}Cl$ values of *β*-HCH remained stable in soil and wheat, revealing no transformation between these systems. Later, Liu et al. (2021) observed that while in wheat leaves, stems and roots, transformation of α-HCH was indicated due to a C-Cl bond cleavage by increasing $\delta^{13}C$ and $\delta^{37}Cl$ compared to the nutrient solution, for β-HCH, in contrast, no transformation was detected. In addition, 1,3,4,5,6-pentachlorocyclohexene (PCCH) was identified as the major metabolite of α-HCH transformation.

Liu et al. (2020b) observed significant chlorine and carbon isotope fractionation of different HCH isomers in samples from a contaminated aquifer in Germany. They used $^{37}Cl/^{35}Cl$ and $^{13}C/^{12}C$ isotope fractionation patterns of different HCH isomers from laboratory experiments diagnostically in a model to characterise microbial dichloroelimination in this field study. At this specific contaminated aquifer it was shown that the transformation of HCH isomers was mainly governed by microbial dichloroelimination transformation.

15.3 Bromine Isotopes in Organic Compounds Found in the Environment

Although a few methods have been developed to analyse the bromine isotope compositions of organic compounds (Sylva et al. 2007; Holmstrand et al. 2010; Gelman and Halicz 2010; Hitzfeld et al. (2011) not many studies have been published showing bromine isotope data in contaminated natural samples.

The first such study was conducted by Carrizo et al. (2011). In this study the bromine isotope compositions of six brominated compounds, produced in the chemical industry, were analysed and compared to the bromine isotope composition of a 2,4-dibromophenol sample that was extracted from whale blubber. Results obtained showed that the anthropogenic brominated organic compounds had a bromine isotope variation from −4.3 to −0.4‰. The 2,4-dibromophenol extracted from the whale blubber (Teuten et al. 2006) had a value of 0.2 ± 1.6‰ (1 SD), which was significantly different from the average value of −1.1 ± 0.9‰ (1 SD) obtained from anthropogenic sources. This difference, about 1.4‰ showed a P of less than 0.05. The results showed that the δ^{81}Br of four out of six industrial compounds was found to be significantly different from that of the natural sample. The study by Carrizo et al. (2011) indicated a baseline for future work employing δ^{81}Br variations between organobromine compounds to study the sources and degradation of organobromine compounds in the environment. They conclude that it is conceivable that ambient microbial and photochemical degradation of brominated flame retardants, for example, induces bromine isotope effects so that bromine isotope analysis can be used to assess the extent of environmental degradation of such chemicals, in analogy to the known isotope fractionation for chlorine in organochlorine compounds.

Chen et al. (2017) studied the importance of both bromine and carbon isotopes in the identification of the source of poly brominated diphenyl ethers (PBDEs). To assess the significance of carbon and bromine isotopes in the source identification of PBDEs, the δ^{13}C and δ^{81}Br values of a few PBDEs from different suppliers were measured. The results showed that individual PBDE samples have unique δ^{13}C and δ^{81}Br values, possibly due to differences in the precursors and manufacturing processes, indicating that the isotope composition is a promising probe to determine the source of PBDEs in the environment.

A summary of the techniques and recent applications on the use of both stable chlorine and bromine isotopes in environmental research was published, in Chinese, by Hu et al. (2021). A few of the bromine isotope studies discussed in this review are by Horst et al. (2014), on the bromine isotope composition of methyl bromide released from plant matter, Kozell et al. (2015) on the transformations of tribromoneopentyl alcohol, Kuntze et al. (2016) on the use of dual carbon-bromine stable isotope analyses to study abiotic transformations in ethylene dibromine, and Golan et al. (2019) on the degradation of 4-bromophenol by *Ochrobactrum sp.* Another review discussing the possibilities and challenges that come with the analyses of bromine isotopes in organic substances was published by Xiong et al. (2021). Their review describes the analytical methods currently in use for measuring carbon, hydrogen, oxygen and bromine stable isotopes in brominated organic compounds by the use of mass spectrometry. Further, they discuss the applications to identify sources, tracing transformation pathways and biological metabolisms. They also discuss the kinetic isotope effects (KIEs), based on the reaction mechanisms. Finally, Xiong et al. (2021) highlight future challenges and prospects that will occur with the bromine isotope analysis of brominated organic compounds and they discuss a perspective on future activities that may arise in this growing field of bromine isotope applications.

References

Aeppli C, Holmstrand H, Andersson P, Gustafsson Ö (2010) Direct compound-specific stable chlorine isotope analysis of organic compounds with quadrupole GC/MS using standard isotope bracketing. Anal Chem 82:420–426

Carrizo D, Unger M, Holmstrand H, Andersson P, Gustafsson Ö, Sylva SP, Reddy CM (2011) Compound-specific bromine isotope compositions of one natural and six industrially synthesised organobromine substances. Environ Chem 8:127–132

Chen LZ, Shouakar-Stash O, Ma T, Wang CL, Liu L (2017) Significance of stable carbon and bromine isotopes in the source identification of PBDEs. Chemosphere 186:160–166

Gelman F, Halicz L (2010) High precision determination of bromine isotope ratio by GC-MC-ICPMS. Int J Mass Spectrom 289:167–169

Golan R, Gelman F, Kuder T et al (2019) Degradation of 4-bromophenol by *Ochrobactrum* sp. HI1 isolated from desert soil: pathway and isotope effects. Biodegradation 30:37–46

Heckel B, Rodríguez-Fernández D, Torrentó C, Meyer A, Palau J, Domènech C, Rosell M, Soler A, Hunkeler D,

Elsner M (2017) Compound-specific chlorine isotope analysis of tetrachloromethane and trichloromethane by gas chromatographyisotope ratio mass spectrometry vs gas chromatography-quadrupolemass spectrometry: method development and evaluation of precision and trueness. Anal Chem 89:3411–3420

Hitzfeld KL, Gehre M, Richnow HH (2011) A novel online approach to the determination of isotopic ratios for organically bound chlorine, bromine and Sulphur. Rapid Commun Mass Spectrom 25:3114–3122

Holmstrand H, Mandalakis M, Zencak Z, Gustafsson Ö, Andersson P (2006) Chlorine isotope fractionation of a semi-volatile organochlorine compound during preparative megabore-column capillary gas chromatography. J Chromatogr A 1103:133–138

Holmstrand H, Mandalakis M, Zencak Z, Andersson P, Gustafsson O (2007) First compound-specific chlorine-isotope analysis of environmentally-bioaccumulated organochlorines indicates a degradation-relatable kinetic isotope effect for DDT. Chemosphere 69:1533–1539

Holmstrand H, Zencak Z, Mandalakis M, Andersson P, Gustafsson Ö (2010) Chlorine isotope evidence for the anthropogenic origin of tris-(4-chlorophenyl)methane. Appl Geochem 25:1301–1306

Holt BD, Sturchio NC, Abrajano TA, Heraty LJ (1997) Conversion of chlorinated volatile organic compounds to carbon dioxide and methyl chloride for isotopic analysis of carbon and chlorine. Anal Chem 69:2727–2733

Holt BD, Heraty LJ, Sturchio NC (2001) Extraction of chlorinated aliphatic hydrocarbons from groundwater at micromolar concentrations for isotopic analysis of chlorine. Environ Pollut 113:263–269

Horst A, Holmstrand H, Andersson P et al (2014) Stable bromine isotopic composition of methyl bromide released from plant matter. Geochim Cosmochim Acta 125:186–195

Hu XX, Hou XW, Liu JY LQ, Jiang GB (2021) Techniques for stable chlorine/bromine isotope analysis and recent applications in environmental research. Environ Chem 40:331–342. (in Chinese with English abstract)

Hunkeler D, Aravena R (2000) Determination of stable carbon isotope ratios of chlorinated methanes, ethanes and ethenes in aqueous samples. Environ Sci Technol 34:2839–2844

Hunkeler D, Chollet N, Pittet X, Aravena R, Cherry JA, Parker BL (2004) Effect of source variability and transport processes on carbon isotope ratios of TCE en PCE in two sandy aquifers. J Contam Hydrol 74:265–282

Hunkeler D, Aravena R, Shouakar-Stash O, Weisbrod N, Nasser A, Netzer L, Ronen D (2011) Carbon and chlorine isotope ratios of chlorinated ethenes migrating through a thick unsaturated zone of a sandy aquifer. Environ Sci Technol 45:8247–8253

Ivdra N, Ficher A, Herrero-Martin S, Giunta T, Bonifacie M, Richnow HH (2017) Carbon, hydrogen and chlorine stable isotope fingerprinting for forensic investigations of Hexachlorocyclohexanes. Environ Sci Technol 51:446–454

Jendrzejewski N, Eggenkamp HGM, Coleman ML (1997) Sequential determination of chlorine and carbon isotopic composition in single microliter samples of chlorinated solvent. Anal Chem 69:4259–4266

Jendrzejewski N, Eggenkamp HGM, Coleman ML (2001) Characterisation of chlorinated hydrocarbons from chlorine and carbon isotopic compositions: scope and application to environmental problems. Appl Geochem 16:1021–1031

Jensen S, Athanasiadou M, Bergman Å (1992) A technique for separation of xenobiotics from large amounts of lipids. Organohalogen Compd 8:79–80

Kannath S, Adamcyk P, Wu LP, Richnow HH, Dybala-Defratyka A (2019) Can alkaline hydrolysis of γ-HCH serve as a model reaction to study its aerobic enzymatic dehydrochlorination by LinA? Int J Mol Sci 20:5955

Kaown D, Shouakar-Stash O, Yang J, Hyun YJ, Lee KK (2014) Identification of multiple sources of groundwater contamination by dual isotopes. Ground Water 52:875–885

Kozell A, Yecheskel Y, Balaban N et al (2015) Application of dual carbon-bromine isotope analysis for investigating abiotic transformations of tribromoneopentyl alcohol (TBNPA). Environ Sci Technol 49:4433–4440

Kuntze K, Kozell A, Richnow HH et al (2016) Dual carbon-bromine stable isotope analysis allows distinguishing transformation pathways of ethylene dibromide. Environ Sci Technol 50:9855–9863

Liu X, Wu LP, Kümmel S, Merbach I, Lal R, Richnow HH (2020a) Compound-specific isotope analysis and enantiomer fractionation to characterize the transformation of hexachlorocyclohexane isomers in a soil–wheat pot system. Environ Sci Technol 54:8690–8698

Liu YQ, Kümmel S, Yao J, Nijenhuis I, Richnow HH (2020b) Dual C-cl isotope analysis for characterizing the anaerobic transformation of α, β, γ, and δ-hexachlorocyclohexane in contaminated aquifers. Water Res 184:116128

Liu X, Wu LP, Kümmel S, Richnow HH (2021) Characterizing the biotransformation of hexachlorocyclohexanes in wheat using compound-specific stable isotope analysis and enantiomer fraction analysis. J Hazard Mat 406:124301

Lojkasek-Lima P, Aravena R, Parker BL, Chrry JA (2012) Fingerprinting TCE in a bedrock aquifer using compound-specific isotope analysis. Ground Water 50:754–764

Murray AM, Ottosen CB, Maillard J, Holliger C, Johansen A, Brabæk L, Kristensen IL, Zimmermann J, Hunkeler D, Broholm MM (2019) Chlorinated ethene plume evolution after source thermal remediation: determination of degradation rates and mechanisms. J Contam Hydrol 227:103551

Palau J, Marchesi M, Chambon JCC, Aravena R, Canals À, Binning PJ, Bjerg PL, Otero N, Soler A (2014) Multi-isotope (carbon and chlorine) analysis for fingerprinting and site characterization at a fractured bedrock aquifer contaminated by chlorinated ethenes. Sci Total Environ 475:61–70

Palau J, Yu R, Hatijah Mortan S, Shouakar-Stash O, Rosell M, Freedman DL, Sbarbati C, Fiorenza S, Aravena R, Marco-Urrea E, Elsner M, Soler A, Hunkeler D (2017) Distinct dual C-cl isotope fractionation patterns during anaerobic biodegradation of 1,2-dichloroethane: potential to characterize microbial degradation in the field. Environ Sci Technol 51:2685–2694

Reddy CM, Heraty LJ, Holt BD, Sturchio NC, Eglinton T, Drenzek NJ, Xu L, Lake JL, Maruya KA (2000) Stable chlorine isotopic compositions of aroclors and aroclor-contaminated sediments. Environ Sci Technol 34:2866–2870

Reddy CM, Xu L, Drenzek NJ, Sturchio NC, Heraty LJ, Kimblin C, Butler A (2002) A chlorine isotope effect for enzyme-catalyzed chlorination. J Am Chem Soc 124:14526–14527

Rodríguez-Fernández D, Torrentó C, Palau J, Marchesi M, Soler A, Hunkeler D, Domènech C, Rosell M (2018a) Unravelling long-term source removal effects and chlorinated methanes natural attenuation processes by C and Cl stable isotopic patterns at a complex field site. Sci Total Environ 645:286–296

Rodríguez-Fernández D, Torrentó C, Guivernau M, Viñas M, Hunkeler D, Soler A, Domènech C, Rosell M (2018b) Vitamin B 12 effects on chlorinated methanes degrading microcosms: dual isotope and metabolically active microbial populations assessment. Sci Total Environ 621:1615–1625

Rodríguez-Fernández D, Heckel B, Torrentó C, Meyer A, Elsner M, Hunkeler D, Soler A, Rosell M, Domènech C (2018c) Dual element (C-cl) isotope approach to characterize abiotic reactions of chlorinated methanes by Fe(0) and by Fe(II) on iron minerals at neutral and alkaline pH. Chemosphere 206:447–456

Ronen D, Magaritz M, Levy I (1986) A multilayer sampler for the study of detailed hydrochemical profiles in groundwater. Water Res 20:311–315

Ronen D, Lev-Wiener H, Graber E, Dahan O, Weisbrod N (2010) Simultaneous counter-flow of chlorinated volatile organic compounds across the saturated-unsaturated interface region of an aquifer. Water Res 29:261–280

Sakaguchi-Söder K, Jager J, Grund H, Matthäus F, Schüth C (2007) Monitoring and evaluation of dechlorination processes using compound-specific chlorine isotope analysis. Rapid Commun Mass Spectrom 21:3077–3084

Shouakar-Stash O, Drimmie RJ, Zhang M, Frape SK (2006) Compound-specific chlorine isotope ratios of TCE, PCE and DCE isomers by direct injection using CF-IRMS. Appl Geochem 21:766–781

Stiber NA, Small MJ, Fischbeck PS (1998) The relationship between historic industrial site use and environmental contamination. J Air Waste Manag Assoc 48:809–818

Sturchio NC, Clausen JL, Heraty LJ, Huang L, Holt BD, Abrajano TA (1998) Chlorine isotope investigation of natural attenuation of trichloroethene in an aerobic aquifer. Environ Sci Technol 32:3037–3042

Sylva SP, Ball L, Nelson RK, Reddy CM (2007) Compound-specific $^{81}Br/^{79}Br$ analysis by capillary gas chromatography/multicollector inductively coupled plasma mass spectrometry. Rapid Commun Mass Spectrom 21:3301–3305

Teuten EL, King GM, Reddy CM (2006) Natural ^{14}C in *Saccoglossus bromophenolosus* compared to ^{14}C in surrounding sediments. Mar Ecol Prog Ser 324:167

Torrentó C, Palau J, Rodríguez-Fernández D, Heckel B, Meyer A, Doménech C, Rosell M, Soler A, Elsner M, Hunkeler D (2017) Carbon and chlorine isotope fractionation patterns associated with different engineered chloroform transformation reactions. Environ Sci Technol 51:6174–6184

Vijgen J (2006) The legacy of Lindane HCH isomer production: a global overview of residue management, formulation and disposal. International HCH & Pesticides Association, Holte, Denmark

Voldner EC, Li YF (1995) Global usage of selected persistent organochlorines. Sci Total Environ 160–161:201–210

Wanner P, Hunkeler D (2015) Carbon and chlorine isotopologue fractionation of chlorinated hydrocarbons during diffusion in water and low permeability sediments. Geochim Cosmochim Acta 157:198–212

Wanner P, Parker BL, Chapman SW, Aravena R, Hunkeler D (2017) Does sorption influence isotope ratios of chlorinated hydrocarbons under field conditions? Appl Geochem 84:348–359

Wiegert C, Aeppli C, Knowles T, Holmstrand H, Evershed R, Pancost RD, Macháčková J, Gustafsson O (2013) Dual Carbon-chlorine stable isotope investigation of sources and fate of chlorinated ethenes in contaminated groundwater. Environ Sci Technol 46:10918–10925

Wu LP, Moses S, Liu YQ, Renpenning J, Richnow HH (2019) A concept for studying the transformation reaction of hexachlorocyclohexanes in food webs using multi-element compound-specific isotope analysis. Anal Chim Acta 1064:56–64

Xiao H, Li NQ, Wania F (2004) Compilation, evaluation, and selection of physical-chemical property data for α-, β-, and γ-hexachlorocyclohexane. J Chem Eng Data 49:173–185

Xiong JK, Li GY, An TC (2021) New advance in the application of compound-specific isotope analysis (CSIA) in identifying sources, transformation mechanisms and metabolism of brominated organic compounds. Crit Rev Environ Sci Technol 50:3973–3996

Zimmermann J, Halloran LJS, Hunkeler D (2020) Tracking chlorinated contaminants in the subsurface using compound-specific chlorine isotope analysis: a review of principles, current challenges and applications. Chemosphere 244:125476

16 Chlorine and Bromine Isotope Characteristics of the Atmosphere

The atmosphere is the only significant reservoir on earth where a large part of chlorine and bromine are present in the form of oxidised and organic molecules. The existence of these compounds, in which chlorine and bromine are present in higher oxidation states than −I is partly due to natural circumstances such as oxidation of chlorides and bromides by ozone or ultraviolet radiation present in the atmosphere. For a large part however chlorine and bromine are also present in the form of man made compounds such as chlorofluorocarbons and related organic chlorine compounds such as carbon tetrachloride, chloroform and dichloromethane, the concentrations of which have decreased considerably since the signing of the Montreal Protocol in 1987. Although most chlorine and bromine isotope studies on organic chlorine compounds (but focussed on their occurrence in soils) were described in Chap. 15 a short summary is given here to show the isotope characteristics of this small but important reservoir and to easily compare it to the other reservoirs.

In their letter to Nature Tanaka and Rye (1991) suggest a method that should be able to analyse organochlorine compounds that were expected to exist in the stratosphere. They stated however that, due to the very low concentrations of these compounds in the stratosphere, they expected that it would be very difficult to actually measure them. It was only around 2010, almost 20 years later, that the first methods were developed with which it was possible to extract amounts of chlorine and bromine compounds from atmospheric samples that were sufficient for isotope analyses.

16.1 Chlorine

16.1.1 Chloromethane in the Environment

Chloromethane is the most abundant natural chlorinated organic compound present in the atmosphere and contributes about 17% of all chlorine present in the atmosphere (Engel et al. 2019). In order to study the sources and sinks of this compound many studies have been applied to determine the chlorine isotope fractionation as the result of formation or degradation processes of chloromethane in the environment, from which some recent studies are e.g. Horst et al. (2019), Keppler et al. (2020), and Hartmann et al. (2023). Hartmann et al. (2023) analysed the chlorine isotope composition of chloromethane emitted from the royal fern *Osmunda regalis* and found that the δ^{37}Cl of chloromethane was −0.91 ± 0.84‰. In spite of this, there seem to be no studies in which chloromethane from the atmosphere was sampled and analysed comprehensively. In light of the studies on isotope fractionation that is observed during processes that produce and degrade chloromethane in the environment it is hoped that such a study will be applied in due course.

H. Eggenkamp, *The Geochemistry of Stable Chlorine and Bromine Isotopes*, Advances in Isotope Geochemistry, https://doi.org/10.1007/978-3-031-75633-7_16

16.1.2 Chlorine Isotope Variations in CFCs

The first real chlorine isotope data from stratospheric gas were presented by Laube et al. (2010) who analysed the chlorine isotope composition of atmospheric CF_2Cl_2 (CFC-12). They found that with increasing altitude the $\delta^{37}Cl$ of the CF_2Cl_2 increased from 0‰ at 14 km to +27‰ at 34 km. They explained this increase due to the faster decomposition of $CF_2{}^{35}Cl_2$ relative to $CF_2{}^{37}Cl^{35}Cl$ and $CF_2{}^{37}Cl_2$ by photolysis and the reaction with O^1D. These data indicate that, within the fairly large errors in the measurements of up to ±4‰, the lower atmosphere has a chlorine isotope composition close to SMOC, which increases to very positive values in the stratosphere due to prevalent decomposition of the molecules with ^{35}Cl bonds as compared to molecules with ^{37}Cl bonds. Based on the preliminary results published by Laube et al. (2010) further chlorine isotope research was done to measure the chlorine isotope composition of CCl_3F (CFC-11) and $CClF_2CCl2_F$ (CFC-113) on archived air samples (Allin et al. 2015). This way it was possible to obtain a long term chlorine isotope record that could help to identify and quantify the sources and sinks of these compounds. Allin et al. (2015) observed that significant stratospheric chlorine isotope fractionation exists for CFC-11 and CFC-113, just like was the case for CFC-12. Stratospheric data indicate a decrease of $\delta^{37}Cl$ over time for all three CFCs, leading to apparent isotope fractionation (ε_{app}) values for mid- and high-latitude stratospheric samples of respectively −2.4 ± 0.5 and −2.3 ± 0.4‰ for CFC-11, −12.2 ± 1.6 and −6.8 ± 0.8‰ for CFC-12 and -3.5 ± 1.5 and 3.3 ± 1.2‰ for CFC-113. In the troposphere the $\delta^{37}Cl$ variations are less and vary roughly between −4 and + 4‰ for all three CFCs. The observations made by Allin et al. (2015) are consistent with an isotopically invariant source signature, but the high measurement uncertainty, which is ±2.7‰ to ±3.8‰, does not allow the authors to preclude the possibility that it has changed over time.

16.2 Bromine

The first bromine isotope measurements of atmospheric CH_3Br samples were reported by Horst et al. (2013) and Horst (2013). They used the method developed by Thornton et al. (2013) to extract sufficient amounts of CH_3Br from atmospheric samples from sites near Stockholm (an urban site) and Abisko (a rural site) in Sweden. Further, at the Abisko site also soil-air samples were collected to determine isotopic fingerprinting and the importance of this source. The results obtained by Horst et al. (2013) and Horst (2013) indicated that, overall, the air samples showed similar $\delta^{81}Br$ values for both field studies ranging from −0.47 to +1.75‰ vs. SMOB for Stockholm samples and −0.14 to +0.48‰ vs. SMOB for Abisko samples. However, in Stockholm the $\delta^{81}Br$ range was larger and the mixing ratios of CH_3Br were higher (14.4 ± 2.1 pptv) whereas for the Abisko samples the mixing ratios (9.8 pptv ±3.1 pptv) were statistically indistinguishable from global background (8 pptv). The Stockholm samples showed a clear trend of depleted bromine isotope signatures (i.e. negative $\delta^{81}Br$ values) with higher concentrations towards enriched $\delta^{81}Br$ values with lower concentrations. This trend is suggestively due to decomposition of CH_3Br in the atmosphere and/or mixing between potential sources and the background methyl bromide. The few soil-air samples that were taken at Abisko showed lower CH_3Br concentrations than the air at the site and the $\delta^{81}Br$ values seemed to be higher than in the air samples. Due to the small number and premature character of these measurements Horst et al. (2013) and Horst (2013) were not yet able to give an explanation for these results.

In order to understand $\delta^{81}Br$ values measured in atmospheric samples, Horst et al. (2014) studied the bromine isotope composition of bromomethane that was released from plant matter into the atmosphere. Horst et al. (2014) reported the measurements of stable bromine isotopes in bromomethane from abiotic plant emissions. They measured the isotope enrichment relative to a salt substrate in KBr-fortified pectin, a ubiquitous cell-stabilizing macromolecule, and of a natural halophyte (*Salicornia fruticosa*). They found that isotope enrichment was negative for both the KBr-fortified pectin and the halophyte, −2.00 ± 0.23‰ and −1.82 ± 0.02‰ respectively at all tested temperatures if only small amounts of the initial salt content was consumed by the reaction (<10%). In experiments where larger amounts of the initial salt were consumed isotope enrichment was closer to 0, so that the isotope difference between the original salt and the produced bromomethane became less. These experiments show that negative $\delta^{81}Br$ values for atmospheric bromomethane can originate from bromomethane release from plants. To explain positive $\delta^{81}Br$ values in the atmosphere however different explanations still need to be developed.

References

Allin SJ, Laube JC, Witrant E, Kaiser J, McKenna E, Dennis P, Mulvaney R, Capron E, Martinerie P, Röckmann T, Blunier T, Schwander J, Fraser PJ, Langenfelds RL, Sturges WT (2015) Chlorine isotope composition in chlorofluorocarbons CFC-11, CFC-12 and CFC-113 in firn, stratospheric and tropospheric air. Atmos Chem Phys 15:6867–6877

Engel A, Rigby MLA, Burkholder JB, Fernandez RP, Froidevaux L, Hall BD et al (2019) Update on ozone-depleting substances (ODSs) and other gases of interest to the Montreal protocol. In: Global ozone research and monitoring project — Report No. 58. In Chapter 1 in scientific assessment of ozone depletion. World Meteorological Organization

Hartmann SC, Keppler F, Greule M, Lauer R, Horst A (2023) Triple-element stable isotope analysis of chloromethane emitted by Royal Fern and Degraded by Club Moss. J Geophys Res Biogeosci 128:e2022JG007256

Horst A (2013) Stable bromine isotopic composition of methyl bromide. Method development and applications. PhD thesis. Stockholm University

Horst A, Thornton BJ, Holmstrand H, Andersson P, Crill PM, Gustafsson Ö (2013) Stable bromine isotopic composition of atmospheric CH_3Br. Tellus Ser B Chem Phys Meteorol 65:21040

Horst A, Holmstrand H, Andersson P, Thornton BF, Wishkerman A, Keppler F, Gustafsson Ö (2014) Stable bromine isotopic composition of methyl bromide released from plant matter. Geochim Cosmochim Acta 125:186–195

Horst A, Bonifacie M, Bardoux G, Richnow HH (2019) Isotopic characterisatization (^{2}H, ^{13}C, ^{37}Cl, ^{81}Br) of abioticdegradation of methyl bromide and methyl chloride in water and implications for future studies. Environ Sci Technol 53:8813–8822

Keppler F, Barnes JD, Horst A, Bahlmann E, Luo J, Nadalig T, Greule M, Hartmann SC, Vuilleumier S (2020) Chlorine isotope fractionation of the major chloromethane degradation processes in the environment. Environ Sci Technol 54:1634–1645

Laube JC, Kaiser J, Sturges WT, Bönisch EA (2010) Chlorine isotope fractionation in the stratosphere. Science 329:1167

Tanaka N, Rye DM (1991) Chlorine in the stratosphere. Nature 353:707

Thornton B, Horst A, Carrizo D, Holmstrand H, Andersson P, Crill PM, Gustafsson Ö (2013) A high-volume cryosampler and sample purification system for bromine isotope studies of methyl bromide. J Atmos Ocean Technol 30:2095–2107

Part V

Extraterrestrial Chlorine and Bromine Isotope Ratio Variations

17 Chlorine Isotope Characteristics of the Moon

There is only limited opportunity to measure the chlorine isotope composition of lunar material. Samples exist of lunar meteorites, that have the risk that they are contaminated with terrestrial material, and of Apollo return samples. Both sample types are only available in small quantities, making it necessary that chemical and isotopic analyses are done with utmost care on as little as possible sample material. For that reason, it is seen that many chlorine isotope studies on lunar material are applied by SIMS on lunar apatite (both meteorites and return samples). A smaller number of studies have been done by IRMS on larger (bulk) samples. An important difference exist between the two types of analyses and for that reason it still is important that, in spite of the relatively large amount of sample necessary per measurement, that bulk samples are continued to be measured. Because terrestrial samples are still mostly measured by IRMS techniques the measurement of lunar bulk samples by IRMS is important to facilitate proper comparisons between samples from Earth and the Moon.

17.1 Early Lunar Sample Measurements

The first systematic stable chlorine isotope study on lunar material was published by Sharp et al. (2010a). Sharp et al. (2010a) studied the isotope composition of 12 lunar samples, which consisted of basalts and glasses. Three of these samples were apatite from basalt and these were measured by SIMS, the other samples were water-soluble leachates and silicate residues and these samples were measured by gas source IRMS. They observed that the total $\delta^{37}Cl$ range in these samples was much larger than on Earth and it ranged from about −1 to +24‰ vs. SMOC. The variations were supposedly related to fractionation due to devolatilization of anhydrous lunar basalts. The lowest $\delta^{37}Cl$ values measured in the study were close to SMOC and overlap with the range of terrestrial samples, indicating a homogeneous source for the Earth and Moon system. The samples with more positive $\delta^{37}Cl$ values could then be explained by fractionation due to degassing where the light isotope got lost. On Earth chlorine isotopes do not fractionate considerable during degassing of magmas and as a result the $\delta^{37}Cl$ variation in basalts on Earth is not very large. In the lunar samples this is explained as the result of the very low hydrogen content of lunar magmas. On Earth chlorine isotope fractionation is controlled by two factors. Firstly by the preferential loss of lighter ^{35}Cl to the vapour phase due to its higher translational velocity and vapour pressure, and secondly by the preferential incorporation of the heavier ^{37}Cl into hydrogen chloride gas due to the covalent character of the bond (Schauble et al. 2003; Sharp et al. 2010b). These two effects appear to cancel each other out due to circumstances that exist on Earth, but apparently

H. Eggenkamp, *The Geochemistry of Stable Chlorine and Bromine Isotopes*, Advances in Isotope Geochemistry, https://doi.org/10.1007/978-3-031-75633-7_17

not on the moon, where it appears that the second, hydrogen chloride fractionating effect is much smaller than on earth. If no hydrogen (in the form of water) is present in a melt the chloride can be volatilised in the form of metal chlorides such as NaCl, $ZnCl_2$ and $FeCl_2$ (Fegley 1991; Colson 1992). The bonds in the melt and in the vapour phase are similar, so that the kinetic fractionation that occurs during loss of volatiles is cancelled out by a different fractionating effect, resulting in evaporation of the lighter isotope, and thus the increase of δ^{37}Cl in the remaining magma phase. Metal chlorides have been observed as surface coatings on lunar volcanic materials (Chou et al. 1975; McKay and Wentworth 1993), consistent with deposition of metal chloride vapour phases. Thermodynamic calculations also support that metal halide vapours deliver the dominant chlorine species when the Cl/H ratio is substantially greater than 1 (Fegley and Swindle 1993). Sharp et al. (2010a) calculated the amount of chloride loss by volatilization of metal chlorides by assuming a Rayleigh type fractionation and found that a fractionation of 20‰ could be reached when 63 to 98% of chloride is lost assuming fractionation factors between 0.983 and 0.992. The observed fractionation of chloride in lunar rocks could only be possible if the hydrogen content in lunar magma was low compared to the chloride content in these magmas. Calculations by Sharp et al. (2010a) indicate that the original lunar mantle would have chloride concentrations of about 400 ppb, and considering that the hydrogen content must be lower than the chloride content in order to explain the large kinetic chloride fractionation the hydrogen content must be in the order of 10(s of) ppb, much lower (4 orders of magnitude) than the bulk Earth value of 260 ppm. Shearer et al. (2014) applied an in-depth study on some of the samples reported by Sharp et al. (2010a), although with almost no comment or interpretation, "rusty rock" 66,095 and lunar soils (samples 64,501,232 and 61,220,39) collected during the Apollo 16 mission. Based on the chemical compositions these samples contain a substantial volatile component. The chlorine isotope compositions of these volatile-rich samples are enriched in ^{37}Cl. In sample 66,095, δ^{37}Cl ranges from +14.0 to +15.6‰, whereas δ^{37}Cl for the volatile-rich Apollo 16 soils ranges from +5.6 to +15.7‰ (see also Sharp et al. 2010a). Based on the δ^{37}Cl measurements it seems likely that the volatile element enrichments and high Cl isotope values observed in both sample 66,095 and the Apollo 16 soils did not result from extra-lunar additions, but are most likely indigenous to the Moon (Shearer et al. 2014).

17.2 Chlorine Isotope Variation in Lunar Rocks as Determined by SIMS

Most studies of chlorine isotope compositions of lunar material have been done on chloride present in lunar apatite and was measured by SIMS. The information obtained from the measurement of chloride in apatite by SIMS is different from the information obtained from bulk samples, mostly because of the much smaller scale on which information is gathered. The range of δ^{37}Cl values measured in apatite is huge, showing values between −4 to +81‰ (Wang et al. 2012, 2019; Boyce et al. 2015) observed in both lunar meteorites and Apollo return samples. Interestingly, just like in bulk samples only few samples show negative values, while the range towards positive samples is largely extended. Tartèse et al. (2014) evaluated the chlorine characteristics of apatite from two lunar meteorites found in hot-desert environments (NWA 4472 and SaU 169). In these meteorites δ^{37}Cl varied from about 5 to 20‰ vs. SMOC, consistent with δ^{37}Cl variations in Mare basalts. The elevated δ^{37}Cl values suggest that apatite in these hot-desert meteorites have retained their original lunar isotopic signatures. Boyce et al. (2018) summarised all apatite chlorine isotope data published until 2018. The data were taken from Sharp et al. (2010a), Boyce et al. (2015), Barnes et al. (2016) and Barnes et al. (2019). The samples consisted of different Apollo 12, 14 and 15 high- and low-Ti and KREEP (potassium/rare earth elements/phosphorus) basalts. To these data were added chlorine isotope values from low-Ti basalts from meteorites analysed by Wang et al. (2012). The data obtained in these studies suggested a different model than anhydrous degassing of metal chlorides as proposed by Sharp et al. (2010a). Elevated $\delta^{37}Cl$ measurements were attributed to mixing between light and heavy isotopic chlorine reservoirs within the Moon. A light reservoir ($\delta^{37}Cl \leq 0‰$) was suggested to be the uncontaminated signal of mare basalts, whereas the heavy reservoir ($\delta^{37}Cl > +25‰$) was attributed to the urKREEP (primeval KREEP), the last vestige of the lunar magma ocean. In order to explore the conditions under which chlorine isotope fractionation occurred in lunar basaltic melts, Potts et al. (2018) selected five Apollo 14 crystalline samples and one basaltic clast to explore the conditions under which chlorine isotope fractionation occurred in lunar basaltic melts. Chlorine isotopes were determined from apatite with secondary ion mass spectrometry. In these samples, $\delta^{37}Cl$ ranged from +14.6 to +40.0‰. In this set of samples it was difficult to interpret the results. Potts et al. (2018) explained the small scale heterogeneity in their samples due to a late-stage interaction with a volatile rich gas phase that originated from devolatilisation of lunar surface regolith rocks ~4 billion years ago. Another study on apatite from Apollo return samples was done by McCubbin and Barnes (2020). They measured the chlorine isotope composition within intercumulus regions and within a holocrystalline olivine-hosted melt inclusion in a magnesian-suite troctolite from Apollo 17. The apatite in the olivine-hosted melt inclusion yielded a $\delta^{37}Cl$ of 28.3 ± 0.9‰ and within the intercumulus regions of the sample $\delta^{37}Cl$ was 28.3 ± 0.9‰. The similarity of the chlorine isotope values from both textural regimes indicated, according to McCubbin and Barnes (2020), that the chlorine isotope composition of apatite in their sample represents the chlorine isotope composition of the KREEP-rich magnesian-suite magmas. These results also imply that the KREEP reservoir had a chlorine isotope composition of about +28 to +30‰ relative to SMOC by at least 4.31 Ga, the approximate age of the samples. Hayden (2023) also studied the chlorine isotope composition from a set of lunar meteorites. In these meteorites the $\delta^{37}Cl$ range in apatite was from −1 to +53‰. Especially the higher $\delta^{37}Cl$ values bridge the gap between earlier data on basaltic lunar meteorites and Apollo return samples, that have $\delta^{37}Cl$ values up to +40‰, and the extreme $\delta^{37}Cl$ values observed in a single lunar meteorite (Wang et al. 2012, 2019) which were between +70 and + 80‰. This new information confirms the large heterogeneity of the existing lunar chlorine isotope variations.

To better understand the large chlorine isotope difference between the Earth and Moon Stephant et al. (2019) determined the chlorine content and isotope composition in olivine and pyroxene hosted melt intrusions from five Apollo 14 basalts. The chlorine isotope compositions of melt inclusions in these five samples showed (weighted) averages of +12.8 ± 2.4‰ for olivine- and + 10.1 ± 3.2‰ for pyroxene-hosted melt inclusions, values that were statistically indistinguishable. These values were also very similar to chlorine isotope compositions of apatites measured in the same samples. Thus, Stephant et al. (2019) concluded that no or little chlorine isotope fractionation occurred after entrapment of the melt inclusions. They concluded that in most cases the chlorine isotope composition of both melt inclusions and apatite reflect the chlorine isotope signature of the source region of the lunar basalts.

For the first time in over 40 years, the Chinese sample-return mission Chang'e 5, returned lunar soil material from the Moon to Earth in 2020. The landing site from this mission was far from the landing sites of previous Apollo and Luna sample-return landing sites and the basalt samples from the Chang'e 5 mission are dated to about 2 billion years, about 1 billion years younger than previously reported lunar basalts. The mission returned 1731 grams of lunar soil material. Interestingly, the chlorine isotope composition of part of this material is one of the first analyses done. Because chlorine isotope variations from lunar material are so much larger than chlorine isotope variations from terrestrial material, special interest in the chlorine isotope composition from material from this mission is to be expected. Ji et al. (2022) performed chlorine isotope measurements by SIMS of chlorine from apatite from the CE5 basalt previously described by Hu et al. (2021), Tian et al. (2021), Li et al. (2021) and Zhang et al. (2022). Ji et al. (2022) observed that chlorine isotope compositions of CE5 apatite vary from 4.5 to 18.9‰ and are positively correlated with the chloride abundances. These data suggest that magmatic degassing of chlorine bearing species during crystallisation of the apatite at or near the lunar surface could have resulted in a large chlorine isotope fractionation. Because the δ^{37}Cl range in the Chang'e 5 samples is similar to the δ^{37}Cl range obtained for the much older Apollo return samples this indicates that the chlorine isotope fractionation in lunar mare basalts is independent from the time when they formed. The general conclusions from the study by Ji et al. (2022) on chlorine isotopes are in line with the conclusions made on previously measured apatite grains form Apollo return samples and lunar meteorites. The new observations made by Ji et al. (2022) confirm an important role for magmatic fractionation of chlorine isotopes during crystallisation of lunar mare lavas and provide information to determine the primordial chlorine isotopic signature of the Moon.

17.3 Overview of Chlorine Isotope Variations in Apollo-Return Bulk Samples

Gargano et al. (2020) presented bulk δ^{37}Cl values from the structural and water-soluble chloride in a large set of Apollo-return samples. Water-soluble chloride constitutes about 65 ± 10% of total the total chlorine in lunar material and has δ^{37}Cl values that are on average 3.0 ± 4.3‰. The rest of the chloride is structurally bound chloride and has δ^{37}Cl values that are on average 7.3 ± 3.5‰. These chlorine isotope values are high but distinctly different from each other. The large difference between the two types of chloride are inconsistent with equilibrium fractionation processes and instead suggest early and extensive degassing of an isotopically light vapour. δ^{37}Cl values of apatite grains, determined by SIMS, are generally higher than the structurally bound chloride. Gargano et al. (2020) conclude that apatite grains are not representative of the bulk rock, and instead record localized degassing during the final stages of lunar magma ocean or later melt crystallization. The large variability of δ^{37}Cl values from apatite within individual thin sections supports this conclusion. While urKREEP has been proposed as the source of the Moon's high Cl isotope values (Boyce et al. 2018), ferroan anorthosites have the highest δ^{37}Cl values and show a positive correlation with chloride content, yet they do not contain apatite, nor evidence of a KREEP component. Gargano et al. (2020) explained the high δ^{37}Cl values in ferroan anorthosites by the incorporation of a more than 30‰ HCl vapour from a highly evolved lunar magma ocean.

References

Barnes JJ, Tartèse R, Anand M, McCubbin FM, Neal CR, Franchi IA (2016) Early degassing of lunar urKREEP by crust-breaching impact(s). Earth Planet Sci Lett 447:84–94

Barnes JJ, Franchi IA, McCubbin FM, Anand M (2019) Multiple volatile reservoirs in the lunar interior revealed by the isotopic composition of chlorine in lunar basalts. Geochim Cosmochim Acta 266:144–162

Boyce JW, Treiman AH, Guan Y, Ma C, Eiler JM, Gross J, Greenwood JP, Stolper EM (2015) The chlorine isotope fingerprint of the lunar magma ocean. Sci Adv 1:1–8

Boyce JW, Kanee SA, McCubbin FM, Barnes JJ, Bricker H, Traiman AH (2018) Early loss, fractionation, and redistribution of chlorine in the moon as revealed by the low-Ti lunar mare basalt suite. Earth Planet Sci Lett 500:205–214

Chou CL, Boynton WV, Sundberg LL, Wasson JT (1975) Volatiles on the surface of Apollo 15 green glass and trace-element distributions among Apollo 15 soils. In: Proceedings of Lunar and Planetary Science Conference, vol 2, pp 1701–1727

Colson RO (1992) Mineralization on the moon? Theoretical considerations of Apollo 16 'rusty rocks', sulfide replacement in 67016, and surface-correlated volatiles on lunar volcanic glass. Proc Lunar Planet Sci 22:427–436

Fegley BJ (1991) Thermodynamic models of the chemistry of lunar volcanic gases. Geophys Res Lett 18:2073–2076

Fegley BJ, Swindle TD (1993) Lunar volatiles, implications for lunar resource utilization. In: Lewis MSM,

Guerrieri ML (eds) Resource of Near-Earth Space. University of Arizona Press, Tucson, AZ, pp 367–426

Gargano A, Sharp Z, Shearer C, Simon JI, Halliday A, Buckley W (2020) The Cl isotope composition and halogen contents of Apollo-return samples. Proc Natl Acad Sci 117:38

Hayden TS (2023) Assessing the volatile inventory and history of the moon using lunar meteorites. PhD thesis. The Open University

Hu S, He H, Ji J, Lin Y, Hui H, Anand M, Tartese R, Yan Y, Hao J, Li R, Gu L, Guo Q, He H, Ouyang Z (2021) A dry lunar mantle reservoir for young mare basalts of Chang'e-5. Nature 600:49–53

Ji JL, He HC, Hu S, Lin YT, Hui HJ, Hao JL, Li RY, Yang W, Yan YH, Tian HC, Zhang C, Anand M, Tartèse R, Gu LX, Li JH, Zhang D, Mao Q, Jia LH, Chen Y, Wu ST, Wang H, He HY, Li XH, Wu FY (2022) Magmatic chlorine isotope fractionation recorded in apatite from Chang'e-5 basalts. Earth Planet Sci Lett 591:117636

Li QL, Zhou Q, Liu Y, Xiao Z, Lin Y, Li JH, Ma HX, Tang GQ, Guo S, Tang X, Yuan JY, Li J, Wu FY, Ouyang Z, Li C, Li XH (2021) Two-billion-year-old volcanism on the moon from Chang'e-5 basalts. Nature 600:54–58

McCubbin FM, Barnes JJ (2020) The chlorine-isotopic composition of lunar KREEP from magnesian-suite troctolite 76535. Am Mineral 105:1270–1274

McKay DS, Wentworth SJ (1993) Grain surface features of Apollo 17 orange and black glass. In: Lunar Planet Sci Conf, vol 24, pp 961–962

Potts NJ, Barnes JJ, Tartèse R, Franchi IA, Anand M (2018) Chlorine isotopic compositions of apatite in Apollo 14 rocks: evidence for widespread vapor-phase metasomatism on the lunar nearside ~4 billion years ago. Geochim Cosmochim Acta 230:46–59

Schauble EA, Rossman GR, Taylor HP (2003) Theoretical estimates of equilibrium chlorine-isotope fractionations. Geochim Cosmochim Acta 67:3267–3281

Sharp ZD, Shearer CK, McKeagan KD, Barnes JD, Wang YQ (2010a) The chlorine isotope composition of the Moon and implications for an anhydrous mantle. Science 329:1050–1053

Sharp ZD, Barnes JD, Fischer TP, Halick M (2010b) An experimental determination of chlorine isotope fractionation in acid systems and applications to volcanic fumaroles. Geochim Cosmochim Acta 74:264–273

Shearer CK, Sharp ZD, Burger PV, McCubbin FM, Provencio PP, Brearly AJ, Steel A (2014) Chlorine distribution and its isotopic composition in "rusty rock" 66095. Implications for volatile element enrichments of "rusty rock" and lunar soils, origin of "rusty" alteration, and volatile element behavior on the moon. Geochim Cosmochim Acta 139:411–433

Stephant A, Anand M, Zhao XC, Chan QHS, Bonifacie M, Franchi IA (2019) The chlorine isotopic composition of the moon: insights from melt inclusions. Earth Planet Sci Lett 523:115715

Tartèse R, Anand M, Joy KH, Franchi IA (2014) H and Cl isotope systematics of apatite in brecciated lunar meteorites Northwest Africa 4472, Northwest Africa 773, Sayh al Uhaymir 169, and Kalahari 009. Meteorit Planet Sci 2014:1–24

Tian HC, Wang H, Chen Y, Yang W, Zhou Q, Zhang C, Lin HL, Huang C, Wu ST, Jia LH, Xu L, Zhang D, Li XG, Chang R, Yang YH, Xie LW, Zhang DP, Zhang GL, Yang SH, Wu FY (2021) Non-KREEP origin for Chang'e-5 basalts in the Procellarum KREEP terrane. Nature 600:59–63

Wang Y, Guan Y, Hsu W, Eiler JM (2012) Water content, chlorine and hydrogen isotope compositions of lunar apatite. In: 75th Annual Meteoritical Society Meeting. Abs 5170

Wang Y, Hsu WB, Guan YB (2019) An extremely heavy chlorine reservoir in the moon: insights from the apatite in lunar meteorites. Sci Rep 9:7527

Zhang D, Su B, Chen Y, Yang W, Mao Q, Jia LH (2022) Titanium in olivine reveals low-Ti origin of the Chang'E-5 lunar basalts. Lithos 414–415:106639

18 Chlorine Isotope Variations on Mars

Although no samples have yet been brought directly from Mars by any sample return mission, still a significant amount of chlorine isotope research has been done on Martian material and Mars itself. In part this research has been done on Earth on Martian meteorites, but for a large part the research has also been applied directly on the surface and atmosphere from Mars trough landers and satellites that are or were operating on the planet itself.

18.1 Martian Meteorites

As no Martian rocks have been returned to Earth yet, knowledge on the chlorine isotope composition of Mars must be concluded from other means. The first chlorine isotope measurements on Marian material were made on Martian meteorites. Sharp et al. (2011) presented the chlorine isotope compositions of eight Martian meteorites. The $\delta^{37}Cl$ values of these samples varied between −3 and + 2‰. It was observed that shergottites had negative $\delta^{37}Cl$ values while cumulates had positive values. These preliminary results were explained as the result of near-surface contamination or volatile loss.

18.1.1 Bulk Chlorine Isotope Measurements on Martian Meteorites

In later years more Martian meteorites were analysed resulting in a better understanding of the chlorine isotope characteristics from Mars. These meteorites provided a unique opportunity to learn about interactions between the Martian mantle magmas and the Martian crust (Williams et al. 2016). Williams et al. (2016) analysed the chlorine isotope composition of 17 Martian meteorites of very different environments, ages and mantle reservoirs. They found that the Martian mantle, as characterised by shergottites (olivine-phyric and basaltic), has a chlorine isotope value as low as −3.8‰ relative to SMOC, the lowest value encountered for any planetary body at the time of publication of this study. The Martian crust, represented by a regolith breccia, has values that are mostly just above SMOC (values between +0.1 and + 1.5‰, with one measurement as high as +8.6‰. These high values were explained by preferential loss of ^{35}Cl from the crust to space. The other meteorites have $\delta^{37}Cl$ values that are mostly intermediate between these extreme values. This intermediate range was

H. Eggenkamp, *The Geochemistry of Stable Chlorine and Bromine Isotopes*, Advances in Isotope Geochemistry, https://doi.org/10.1007/978-3-031-75633-7_18

explained by a two component mechanical mixing system, where impacts, fluid interaction and assimilation-fractional crystallisation are the main causes of the mixing of both end-members (Williams 2016; Williams et al. 2016; Sharp et al. 2016). Sharp et al. (2016) observed that the very low $\delta^{37}Cl$ values for the Martian mantle is different from Earth and most chondrites, both of which are close to 0‰. Sharp et al. (2016) state that no process is known that chlorine isotope values can decrease to −3.8‰, and for that reason they suggest that this low $\delta^{37}Cl$ value represents the primordial bulk composition that Mars inherited during accretion. They further suggest that the higher $\delta^{37}Cl$ values seen in many chondrites are explained by later incorporation of ^{37}Cl-enriched HCl-hydrate. A large study was made from the Martian meteorite Northwest Africa (NWA) 8159, which is considered a unique Martian basalt (Herd et al. 2017; Shearer et al. 2019). From these studies, Shearer et al. (2019) used, amongst others, the chlorine isotope composition of this meteorite to explore the interaction between mantle-derived magmas and the Martian crust. The bulk $\delta^{37}Cl$ of NWA 8159 is +1.5‰ (Williams et al. 2016). Previous studies indicated that the Martian mantle is isotopically light (−3 to −6‰), whereas the Martian crust is isotopically heavy (>0‰), while intermediate values were produced by the interaction between these two reservoirs (Williams 2016; Williams et al. 2016; Sharp et al. 2016; Shearer et al. 2018). The high $\delta^{37}Cl$ value found in meteorite NWA 8159 clearly indicates a large crustal and atmospheric imprint on the composition of this meteorite.

18.1.2 SIMS Measurements on Martian Meteorites

Compared to IRMS, with the use of SIMS much smaller samples can be analysed, while also small scale isotope variations within individual specimens can easily be studied. The first comprehensive study on Martian meteorites with SIMS analyses was presented by Belluci et al. (2017). Belluci et al. (2017) measured the chlorine isotope compositions of apatite [$Ca_5(PO_4)_3(F,Cl,OH)$] and merrillite [$Ca_9NaMg(PO_4)_7$] from eight Martian meteorites. In apatite the chloride content can be very high in chlor-apatite, but also very low in fluor- and hydroxyl-apatites. In merrillite the chloride content is normally rather low, and only rarely it is above 100 ppm. Distinct differences in chlorine isotope compositions of phosphates were observed between very old (>4 Ga) and young (<600 Ma) basaltic meteorites. The old meteorites show positive $\delta^{37}Cl$ values, between +1.1 and + 2.5‰, while the younger meteorites have negative $\delta^{37}Cl$ values, between −5.6 and −0.2‰. The most negative $\delta^{37}Cl$ values are observed in zoned grains, where the lowest $\delta^{37}Cl$ values were found at the rims suggesting interaction with the surface of Mars during the latest stages of basalt crystallisation. Belluci et al. (2017) observed that phosphates with no textural, major element, or halogen enrichment evidence for mixing with this surface reservoir have an average $\delta^{37}Cl$ of −0.6‰. This supports the hypothesis for a similar initial Cl isotope composition for Mars, the Earth, and the Moon. At the surface of Mars perchlorate has been detected at weight percent concentrations (Hecht et al. 2009; Catling et al. 2010; Carrier and Kounaves 2015), and the redox reactions responsible for the formation of perchlorate (from either chloride or chlorine) are the only known reactions that can fractionate chlorine isotopes substantially. The chlorine isotope data obtained by Belluci et al. (2017), with a range that is larger than exists on Earth, suggest that perchlorate formation and halogen cycling via brines, which have been documented on the Martian surface, has been active throughout Martian history. Further chlorine isotope research on apatite from a single Martian meteorite (Chassigny) also showed a large chlorine isotope range. The range in this single meteorite showed to be about 10‰, which was only little less than the total range in all meteorites measured before then (Shearer et al. 2018). Shearer et al. (2018) concluded that this range was the result of the mixing of two reservoirs, where representatives of both reservoirs were present in this single meteorite. The two reservoirs were an isotopically light mantle reservoir ($-6‰ < \delta^{37}Cl < -4‰$) with little variability, and an isotopically heavy crustal reservoir ($\delta^{37}Cl > 0‰$) with significant isotope variability. Shearer et al. (2018) noticed that the low $\delta^{37}Cl$ value of the Martian mantle is different from that of the Earth, but also from the Moon and many chondritic meteorites, that show values much closer to 0‰. They suggest that the low $\delta^{37}Cl$ values from the Martian mantle represent the $\delta^{37}Cl$ of the primordial solar nebula from which Mars accreted, while the higher values observed on Earth, the Moon and many chondritic meteorites do not represent the primordial solar nebula, but rather the later incorporation of ^{37}Cl-enriched HCl-hydrates into the accreting material. The high $\delta^{37}Cl$ values of the Martian crust are explained by Shearer et al. (2018) though protracted loss of ^{35}Cl into space, a process that must have started already very early in the Martian history.

Hu et al. (2019) studied the chlorine isotope composition of a single apatite grain from Martian meteorite NWA 7034. They found low $\delta^{37}Cl$ values at the rim of this grain (approximately −2‰) and high $\delta^{37}Cl$ values in the centre of this grain (up to +4‰), and this effect was negatively correlated with the water content in this grain (0.09% at the rim and 0.02% in the centre). While the U-Pb age of zircons in this meteorite is about 4.5 Ga, the U-Pb age of the apatites is about 1.5 Ga. It was suggested that a metamorphic event took place about 1.5 Ga ago, that reset the U-Pb system and that introduced a very ^{37}Cl rich fluid that was responsible for the formation of the apatite grain. The low $\delta^{37}Cl$ values at the rim of the grain, with low water content, then suggest degassing of water at a later time.

With nanoSIMS it is possible to go to even lower spatial resolutions than with ordinary SIMS with measurement areas as small as 5x5 μm^2. The lowest spatial resolution can be reached with Atom Probe Tomography (APT; Gordon et al. 2012) and with individual atom counts (^{35}Cl and ^{37}Cl) and calculated atom ratios it was possible to measure the chlorine isotope composition at nanoscale boundaries (Darling et al. 2021). While Darling et al. (2021) observed rather normal $\delta^{37}Cl$ values in apatite and merrilite from the Martian meteorite NWA 5298 (−3.15 to +1.08‰) with nanoSIMS, much larger variations were observed with the use of APT at boundaries. APT measurements at bulk locations revealed $\delta^{37}Cl$ values between −12.5 and + 6.4‰, a range that, although significantly larger than the range measured by SIMS, is still within values that are rarely obtained with other analytical methods. At the nanoscale boundaries however, $\delta^{37}Cl$ values between −45.3 and −12.0‰ were measured. Especially the lowest of these values are rather extreme. Darling et al. (2021) explain these data as the result of shock metamorphism, principally as a result of chlorine mobilisation during shock melting and recrystallization. Overall however the Martian meteorite NWA 5298 has an average $\delta^{37}Cl$ value of $-0.3 \pm 0.6‰$.

18.2 In-Situ Mass Spectrometric Analyses of Martian Sediments

Onboard the Curiosity Rover, that traverses through the Gale Crater on Mars, the Sample Analysis at Mars instrument (SAM; Mahaffy et al. 2012) is present for chemical analysis. Part of this instrument is a quadrupole mass spectrometer that measures individual mass intensities. HCl is one of the most abundant species that is observed during evolved gas analysis (EGA) of every sample, typically at levels of approximately 10 μmol/g. Most and possibly all of this chlorine derives from the decomposition of oxidised chlorine species, of which the perchlorate ion is the most obvious candidate (Glavin et al. 2013). This perchlorate accounts for a substantial fraction of the total Cl in these rocks (Archer et al. 2015) and is thought to be an ancient constituent (Ming et al. 2014). Although it is possible that additional HCl could be formed from other (chloride) minerals in the sample (such as halite or akaganeite), these minerals are present in low quantities only and it is unknown whether these minerals can release HCl under the analytical conditions adopted for the SAM (Glavin et al. 2013). The presence of oxychlorine compounds including perchlorate was recently reported in a

Martian meteorite (Kounaves et al. 2014). The isotopic composition of chlorine can be measured in the evolved HCl in the SAM quadrupole mass spectrometer (QMS). Although corrections for isobaric species preclude high precision isotope ratio determinations, variability is nevertheless apparent. Farley et al. (2016) reported unexpectedly light and highly variable $^{37}Cl/^{35}Cl$ ratios measured by the quadrupole mass spectrometer of the SAM in this evolved HCl that, as stated above, most probably originates from perchlorates present in the Martian sediment. The total range of $\delta^{37}Cl$ values measured was from −51 to −1‰ but standard deviations of the measurements are generally rather large. In most cases however they are still much smaller than the chlorine isotope variations. Clear differences are observed for the chlorine isotope composition at different sample locations in the Gale Crater. While $\delta^{37}Cl$ values in the Yelloknife Bay area are modestly negative with values between −20 and −1‰ (and are analytically indistinguishably with a mean $\delta^{37}Cl$ of −11 ± 7‰), in the Pahrup and Kimberly areas they are much more negative with values between −41 and −37‰ in the Pahrup area and between −51 and −49‰ in the Kimberley area. Taking these last two areas together they show analytically indistinguishable values with a mean $\delta^{37}Cl$ of −43 ± 6‰. Farley et al. (2016) consider two hypotheses to explain these highly negative $\delta^{37}Cl$ values. (1) SAM is detecting HCl from perchlorate (or chlorate) that is produced from volcanic gas emissions by atmospheric chemical reactions. Similar reactions may be responsible on Earth for the isotopically lightest known chlorine from Earth in perchlorate from the Atacama Desert. Some of the Gale Crater $\delta^{37}Cl$ values are lower than those in Atacama perchlorate, but at the moment of writing their study, Farley et al. (2016) were unable to assess whether this difference is prohibitive. If the negative $\delta^{37}Cl$ signal is produced due to volcanic emissions followed by oxidation processes, the isotope variations between samples could arise from variations in the relative sizes of the reactant chloride and product perchlorate reservoirs, or from variations in the fraction of perchlorate that is reduced back to chloride after its deposition. This reduction process enriches ^{37}Cl in the residual perchlorate significantly. (2) An alternative end-member model only takes into account perchlorate reduction. This model can explain the observed data if SAM measured HCl derived from chloride. In this model perchlorate with a $\delta^{37}Cl$ close to 0‰ and that was produced by an as yet unknown mechanism is reduced to chloride. Depending on the size of the reduced reservoir, the produced chloride can vary in isotope composition from approximately −70‰ in the first increment (Urey 1947) all the way to the starting $\delta^{37}Cl$ value when the perchlorate is completely reduced. Thus, variable degrees of perchlorate reduction can produce chloride with the whole observed $\delta^{37}Cl$ range. Of course, a combination of both end-member models is also possible. Farley et al. (2016) state that the determination of the phase that hosts the chlorine measured by SAM, either perchlorate, chlorate or chloride, is critical for selecting between the models. At the moment of writing their paper, Farley et al. (2016) were not able to establish which of the models (or both) they proposed was responsible for the observed chlorine isotope variations. Based on the very limited evidence they had seen they believe however that the chlorine from which they measured the chlorine isotope composition derived for the most part from perchlorate (or chlorate).

18.3 The Chlorine Isotope Composition of the Martian Atmosphere

The Trace Gas Orbiter Atmospheric Chemistry Suite (ACS) and the Nadir Occultation for MArs Discovery (NOMAD) spectrometers reported recently for the first time the presence of HCl in the Martian atmosphere (Aoki et al. 2021; Korablev et al. 2021). This discovery immediately led to attempts to measure the chlorine isotope composition of this HCl. Two such studies appeared independently of each other in early 2021 (Liuzzi et al. 2021; Trokhimovskiy et al. 2021). Liuzzi et al. (2021) determined a chlorine isotope value of −6 ± 78‰ while Trokhimovskiy et al. (2021) determined a value of −7 ± 20‰. Because of the large uncertainty in the value, both determinations are indistinguishably from a terrestrial chlorine isotope composition. However, as both determinations appear to lend to slightly negative values it is possible that the HCl in the Martian atmosphere descends from the low δ^{37}Cl perchlorate present at the Martian surface as determined earlier by Farley et al. (2016).

References

Aoki S, Daerden F, Viscardy S, Thomas IR, Erwin JT, Robert S, Trompet L, Neary GL, Villanueva G, Liuzzi M, Crismani MJ, Lancy RT, Whiteway J, Schmidt F, Lopez-Valverde MA, Ristic B, Patel MR, Bellucci G, Lopez-Moreno JJ, Olsen KS, Lefèvre F, Montmessin F, Trokhimovskiy A, Fedorova AA, Korablev O, Vandaele AC (2021) Annual appearance of hydrogen chloride on Mars and a striking similarity with the water vapor vertical distribution observed by TGO/NOMAD. Geophys Res Lett 48:e2021GL092506

Archer PD, Ming DW, Sutter B, Morris RV, Clark BC, Mahaffy PH, Wray JJ, Fairen AG, Gellert R, Yen AS, Blake DF, Vaniman DT, Glavin DT, Eigenbrode JL, Trainer MG, Navarro-Gonzalez R, McKay CP, Freissinet C (2015) Oxychlorine species on mars: the gale crater story. In: 46th lunar and planetary science conference, 2971

Belluci JJ, Whitehouse MJ, John T, Nemchin AA, Snape JF, Bland PA, Benedix GK (2017) Halogen and Cl isotopic systematics in Martian phosphates: implications for the Cl cycle and surface halogen reservoirs on Mars. Earth Planet Sci Lett 458:192–202

Carrier BL, Kounaves SP (2015) The origins of perchlorate in the Martian soil. Geophys Res Lett 42:3739–3745

Catling DC, Claire MW, Zahnle KJ, Quinn RC, Clark BC, Hecht MH, Kounaves S (2010) Atmospheric origins of perchlorate on Mars and in the Atacama. J Geophys Res 115:E00E11

Darling JR, White LF, Kizovski T, Černok A, Moser DE, Tait KT, Dunlop J, Langelier B, Douglas JO, Zhao X, Franchi IA, Anand M (2021) The shocking state of apatite and merrillite in shergottite Northwest Africa 5298 and extreme nanoscale chlorine isotope variability revealed by atom probe tomography. Geochim Cosmochim Acta 293:422–437

Farley KA, Martin Jr P, Archer PD, Atreya SK, Conrad PG, Eigenbrode JL, Fairén AG, Franz HB, Freissinet C, Glavin DP, Mahaffy PR, Malespin C, Ming DW, Navarro-Gonzalez R, Sutter B (2016) Light and variable ^{37}Cl/^{35}Cl ratios in rocks from gale crater, Mars: possible signature of perchlorate. Earth Planet Sci Lett 438:14–24

Glavin DP, Freissinet C, Miller KE, Eigenbrode JL, Brunner AE, Buch A, Sutter B, Archer PD, Atreya SK, Brinckerhoff WB, Cabane M, Coll P, Conrad PG, Coscia D, Dworkin JP, Franz HB, Grotzinger JP, Leshin LA, Martin MG, McKay C, Ming DW, Navarro-Gonzalez R, Pavlov A, Steele A, Summons RE, Szopa C, Teinturier S, Mahaffy PR (2013) Evidence for perchlorates and the origin of chlorinated hydrocarbons detected by SAM at the Rocknest aeolian deposit in gale crater. J Geophys Res Planets 118:1955–1973

Gordon LM, Tran L, Joester D (2012) Atom probe tomography of apatites and bone-type mineralized tissues. ACS Nano 6:10667–10675

Hecht MH, Kounaves SP, Quinn RC, West SJ, Young SMM, Ming DW, Catling DC, Clark BC, Boynton WV, Hoffman J, DeFlores LP, Gospodinova K, Kapit J, Smith PH (2009) Detection of perchlorate and the soluble chemistry of Martian soil at the Phoenix lander site. Science 325:64–67

Herd CDK, Walton EWL, Agee CB, Muttik N, Ziegler K, Shearer CK, Bell AS, Santos AS, Burger PV, Simon JI, Tappa MJ, McCubbin FM, Gattacceca J, Lagroix F, Sanborn ME, Yin QZ, Cassata WS, Borg LE, Lindvall RE, Kruijer TS, Brennecka GA, Kleine T (2017) The Northwest Africa 8159 Martian meteorite: expanding the Martian sample suite to the early Amazonian. Geochim Cosmochim Acta 218:1–26

Hu S, Lin YT, Zhang JC, Hao JL, Xing WF, Zhang T, Yang W, Changela H (2019) Ancient geologic events on Mars revealed by zircons and apatites from the Martian regolith breccia NWA 7034. Meteorit Planet Sci 54:850–879

Korablev O, Olsen KS, Trokhimovskiy A, Lefèvre F, Montmessin F, Fedorova AA, Toplis MJ, Alday J, Belyaev DA, Patrakeev A, Ignatiev NI, Shakun AV, Grigoriev AV, Baggio L, Abdenour I, Lacombe G, Ivanov YS, Aoki S, Thomas IR, Daerden F, Ristic B, Erwin JT, Patel M, Bellucci G, Lopez-Moreno JJ, Vandaele AC (2021) Transient HCl in the atmosphere of Mars. Sci Adv 7:eabe4386

Kounaves SP, Carrier BL, O'Neil GD, Stroble ST, Claire MW (2014) Evidence of martian perchlorate, chlorate, and nitrate in Mars meteorite EETA79001: implications for oxidants and organics. Icarus 229:206–213

Liuzzi G, Villanueva GL, Viscardy S, Mège D, Crismani MMJ, Aoki S, Gurgurewicz J, Tesson PA, Mumma MJ, Smith MD, Faggi S, Kofman SV, Knutsen EW, Daerden F, Neary L, Schmidt F, Trompet L, Erwin JT, Robert S, Thomas IR, Ristic B, Bellucci G, Lopez-Moreno JJ, Patel MR, Vandaele AC (2021) Probing the atmospheric Cl isotopic ratio on mars: implications for planetary evolution and atmospheric chemistry. Geophys Res Lett 48:e2021GL092650

Mahaffy PR, Webster CR, Cabane M, Conrad PG, Coll P, Atreya SK, Arvey R, Barciniak M, Benna M, Bleacher L, Brinckerhoff WB, Eigenbrode JL, Carignan D, Cascia M, Chalmers RA, Dworkin JP, Errigo T, Everson P, Franz H, Farley R, Feng S, Frazier G, Freissinet C, Glavin DP, Harpold DN, Hawk D, Holmes V, Johnson CS, Jones A, Jordan P, Kellogg J, Lewis J, Lyness E, Malespin CA, Martin DK, Maurer J, McAdam AC, McLennan D, Nolan TJ, Noriega M,

Pavlov AA, Prats B, Raaen E, Sheinman O, Sheppard D, Smith J, Stern JC, Tan F, Trainer M, Ming DW, Morris RV, Jones J, Gundersen C, Steele A, Wray J, Botta O, Leshin LA, Owen T, Battel S, Jakosky BM, Manning H, Squyres S, Navarro-Gonzalez R, McKay CP, Raulin F, Sternberg R, Buch A, Sorensen P, Kline-Schoder R, Coscia D, Szopa C, Teinturier S, Baffes C, Feldman J, Flesch G, Forouhar S, Garcia R, Keymeulen D, Woodward S, Block BP, Arnett K, Miller R, Edmonson C, Gorevan S, Mumm E (2012) The sample analysis at Mars investigation and instrument suite. Space Sci Rev 170:401–478

Ming DW, Archer PD, Glavin DP, Eigenbrode JL, Franz HB, Sutter B, Brunner AE, Stern JC, Freissinet C, AC MA, Mahaffy PR, Cabane M, Coll P, Campbell JL, Atreya SK, Niles PB, Bell JF, Bish DL, Brinckerhoff WB, Buch A, Conrad PG, Des Marais DJ, Ehlmann BL, Fairen AG, Farley K, Flesch GJ, Francois P, Gellert R, Grant JA, Grotzinger JP, Gupta S, Herkenhoff KE, Hurowitz JA, Leshin LA, Lewis KW, McLennan SM, Miller KE, Moersch J, Morris RV, Navarro-Gonzalez R, Pavlov AA, Perrett GM, Pradler I, Squyres SW, Summons RE, Steele A, Stolper EM, Sumner DY, Szopa C, Teinturier S, Trainer MG, Treiman AH, Vaniman DT, Vasavada AR, Webster CR, Wray JJ, Yingst RA, MSL Science Team (2014) Volatile and organic compositions of sedimentary rocks in Yellowknife Bay, gale crater. Mar Sci 343:1245267

Sharp ZD, Shearer CK, FM MC, Agee CB (2011) The Cl isotope composition of the Moon and Mars. In: 21th Annual V.M. Goldschmidt conference, vol 75. Mineral Mag, Prague, Czech Republic, p 1848

Sharp Z, Williams J, Shearer C, Agee C, McKeegan K (2016) The chlorine isotope composition of Martian meteorites 2. Implications for the early solar system and the formation of Mars. Meteorit Planet Sci 51:2111–2126

Shearer CK, Messenger S, Sharp ZD, Burger PV, Nguyen AN, McCubbin FM (2018) Distinct chlorine isotopic reservoirs on Mars. Implications for character, extent and relative timing of crustal interactions with mantle-derived magmas, evolution of the martian atmosphere, and the building blocks of an early Mars. Geochim Cosmochim Acta 234:24–36

Shearer CK, Bell AS, Herd CDK, Burger PV, Provencio P, Sharp ZD, Papike JJ (2019) The Northwest Africa 8159 (NWA 8159) Martian meteorite part 2. Spinel-orthopyroxene intergrowths. A record of fO_2 and crust-basalt interactions. Geochim Cosmochim Acta 258:242–257

Trokhimovskiy A, Fedorova AA, Olsen KS, Alday J, Korablev O, Montmessin F, Lefèvre F, Patrakeev A, Belyaev D, Shakun AV (2021) Isotopes of chlorine from HCl in the Martian atmosphere. Astron Astrophys 651:A32

Urey HC (1947) The thermodynamic properties of isotopic substances. J Chem Soc 1947:562–581

Williams JT (2016) Cl in the early solar system. MSc thesis Univ New Mexico

Williams JT, Shearer CK, Sharp ZD, Burger PV, McCubbin FM, Santos AR, Agee CB, McKeegan KD (2016) The chlorine isotopic composition of Martian meteorites 1: Chlorine isotope composition of Martian mantle and crustal reservoirs and their interactions. Meteorit Planet Sci 51:2092–2110

19 Chlorine Isotope Characteristics of Other Solar System Bodies

Apart from the Earth, the Moon and Mars chlorine isotope measurements have been made on a variety of other solar system bodies. Most significantly, these measurements were made on various types of meteorites found on Earth. Apart from general meteorites, chlorine isotopes are also determined on meteorites that originated from the planetoid 4 Vesta that were found on Earth, in situ in the coma of comet 67P/Churyumov-Gerasimenko and in the atmosphere of Venus with ground-based telescopy from Earth.

19.1 Meteorites

Meteorites are solid bodies found on Earth and that originate from outer space. They can have various chemical compositions, varying from mostly rock, to mostly nickel-iron. Intermediate compositions are also known. The classification of meteorites is, due to the large variation in chemical compositions very complex (Krot et al. 2007). Most meteorites, about 90% of the total, are stony. All iron and stony-iron meteorites, and a small number of the stony meteorites, are differentiated. This means that they are mixtures of different components that are produced by crystallisation from a melt in planetary or asteroidal interiors (Hutchinson 1989). All other meteorites, the chondrites, have not undergone melting on a planetary or asteroidal scale since their formation about 4.5 Ga ago. As such these meteorites are considered as representative of the original undifferentiated composition of the solar nebula. Their chlorine isotope composition thus is also considered as representative of the original, which is currently the average, chlorine isotope value of the solar nebula. For that reason the determination of the chlorine isotope composition of chondrites has been considered as very important with regard to our understanding of the chorine isotope composition of the solar system in general.

19.1.1 Chondrites (Undifferentiated Stony Meteorites)

Already shortly after the discovery of the chlorine isotopes and the determination of the isotope ratio between ^{35}Cl and ^{37}Cl interest arose to determine this ratio in meteorites too in order to test whether the extraterrestrial chlorine isotope ratio is different from it on Earth. While it was expected that no chlorine isotope ratio variations would exist on Earth, because on Earth thorough mixing of chlorine exists, in meteoric chlorine this was not expected, especially that it was not mixed with terrestrial chlorine (Harkins and Stone 1926). Harkins and Stone (1925a, 1925b, 1926) measured the chlorine isotope ratio in a meteorite that was obtained from the National Museum, and from which the Museum stated that the chlorine was certainly of meteoric origin. Harkins

H. Eggenkamp, *The Geochemistry of Stable Chlorine and Bromine Isotopes*, Advances in Isotope Geochemistry, https://doi.org/10.1007/978-3-031-75633-7_19

and Stone (1925a, 1925b, 1926) found that the atomic weight of meteoric chlorine was (within error) exactly equal to the terrestrial chlorine atomic weight. The most detailed results were presented in Harkins and Stone (1926) and show that the atomic weight of chlorine from this meteorite was 35.457 ± 0.002 (1SD) and for terrestrial materials (HCl, sodalite, wernerite and apatite) it also was 35.457 ± 0.002 (1SD). Data are given here in three decimal places as this was the precision with which Harkins and Stone (1925a, 1925b, 1926) could determine the atomic weight of chlorine. Hoering and Parker (1961) also determined the chlorine isotope composition of four stony meteorites. The results, between −0.7 and + 0.7‰, were in line with their analyses of terrestrial materials. Because all data were within 1‰ from the standard, and very close to the error of the measurement (defined in the meteorite samples as ±0.5‰), the chlorine isotope variation was, at the time of Hoering and Parker (1961), not considered to be significant.

The earliest chlorine isotope variations in meteorites were observed by Magenheim et al. (1994, 1995). Their data, with δ^{37}Cl values in the range between +2.7 and + 4‰ were at the time of publication considered as being intermediate between mantle and crust values and were applied to determine differentiation of chlorine isotopes between the mantle and the crust of the Earth. This, because it was considered that the chlorine isotope composition in these meteorites represented primordial chloride. These measurements, which were determined by the method developed by Magenheim et al. (1994), are currently not considered as valid anymore due to possible artefacts of the method, as discussed earlier in this book. Other early chlorine isotope measurements were done on water soluble chloride and halite from the ZAG-II chondrite (Bridges et al. 2004). These samples, also analysed by P-TIMS (Cs_2Cl^+) but at the University of Leeds, showed negative values between −2.8 and −1.4‰. These values were explained as either a second reservoir from the early solar system, or large secondary fractionation due to low-temperature fluid activity. Nakamura et al. (2007) presented an overview of 24 bulk chondrites and observed that O-chondrites and E-chondrites have a negative correlation between the chloride concentration and chlorine isotope compositions, while C-chondrites had a positive correlation. They found a very large range in chlorine isotope compositions, from −4.6 to +5.0‰, analysed with (P-)TIMS. As this technique was later considered as possibly delivering false results it is not known if these data would still stand.

Sharp et al. (2007) analysed the chlorine isotope composition of chondrites by gas-IRMS (CH_3Cl) analysis. Larger samples were analysed by dual inlet, smaller by continuous flow. They analysed five chondrites, and found that their δ^{37}Cl values were relatively close to 0‰, although with a fairly large spread (−0.71 to +1.21‰), from which they concluded that the average δ^{37}Cl value of meteorites was indistinguishable from the Earth's average chlorine isotope composition. Sharp et al. (2007) also analysed 10 sodalite samples from the meteorite Allende. These samples were analysed by Secondary Ion Mass Spectrometry (SIMS) using a Cameca 1270 ion microprobe at CNRS, CRPG, Vandoeuvre-les-Nancy, France. These samples were all significantly negative relative to SMOC with an average value of −1.3‰ (the range of values was between −2.09 and −0.39‰). Two possible explanations were given for these low values. Either a second nebular chlorine reservoir, or, because of equilibrium isotope fractionation during the formation of sodalite from nepheline and sodium chloride, a secondary isotope fractionation. The experimentally determined fractionation between sodalite and sodium chloride indeed suggested negative δ^{37}Cl values for sodalite relative to sodium chloride. As this fractionation ($10^3 \ln\alpha$) was experimentally determined as −0.3 at 825 °C, which would logically increase at lower temperatures, values around −1‰ for sodalite in the meteorite Allende, seemed reasonable values obtained due to this equilibrium fractionation process.

A further set of 19 chondrites were analysed by Sharp et al. (2013). They analysed carbonaceous, ordinary and enstatite chondrites, both in altered and unaltered form. They also analysed the structurally bound and water soluble chloride separately. It was found that the bulk chlorine isotope composition of the least altered chondrites showed average values of −0.3 ± 0.5‰. The structurally bound values of the chloride in these chondrites is −0.3 ± 0.3‰. These values are indistinguishable from the bulk chlorine isotope composition of the Earth as determined by Sharp et al. (2007, 2013) and are comparable to the lowest $\delta^{37}Cl$ values that are determined on the Moon (Sharp et al. 2010). For this reason Sharp et al. (2013) concluded that the inner regions of the early solar nebula had a homogeneous chlorine isotope reservoir. Sharp et al. (2013) did find larger variations in meteorites that had undergone secondary processing. Samples that were metamorphosed showed bulk $\delta^{37}Cl$ values between −1.2 and + 0.8‰. Samples that were aqueously altered showed values between −0.3 and + 0.3‰. The averages of both groupings are very close to 0‰ vs. SMOC. These larger variations are explained by fractionation processes which are similar to processes that occur on Earth. Sharp et al. (2013) also didn't observe any relationship between the chloride concentrations and the chlorine isotope compositions. Their general conclusion is that no chlorine isotope variations existed in the primary solar nebula (thus that it was homogeneous) and that all chlorine isotope variations in meteorites, but also on Earth are the result of secondary fractionation. While $\delta^{37}Cl$ values in most chondrites measured in Sharp et al. (2013) were close to SMOC, one of the measured chondrites, Parnallee, showed an exceptionally low $\delta^{37}Cl$ value of −4.1 ± 0.1‰. Sharp et al. (2013) explained this low value as follows; a fraction of HCl hydrate was formed in a cold region of the nebula at about 140–160 K. This fraction was subsequently removed from the system, leaving a residual HCl gas that was depleted in ^{37}Cl. Incorporation of this light HCl gas in the host of Parnallee could explain the low $\delta^{37}Cl$ value of this sample. A comparatively low $\delta^{37}Cl$ value of −4.5‰ was observed in the Marian meteorite NWA 8276 (Sharp et al. 2016; Williams et al. 2016). These low values were interpreted as representative for the Martian mantle, potentially a primary signature of Martian source material, and the solar nebula. Williams (2016) analysed the chlorine isotope composition of all other 3.00 LL chondrites, a very primitive subgroup of chondrites (NWA 7731, NWA 10061 and Semarkona). All three chondrites showed $\delta^{37}Cl$ values close to 0 (−0.75, −0.52 and + 0.29‰ respectively). Based on petrographic evidence in the meteorite Semarkona that chloride was incorporated into chondrules prior to parent body accretion, which was attributed to HCl-clathrate formation that is predicted to have higher $\delta^{37}Cl$ values than the nebular gas by 3 to 6‰, resulted in $\delta^{37}Cl$ values close to 0‰ in these other meteorites. It was then concluded that NWA 8276 had escaped this clathrate incorporation and consequently represented the primitive chlorine isotope composition of the solar nebula.

To assess the chlorine isotope variation in secondary processed chondrites Schauble and Sharp (2011) modelled the chlorine isotope fractionation between gas phase HCl and crystalline phases $HCl.3H_2O$, NaCl, KCl and $Na_4Al_3(SiO_4)_3Cl$. They applied this exercise using a combination of experimental vibrational frequencies and electronic structure models. It was calculated that sodalite ($Na_4Al_3(SiO_4)_3Cl$) would have a $\delta^{37}Cl$ that would be about ±0.7‰ lower than HCl at 950 K, which is in line, although somewhat smaller, with the chlorine isotope composition of sodalite inclusions in the meteorite Allende (Sharp et al. 2007). The fractionation between solid $HCl.3H_2O$ and gaseous HCl is predicted to be between +3 and + 6‰ at temperatures that are relevant to low-pressure gas/crystal partitioning, which is about 140 to 160 K or −133 to −113 °C. This fractionation, with isotopically heavier HCl in the crystalline form is opposite the expected fractionation that this crystalline form would be lighter. These results led to the conclusion that it may be possible to distinguish between icy and silicate sources of chloride in planets and planetary precursors (Zolotov and Mironenko 2007), at least as long as the chloride condensation was not quantitatively complete. Very positive $\delta^{37}Cl$ values which are derived from acid hydrates could, if they were to be found, indicate a post-condensation origin of chlorine bearing minerals and would be consistent with a correlation of high chloride concentration and hydration of chloride in CI chondrites.

19.1.2 Achondrites (Partially Differentiated Meteorites)

Tartèse et al. (2019) determined the chlorine isotope composition of phosphates from the achondrite Graves Nunataks (GRA) 06128 in order to better constrain the origin and processing of volatiles in the early Solar System. The meteorite GRA 06128 originates from an asteroid that underwent partial melting within a few million years after the start of Solar System formation, and as such is not undifferentiated. δ^{37}Cl in the apatite from this meteorite is homogeneous and has a high value of +3.2 ± 0.7‰, very different from chondrite values. Tartèse et al. (2019) suggest that it may be possible that the chlorine isotope value results from interaction of GRA 06128 with fumarole-like fluids derived from HCl-rich ice that is delivered to the parent body of GRA 06128 by an ice-rich impactor. More recently Stephant et al. (2024) evaluated the chlorine isotope compositions of a series of achondrites, acapulcoites and lodranites. Stephant et al. (2024) observed that no chlorine isotope fractionation was recorded in apatite from the evaluated meteorites. For that reason they concluded that these achondritic meteorites had retained their chondritic precursor chlorine isotope signature.

19.1.3 Iron Meteorites

Iron meteorites are fully differentiated and represent the iron core of solar system bodies that sank to the centre of the body after differentiation form the much lighter rock facies. A large set of different types of iron meteorites were measured for their chlorine isotope composition by Gargano and Sharp (2019). Chlorine isotope compositions vary considerable with a large range compared to other types of meteorites. The δ^{37}Cl values of magmatic irons range from −7.2 to +18.0‰ versus SMOC and the isotope compositions are unrelated to the chloride concentrations, that range from 0.3 to 161 ppm. The non-magmatic irons (IAB) are relatively chlorine rich and contain >161 ppm and show δ^{37}Cl values with a ranging from −6.1 to −3.2‰.The most extreme δ^{37}Cl values are inconsistent with a terrestrial source, and because the chloride contents in magmatic iron meteorites are largely consistent with derivation from a chondrite-like silicate complement, Gargano and Sharp (2019) suggest that their chlorine is indigenous to iron meteorites. Very high δ^{37}Cl values in two of these meteorites (+13.4 and + 18.0‰) are explained by degassing during a disruption event on their parent body. Because no processes are known to decrease δ^{37}Cl values in meteorites (degassing, when light ^{35}Cl escapes faster from the body than heavy ^{37}Cl, tends to increase the δ^{37}Cl), the low δ^{37}Cl values were interpreted by Gargano and Sharp (2019) to be representative for the primordial chlorine isotope composition of the solar nebula. Consequently, Gargano and Sharp (2019) state that the poor retention of low δ^{37}Cl values in many differentiated planetary materials suggest that extensive devolatilisation occurred during the formation of the planets, and this can explain the relatively high average δ^{37}Cl value on Earth due to the loss of approximately 60% of the initial chlorine content that was initially present.

19.2 Asteroid 4 Vesta

A number of differentiated meteorites, the so called eucrites, howardites and diogenites are genetically linked to each other and most of them are thought to derive from the asteroid 4 Vesta, although a few may descend from other, geochemically similar asteroids (Takeda and Graham 1991; Scott et al. 2009; McSween et al. 2012; Sarafian et al. 2013, 2014). As a result, these meteorites show a nice view on the origin of this asteroid. Sarafian et al. (2017) measured the chlorine isotope composition of apatite and merrilite in seven basaltic eucrites with SIMS. Chlorine isotope values expressed a relatively large range in these eucrite samples from −3.8 to +7.7‰. These isotope values correlate positively with magnesium and scandium, while they correlate negatively with potassium, vanadium and chromium. Sarafian et al. (2017) explain these correlations with degassing of 4 Vesta's magma ocean. It is suggested that the lowest observed value agrees with the $\delta^{37}Cl$ of the accreted parent body of 4 Vesta. The low value suggests that at least some parts of the solar nebula have been isotopically light compared to the Earth, which has on average a $\delta^{37}Cl$ of 0‰. This low value seems to agree with suggestions of a low $\delta^{37}Cl$ in (parts of) the solar nebula as discussed earlier in this chapter based on other meteorites from various sources (Sharp et al. 2016; Williams et al. 2016; Williams 2016; Gargano and Sharp 2019).

19.3 Comet 67P/Churyumov-Gerasimenko

The ESA's Rosetta mission was aimed at studying a comet at close range. Rosetta orbited around the comet 67P/Churyumov-Gerasimenko and one of the instruments it contained was a Double Focusing Mass Spectrometer (DFMS; Balsiger et al. 2007). With this instrument it was possible to measure the intensity of chlorine bearing species, mostly Cl^+ and HCl^+ of both stable chlorine isotopes in the coma of the comet (Dhooghe et al. 2017, 2021). Due to the low concentrations of chlorine in the coma the concentrations for Cl^+ and HCl^+ ware taken together to calculate an isotope ratio value. Dhooghe et al. (2017) measured abundances of ^{37}Cl and ^{35}Cl fragments in the coma of 67P/Churyumov-Gerasimenko for the first time. They correlated the abundances of ^{37}Cl and ^{35}Cl fragments and this correlation resulted in a $^{37}Cl/^{35}Cl$ chlorine isotope ratio of 0.29 ± 0.02. While Dhooghe et al. (2017) state that this ratio is compatible with the ratio for the early Sun, Earth, Moon, Mars, asteroids and meteorites, within the measurement uncertainties, Liuzzi et al. (2021), who plotted the obtained value relative to other Solar system chlorine isotope ratio determinations, conclude that the coma of 67P/Churyumov-Gerasimenko, with a $\delta^{37}Cl$ value recalculated from a ratio of 0.29 ± 0.02 of −93 ± 63‰ relative to SMOC consists of the isotopically lightest chlorine in the solar system. In their study, Dhooghe et al. (2017) only used data from a short observation period between 1 and 31 October 2014, with a total of 341 spectra used. In a later study, Dhooghe et al. (2021) presented the chlorine isotope ratio based on the data analyses from the entire Rosetta mission, which lasted from 6 August 2014 until 30 September 2016. Compared to their original study, taking into account data from the entire mission decreased the random error sufficiently resulting in uncertainty margins on the ratios that become bounded by the systematic errors. Fortunately, between the two studies, there had been several improvements made regarding the calibration of the instrument and the data reduction, which helped Dhooghe et al. (2021) to reduce systematic errors in the dataset. The number of spectra used in the second study was well over 10,000 from two channels. Because the errors of the measurements on both channels are statistically independent and the correction had been done independently for both channels, it was justifiable by Dhooghe et al. (2021) to combine the results for both channels and for both Cl and HCl by taking their logarithmically weighted averages. Overall, this resulted in a $^{37}Cl/^{35}Cl$ ratio in the coma of 67P/Churyumov-Gerasimenko of 0.336 ± 0.017. This agrees with the error to the $^{37}Cl/^{35}Cl$ ratio of the oceans on Earth of 0.320. Recalculated to $\delta^{37}Cl$ the chlorine isotope value is +51 ± 55 (1σ). This value is considerably different from the value reported in Dhooghe et al. (2017). The difference is explained by the fact that Dhooghe et al. (2017) calculated the chlorine isotope ratio from the integrated areas of the ^{37}Cl and ^{35}Cl peaks. Dhooghe et al. (2021) on the contrary calculated the isotope ratio from the total number of ions, which, because these are absolute numbers, reflect the total chlorine and isotope ratio much better. The ratio value however is about 18% higher when the total number of ions in taken into account and this then results in a chlorine isotope ratio that appears to be slightly higher, instead of slightly lower than the isotope value encountered on Earth. While the isotope value is not significantly different from the isotope value on Earth, the value seems to tend to higher values than observed on Earth and on most other Solar system bodies. This may be explained due to degassing or devolatilization, releasing isotopically light chlorine from the comet, in a way similar to what was concluded as the origin of high $\delta^{37}Cl$ values that were found on lunar rocks (Sharp et al. 2010; Wang et al. 2012, 2019; Boyce et al. 2015).

19.4 The Chlorine Isotope Ratio in the Atmosphere of Venus

It had been observed that HCl is probably the most important chlorine species in the atmosphere of Venus (Yung and DeMore 1982; Krasnopolsky and Parshev 1983). Iwagami et al. (2008), with the use of infra-red spectroscopy, were able to determine the HCl content and the $^{35}Cl/^{37}Cl$ ratio in the atmosphere of Venus. They observed an average value of 0.74 ± 0.06 ppm for HCl above the clouds of Venus' atmosphere. The $^{35}Cl/^{37}Cl$ isotope ratio they obtained was 3.1 ± 0.4, a value that agrees fully with the value of 3.1 normally observed on Earth.

References

Balsiger H, Altwegg K, Bochsler P, Eberhardt P, Fischer J, Graf S, Jäckel A, Kopp E, Langer U, Mildner M, Müller J, Riesen T, Rubin M, Scherer S, Wurz P, Wüthrich S, Arijs E, Delanoye S, De Keyser J, Neefs E, Nevejans D, Rème H, Aoustin C, Mazelle C, Médale JL, Sauvaud JA, Berthelier JJ, Bertaux JL, Duvet L, Illiano JM, Fuselier SA, Ghielmetti AG, Magoncelli T, Shelley EG, Korth A, Heerlein K, Lauche H, Livi S, Loose A, Mall U, Wilken B, Gliem F, Fiethe B, Gombosi TI, Block B, Carignan GR, Fisk LA, Waite JH, Young DT, Wollnik H (2007) Rosina – Rosetta orbiter spectrometer for ion and neutral analysis. Space Sci Rev 128:745–801

Boyce JW, Treiman AH, Guan Y, Ma C, Eiler JM, Gross J, Greenwood JP, Stolper EM (2015) The chlorine isotope fingerprint of the lunar magma ocean. Sci Adv 1:1–8

Bridges JC, Banks DA, Smith M, Grady MM (2004) Halite and stable chlorine isotopes in the Zag H3-6 breccia. Meteorit Planet Sci 39:657–666

Dhooghe F, De Keyser J, Altwegg K, Briois C, Balsiger H, Berthelier JJ, Calmonte U, Cessateur G, Combi MR, Equeter E, Fiethe B, Fray N, Fuselier S, Gasc S, Gibbons A, Gombosi T, Gunell H, Hässig M, Hilchenbach M, Le Roy L, Maggiolo R, Mall U, Marty B, Neefs E, Rème H, Rubin M, Sémon T, Tzou CY, Wurz P (2017) Halogens as tracers of protosolar nebula material in comet 67P/Churyumov–Gerasimenko. Monthly Not R Astron Soc 472:1336–1345

Dhooghe F, De Keyser J, Hänni N, Altwegg K, Cessateur G, Jehin E, Maggiolo R, Rubin M, Wurz P (2021) Chlorine-bearing species and the $^{37}Cl/^{35}Cl$ isotope ratio in the coma of comet 67P/Churyumov–Gerasimenko. Monthly Not R Astron Soc 508:1020–1032

Gargano A, Sharp Z (2019) The chlorine isotope composition of iron meteorites: evidence for the Cl isotope composition of the solar nebula and implications for extensive devolatilization during planet formation. Meteorit Planet Sci 54:1619–1631

Harkins WD, Stone SD (1925a) The isotopic composition and the atomic weight of chlorine in meteorites. Nature 116:426

Harkins WD, Stone SB (1925b) The isotopic composition of the element chlorine in meteorites: the atomic weight of meteoric and terrestrial chlorine. Proc Natl Acad Sci 11:643–645

Harkins WD, Stone SB (1926) The isotopic composition and atomic weight of chlorine from meteorites and from minerals of non-marine origin. J Am Chem Soc 48:938–949

Hoering TC, Parker PL (1961) The geochemistry of the stable isotopes of chlorine. Geochim Cosmochim Acta 23:186–199

Hutchinson R (1989) Meteorites-petrology. In: Petrology. Encyclopedia of earth science. Springer, Boston, MA, pp 362–367

Iwagami N, Ohtsuki S, Tokuda K, Ohira N, Kasaba Y, Imamurac T, Sagawa H, Hashimoto GL, Takeuchi S, Ueno M, Okumura S (2008) Hemispheric distributions of HCl above and below the Venus' clouds by ground-based 1.7 mm spectroscopy. Planet Space Sci 56:1424–1434

Krasnopolsky VA, Parshev VA (1983) Photochemistry of the Venus atmosphere. In: Hunten DM, Colin L, Donahue TM, Moroz VI (eds) Venus. Univ Ariz Press, Tucson, AZ, USA, pp 431–483

Krot AN, Keil K, Scott ERD, Goodrich CA, Weisberg MK (2007) 1.05 – classification of meteorites. Treatise Geochem 1:1–52

Liuzzi G, Villanueva GL, Viscardy S, Mège D, Crismani MMJ, Aoki S, Gurgurewicz J, Tesson PA, Mumma MJ, Smith MD, Faggi S, Kofman SV, Knutsen EW, Daerden F, Neary L, Schmidt F, Trompet L, Erwin JT, Robert S, Thomas IR, Ristic B, Bellucci G, Lopez-Moreno JJ, Patel MR, Vandaele AC (2021) Probing the atmospheric cl isotopic ratio on mars: implications for planetary evolution and atmospheric chemistry. Geophys Res Lett 48:e2021GL092650

Magenheim AJ, Spivack AJ, Volpe C, Ransom B (1994) Precise determination of stable chlorine isotopic ratios in low-concentration natural samples. Geochim Cosmochim Acta 58:3117–3121

Magenheim AJ, Spivack AJ, Michael PJ, Gieskes JM (1995) Chlorine stable isotope composition of the oceanic crust: implications for Earth's distribution of chlorine. Earth Planet Sci Lett 131:427–432

McSween Jr HY, Mittlefehldt DW, Beck AW, Mayne RG, McCoy TJ (2012) HED meteorites and their relationship to the geology of Vesta and the Dawn mission. In: Russell C, Raymond C (eds) The Dawn Mission to minor planets 4 Vesta and 1 Ceres. Springer, New York, NY, pp 141–174

Nakamura N, Fujitani T, Kimura M (2007) A new isotope tracer for the early solar system processes: stable chlorine isotopes and the distribution of Cl-bearing phases in chondrites. 38th Lunar Planet Sci Conf, abs 1707

Sarafian AR, Roden MF, Patiño Douce AE (2013) The nature of volatiles in eucrites: clues from apatite. Meteorit Planet Sci 48:2135–2154

Sarafian AR, Nielsen SG, Marschall HR, McCubbin FM, Monteleone BD (2014) Early accretion of water in the inner solar system from a carbonaceous chondrite-like source. Science 346:623–626

Sarafian AR, John T, Roszjar J, Whitehouse MJ (2017) Chlorine and hydrogen degassing in Vesta's magma ocean. Earth Planet Sci Lett 459:311–319

Schauble EA, Sharp ZD (2011) Modeling isotopic signatures of nebular chlorine condensation. In: 21th Annual V.M. Goldschmidt conference, vol 75. Mineral Mag, Prague, Czech Republic, p 1810

Scott ERD, Greenwood RC, Franchi IA, Sanders IS (2009) Oxygen isotopic constraints on the origin and parent bodies of eucrites, diogenites, and howardites. Geochim Cosmochim Acta 73:5835–5853

Sharp ZD, Barnes JD, Brearly AJ, Chaussidon M, Fisher TP, Kamenetsky VS (2007) Chlorine isotope homogeneity of the mantle, crust and carbonaceous chondrites. Nature 446:1062–1065

Sharp ZD, Shearer CK, McKeagan KD, Barnes JD, Wang YQ (2010) The chlorine isotope composition of the Moon and implications for an anhydrous mantle. Science 329:1050–1053

Sharp ZD, Mercer JA, Jones RH, Brearley AJ, Selverstone J, Bekker A, Stachel T (2013) The chlorine isotope composition of chondrites and earth. Geochim Cosmochim Acta 107:189–204

Sharp Z, Williams J, Shearer C, Agee C, McKeegan K (2016) The chlorine isotope composition of Martian meteorites 2. Implications for the early solar system and the formation of Mars. Meteorit Planet Sci 51:2111–2126

Stephant A, Anand M, Carli C, Zhao X, Davidson J, Cuppone T, Pratesi C, Franchi IA (2024) Chondritic chlorine isotope composition of acapulcoites and lodranites. Geochem Perspect Lett 29:14–19

Takeda H, Graham AL (1991) Degree of equilibration of eucritic pyroxenes and thermal metamorphism of the earliest planetary crust. Meteoritics 26:129–134

Tartèse R, Anand M, Franchi IA (2019) H and Cl isotope characteristics of indigenous and late hydrothermal fluids on the differentiated asteroidal parent body of Grave Nunataks 06128. Geochim Cosmochim Acta 266:529–544

Wang Y, Guan Y, Hsu W, Eiler JM (2012) Water content, chlorine and hydrogen isotope compositions of lunar apatite. In: 75th Annual Meteoritical society meeting. Abs 5170

Wang Y, Hsu WB, Guan YB (2019) An extremely heavy chlorine reservoir in the Moon: insights from the apatite in lunar meteorites. Sci Rep 9:7527

Williams JT (2016) Cl in the early solar system. MSc thesis Univ New Mexico

Williams JT, Shearer CK, Sharp ZD, Burger PV, McCubbin FM, Santos AR, Agee CB, McKeegan KD (2016) The chlorine isotopic composition of Martian meteorites 1: chlorine isotope composition of Martian mantle and crustal reservoirs and their interactions. Meteorit Planet Sci 51:2092–2110

Yung YL, DeMore WB (1982) Photochemistry of the stratosphere of Venus: implications for atmospheric evolution. Icarus 51:199–247

Zolotov MYu, Mironenko MV (2007) Hydrogen chloride as a source of acid fluids in parent bodies of chondrites. 38th Lunar Planet Sci Conf, abs 2340

20 Chlorine Isotope Ratios in Stars and Interstellar Space

20.1 Possible Rates of Nucleosynthesis for ^{35}Cl and ^{37}Cl

The nucleosynthesis of ^{35}Cl and ^{37}Cl is different from each other as discussed in Chap. 1. This results in different amounts of ^{35}Cl and ^{37}Cl produced in different stars as the evolution of stars can be vastly different. As a result, in stars and interstellar space the chlorine isotope ration can be significantly different from the ratio in the solar system. Because stable isotopes from individual elements opened a new window into the study of the details of stellar evolution, supernovae and galactic chemical evolution (Kobayashi et al. 2011) a significant number of studies have been published in which the different production rates of individual isotopes were estimated. Rauscher et al. (2002) made a very comprehensive study on the nucleosynthesis in stars with solar masses between 15 and 25. In principle production yields for ^{35}Cl and ^{37}Cl are in the same order of magnitude, and obtain $^{35}Cl/^{37}Cl$ production factors varying between 1.52 and 2.82, except for the model with solar mass 20, where production factor ratios are between 10.59 and 10.84. The discrepancies at solar mass 20, characterised by much higher yields of ^{35}Cl as compared to ^{37}Cl, are a consequence of the merging of the oxygen, neon, and carbon shells about a day prior to core collapse (Rauscher et al. 2002). Hoppe et al. (2023) showed that the model by Rauscher et al. (2002) provided good fits for the isotope compositions of many elements in presolar grains from SNe (e.g., Hoppe et al. 2010) and for the S-isotopic compositions of volatile species from 67P/Churyumov-Gerasimenko (Hoppe et al. 2018). Maas and Pilachowski (2018) summarised various other chlorine nucleosynthesis predictions (Kobayashi et al. 2006, 2011; Chieffi and Limongi 2013; Travaglio et al. 2004; Karakas and Lugaro 2016; Cristallo et al. 2015) In general, it appears that that the ^{37}Cl yield is increasing (resulting in decreasing $^{35}Cl/^{37}Cl$ ratios) with increasing mass of the stars in which the isotopes are produced. Also the metallicity in a star can impact the $^{35}Cl/^{37}Cl$ production ratio. Maas and Pilachowski (2018) noticed that different published $^{35}Cl/^{37}Cl$ production ratios range from 1.20 to 4.23, compared to 3.13 for the observed solar system $^{35}Cl/^{37}Cl$ ratio (Lodders et al. 2009). ^{37}Cl yields can increase in principle to extreme values in He-core masses and neutron excess (Prantzos et al. 1990). For example, in a 25 times solar mass star, ^{37}Cl can be over-abundant by a factor of nearly 50 compared to the solar system ^{37}Cl abundance (Pignatari et al. 2010). These resulting extremely low $^{35}Cl/^{37}Cl$ isotope ratios of ~0.06 have in practice not yet been observed in the universe.

H. Eggenkamp, *The Geochemistry of Stable Chlorine and Bromine Isotopes*, Advances in Isotope Geochemistry, https://doi.org/10.1007/978-3-031-75633-7_20

20.2 Measured Chlorine Isotope Ratios in the Galaxy

Chlorine is a relatively rare element in stars, with abundances about 10^6 to 10^8 times lower than hydrogen depending on the species observed as calculated by Tsuji (1973). Chlorine abundance measurements in stellar spectra are difficult due to their low abundances and because no strong optical absorption features are present. Suess and Urey (1956) had already suggested that the cosmic abundances of chlorine would be relatively high and due to the abundances of metal ions it was expected (Tsuji 1973) that AlCl, CaCl, NaCl and KCl could be important metal-chloride species in stellar atmospheres. In their first determination of chlorine isotope ratios in IRC + 10216 Carnicharo and Guélin (1987) were able to determine abundances of AlCl, NaCl and KCl and the isotope ratios of NaCl and AlCl. Apart from metal chlorides the most common chlorine species in stars is HCl (Maas and Pilachowski 2018). Also chloronium (H_2Cl^+) has been observed in interstellar material (Lis et al. 2010). Right from the start of the detection of chlorine bearing chemical compounds in stars and interstellar matter it was possible to detect, to a large extent, the minor isotope ^{37}Cl too, and for that reason it was possible to calculate chlorine isotope ratios and compare the results to the chlorine isotope ratio in the solar system. The reason that the minor isotope could be detected from early on is of course the fact that the ^{35}Cl/^{37}Cl ratio is generally somewhere between 2 and 4, indicating that the minor isotope is still rather common, contrary to other isotope systems such as carbon of oxygen where the ratio between the major and minor isotope is about 100.

Since the early study by Carnicharo and Guélin (1987) a significant number of studies have been done to study chlorine compounds and stable chlorine isotope ratios in stars and interstellar matter. Maas and Pilachowski (2018) measured the chlorine isotope ratios in a large set of stars, and also summarised all data measured before their study. They divide the data into three groups, based on the stellar origin of the measurements. The ratio determination could also be done on different chemical species (metal halides, hydrogen chloride and chloronium). Overall it was found that chlorine isotope ratio determinations were done on 40 stars and interstellar matter.

The earliest chlorine isotope studies were made on metal chlorides in the interstellar envelope IRC + 10216 (Carnicharo and Guélin 1987; Carnicharo et al. 2000; Kahane et al. 2000). At a later stage chlorine isotope ratio on this system were also determined on HCl (Agundez et al. 2011). Chlorine isotope ratios were also determined on NaCl from a different post-AGB star, CRL 2688, by Highberger et al. (2003) and on AgCl from W Aquilae by Danilovich et al. (2021). Quite a few chlorine isotope ratios have been determined on interstellar matter (ISM). The earliest determinations were made by Salez et al. (1996) on the Orion A giant molecular cloud, a region of intense star formation, that is characterised by predominantly very massive stars. This was followed by a very comprehensive study Peng et al. (2010) who studied HCl in a large set of 27 galactic star-forming regions of which it was possible to determine the chlorine isotope ratio in 11 of them. Further studies on chlorine isotope ratios on hydrogen chloride in ISM were published later by Lis et al. (2010), Cernicharo et al. (2010), De Luca et al. (2012), Goto et al. (2013), Monje et al. (2013) and Kama et al. (2015). Chlorine isotope ratios on the chloronium ion in ISM were published by Lis et al. (2010), Neufeld et al. (2012), Muller et al. (2014) and Kama et al. (2015). Finally a large survey of chlorine and it isotope ratios has been conducted by Maas et al. (2016) and Maas and Pilachowski (2018), leading to the PhD thesis of Zachary Maas in 2020 (Maas 2020).

Table 20.1 summarises the measured ^{35}Cl/^{37}Cl ratios in the galaxy cited above. Maas and Pilachowski (2018) summarised this set and split the data into three groups, (1) carbon and (post) AGB stars, (2) ISM and (3) red giants. The isotope measurements are made on five different chemical species, HCl, H_2Cl^+, NaCl, KCl or AlCl. Figure 20.1 shows the published data in a histogram with bin size 0.2. The ^{35}Cl/^{37}Cl ratio in the solar system is about 3.1.

Table 20.1 Overview of known chlorine isotope ratios is stars and interstellar matter

#	Object	Type	Species	$^{35}Cl/^{37}Cl$ ratio	Reference
1	IRC + 10216	C / (post)AGB stars	HCl	3.3 ± 0.3	Agundez et al. (2011)
2	IRC + 10216	C / (post)AGB stars	NaCl	2.0 ± 1.0	Carnicharo and Guélin (1987)
3	IRC + 10216	C / (post)AGB stars	NaCl	3.8 ± 1.0	Carnicharo et al. (2000)
4	IRC + 10216	C / (post)AGB stars	NaCl	2.3 ± 0.5	Kahane et al. (2000)
5	CRL 2688	C / (post)AGB stars	NaCl	2.1 ± 0.8	Highberger et al. (2003)
6	IRC + 10216	C / (post)AGB stars	AlCl	2.4 ± 0.6	Carnicharo and Guélin (1987)
7	IRC + 10216	C / (post)AGB stars	AlCl	2.9 ± 0.5	Carnicharo et al. (2000)
8	IRC + 10216	C / (post)AGB stars	AlCl	2.2 ± 0.3	Kahane et al. (2000)
9	W Aquilae	C / (post)AGB stars	AlCl	1.2	Danilovich et al. (2021)
10	W Aquilae	C / (post)AGB stars	AlCl	2.4	Danilovich et al. (2021)
11	IRC + 10216	C / (post)AGB stars	KCl	2.7 ± 2.0	Carnicharo et al. (2000)
12	Orion A (OMC-1)	ISM	HCl	4.0 ± 0.3	Salez et al. (1996)
13	Orion A (OMC-1)	ISM	HCl	6.5 ± 2.2	Salez et al. (1996)
14	W31RS4	ISM	HCl	0.8 ± 0.3	Peng et al. (2010)
15	W31RS4	ISM	HCl	1.2 ± 0.5	Peng et al. (2010)
16	W3OH	ISM	HCl	2.3 ± 1.0	Peng et al. (2010)
17	OMC-1(0,-60)	ISM	HCl	2.5 ± 0.8	Peng et al. (2010)
18	OMC-1(0,-100)	ISM	HCl	2.1 ± 0.4	Peng et al. (2010)
19	Monoceros R2	ISM	HCl	2.3 ± 0.9	Peng et al. (2010)
20	IRC + 10216	ISM	HCl	2.6 ± 0.8	Peng et al. (2010)
21	IRAS16293	ISM	HCl	2.1 ± 1.2	Peng et al. (2010)
22	G5 89-0.39	ISM	HCl	2.1 ± 0.6	Peng et al. (2010)
23	G5 89-0.39	ISM	HCl	2.5 ± 0.7	Peng et al. (2010)
24	DR21(OH)	ISM	HCl	5.0 ± 3.0	Peng et al. (2010)
25	NGC6334I	ISM	HCl	2.7	Lis et al. (2010)
26	W3 A*	ISM	HCl	2.1 ± 0.5	Carnicharo et al. (2010)
27	CRL 2136	ISM	HCl	2.3 ± 0.4	Goto et al. (2013)
28	W31C (G10.6-0.4)	ISM	HCl	2.9	Monje et al. (2013)
29	OMC-2-FIR-4	ISM	HCl	3.2 ± 0.1	Kama et al. (2015)
30	W31C	ISM	HCl	2.1 ± 1.5	De Luca et al. (2012)
31	NGC6334I	ISM	H_2Cl^+	3.0	Lis et al. (2010)
32	W31C	ISM	H_2Cl^+	2.6 ± 0.5	Neufeld et al. (2012)
33	Sgr A	ISM	H_2Cl^+	3.4 ± 1.2	Neufeld et al. (2012)
34	PKS-18390-211 (SW)	ISM	H_2Cl^+	3.1 ± 0.3	Muller et al. (2014)
35	W49N	ISM	H_2Cl^+	3.5 ± 0.4	Neufeld et al. (2015)
36	OMC-2 foreground	ISM	H_2Cl^+	4.3 ± 0.8	Kama et al. (2015)
37	RZ Ari	Red giant	HCl	2.2 ± 0.4	Maas and Pilachowski (2018)
38	GN Cet	Red giant	HCl	1.76 ± 0.17	Maas and Pilachowski (2018)
39	AH Phe	Red giant	HCl	2.20 ± 0.31	Maas and Pilachowski (2018)
40	BD-17491	Red giant	HCl	2.42 ± 0.30	Maas and Pilachowski (2018)
41	AV Eri	Red giant	HCl	3.42 ± 0.50	Maas and Pilachowski (2018)
42	BQ Mon	Red giant	HCl	2.92 ± 0.31	Maas and Pilachowski (2018)
43	KO Mon	Red giant	HCl	3.22 ± 0.42	Maas and Pilachowski (2018)
44	NML Cygni	Red giant	NaCl	3.9 ± 0.3	Singh et al. (2022)

In this histogram it is visible that most reported $^{35}Cl/^{37}Cl$ ratios are between 2.0 and 2.4. Because of the large errors that are involved in the measurement of chlorine isotope ratios in stars and ISM the reported values are considered, in most cases, as consistent with the solar value of 3.1. To be overall consistent with the solar ratio however it would be expected that the median value of a large number of observations (even with large errors) would be equal, or at least much closer, to the solar ratio. From the total sample set of 44 ratio observations however, the median value is

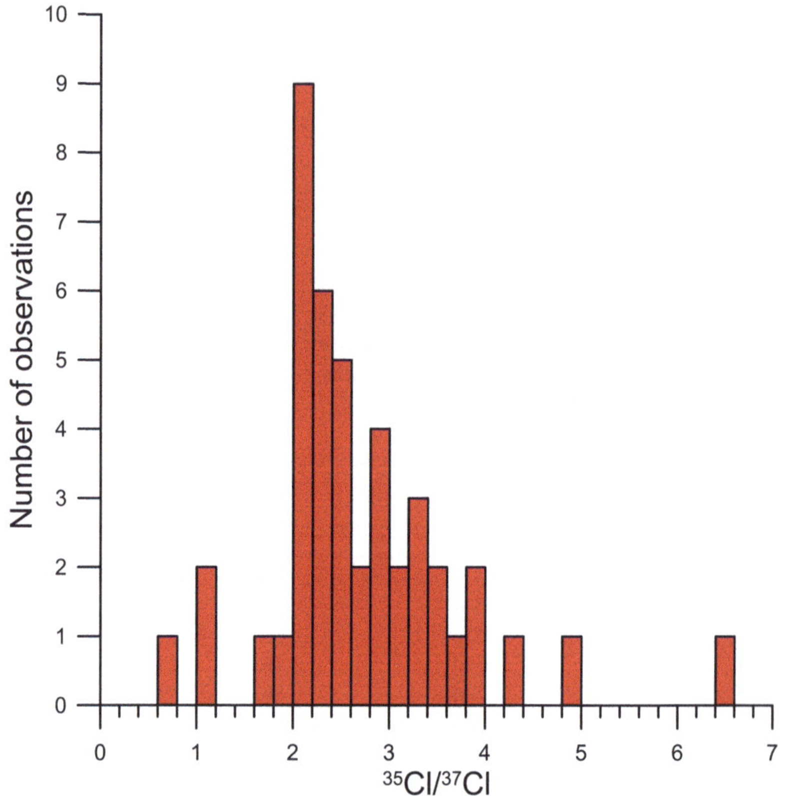

Fig. 20.1 Histogram of the known chlorine isotope ratios in stars and interstellar matter, shown with 0.2 ratio bins. The chlorine isotope ratio in the solar system is 3.13, as is visible in the histogram most reported ratios are below the ratio of the solar system

only 2.5, or 0.6 ratio units lower than the ratio in the solar system. From this overview it appears that overall in the galaxy the chlorine isotope ratio is much lower than in the solar system and that $^{35}Cl/^{37}Cl$ ratios between 2.0 and 2.5 are the most common.

An overview of the published $^{35}Cl/^{37}Cl$ ratios in the galaxy with the associate errors as tabulated in Table 20.1 is given in Fig. 20.2. Individual determinations are coloured by the type of object they were measured from and the form of the dots indicate the chemical species the isotope ratio was measured on. From the figure it is clear that the most common isotope ratios are between 2 and 3, with a few lower than 2, down to values as low as 0.8. Values above 3 are more common, although values above 4 are very rare and most of them have very large errors associated with them. From the figure it is also visible that there are no significant chlorine isotope ratios differences between different object types and chemical species. Perhaps the chlorine isotope ratios in interstellar chloronium seems higher than in interstellar HCl. Also in the red giants about half of the observations are very clearly close to the solar chlorine isotope ratio. Overall however, from the published chlorine isotope ratio determinations, it is clear that observed chlorine isotope ratios in the galaxy are in general lower than the chlorine isotope ratio in the solar system.

It has to be realised however that the determination of chlorine in stars and ISM is very difficult. This is due to its low abundance and because there are no strong optical absorption features. Hotter stars with temperatures above about 4000 K have no HCl features in theirs spectra because of the low dissociation energy of this molecule (Maas and Pilachowski 2018). For that reason chlorine can only be detected in cooler stars. The total variability of chlorine isotope ratios in stars and ISM is well discussed by Maas and Pilachowski (2018). Maas and Pilachowski (2018) state that the dominant source for the production of chlorine in the galaxy are core

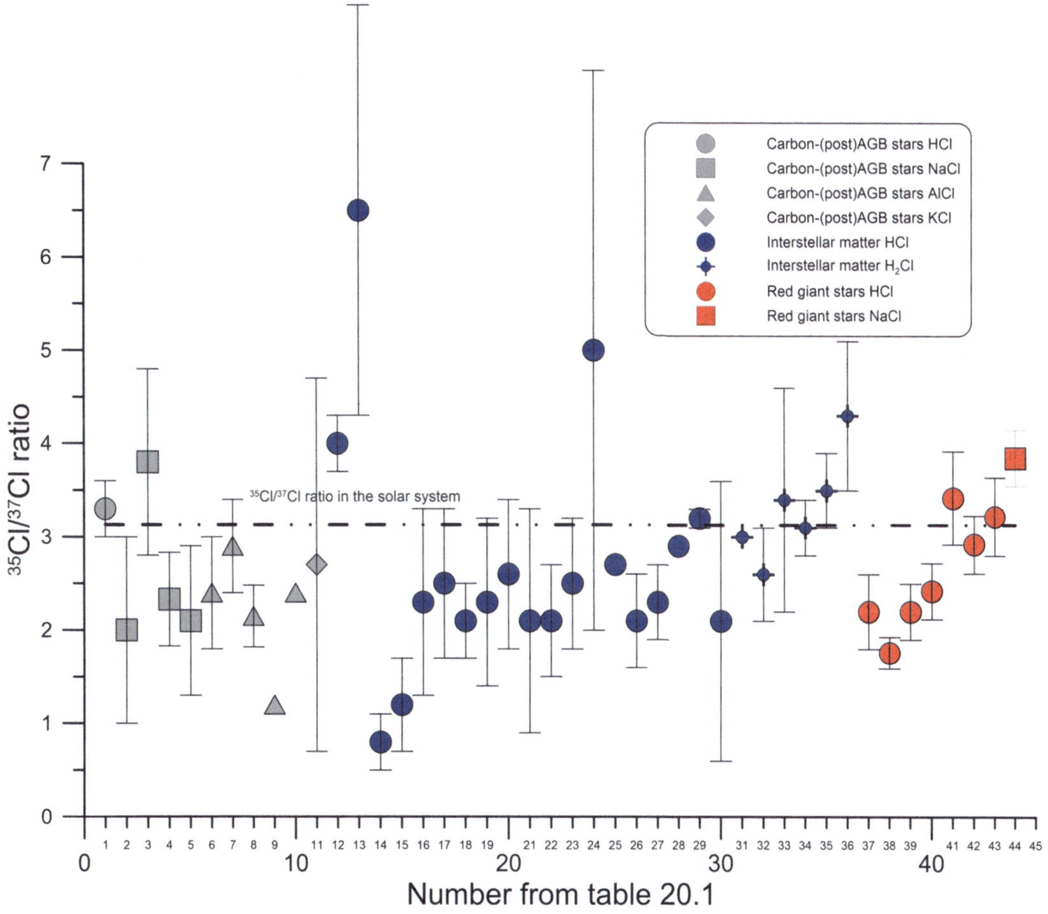

Fig. 20.2 All known chlorine isotope ratios of stars and interstellar matter. The horizontal line at $^{35}Cl/^{37}Cl = 3.13$ indicates the chlorine isotope ratio of the solar system. The data shown in this figure are taken from Table 20.1

collapsing super novae (CCSN). The isotope ratio varies depending on the progenitor star mass, metallicity, and explosion energy, and results in lower chlorine isotope ratios with larger progenitors and higher metallicities. According to Maas and Pilachowski (2018) this may explain the higher isotope ratios in certain red giants from smaller ("low-mass") progenitors. In circumstellar material, most of which is studied on IRC + 10216, it is suggested that the s-process produces ^{37}Cl and reduces the chlorine isotope ratio as suggested by Kahane et al. (2000) and Highberger et al. (2003). Other studies however suggest much smaller ^{37}Cl production (Cristallo et al. 2015; Karakas and Lugaro 2016) and can explain also in the circumstellar material chlorine isotope ratios close to the solar value. Overall it can be concluded that, in spite of the large number of studies the average chlorine isotope ratio in the universe can not be concluded upon yet. The main reason is because chlorine abundances can not be measured in important subsets of stars, especially the hotter ones. On the other hand, it is interesting to note that it is clear that the chlorine isotope ratio in the universe is not constant and that $^{35}Cl/^{37}Cl$ isotope ratios ranging from below 1 to above 6 have been observed. Further studies will definitely improve our understanding of the origin and evolution of chlorine in the universe and result in a better understanding of secondary nucleosynthesis processes (see e.g. Nomoto et al. 2013). Important to that aspect is of course that chlorine isotope ratios must be measured in a larger array of galactic material in the future, so that a better understanding of the overall universal chlorine isotope ratio variation will become available.

20.3 The Chlorine Isotope Composition of Cosmic Rays

A few studies have been made to attempt to determine the chlorine isotope ratio in cosmic rays. Wiedenbeck (1985) presented measurements of ^{35}Cl, ^{36}Cl and ^{37}Cl abundances in galactic cosmic rays by the ISEE-3 spacecraft. The ratio $^{35}Cl/^{37}Cl$ was found to be 2.0 with rather large standard deviations suggesting that the actual ratio must be somewhere between 1.2 and 3.3. It was concluded that the abundances of the two isotopes are consistent with secondary production as expected from a propagation model that was developed to account for both light and sub-iron secondary isotopes. The isotope composition of chlorine as determined from Voyager measurements was presented by Lukasiak et al. (1996) at the Joint Meeting of the APS and the AAPT. Unfortunately they did not present an isotope ratio in their abstract, nor were the results at a later date formally published.

References

Agúndez M, Cernicharo J, Waters LBFM, Decin L, Encrenaz P, Neufeld D, Teyssier D, Daniel F (2011) HIFI detection of hydrogen fluoride in the carbon star envelope IRC +10216. Astron Astrophys 533:L6

Carnicharo J, Guélin M (1987) Metals in IRC+10216: detection of NaCl, AlCl, and KCl, and tentative detection of AlF. Astron Astrophys 183:L10–L12

Carnicharo J, Guélin M, Kahane C (2000) A λ2 mm molecular line survey of the C-star envelope IRC+10216. Astron Astrophys Suppl Ser 142:181–215

Cernicharo J, Goicoechea JR, Daniel F, Agúndez M, Caux E, de Graauw T, De Jonge A, Kester D, Leduc HG, Steinmetz E, Stutzki J, Ward JS (2010) The $^{35}Cl/^{37}Cl$ isotopic ratio in dense molecular clouds: HIFI observations of hydrogen chloride towards W3 a. Astron Astrophys 518:L115

Chieffi A, Limongi M (2013) Pre-supernova evolution of rotating solar metallicity stars in the mass range 13–120 $M_{\odot}$ and their explosive yields. Astrophys J 764:21

Cristallo S, Straniero O, Piersanti L, Gobrecht D (2015) Evolution, nucleosynthesis, and yields of AGB stars at different metallicities. III. Intermediate-mass models, revised low-mass models, and the pH-fruity interface. Astrophys J 219:2

Danilovich T, Van de Sande M, Plane JMC, Millar TJ, Royer P, Amor MA, Hammami K, Decock L, Gottlieb CA, Decin L, Richards AMS, De Beck E, Baudry A, Bolte J, Cannon E, De Ceuster F, de Koter A, Etoka S, Gobrecht D, Gray M, Herpin F, Homan W, Jeste M, Kervella P, Khouri T, Lagadec E, Maes S, Malfait J, McDonald I, Menten KM, Montargès M, Müller HSP, Pimpanuwat B, Sahai R, Wallström SHJ, Waters LBFM, Wong KT, Yates J, Zijlstra A (2021) ATOMIUM: halide molecules around the S-type AGB star W Aquilae. Astron Astrophys 655:A80

De Luca M, Gupta H, Neufeld D, Gerin M, Teyssier D, Drouin BJ, Pearson JC, Lis DC, Monje R, Phillips TG, Goicoechea JR, Godard B, Falgarone E, Coutens A, Bell TA (2012) *Herschel*/HIFI discovery of HCl^+ in the interstellar medium. Astrophys J Lett 751:L37

Goto M, Usuda T, Geballe TR, Menten KM, Indriolo N, Neufeld DA (2013) Fundamental vibrational transitions of hydrogen chloride detected in CRL 2136. Astron Astrophys 558:L5

Highberger JL, Thonson KJ, Young PA, Arnett D, Ziurys LM (2003) The salty scrambled egg: detection of NaCl toward CRL 2688. Astron J 593:393–401

Hoppe P, Leitner J, Gröner E, Marhas KK, Meyer BS, Amari S (2010) NanoSIMS studies of small presolar SiC grains: new insights into supernova nucleosynthesis, chemistry, and dust formation. Astrophys J 719:1370–1384

Hoppe P, Rubin M, Altwegg K (2018) Presolar isotopic signatures in meteorites and comets: new insights from the Rosetta mission to comet 67P/Churyumov-Gerasimenko. Space Sci Rev 214:106

Hoppe P, Rubin M, Altwegg K (2023) A comparison of presolar isotopic signatures in laboratory-studied primitive solar system materials and comet 67P/Churyumov-Gerasimenko: new insights from light elements, halogens, and Noble gases. Space Sci Rev 219:32

Kahane C, Dufour E, Busso M, Gallino R, Lugaro M, Forestini M, Straniero O (2000) Improved isotopic ratio determinations in IRC+10216, the progenitor mass and the s process. Astron Astrophys 357:669–676

Kama M, Caux E, López-Sepulcre A, Wakelam V, Dominik C, Ceccarelli C, Lanza M, Lique F, Ochsendorf BB, Lis DC, Caballero RN, Tielens AGGM (2015) Depletion of chlorine into HCl ice in a protostellar core. The CHESS spectral survey of OMC-2 FIR 4. Astron Astrophys 574:A107

Karakas AI, Lugaro M (2016) Stellar yields from metal-rich asymptotic giant branch models. Astrophys J 825:26

Kobayashi C, Umeda H, Nomoto K, Tominaga N, Ohkubo T (2006) Galactic chemical evolution: carbon through zinc. Astrophys J 653:1145–1171

Kobayashi C, Karakas AI, Umeda H (2011) The evolution of isotope ratios in the milky way galaxy. Mon Not R Astron Soc 414:3231–3250

Lis DC, Pearson JC, Neufeld DA, Schilke P, Müller HSP, Gupta H, Bell TA, Comito C, Phillips TG, Bergin EA, Ceccarelli C, Goldsmith PF, Blake GA, Bacmann A, Baudry A, Benedettini M, Benz A, Black J, Boogert A, Bottinelli A, Cabrit S, Caselli P, Castets A, Caux E, Cernicharo J, Codella C, Coutens A, Crimier N, Crockett NR, Daniel F, Demyk K, Dominic C, Dubernet ML, Emprechtinger M, Encrenaz P, Falgarone E, Fuente A, Gerin M, Giesen TF, Goicoechea JR, Helmich F, Hennebelle P, Henning T, Herbst E, Hily-Blant P, Hjalmarson Å, Hollenbach D, Jack T, Joblin C, Johnstone D, Kahane C, Kama M, Kaufman M, Klotz A, Langer WD, Larsson B, Le Bourlot J, Lefloch B, Le Petit F, Li D, Liseau R, Lord SD, Lorenzani A, Maret S, Martin PG, Melnick GJ, Menten KM, Morris P, Murphy JA, Nagy Z, Nisini B, Ossenkopf V, Pacheco S, Pagani L, Parise B, Pérault M, Plume R, Qin SL, Roueff E, Salez M, Sandqvist A, Saraceno P, Schlemmer S, Schuster K, Snell R, Stutzki J, Tielens A, Trappe N, van der Tak FFS, van der Wiel MHD, van Dishoeck E, Vastel C, Viti S, Wakelam V, Walters A, Wang S, Wyrowski F, Yorke HW, Yu S, Zmuidzinas J, Delorme Y, Desbat JP, Güsten R, Krieg JM, Delforge B (2010) Herschel/HIFI discovery of interstellar chloronium (H_2Cl^+). Astron Astrophys 521:L9

Lodders K, Palme H, Gail HP (2009) Abundances of the elements in the solar system. In: Solar system, Landolt-Börnstein – Group VI astronomy and astrophysics, vol 4B. Springer-Verlag, Berlin Heidelberg, p 712. https://arxiv.org/abs/0901.1149

Lukasiak A, McDonald FB, Webber WR (1996) Voyager measurements of the isotopic composition of cosmic-ray chlorine and implications for the propagation of cosmic rays. Am Phys Soc, APS/AAPT Joint Meeting, May 2–5, 1996, abstract id K10.01

Maas ZG (2020) The galactic chemical evolution of chlorine and phosphorus. PhD Thesis. Indiana University Bloomington

Maas ZG, Pilachowski CA (2018) Chlorine isotope ratios in M giants. Astron J 156:2

Maas ZG, Pilachowski CA, Hinkle K (2016) Chlorine abundances in cool stars. Astron J 152:196

Monje RR, Lis DC, Roueff E, Gerin M, De Luca M, Neufeld DA, Godard B, Phillips TG (2013) Hydrogen chloride in diffuse interstellar clouds along the line of sight to W31C (G10.6-0.4). Astrophys J 767:81

Muller S, Black JH, Guélin M, Henkel C, Combes F, Gérin M, Aalto S, Beelen A, Darling J, Horellou C, Martín S, Menten KM, Trung DV, Zwaan MA (2014) Detection of chloronium and measurement of the $^{35}Cl/^{37}Cl$ isotopic ratio at z = 0.89 toward PKS 1830–211. Astron Astrophys 566:L6

Neufeld DA, Roueff E, Snell RL, Lis D, Benz AO, Bruderer S, Black JH, De Luca M, Gerin M, Goldsmith PF, Gupta H, Indriolo N, Le Bourlot J, Le Petit F, Larsson B, Melnick GJ, Menten KM, Monje R, Nagy Z, Phillips TG, Sandqvist A, Sonnentrucker P, van der Tak F, Wolfire MG (2012) Herschel observations of interstellar chloronium. Astrophys J 748:37

Neufeld DA, Black JH, Gerin M, Goicoechea JR, Goldsmith PF, Gry C, Gupta H, Herbst E, Indriolo N, Lis D, Menten KM, Monje R, Mookerjea B, Persson C, Schilke P, Sonnentrucker P, Wolfire MG (2015) *Herschel* observations of interstellar chloronium. II. Detections toward G29.96-0.02, W49N, W51, and W3(OH), and determinations of the ortho-to-para and $^{35}Cl/^{37}Cl$ isotopic ratios. Astrophys J 807:54

Nomoto K, Kobayashi C, Tominaga N (2013) Nucleosynthesis in stars and the chemical enrichment of galaxies. Annu Rev Astron Astrophys 51:457–509

Peng RS, Yoshida H, Chamberlin RA, Phillips TG, Lis DC, Gerin M (2010) A comprehensive survey of hydrogen chloride in the galaxy. Astrophys J 723:218–228

Pignatari M, Gallino R, Heil M, Wiescher M, Käppeler F, Herwig F, Bisterzo S (2010) The weak *s*-process in massive stars and its dependence on the neutron capture cross sections. Astrophys J 710:1557–1577

Prantzos N, Hashimoto M, Nomoto K (1990) The *s*-process in massive stars – yields as a function of stellar mass and metallicity. Astron Astrophys 234:211–229

Rauscher T, Heger A, Hoffmann RD, Woosley SE (2002) Nucleosynthesis in massive stars with improved nuclear and stellar physics. Astrophys J 576:323

Salez M, Frerking MA, Langer WD (1996) Hydrogen chloride detection in Orion a and Monoceros R2 and derivation of the $H^{35}Cl/H^{37}Cl$ isotopic ratio. Astrophys J 467:708–716

Singh AP, Edwards JL, Ziurys LM (2022) The Arizona Radio Observatory 1 mm Spectral Survey of the Hypergiant Star NML Cygni (215–285 GHz). Astron J 164:230

Suess HE, Urey HC (1956) Abundances of the elements. Rev Mod Phys 28:53–74

Travaglio C, Hillebrandt W, Reinecke M, Thielemann FK (2004) Nucleosynthesis in multi-dimensional SN Ia explosions. Astron Astrophys 425:1029–1040

Tsuji T (1973) Molecular abundances in stellar atmospheres. II. Astron Astrophys 23:411–431

Wiedenbeck ME (1985) The isotopic composition of cosmic ray chlorine. NASA, Goddard Space Flight Center 19th Intern Cosmic Ray Conf, Vol 2, pp 84–87

21 Bromine and Bromine Isotope Ratios in Extraterrestrial Environments

The cosmic abundance of bromine is very low. According to Lodders (2003) the bromine abundance in the solar system is $10^{2.59}$, compared to 10^{12} for hydrogen. On the same scale the chlorine abundance is $10^{5.26}$. This abundance is not directly measured in the sun, but based on determinations on chondrites. No spectroscopic measurements for bromine are known. As a result of these low abundances very little information is available on bromine concentrations, and consequently even less, on bromine isotope ratios. Still, a few studies are known where the ratio of the stable bromine isotopes in extraterrestrial material have been determined.

21.1 The Bromine Isotope Ratio in Stony Meteorites

The only currently known study in which the bromine isotope composition in meteorites was determined was published by Wyttenbach et al. (1964). Wyttenbach et al. (1964) determined bromine concentrations on 11 and the isotope compositions on five stony meteorites with the use of neutron activation analyses. While large variations were found in the concentration of bromine in these meteorites, from 0.067 to 1.56 ppm bromine, the isotope values were, within the errors of the measurement, the same. Unfortunately the errors of the measurement were rather large with this technique, and considered to be ±20‰ for the standard and ± 30‰ for the meteorite samples, recalculated from the bromine isotope ratios as reported by Wyttenbach et al. (1964). The reported isotope values were between −10 and + 30‰ compared to the used standard, a sample of NH_4Br obtained from AnalaR, and this variation was not considered significant. The conclusion reached by Wyttenbach et al. (1964) was, that within the limits of accuracy of the method used, the isotope composition of bromine in the five samples was the same as the isotope composition of terrestrial bromine.

21.2 The Bromine Isotope Composition in the Coma of Comet 67P/Churyumov-Gerasimenko

Dhooghe et al. (2017) were not only able to determine the chlorine abundance in the coma of comet 67P/Churyumov–Gerasimenko, but also the bromine abundance, which is about 50 times less common than chlorine in this coma. From the determinations they were able to determine the bromine isotope ratio, $^{81}Br/^{79}Br$ too. The isotope ratio based on 10 days of measurements, obtained during the period from 18 to 28 October 2014, was 0.95 ± 0.07 which agrees on the delta scale to a value of −25 ± 72‰, obviously within the range that is also observed on Earth.

H. Eggenkamp, *The Geochemistry of Stable Chlorine and Bromine Isotopes*, Advances in Isotope Geochemistry, https://doi.org/10.1007/978-3-031-75633-7_21

This bromine isotope ratio is calculated, just like the chlorine isotope ratio by Dhooghe et al. (2017), from the bromine isotope ratio of the integrated areas of the ^{81}Br and ^{79}Br peaks. In their second study on the chlorine isotope ratio in the coma from 67P/Churyumov-Gerasimenko Dhooghe et al. (2021) determined the chlorine isotope ratio from the total number of ions, an approach that reflects the chlorine isotope ratio better than the earlier approach, and also reached to a different, significantly higher chlorine isotope ratio. Unfortunately, Dhooghe et al. (2021) did not present bromine abundances and isotope ratios in their second paper. For that reason it is not possible to tell whether the bromine isotope ratio, if determined from the total number of ions, will be different from the approach to determine it from the integrated areas.

21.3 Bromine Beyond the Solar System

21.3.1 Predicted Bromine Isotope Ratio from Nucleosynthesis Ratios

Calculations for the nucleosynthesis of bromine have also been made by Rauscher et al. (2002). $^{79}Br/^{81}Br$ isotope ratios in nucleosynthesis were calculated as being between 1.13 and 1.71 for most solar between 15 and 25, except, just like for chlorine at solar mass 20. At solar mass 20 the predicted ratio is lower than unity with values of 0.90 and 0.87. Hoppe et al. (2023) showed that the models developed by Rauscher et al. (2002) provided good fits for the isotope compositions of many elements in presolar grains from SNe (e.g., Hoppe et al. 2010) and for the S-isotopic compositions of volatile species from 67P/Churyumov–Gerasimenko (Hoppe et al. 2018). The predicted production ratios are mostly higher than the solar $^{79}Br/^{81}Br$ ratio which is 1.028 (Catanzaro et al. 1964), except for predictions at solar mass 20, which are a consequence of the merging of the oxygen, neon, and carbon shells about a day prior to core collapse (Rauscher et al. 2002). The predictions by Rauscher et al. (2002) do not suggest a clear relationship between mass and isotope ratio in the nucleosynthesis of the bromine isotopes, although they are mostly higher than the solar system ratio, and it thus is not clear yet at which circumstances the solar bromine isotope ratio is expected to be produced.

21.3.2 Bromine in Stars and Interstellar Matter

Until recently, bromine or bromine containing molecules were not observed in stars or interstellar matter. Ligterink and Kama (2018) made a search for bromine on archived *Herschel*/HIFI data from Orion KL, Sgr B2(N), and NGC 6334I. Unfortunately, no HBr or HBr^+ features were observed in these data. Based on the fact that no features were found Ligterink and Kama (2018) were able to fix upper limits on the bromine abundances in the observed stars. Most of these upper limits lie above the values expected from a simple scaling down of HCl emission using the chlorine bromine elemental ratios. Ligterink and Kama (2018) compared the upper limits of the bromine abundances values with the observed abundance in the coma of comet 67P/Churyumov–Gerasimenko (Dhooghe et al. 2017). They observed that the upper limits for the bromine abundance that they found for Orion KL were about an order of magnitude lower than the values observed in 67P/Churyumov-Gerasimenko. They concluded that these low values were consistent with the modelling of the possible abundance of bromine in the gas phase and for that reason, Ligterink and Kama (2018) concluded that the high bromine abundance in 67P/Churyumov-Gerasimenko was formed in icy grain mantles. About the same time, Werner et al. (2018) also searched for bromine in the galaxy. Werner et al. (2018) looked for bromine (and antimony) in hot stars with effective temperatures between 49,500 and 70,000 K, so called helium-rich white dwarfs. Werner et al. (2018) found bromine in the spectra of RE 0503-289 and HD149499 B at the correct wavelength and therefore conclude that they really observed bromine, contrary to the observations of Chayer et al. (2005) who unfortunately erroneously identified bromine, although preliminary, in the cool star DO HD149499 B. Werner et al. (2018) were able to detect bromine in RE 0503-289 and HD149499 B because the bromine abundances in these stars were about 1600 and 2500 times higher than the solar bromine abundance. Still, it was not yet possible to detect ^{81}Br and for that reason no isotope ratio estimations could be made. So, in spite of the efforts, no bromine isotope ratio determinations from outside the solar system are known at this moment.

References

Catanzaro EJ, Murphy TJ, Garner EL, Shields WR (1964) Absolute isotopic abundance ratio and the atomic weight of bromine. J Res Natl Bur Stand 68A:593–599

Chayer P, Vennes S, Dupius J, Kruk JW (2005) Abundance of elements beyond the iron Group in Cool DO white dwarfs. Astrophys J 630:L169

Dhooghe F, De Keyser J, Altwegg K, Briois C, Balsiger H, Berthelier JJ, Calmonte U, Cessateur G, Combi MR, Equeter E, Fiethe B, Fray N, Fuselier S, Gasc S, Gibbons A, Gombosi T, Gunell H, Hässig M, Hilchenbach M, Le Roy L, Maggiolo R, Mall U, Marty B, Neefs E, Rème H, Rubin M, Sémon T, Tzou CY, Wurz P (2017) Halogens as tracers of protosolar nebula material in comet 67P/Churyumov–Gerasimenko. Mon Not R Astron Soc 472:1336–1345

Dhooghe F, De Keyser J, Hänni N, Altwegg K, Cessateur G, Jehin E, Maggiolo R, Rubin M, Wurz P (2021) Chlorine-bearing species and the $^{37}Cl/^{35}Cl$ isotope ratio in the coma of comet 67P/Churyumov–Gerasimenko. Mon Not R Astron Soc 508:1020–1032

Hoppe P, Leitner J, Gröner E, Marhas KK, Meyer BS, Amari S (2010) NanoSIMS studies of small presolar SiC grains: new insights into supernova nucleosynthesis, chemistry, and dust formation. Astrophys J 719:1370–1384

Hoppe P, Rubin M, Altwegg K (2018) Presolar isotopic signatures in meteorites and comets: new insights from the Rosetta mission to comet 67P/Churyumov-Gerasimenko. Space Sci Rev 214:106

Hoppe P, Rubin M, Altwegg K (2023) A comparison of Presolar isotopic signatures in laboratory-studied primitive solar system materials and comet 67P/Churyumov-Gerasimenko: new insights from light elements, halogens, and noble gases. Space Sci Rev 219:32

Ligterink MFW, Kama M (2018) Interstellar bromine abundance is consistent with cometary ices from Rosetta. Astron Astrophys 614:A112

Lodders K (2003) Solar system abundances and condensation temperatures of the elements. Astrophys J 591:1220–1247

Rauscher T, Heger A, Hoffmann RD, Woosley SE (2002) Nucleosynthesis in massive stars with improved nuclear and stellar physics. Astrophys J 576:323

Werner K, Rauch T, Knörzer M, Kruk JW (2018) First detection of bromine and antimony in hot stars. Astron Astrophys 614:A96

Wyttenbach A, van Gunten HR, Scherle W (1964) Determination of bromine content and isotopic composition of bromine in stoney meteorites by neutron activation. Geochim Cosmochim Acta 29:467–474

Part VI

Historical and Future Developments of Halogen Isotope Geochemistry

22 Halogen Isotope Studies in Historical Perspective and Future Outlook

In 2024 it was 40 years ago that the first major work on chlorine isotopes was published (Kaufmann 1984). In this PhD thesis Ronald Kaufmann described the first method with which it was possible to measure the small natural variations of chlorine stable isotopes that appear in groundwaters and applied the technique to better understand the hydrological cycle of the chloride ion. Considering the fact that no chlorine isotope variations in natural samples were known before Ronald Kaufmann started his PhD studies the research focussed on ground water for the following three reasons: (1) The only samples that (possibly) showed natural variations in the study by Hoering and Parker (1961) were ground waters; (2) Groundwater environments are diverse and many locations might have fractionated chlorine due to processes that could be understood at that time; (3) Chloride could be removed quantitatively from water samples through a relatively simple technique by precipitating chloride with silver ions as silver chloride. Kaufmann (1984) studied chlorine isotope variations in different types of water samples, such as hydrothermal waters, formation waters and aquifer samples. Kaufmann (1984) also analysed a few halite samples. These early measurements showed clearly that the chlorine isotope ratio varied by at least 2.1‰, with the observation of values between −1.3 and + 0.8‰ relative to SMOC, a much larger range than was seen before and probably also a much larger variation than was expected before the research started (Kaufmann et al. 1984). Interestingly in oil field brine samples the $\delta^{37}Cl$ values showed an increasing trend with depth and it was suggested that this enrichment with depth might reflect an isotope fractionation mechanism related to the electrostatic properties of subsurface clays. These early studies further showed positive evidence that chlorine isotope variations could be observed in studies on groundwater diffusion (Desaulniers et al. 1986) and ion-filtration (Campbell 1985).

The outlook for the future of chlorine isotope geochemistry in 1984 was very optimistic. As it was finally shown that natural chlorine isotope variations could be measured in natural samples without problems it was expected that other laboratories would step in to start analysing chlorine isotopes. It was expected that most research would be done towards the three well understood fractionation processes known at that time. These processes were diffusion, and equilibrium and kinetic fractionation during precipitation. Both laboratory and field experiments were anticipated. It was expected that counter-current diffusion experiments would be more simple to simulate than simple diffusion due to larger sample size and longer time constraints. Fractionation of chlorine isotopes during halite precipitation was also anticipated as a valid experimental proposal. It was expected that isotope fractionation in silicate melts would be the most difficult to test experimentally, which indeed was shown later, as

H. Eggenkamp, *The Geochemistry of Stable Chlorine and Bromine Isotopes*, Advances in Isotope Geochemistry, https://doi.org/10.1007/978-3-031-75633-7_22

only in 2017 the first chlorine isotope diffusion study from a melt was published (Fortin et al. 2017).

It was anticipated that an inventory of chlorine isotope data in different kinds of geological materials, waters, rocks and gases would be made as a first step to answer the main geochemical questions considering the geochemical cycle of chlorine such as the origin of the distribution of chlorine between the major reservoirs and the possible mechanisms of mass exchange between these reservoirs about which they could only speculate at the time of writing of Kaufmann's thesis. Some of the questions that were expected to be solved in the near future (counted from 1984) were questions regarding the variability of the chlorine isotope compositions through time by analysing their variations in salt deposits and the distribution of chlorine isotopes between mantle and surface reservoirs by analysis of chlorine isotope compositions in mantle derived rocks. It is very interesting to realise that scientific discussions are still being held about these processes as they, in spite of the numerous studies on their chlorine isotope compositions, are not completely understood yet. These examples indicate that in many cases it was not trivial to analyse the chlorine isotope compositions (such as in rocks) and secondary fractionation that could appear can also hinder the understanding of the variations that are found in these samples.

Before the research by Kaufmann (1984) most studies on chlorine isotopes were just "one off" studies. With this I mean that these studies were apparently conducted to test the hypothesis that natural chlorine isotope variations could be determined and which was abandoned after significant (relative to the precision of the methodology used at the time of the research) chlorine isotope variations were not observed. Some important studies in this category were Curie (1921), Harkins and Stone (1926), Owen and Schaeffer (1955) and Hoering and Parker (1961). One exception did exist in the earliest days of isotope research. Even before it was known and shown that chlorine existed of two stable isotopes William D. Harkins developed research to show that chlorine must exist of (at least) two stable isotopes, based on the hypothesis that the atomic mass of chlorine was not an integer number. Harkins hypothesised that with an atomic mass of 35.5 chlorine must exist of at least two isotopes. To show this hypothesis he and his co-workers applied, over a number of years, diffusion experiments to change the atomic mass of chlorine and they finally succeeded with this approach (e.g. Harkins and Wilson 1915; Harkins and Hall 1916; Harkins 1920a, 1920b, 1920c; Harkins and Hayes 1922; Harkins and Jenkins 1926). In the period between 1925 and 1980 only two research groups focussed on chlorine isotope studies in longer term projects. A collaborative research project between the Max-Planck Institute for Chemistry in Mainz and the Physical Institute of the Chalmers University of Technology in Göteborg initiated by Prof Alfred Klemm resulted in several publications between 1955 and 1960 on the experimentally determined isotope enrichment of chlorine and bromine during diffusion in molten salts (e.g. Klemm and Lundén 1955; Lundén and Herzog 1956; Cameron et al. 1956; Herzog and Klemm 1958; Lundén and Lodding 1960). A research group led by James W Taylor published several studies on the kinetic isotope effects of chlorine and bromine in S_N1 and S_N2 systems in the period between 1969 until 1980 (e.g. Taylor and Grimsrud 1969; Grimsrud and Taylor 1970; Turnquist et al. 1973; Graczyk and Taylor 1974; Willey and Taylor 1978, 1980). Because of continued interest in chlorine (and bromine) isotopes for so many years by these research groups they also contributed significantly to the development of analytical methods for chlorine and bromine isotopes. As a result the analytical precision of the measurements improved significantly during these periods (e.g. Herzog and Dörnenburg 1958; Taylor and Grimsrud 1969; Willey and Taylor 1978).

After the publication of his PhD thesis by Ron Kaufmann (1984), the research group from Tucson, Arizona continued with chlorine isotope research during the 1980s in which period several MSc students finished their course with chlorine isotope research (e.g. Campbell 1985; Wirt 1988; Martin 1993) until they published their improved method for chlorine isotope analysis (Long et al.

1993). Apparently interest in chlorine isotope research faded after the retirement of Austin Long, although Chris Eastoe published numerous important papers on evaporites and formation brines (e.g. Eastoe and Peryt 1999; Eastoe et al. 1999, 2001, 2007; Eastoe 2016) until the breakdown of the mass spectrometer rendered further local chlorine isotope research impossible. Although retired now, Chis Eastoe is still very active as a chlorine isotope geochemist, now in various collaborations with mostly Chinese scientists (e.g. Zhao et al. 2021; Liu et al. 2021a, 2021b, 2021c, 2021d; Zhang et al. 2024). Especially interesting in this respect is his suggestion that the chlorine isotope fractionation associated with salt precipitation depends on the seawater chemistry (Eastoe and Meng 2024). This suggestion was based on the observation that the chlorine isotope composition of halite regularly falls outside the expected range assuming an ε^{37}Cl of +0.3‰. For that reason Eastoe and Meng (2024) suggest that the isotope fractionation can vary, depending on the sea water chemistry, between −0.4 and + 0.3‰. This hypothesis is not yet confirmed by experiments, and it is hoped that these experiments can be organised in due course. Sadly Ron Kaufmann had already passed away in 2005, and Austin Long in 2010.

The second research group to take interest in chlorine isotope research after the research group from Tucson was from the Department of Earth Sciences of the University of Waterloo in Canada. Shaun Frape and Peter Fritz were already collaborating with their colleagues from Tucson in their study on Canadian Shield brines (Kaufmann et al. 1987) and later on other formation waters (Kaufmann et al. 1993). This research group also was the first to explore the possibilities to analyse chlorine isotopes in organic compounds for potential environmental study purposes (Van Warmerdam et al. 1995; Beneteau et al. 1999). During the early twenty-first century the stable isotope laboratory of Waterloo University became the most productive chlorine and bromine isotope laboratory in the world especially after Orfan Shouakar-Stash published his methodology to analyse chlorine isotope variations in organic compounds (Shouakar-Stash et al. 2003) and after they published methods to analyse chlorine and bromine isotopes online using GC-IRMS (Shouakar-Stash et al. 2005a, 2005b; Shouakar-Stash 2008). Later they also developed methods to analyse the chlorine isotope composition of organic compounds by direct injection of these compounds into a mass spectrometer (Shouakar-Stash et al. 2009). This research group continues to be very productive to the present day, among which several studies in which stable chlorine and bromine isotope measurements were provided for third party research (e.g. Stotler et al. 2010; Boschetti et al. 2011; Audo-Miró et al. 2013; Gwynne et al. 2013; Le Gal La Salle et al. 2013; Rebeix et al. 2014; Kaown et al. 2014; Hanlon et al. 2017; Henkemans et al. 2018; Stotler et al. 2021; Saeed et al. 2021).

August Koster van Groos (University of Illinois at Chicago) had learned of the appearance of Ron Kaufmann's thesis early on and a few years later he had the opportunity to test the application of the technique during a sabbatical at Utrecht University in 1986/1987. Tests were promising and together with Rob Kreulen a PhD project was set up at Utrecht University, first with Math Kohnen and later Hans Eggenkamp as students. This resulted in the publication of a PhD thesis and several papers (Eggenkamp 1994; Eggenkamp et al. 1994, 1995; Eggenkamp and Koster van Groos 1997). After Hans Eggenkamp left Utrecht the equipment continued to be used at a lower frequency for chlorine isotope analyses for third parties (such as Groen et al. 2000 and Reitz et al. 2011) and the post-doctoral work from Masaaki Musashi (e.g. Musashi et al. 1998, 2004, 2008; Markl et al. 1997). Although the amount of chlorine isotope work done in Utrecht was relatively limited between 2000 and 2010 the Iranian PhD student Rahim Bagheri conducted part of his research at this laboratory (Bagheri et al. 2014a, 2014b) and from 2012 to 2016 a PhD study in which chlorine isotope measurements play a crucial role was again conducted (e.g. Rodriguez et al. 2016, 2018). Unfortunately, no further research on chlorine isotopes was conducted in Utrecht after this study was finished.

After Hans Eggenkamp left Utrecht he built the stable chlorine isotope laboratory at the

University of Reading under the supervision of Prof Max Coleman. After this was done successfully several post-doctoral studies were accomplished in Reading (e.g. Jendrzejewski et al. 1997, 2001; Loomis et al. 1997; Rosenbaum et al. 2000; Coleman et al. 2001, 2003; Ader et al. 2001, 2008; Van Acker et al. 2006; Eggenkamp and Coleman 2009) among which the first study presenting a method that showed natural bromine isotope variations (Eggenkamp and Coleman 2000). After Max Coleman left Reading for the Jet Propulsion Laboratory unfortunately the possibility to analyse chlorine isotopes in Reading ceased towards the end of the first decade of the twenty-first century.

After Nathalie Jendrzejewski, one of Max Coleman's post-doctoral fellows, left Reading she set up a facility to analyse chlorine isotopes at the Institut de physique du globe de Paris (IPGP). Once build it led to a series of interesting chlorine isotope papers (e.g. Godon et al. 2004; Lavastre et al. 2005; Woulé Ebongué et al. 2005). Especially interesting were the papers written by Magali Bonifacie in which important progress was shown to our understanding of the global chlorine cycle (Bonifacie et al. 2005, 2007, 2008). After a few quite years research has taken up again with the PhD project of Thomas Giunta that mostly focussed on different applications of chlorine isotopes in CO_2 sequestration studies. In this period the experimental determination of isotope fractionation factors during chloride oxidation (Giunta et al. 2013, 2017) and a study on the isotope fractionation for different chloride and bromide salts (Eggenkamp et al. 2016) were also conducted. From 2015 chlorine and bromine isotope research was revived considerably at the IPGP. Under the guidance of Pierre Agrinier and later Magali Bonifacie the extraction lines to measure chlorine isotopes were modernised. Also, in collaboration with the ICP-MS laboratory at the IPGP a method to measure high quality bromine isotopes was set up (Louvat et al. 2016), a method that was later on made applicable to low Br/Cl ratios such that also bromine isotopes in halite were measurable (Eggenkamp and Louvat 2018). During this period Hans Eggenkamp was able to extend his research on chlorine and bromine isotope variations in salt deposits (Eggenkamp et al. 2019a, 2019b). The chlorine isotope laboratory at the IPGP continues to be very active, and this includes significant collaboration with other institutions, in part also to study the correlations between different instrumental techniques to measure chlorine isotopes (e.g. Agrinier et al. 2019, 2021, 2023; Horst et al. 2019; Benard et al. 2020; Eggenkamp et al. 2021, 2022; Wudarska et al. 2021; Anderson et al. 2022; Ciu et al. 2021; Strydom et al. 2022; Lenoir et al. 2024).

The first major studies showing chlorine isotope variations using thermal ionization mass spectrometry were already done during the early 1990s. It was discovered by Ying-Kai Xiao that treatment of the filament in a thermal ionization mass spectrometer with graphite improved the precision of boron isotope measurements considerably (Xiao et al. 1988). After setting up a laboratory at the Qinghai Institute of Salt Lakes in Xining he discovered that this treatment also improved the measurement of chlorine isotope ratios considerably (Xiao et al. 1991; Xiao and Zhang 1992). Only slightly later it was also shown that this method works very well for bromine isotopes too (Xiao et al. 1993). These discoveries led to many studies in which especially the isotope behaviour of chlorine during processes that exist in continental salt lakes were studied and these studies continue until the present day (e.g. Xiao et al. 1997, 2000; Liu et al. 1997; Tan et al. 2005, 2006, 2009; Luo et al. 2012, 2014, 2015, 2016). Besides these studies Prof. Xiao was also involved in studies on possible isotope variations in seawater chloride (Shirodkar et al. 2003, 2006) and the development of an international secondary reference standard for chlorine isotopes (Xiao et al. 2000). After the retirement of Ying-Kai Xiao isotope research continued under the direction of Yun-Qi Ma. While most research focussed on the isotope chemistry of boron in salt lakes chlorine and bromine isotope techniques and applications continued to be developed (e.g. Ma et al. 2016a, 2016b; Gui et al. 2022; Yu et al. 2023). One of the students of Ying-Kai Xiao, Hai-Zhen Wei moved to Nanjing University around 2010. While focussing mostly on boron isotopes in the first few years (e.g. Yan et al. 2012; Wei et al. 2014a, 2014b, 2014c, 2015a), chlorine and bromine isotope research was developed from the beginning (e.g. Wei et al. 2011, 2012, 2015b). After a few quite years new chlorine isotope studies were published starting with a review (Liu et al. 2020a, 2020b, 2020c) and a study on evaporites in the Carpathian Mountains (Zhao et al. 2021). Most recently some very important papers on important experimental and theoretical principles of chlorine isotope geochemistry were published (e.g. Liu et al. 2021a, 2021b; Han et al. 2023; Zhang et al. 2024), that improved our understanding on the principle theoretical aspects on equilibrium isotope fractionation of chlorine considerably.

The second laboratory where the method to analyse chlorine isotopes with thermal ionization mass spectrometry was successfully applied was at Scripps Institution of Oceanography in San Diego. At this group the focus was on the chlorine isotope variations during subduction of pore fluids (Ransom et al. 1995) and marine aerosols (Volpe and Spivack 1994; Volpe et al. 1998). Later they managed to develop methods to analyse chlorine isotope composition in rocks and made the first estimates of chlorine isotope composition of the mantle (Magenheim et al. 1994, 1995). Due to possible analytical problems of their technique later studies by other laboratories using different techniques were not able to come to the same conclusions on the mantle chlorine isotope composition (Sharp et al. 2007; Bonifacie et al. 2008). Further research on chlorine isotopes seemed to vanish, probably due to the move of Art Spivack to North Carolina and later to Rhode Island. Only during 2008 one more paper appeared in which the pyrohydrolysis method apparently was improved enough so that is was possible to present new rock data using pyrohydrolysis extraction of chloride followed by thermal ionization mass spectrometry (Wei et al. 2008a, 2008b).

Another laboratory where chlorine isotopes were already studied at an early date was at Argonne National Laboratory. Under the directions of Neil Sturchio a method was developed to analyse both the chlorine and carbon isotope compositions of chlorinated organic compounds independently from the group at Reading and published it a few months earlier (Holt et al. 1997). They continued their research afterwards and were the first to develop methods to extract large enough quantities of chlorinated organic compounds from environmental samples to be able to analyse them for chlorine and carbon isotopes (Holt et al. 2001). After Neil Sturchio moved to the University of Illinois at Chicago his focus on chlorine isotope research changed to isotope studies of perchlorates. They developed methods to analyse perchlorates in very small amounts and to extract minute amounts of perchlorates from environmental samples (e.g. Böhlke et al. 2005, 2009; Sturchio et al. 2007). Their methods were improved considerably over this period and were ultimately good enough to be able to map the perchlorate isotope compositions in groundwaters for forensic purposes (Sturchio et al. 2012). While Neil Sturchio moved to the University of Delaware in 2014 isotope research on perchlorate and later chlorate actively continued to be applied by a relatively small group of researchers (among which John Karl Böhlke, Baohua Gu, Paul Hatzinger and Andrew Jackson). While the lead scientists were based at different institutions the isotope measurements were still done at Oak Ridge National Laboratory, and (per)chlorate research extended in directions such as the Great Lakes, ground water in arid environments, and human urine (e.g. Poghosyan et al. 2014, 2016; Sturchio et al. 2014; Hatzinger et al. 2017, 2022; Cao et al. 2020; Estrada et al. 2017).

A considerable amount of research on chlorine isotopes in mantle derived rocks, lunar and other extraterrestrial material has been performed by the group around Zach Sharp. They have made great progress in our understanding of the chlorine cycling on Earth (e.g. Sharp and Barnes 2004; Barnes and Sharp 2006; Barnes et al. 2006, 2008, 2009; Selverstone and Sharp 2011), the processes that are responsible for the chlorine isotope varability during chlorine (re)cycling (e.g. Sharp et al. 2010a) and the composition of the mantle, the moon and further extraterrestrial material (e.g.

Sharp et al. 2007, 2010b, 2011, 2013). In 2009 Jaime Barnes left the laboratory in New Mexico to set up a new chlorine isotope facility in Austin, Texas. This move has resulted in a significant number of new chlorine isotope studies (e.g. Amundson et al. 2012; Barnes and Cisneros 2012; Rizzo et al. 2013; Barnes et al. 2013, 2019a, 2019b; Stefánsson and Barnes 2016; Liotta et al. 2017; Cullen et al. 2019, 2021; Reich et al. 2019; Ranta et al. 2021; Wudarska et al. 2021; Gil et al. 2022; Segee-Wright et al. 2024) mostly focussed on mantle-lithosphere interaction. In the meantime also Zach Sharp continued his research on chlorine isotopes, although it seems that interest in this research field is fading and only few studies were published in recent years (e.g. Selverstone and Sharp 2015; Sharp et al. 2016; Williams et al. 2016; Shearer et al. 2018; Gargano and Sharp 2019; Gargano et al. 2020, 2022) suggesting that interest in chlorine isotope research is decreasing, possibly due to changing the research interests of the institute more towards more biological research.

In the same period (between 2000 and 2005) several other laboratories solely involved in the study of organic compounds started to analyse chlorine and bromine isotope ratios. The first of these was the laboratory of the University of Stockholm under the direction of Henry Holmstrand were at first methods were developed to analyse chlorine and later bromine isotopes in environmental samples (Holmstrand et al. 2004, 2010a; Mandalakis et al. 2008) and to apply these methods to natural samples (e.g. Holmstrand et al. 2006, 2007, 2010b; Carrizo et al. 2011). At a later stage they were even able to improve their methods to the level that they were able to extract bromomethane from atmospheric samples and analyse them for bromine isotopes (Horst 2013). After this productive period halogen isotope research has unfortunately been de-emphasised at the University of Stockholm towards research on carbon isotopes on aerosols and methane from subsea permafrost and hydrates.

Another laboratory that is progressively more involved in chlorine and bromine isotope studies on organic compounds is led by Daniel Hunkeler in Neuchâtel who has already published several studies on chlorine isotopes (e.g. Hunkeler et al. 2009, 2011). In the years before the publication of the first edition of this book several more laboratories have started to study chlorine isotopes in organic compounds, which at an early date already resulted in a large interlaboratory comparison study (Bernstein et al. 2011). In recent years a significant number of mostly chlorine isotope studies have been published regularly on subjects like diffusion, biodegradation, reductive dehalogenation (e.g. Palau et al. 2016, 2017, 2023; Wanner and Hunkeler 2015, 2019; Heckel et al. 2017; Rodriguez-Fernandez et al. 2018; Ponsin et al. 2018, 2019; Lihl et al. 2020; Zimmerman et al. 2020; Halloran et al. 2021; Torrentó et al. 2017, 2021).

At the Helmholz Zentrum in Leipzig chlorine isotope research has focussed on the study of isotope effects during production and decomposition in the environment of hexachlorocyclohexane (HCH, $C_6H_6Cl_6$) isomers. This research is conducted under the direction of Hans-Hermann Richnow and was initiated with the development of analytical techniques and comparison with more classic IRMS (Gilevska et al. 2015; Nijenhuis et al. 2015; Ivdra et al. 2017; Horst et al. 2017, 2019). They studied the isotope effects of HCHs in biomass (Lian et al. 2018), alkaline hydrolysis (Kannath et al. 2019), transformations in plants (Wu et al. 2019; Liu et al. 2019, 2021c, 2021d, 2023), soil systems (Kohli et al. 2017; Liu et al. 2020c, 2021c), anaerobic transformation in contaminated aquifers (Liu et al. 2020b), and fractionation during synthesis of HCH (Liu et al. 2022a, 2022b). Except for studies on HCHs also other chlorine isotope systems on chlorinated organic compounds were studied (e.g. Filippini et al. 2018).

While early studies on chlorine and bromine isotopes in China were applied with the use of TIMS, IRMS methods were first developed at the isotope laboratory of the Chinese University of Geosciences in Wuhan, led by Prof. Teng Ma. They developed methods to analyse chlorine and bromine isotopes using a GasBench to separate chloromethane or bromomethane from the other gases present after their preparation from both inorganic chloride or bromide (Liu et al. 2013; Du et al. 2013) and chlorine in organic compounds

(Gan et al. 2013). At about the same time the first studies on the chlorine and bromine isotope geochemistry in groundwater systems in the north east of China (Li et al. 2012; Chen et al. 2014, 2016, 2017; Du et al. 2013, 2016; Yu et al. 2019) were published at a relatively fast speed. After 2019 it seems that unfortunately no chlorine or bromine isotope studies have been published any more. Hopefully this is just a temporal brake and will new halogen stable isotope studies be published in the near future.

SIMS techniques to measure chlorine isotopes were already developed in 2004 at Woods Hole Oceanographic Institution (Layne et al. 2004) and were applied in a few studies on crust mantle interaction (Layne et al. 2009; John et al. 2010). It was not until 2014 that SIMS was applied more regularly for chlorine isotope research. The main reason was that SIMS measurements do have relatively large errors associated with them so that only large isotope variability could be recognised with this technique. These large errors have decreased considerably in more recent years. SIMS techniques can now be applied with relative ease in any laboratory that possesses a SIMS instrument and studies on chlorine isotopes have been published by several institutions now. For high quality SIMS measurements it is important that good quality reference material is available and those are now being more and more developed (e.g. Wudarska et al. 2021). Most SIMS studies are done on the chlorine rich mineral apatite, from which now reference material is available, in extraterrestrial material such as meteorites and lunar return samples. From this type of samples mostly very small amounts of material is available, and only with SIMS techniques it is often possible to obtain proper chlorine isotope data on the very small amounts of sample material that is available. A few laboratories have been relatively active in the field of SIMS measurements of chlorine isotopes. The first of these was the NORDSIMS laboratory in Stockholm where both terrestrial (John et al. 2010; Kusebauch et al. 2015) and meteorite (Belluci et al. 2017; Sarafian et al. 2017) research was done. Most active SIMS laboratories in the field of chlorine isotope studies seem to be Milton Keynes (Barnes et al. 2016, 2019a, 2019b; Barrett 2018; Barrett et al. 2019; Potts et al. 2018; Stephant et al. 2019, 2024; Darling et al. 2021; Hayden 2022) and Lausanne (Manzini et al. 2017; Bouvier et al. 2019, 2022a, 2022b). Other laboratories that produced chlorine isotopes studies with their SIMS instruments include Edmonton (Fortin et al. 2017), Houston (Shearer et al. 2018), Beijing (Hu et al. 2019; Liu et al. 2022b), Manchester (Tartese et al. 2019), Potzdam (Wudarska et al. 2020, 2021), and Guangzhou (Ciu et al. 2021). Considering that chlorine isotopes can now be measured with SIMS with good precision (better than ±0.3‰ in most cases), the opportunity to measure very small samples and to measure profiles over individual crystals there are a lot of possibilities for future growth to study chlorine isotope variations by SIMS. Not only applied to extraterrestrial material on apatite, but also on other chloride rich minerals in terrestrial, for example crustal, material. For example, in these samples variations in fractionation between melt and different minerals and zonation within individual crystals can be studied on a much smaller scale than with bulk samples. These opportunities are very exciting and further development is well expected in the near future.

Aside from these research groups there are several other research groups and laboratories where chlorine isotopes were studied in a less continuous manner. An important example of this type of research groups is found in the laboratory of David Banks (Leeds) were several important chlorine isotope studies were produced (e.g. Richard et al. 2011; Chiaradia et al. 2006; Bridges et al. 2004; Rosenbaum et al. 2000; Banks et al. 2000a, 2000b) in which chlorine isotopes were measured by thermal ionization mass spectrometry. Another important example was the laboratory of the National Water Research Institute in Saskatoon, Canada where methodology for online techniques were developed (Wassenaar and Koehler 2004) and who did very important research on the chlorine isotope composition of rainwater (Kohler and Wassenaar 2009). In this laboratory the chlorine isotope compositions for some third party studies were also analysed (e.g. Gleeson and Smith 2009; Bernal et al. 2014).

Other labs published the methodology they developed for chlorine and/or bromine isotope measurements, such as Gelman and Halicz (2010, 2011) at the Hebrew University in Jerusalem, Israel, and Hałas and Pelc (2009) at the Maria Curie-Skłodowska University in Lublin, Poland. While these research groups started to publish case studies they only did this on a relatively limited scale, such as Zakon et al. (2013, 2014, 2016) and Yankelson (2019) from the Hebrew University in Jerusalem, and Baran et al. (2012, 2015) and Dembska-Sieka et al. (2020) from the Maria Curie-Skłodowska University in Lublin. It is hoped that the number of papers by other institutions such as those mentioned in this section, and also others, will be increased considerably in the near future.

Finally I want to mention studies that have contributed very importantly at a theoretical level towards chlorine and bromine stable isotope geochemistry. First of course the classic paper by Urey (1947) where isotope fractionation between species of different oxidation states for both chlorine and bromine were calculated before any significant variations were measured. The same was actually the case with the study by Richet et al. (1977) who reviewed fractionation calculations between gaseous molecules HCl and Cl_2. Once isotope fractionation was known to exist in natural samples Schauble et al. (2003), Bourg and Sposito (2007), Tang and Peng (2017), Balan et al. (2019), Xu (2020), Gao and Liu (2021) and Pekora et al. (2022) published a new series of papers investigating the theoretical fractionation of chlorine isotopes. These studies were published from institutions that did not normally publish studies on natural samples, although cooperation with scientists that study natural samples was regularly shown. Further important theoretical studies were published by scientists from Nanjin University (Liu et al. 2021a; Zhang et al. 2024) showing that also institutions that are very active in isotope measurements can produce high quality theoretical papers and as such show they have the best of both worlds within a single institute.

This list of laboratories and research groups that study (or studied in the past) chlorine and/or bromine isotopes is certainly not complete, and chlorine and bromine isotopes have been studied, and are actually studied at this moment at more laboratories, but it is believed that the laboratories that have contributed the most important contributions to the study of the 'stable halogen isotopes'have been cited above.

However, considering the fact that the development of halogen isotope applications in geology has now reached the mature age of 40 years, it is some kind of surprising, for some of us perhaps even disappointing that this whole development can be summarised in a small volume like this book. The bibliography of chlorine and bromine isotope geochemistry, counted a total of just under 300 chlorine and bromine isotope papers when the first edition of this book was published in 2014. The rate at which chlorine and bromine isotope papers have been published has increased considerably in the last 10 years. At this moment the total count of published papers on chlorine and bromine isotopes has more than doubled, and since 2014 more than 300 new halogen isotope papers have been published. This definitively shows that the interest in halogen isotope geochemistry is increasing, albeit relatively slowly, considerably in the last decade, which is of course fantastic to note.

Figure 22.1 shows the development of the number of chlorine and bromine isotope papers over time, from 1915 until 2023. The number of papers published each year remained very low from the first paper suggesting that chlorine should have more than one stable isotope in 1915 until 1994, when, except for 1920 and 1921, never more than 5 papers per year were published. This indicates that, even after it was shown that chlorine isotope variations could be measured in natural samples (Kaufmann et al. 1984) it took about 10 years before the number of publications published per annum started to increase significantly. Even after this the number increased only slowly until in 2013 for the first time more than 20 papers were published. Since the publication of the first edition of this book the annual number of halogen isotope papers has increased considerably and appears to be at an average of approximately 30 papers annually since 2014.

The development of stable halogen (chlorine and bromine) isotope research is, as can be viewed from Fig. 22.1, quite regular, in a sense that the number of papers published grew slowly over the last 4 decades. This is quite the opposite as when compared to the development of other light isotope systems such as hydrogen, oxygen, carbon or sulphur, which developed much faster after their initial discoveries. Even after only a short amount of time after their initial development, within about 20 or 30 years, these other isotope systems were well known and commonly studied systems in a much larger number of laboratories than the number of laboratories that are currently studying halogen isotopes. The reasons for this slow development were probably related to the (geochemical) processes chlorine and bromine are involved in and the relatively limited isotope variation that is observed in halogen isotope systems. Also, in IRMS techniques off-line preparation of the samples involves a considerable amount of labour which makes it more difficult to analyse a large number of samples. In the case of bromine isotopes the lack of interest may in addition be related to the difficulties to extract bromine from samples with a large excess of chlorine (which is the case in all natural inorganic samples) and the related problem of the small amounts of bromine that are generally present in those samples which makes it more difficult to produce high quality analyses. Combined with the relatively small variations that are observed for chlorine and bromine isotopes it is also concluded that the variations that are found are often difficult to interpret. As has been stated earlier in this volume the variations that are observed within single reservoirs are often larger that the variations that are found between different reservoirs. As a result of this overlap it never is very easy to determine what processes have taken place, and in virtually all cases additional observations and measurements of other chemical and isotopic parameters need to be taken into account for a proper interpretation of the results. This indicates that analysing chlorine isotopes as the only parameter in a system will never result in a conclusive result, unlike other isotope systems where single measurements can sometimes point to clear conclusions. On the other hand, it has been shown that in several environments the additional information that is obtained from chlorine isotope measurements actually give just the last bit of information that is necessary to solve a problem, such as has been shown for example in the complex diffusion system that is found in the Dutch IJsselmeer

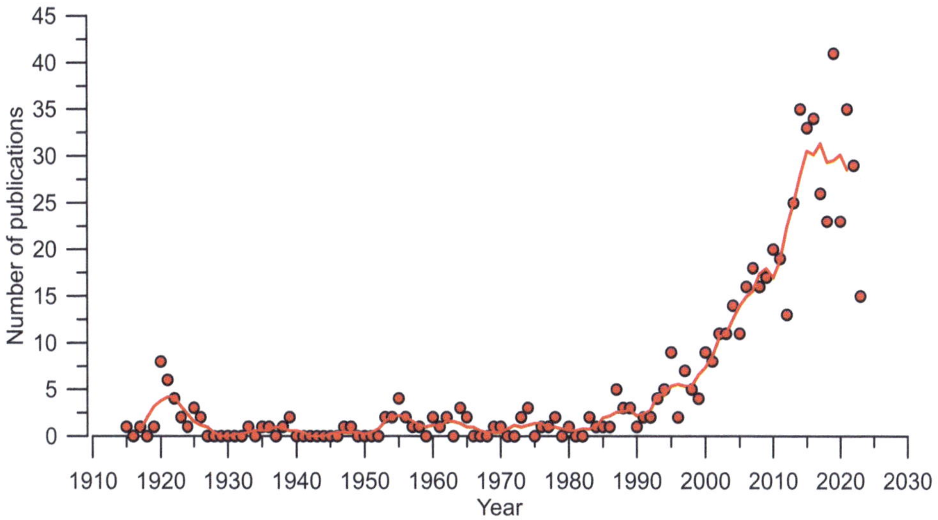

Fig. 22.1 Annual number of publications on chlorine and/or bromine isotopes. The red line represents the 5-year moving average of this number

(Beekman et al. 2011) where alternating freshwater and saline water phases did appear in the past, and where it was shown that only after analysing the chlorine isotope characteristics in combination with the earlier measured parameters it was possible to fully understand the geochemical history of the system, as compared to earlier studies in the same system when chlorine isotope measurements were not available yet (Volker and Van der Molen 1991).

As shown above the number of laboratories that are, or were, involved in chlorine and bromine isotope studies during the last 40 years was small enough to describe most of them in this chapter. It can be concluded from this observation that the expectation that a new isotope system that could in theory be easily picked up by most isotope laboratories that are also active in the other light stable isotopes (as chlorine isotopes can be analysed, just like the other light stable isotopes δ^2H, $\delta^{13}C$, $\delta^{15}N$, $\delta^{18}O$ and $\delta^{34}S$, as a gaseous compound in dual inlet mass spectrometry) did not really came through. Off-line preparation techniques seemed to be too complex to just apply them on an irregular base. As can be concluded from the introduction on the most active groups mentioned above, every now and then halogen isotope studies are developed in new laboratories, but at the same time the studies cease to be done in other laboratories. In many cases it appears that, when the researcher that set up the halogen isotope facilities in a laboratory leaves, interest in the study vanishes. Possibly as a result of the limited number of laboratories where halogen isotopes are studied the total number of papers per years is stabilising at about 30 per year since about 2014.

However, considering the slow but constant increase in published papers on chlorine and bromine stable isotope geochemistry it is safe to expect that the amount of research on chlorine and bromine isotope geochemistry will continue to increase in the near future. Individual laboratories will continue to work, generally in their own niche, and this will continue to improve our knowledge considerably. Development of new techniques, of which SIMS will be very promising because of the ease of measurements and the extreme detailed profiles that can be measured, will continue to grow interest in halogen isotope studies. Halogen isotope research will remain an important niche, although it might never grow into a major research field. It is expected however, due to the enthusiasm of the present leaders in the field, that new applications and further research will continue to be developed and applied to geological environments.

References

Ader M, Coleman ML, Doyle SP, Stroud M, Wakelin D (2001) Methods for the stable isotopic analysis of chlorine in chlorate and perchlorate compounds. Anal Chem 73:4946–4950

Ader M, Chaudhuri S, Coates JD, Coleman M (2008) Microbial perchlorate reduction: a precise laboratory determination of the chlorine isotope fractionation and its possible biochemical basis. Earth Planet Sci Lett 269:604–612

Agrinier P, Destrignevile C, Giunta T, Bonifacie M, Bardoux G, Andre J, Lucazeau F (2019) Strong impact of ion filtration on the isotopic composition of chlorine in young clay-rich oceanic sediment pore fluids. Geochim Cosmochim Acta 245:525–541

Agrinier P, Bonifacie M, Bardoux G, Lucazeau F, Giunta T, Ader M (2021) Chlorine isotope data of chlorides challenge the pore fluid paradigm. Geochim Cosmochim Acta 300:258–278

Agrinier P, Giskes J, Subbarao G, Bardoux G, Bonifacie M (2023) Chloride exchanges between oceanic sediments and seawater: constraints from chlorine isotopes. Geochim Cosmochim Acta 361:10–23

Amundson R, Barnes JD, Ewing S, Heimsath A, Chong G (2012) The stable isotope composition of halite and sulfate of hyperarid soils and its relation to aqueous transport. Geochim Cosmochim Acta 99:271–286

Anderson TS, Boyce JW, McCubbin FM, Turney S, Dudley JM, Haney NC, Bardoux G, Bonifacie M (2022) Development towards stable chlorine isotope measurements of astromaterials using the modified Middleton source of an accelerator mass spectrometer. Int J Mass Spectrom 477:116849

Audí-Miró C, Cretnik S, Otero N, Palau J, Shouakar-Stash O, Soler A, Elsner M (2013) Cl and C isotope analysis to assess the effectiveness of chlorinated ethene degradation by zero-valent iron: evidence from dual element and product isotope values. Appl Geochem 32:175–173

Bagheri R, Nadri A, Raeisi E, Eggenkamp HGM, Kazemi GA, Montaseri A (2014a) Hydrochemical and isotopic ($\delta^{18}O$, δ^2H, $^{87}Sr/^{86}Sr$, $\delta^{37}Cl$ and $\delta^{81}Br$) evidence for the origin of saline formation water in a gas reservoir. Chem Geol 384:62–75

Bagheri R, Nadri A, Raeisi E, Kazemi GA, Eggenkamp HGM, Montaseri A (2014b) Origin of brine in the Kangan Gasfield: isotopic and hydrogeochemical approaches. Environ Earth Sci 72:1055–1072

Balan E, Créon L, Sanloup C, Aléon J, Blanchard M, Paulatto L, Bureau H (2019) First-principles modeling of chlorine isotope fractionation between chloride-bearing molecules and minerals. Chem Geol 525:424–434

Banks DA, Green R, Cliff RA, Yardley BWD (2000a) Chlorine isotopes in fluid inclusions: determination of the origins of salinity in magmatic fluids. Geochim Cosmochim Acta 64:1785–1789

Banks DA, Gleeson SA, Green R (2000b) Determination of the origin of salinity in granite-related fluids: evidence from chlorine isotopes in fluid inclusions. J Geochem Explor 69–70:309–312

Baran A, Pelc A, Chmiel S, Halas S (2012) Pochodzenie mineralnych wód chlorkowo-sodowych w rejonie Rzeszowa w świetle badań izotopowych (δ^{18}O i δD, δ^{13}C, δ^{37}Cl) i chemicznych (in Polish: Origin of saline waters in the vicinity of Rzeszów in light of isotope (δ^{18}O i δD, δ^{13}C, δ^{37}Cl) and hydrochemical studies). Prz Geol 60:657–664

Baran I, Baran A, Chmiel S, Pelc A, Hlas S (2015) Badania izotopowe δ^{18}O, δD oraz δ^{37}Cl w wybranych wodach mineralnych Podkarpacia (in polish: isotopic investigation of δ^{18}O, dD and δ^{37}Cl of selected mineral waters in the Subcarpathian area). Prz Geol 63:172–178

Barnes JD, Cisneros M (2012) Mineralogical control on the chlorine isotope composition of altered oceanic crust. Chem Geol 326–327:51–60

Barnes JD, Sharp ZD (2006) A chlorine isotope study of DSDP/ODP serpentinized ultramafic rocks: insights into the serpentinization process. Chem Geol 228:246–265

Barnes JD, Selverstone J, Sharp ZD (2006) Chlorine isotope chemistry of serpentinites from Elba, Italy, as an indicator of fluid source and subsequent tectonic history. Geochem Geophys Geosyst 7:Q08015

Barnes JD, Sharp ZD, Fischer TP (2008) Chlorine isotope variations across the Izu-Bonin-Mariana arc. Geology 36:883–886

Barnes JD, Sharp ZD, Fisher TP, Hilton DR, Carr MJ (2009) Chlorine isotope variations along the central American volcanic front and back arc. Geochem Geophys Geosyst 10:Q11S17

Barnes JD, Eldam R, Lee CTA, Errico JC, Loewy S, Cisneros M (2013) Petrogenesis of serpentinites from the Franciscan complex, western California, USA. Lithos 178:143–157

Barnes JJ, Tartèse R, Anand M, McCubbin FM, Neal CR, Franchi IA (2016) Early degassing of lunar urKREEP by crust-breaching impact(s). Earth Planet Sci Lett 447:84–94

Barnes JJ, Franchi IA, McCubbin FM, Anand M (2019a) Multiple volatile reservoirs in the lunar interior revealed by the isotopic composition of chlorine in lunar basalts. Geochim Cosmochim Acta 266:144–162

Barnes JD, Penniston-Dorland SC, Bebout GE, Hoover W, Beaudoin GM, Agard P (2019b) Chlorine and lithium behavior in metasedimentary rocks during prograde metamorphism: a comparative study of exhumed subduction complexes (Catalina schist and Schistes Lustrés). Lithos 336–337:40–53

Barrett TJ (2018) The history of water and other associated volatiles in Howardite-Eucrite-Diogenite meteorites. PhD Thesis. Open University, Milton Keyes, UK

Barrett TJ, Barnes JJ, Anand M, Franchi IA, Greenwood RC, Charlier BLA, Zhao X, Moynier F, Grady MM (2019) Investigating magmatic processes in the early solar system using the Cl isotopic systematics of eucrites. Geochim Cosmochim Acta 266:582–597

Beekman HE, Eggenkamp HGM, Appelo CAJ (2011) An integrated modelling approach to reconstruct complex solute transport mechanisms – Cl and δ^{37}Cl in pore water of sediments from a former brackish lagoon in The Netherlands. Appl Geochem 26:257–268

Belluci JJ, Whitehouse MJ, John T, Nemchin AA, Snape JF, Bland PA, Benedix GK (2017) Halogen and Cl isotopic systematics in Martian phosphates: implications for the Cl cycle and surface halogen reservoirs on Mars. Earth Planet Sci Lett 458:192–202

Benard B, Famin V, Agrinier P, Aunay B, Lebeau G, Sanjuan B, Vimaux F, Bardoux G, Dezayes C (2020) Benart et al. 2020 Origin and fate of hydrothermal fluids at piton des Neiges volcano (Réunion Island): a geochemical and isotopic (O, H, C, Sr, Li, Cl) study of thermal springs. J Volcanol Geotherm Res 392:106682

Beneteau KM, Aravena R, Frape SK (1999) Isotopic characterization of chlorinated solvents-laboratory and field results. Org Geochem 30:739–753

Bernal NF, Gleeson SA, Dean AS, Liu XM, Hoskin P (2014) The source of halogens in geothermal fluids from the Taupo Volcanic Zone, North Island, New Zealand. Geochim Cosmochim Acta 126:265–283

Bernstein A, Shouakar-Stash O, Ebert K, Laskov C, Hunkeler D, Jeannottat S, Sakaguchi-Söder K, Laaks J, Jochmann MA, Cretnik S, Jager J, Haderlein SB, Schmidt TC, Aravena R, Elsner M (2011) Compound-specific chlorine isotope analysis: a comparison of gas chromatography / isotope ratio mass spectrometry and gas chromatography / quadrupole mass spectrometry methods in an interlaboratory study. Anal Chem 83:7624–7634

Böhlke JK, Sturchio NC, Gu B, Horita J, Brown GM, Jackson WA, Batista JR, Hatzinger PB (2005) Perchlorate isotope forensics. Anal Chem 77:7838–7842

Böhlke JK, Hatzinger PB, Sturchio NC, Gu B, Abbene I, Mroczkowski SJ (2009) Atacama perchlorate as an agricultural contaminant in groundwater: isotopic and chronologic evidence from Long Island, New York. Environ Sci Technol 43:5619–5625

Bonifacie M, Charlou JL, Jendrzejewski N, Agrinier P, Donval JP (2005) Chlorine isotopic compositions of high temperature hydrothermal vent over ridge axes. Chem Geol 221:279–288

Bonifacie M, Jendrzejewski N, Agrinier P, Coleman M, Pineau F, Javoy M (2007) Pyrohydrolysis-IRMS determination of silicate chlorine stable isotope compositions. Application to oceanic crust and meteorite samples. Chem Geol 242:187–201

Bonifacie M, Jendrzejewski N, Agrinier P, Humler E, Coleman M, Javoy M (2008) The chlorine isotope composition of earth's mantle. Science 319:1518–1520

Boschetti T, Toscani L, Shouakar-Stash O, Iacumin P, Venturelli G, Mucchino C, Frape SK (2011) Salt waters of the northern Apennine Foredeep Basin (Italy): origin and evolution. Aquat Geochem 17:71–108

Bourg IC, Sposito G (2007) Molecular dynamics simulations of kinetic isotope fractionation during the diffusion of ionic species in liquid water. Geochim Cosmochim Acta 71:5583–5589

Bouvier AS, Manzini M, Rose-Koga EF, Nichols ARL, Baumgartner LP (2019) Tracing of Cl input into the sub-arc mantle through the combined analysis of B, O and Cl isotopes in melt inclusions. Earth Planet Sci Lett 507:30–39

Bouvier AS, Portnyagin MV, Flemetakis S, Hoernle K, Klemme S, Berndt J, Mironov NL, John T (2022a) Chlorine isotope behavior in subduction zone settings revealed by olivine-hosted melt inclusions from the Central America Volcanic Arc. Earth Planet Sci Lett 581:117414

Bouvier AS, Rose-Koga EF, Chapuis A (2022b) Deciphering degassing and source effects in Cl isotopes in melt inclusions: the possible role of amphibole in the magma source of Stromboli (Aeolian Island arc). Front Earth Sci 9:793259

Bridges JC, Banks DA, Smith M, Grady MM (2004) Halite and stable chlorine isotopes in the Zag H3-6 breccia. Meteorit Planet Sci 39:657–666

Cameron AE, Herr W, Herzog W, Lundén A (1956) Isotopen-Anreicherung beim Brom durch electrolytische Überführung in geschmolzenem Bleibromid. Z Naturforschng 11a:203–205

Campbell DJ (1985) Fractionation of stable chlorine isotopes during transport through semipermeable membranes. MSc thesis. University of Arizona, Tucson, Ariz

Cao FF, Struchio NC, Ollivier P, Devau N, Heraty LJ, Jaunat J (2020) Sources and behavior of perchlorate in a shallow chalk aquifer under military (World War I) and agricultural influences. J Hazard Mater 398:123072

Carrizo D, Unger M, Holmstrand H, Andersson P, Gustafsson Ö, Sylva SP, Reddy CM (2011) Compound-specific bromine isotope compositions of one natural and six industrially synthesised organobromine substances. Environ Chem 8:127–132

Chen LZ, Ma T, Du Y, Yang J, Liu L, Shan HM, Liu CF, Cai HS (2014) Origin and evolution of formation water in North China plain based on hydrochemistry and stable isotopes (^{2}H, ^{18}O, ^{37}Cl and ^{81}Br). J Geochem Explor 145:250–259

Chen LZ, Ma T, Du Y, Xiao C, Chen XM, Liu CF, Wang YX (2016) Hydrochemical and isotopic (2H, 18O and 37Cl) constraints on evolution of geothermal water in coastal plain of Southwestern Guangdong Province, China. J Volcanol Geotherm Res 318:45–54

Chen LZ, Shouakar-Stash O, Ma T, Wang CL, Liu L (2017) Significance of stable carbon and bromine isotopes in the source identification of PBDEs. Chemosphere 186:160–166

Chiaradia M, Banks D, Cliff R, Marschik R, de Haller A (2006) Origin of fluids in iron oxide-copper-gold deposits: constraints from $\delta^{37}Cl$, $^{87}Sr/^{86}Sr_i$ and Cl/Br. Miner Deposita 41:565–573

Ciu ZX, Yang Q, Xia XP, Wang R, Bonifacie M, Lai CK, Zhang WF, Zhang YQ, Xu J (2021) High-precision apatite $\delta^{37}Cl$ measurement by SIMS with a 10^{12} Ω amplifier faraday cup. J Anal At Spectrom 37:222–238

Coleman ML, Eggenkamp HGM, Aranyossy JF (2001) Chlorine stable isotope characterisation of solute transport in mudrocks. in Etude pour la faisabilité des stockages de déchets radioactifs. Actes Journ Scientif ANDRA, Nancy, 7, 8 et 9 décembre 1999. pp 155–175

Coleman ML, Ader M, Chaudhuri S, Coates JD (2003) Microbial isotopic fractionation of perchlorate chlorine. Appl Environ Microbiol 69:4997–5000

Cullen JT, Hurwitz S, Barnes JD, Lassiter JC, Penniston-Dorland S, Kasemann SA, Thordsen JJ (2019) Temperature-dependent variations in mineralogy, major element chemistry and the stable isotopes of boron, lithium and chlorine resulting from hydration of rhyolite: constraints from hydrothermal experiments at 150 to 350 °C and 25 MPa. Geochim Cosmochim Acta 261:269–287

Cullen JT, Hurwitz S, Barnes JD, Lassiter JC, Penniston-Dorland S, Meixner A, Wilckens F, Kasemann SA, McClesky RB (2021) The systematics of chlorine, lithium, and boron and $\delta^{37}Cl$, $\delta^{7}Li$, and $\delta^{11}B$ in the hydrothermal system of the Yellowstone plateau volcanic field. Geochem Geophys Geosyst 22:e2020GC009589

Curie I (1921) Sur le poids atomique du chlore dans quelques mineraux. Compte Rend Sean 172:1025–1028

Darling JR, White LF, Kizovski T, Černok A, Moser DE, Tait KT, Dunlop J, Langelier B, Douglas JO, Zhao X, Franchi IA, Anand M (2021) The shocking state of apatite and merrillite in shergottite Northwest Africa 5298 and extreme nanoscale chlorine isotope variability revealed by atom probe tomography. Geochim Cosmochim Acta 293:422–437

Dembska-Sięka P, Pelc A, Rajchem L (2020) The origin and chemical composition of Carpathians chloride CO2-rich waters in the light of stable chlorine isotope studies (^{37}Cl and ^{35}Cl). Geol Quart 64:197–204

Desaulniers DE, Kaufmann RS, Cherry JA, Bentley HW (1986) ^{37}Cl-^{35}Cl variations in a diffusion controlled groundwater system. Geochim Cosmochim Acta 50:1757–1764

Du Y, Ma T, Yang J, Liu L, Shan HM, Cai HS, Liu CF, Chen LZ (2013) A precise analytical method for bromine stable isotopes in natural waters by GasBench II-IRMS. Int J Mass Spectrom 338:50–56

Du Y, Ma T, Chen LZ, Xiao C, Liu CF (2016) Chlorine isotopic constraint on contrastive genesis of representative coastal and inland shallow brine in China. J Geochem Explor 170:21–29

Eastoe CJ (2016) Stable chlorine isotopes in arid non-marine basins: instances and possible fractionation mechanisms. Appl Geochem 74:1–12

Eastoe CJ, Meng FW (2024) Is Cl isotope fractionation in the seawater-evaporite system dependent on seawater chemistry? Chem Geol 644:121832

Eastoe CJ, Peryt T (1999) Multiple sources of chloride in Badenian evaporites, Carpathian mountains: stable chlorine isotope evidence. Terra Nova 11:118–123

Eastoe CJ, Long A, Knauth LP (1999) Stable chlorine isotopes in the Palo Duro Basin, Texas: evidence for preservation of Permian evaporite brines. Geochim Cosmochim Acta 63:1375–1382

Eastoe CJ, Long A, Land LS, Kyle JR (2001) Stable chlorine isotopes in halite and brine from the Guld Coast Basin: brine genesis. Chem Geol 176:343–360

Eastoe CJ, Peryt TM, Petrychenko OY, Geisler-Cussey D (2007) Stable chlorine isotopes in Phanerozoic evaporites. Appl Geochem 22:575–588

Eggenkamp HGM (1994) δ^{37}Cl; the geochemistry of chlorine isotopes. Geol Ultraiect 116:1–150

Eggenkamp HGM, Coleman ML (2000) Rediscovery of classical methods and their application to the measurement of stable bromine isotopes in natural samples. Chem Geol 167:393–402

Eggenkamp HGM, Coleman ML (2009) The effect of aqueous diffusion on the fractionation of chlorine and bromine stable isotopes. Geochim Cosmochim Acta 73:3539–3548

Eggenkamp HGM, Koster van Groos AF (1997) Chlorine stable isotopes in carbonatites: evidence for isotopic heterogeneity in the mantle. Chem Geol 140:137–143

Eggenkamp HGM, Louvat P (2018) A simple distillation method to extract bromine from natural water and salt samples for isotope analysis by multi-collector inductively coupled plasma mass spectrometry. Rapid Commun Mass Spectrom 32:612–618

Eggenkamp HGM, Middelburg JJ, Kreulen R (1994) Preferential diffusion of ^{35}Cl relative to ^{37}Cl in sediments of Kau Bay, Halmahera, Indonesia. Chem Geol (Isot Geosci Sect) 116:317–325

Eggenkamp HGM, Kreulen R, Koster van Groos AF (1995) Chlorine stable isotope fractionation in evaporites. Geochim Cosmochim Acta 59:5169–5175

Eggenkamp HGM, Bonifacie M, Ader M, Agrinier P (2016) Experimental determination of stable chlorine and bromine isotope fractionation during precipitation of salt from a saturated solution. Chem Geol 433:46–56

Eggenkamp HGM, Louvat P, Griffioen J, Agrinier P (2019a) Chlorine and bromine isotope evolution within a fully developed upper Permian natural salt sequence. Geochim Cosmochim Acta 245:316–326

Eggenkamp HGM, Louvat P, Agrinier P, Bonifacie M, Bekker A, Krupenik V, Griffioen J, Horita J, Brocks JJ, Bagheri R (2019b) The bromine and chlorine isotope composition of primary halite deposits and their significance for the secular isotope composition of seawater. Geochim Cosmochim Acta 264:13–29

Eggenkamp HGM, Marks MAW, Louvat P, Markl G (2021) Bromine isotope variations in magmatic and hydrothermal sodalite and Tugtupite and the estimation of Br isotope fractionation between melt and sodalite. Minerals 11:370

Eggenkamp HGM, Marks MAW, Bonifacie M, Bardoux G, Agrinier P, Markl G (2022) Cl isotope fractionation in magmatic and hydrothermal eudialyte, sodalite and tugtupite (Ilímaussaq intrusion, South Greenland). Chem Geol 604:120932

Estrada NL, Böhlke JK, Sturchio NC, Gu B, Harvey G, Burkey KO, Grantz DA, MacGrath MT, Anderson TA, Rao B, Sevanthi R, Hatzinger PB, Jackson WA (2017) Stable isotopic composition of perchlorate and nitrate accumulated in plants: hydroponic experiments and field data. Sci Total Environ 595:556–566

Filippini M, Nijenhuis I, Kümmel S, Chiarini V, Crosta G, Richnow HH, Gargini A (2018) Multi-element compound specific stable isotope analysis of chlorinated aliphatic contaminants derived from chlorinated pitches. Sci Tot Environ 640–641:153–162

Fortin MA, Watson EB, Stern R (2017) The isotope mass effect on chlorine diffusion in dacite melt, with implications for fractionation during bubble growth. Earth Planet Sci Lett 480:15–24

Gan YQ, Yu TT, Zhou AG, Liu YD, Liu CF (2013) A technique for carbon and chlorine isotope analyses of chlorinated aliphatic hydrocarbons in groundwater. J Earth Sci 24:274–281

Gao CH, Liu Y (2021) First-principles calculations of equilibrium bromine isotope fractionations. Geochim Cosmochim Acta 297:65–81

Gargano A, Sharp Z (2019) The chlorine isotope composition of iron meteorites: evidence for the Cl isotope composition of the solar nebula and implications for extensive devolatilization during planet formation. Meteorit Planet Sci 54:1619–1631

Gargano A, Sharp Z, Shearer C, Simon JI, Halliday A, Buckley W (2020) The Cl isotope composition and halogen contents of Apollo-return samples. Proc Natl Acad Sci 117:38

Gargano A, Dottin J, Hopkins SS, Sharp Z, Shearer C, Halliday AN, Larner F, Farquar J, Simon JI (2022) The Zn, S, and Cl isotope compositions of mare basalts: implications for the effects of eruption style and pressure on volatile element stable isotope fractionation on the moon. Am Mineral 107:1985–1994

Gelman F, Halicz L (2010) High precision determination of bromine isotope ratio by GC-MC-ICPMS. Int J Mass Spectrom 289:167–169

Gelman F, Halicz L (2011) High-precision isotope ratio analysis of inorganic bromide by continuous flow MC-ICPMS. Int J Mass Spectrom 307:211–213

Gil G, Borowski MP, Barnes JD, Jokubauskas P, Bagiński B, Gunia P, Ilnicki S (2022) Formation of serpentinite-hosted talc in a continental crust setting: petrographic, mineralogical, geochemical, and O, H and Cl isotope study of the Gilów deposit, Góry Sowie Massif (SW Poland). Ore Geol Rev 146:104926

Gilevska T, Ivdra N, Bonifacie M, Richnow HH (2015) Improvement of analytical method for chlorine dual-inlet isotope ratio mass spectrometry of organochlorines. Rapid Commun Mass Spectrom 29:1343–1350

Giunta T, Labadi J, Eggenkamp HGM (2013) Experimental determination of chlorine isotope fractionation in Cl_2-Cl^-_{aq} and $ClOH_{aq}$-Cl^-_{aq}. 23th Annual V.M. Goldschmidt Conference. Firenze, Italy

Giunta T, Labidi J, Eggenkamp HGM (2017) Chlorine isotope fractionation between chloride (Cl^-) and dichlorine (Cl_2). Geochim Cosmochim Acta 213:375–382

Gleeson SA, Smith MP (2009) The sources and evolution of mineralising fluids in iron oxide-copper-gold systems, Norrbotten, Sweden: constraints from Br/Cl ratios and stable Cl isotopes of fluid inclusion leachates. Geochim Cosmochim Acta 73:5658–5672

Godon A, Jendrzejewski N, Eggenkamp HGM, Banks DA, Ader M, Coleman ML, Pineau F (2004) A cross calibration of chlorine isotopic measurements and suitability of seawater as the international reference material. Chem Geol 207:1–12

Graczyk DG, Taylor JW (1974) Chlorine kinetic isotope-effects in nucleophilic-substitution reactions – support for ion-pairs mechanism in reactions of p-methoxybenzyl chloride in 70 percent aqueous acetone. J Am Chem Soc 96:3255–3261

Grimsrud EP, Taylor JW (1970) Chlorine kinetic isotope effects in nucleophilic displacements at a saturated carbon. J Am Chem Soc 92:739–741

Groen J, Velstra J, Meesters AGCA (2000) Salinization processes in paleowaters in coastal sediments from Suriname: evidence from δ^{37}Cl analysis and diffusion modelling. J Hydrol 234:1–20

Gui JY, Chen X, Han XD, Wang Z, Ma YQ (2022) Determination of stable chlorine isotopes by Isotopic Preparative Chromatography (IPC) with positive–thermal ionization mass spectrometry (P-TIMS). Anal Lett 55:2564–2573

Gwynne R, Frape S, Shouakar-Stash O, Love A (2013) ^{81}Br, ^{37}Cl and ^{87}Sr studies to assess groundwater flow and solute sources in the southwestern great Artesian Basin, Australia. Prog Earth Planet Sci 7:330–333

Halas P, Pelc A (2009) New isotope ratio mass spectrometric method of precise δ^{37}Cl determinations. Rapid Commun Mass Spectrom 23:1061–1064

Halloran LJS, Vakili F, Wanner P, Shouakar-Stash O, Hunkeler D (2021) Sorption- and diffusion-induced isotopic fractionation in chloroethenes. Sci Total Environ 788:147826

Han ZY, Hui HJ, Wei HZ, Li WQ (2023) Isotopic fractionation of chlorine and potassium during chloride sublimation under lunar conditions. Geochim Cosmochim Acta 353:112–128

Hanlon C, Stotler R, Frape S, Gwynne R (2017) Comparison of δ81Br and δ37Cl composition of volatiles, salt precipitates, and associated water in terrestrial evaporative saline lake systems. Isot Environ Health Stud 53:446–465

Harkins WD (1920a) The separation of the element chlorine into normal chlorine and meta-chlorine, and the positive electron. Nature 105:230–231

Harkins WD (1920b) The separation of the element chlorine into chlorine and meta-chlorine. Science 51:289–291

Harkins WD (1920c) The nuclei of atoms and the new periodic system. Phys Rev 15:73–94

Harkins WD, Hall RE (1916) The periodic system and the properties of the elements (Fifth paper on atomic structure). J Am Chem Soc 38:169–221

Harkins WD, Hayes A (1922) The separation of the element chlorine into isotopes (isotopic atomic species). Phys Rev 19:403–404

Harkins WD, Jenkins FA (1926) The separation of the element chlorine into isotopes. The light fraction. J Am Chem Soc 48:58–69

Harkins WD, Stone SB (1926) The isotopic composition and atomic weight of chlorine from meteorites and from minerals of non-marine origin. J Am Chem Soc 48:938–949

Harkins WD, Wilson ED (1915) The changes of mass and weight involved in the formation of complex atoms. J Am Chem Soc 37:1367–1383

Hatzinger PB, Harvey G, Jackson WA, Böhlke JK, Sturchio NC, Gu BH, Grantz D, Burkey K, McGrath M (2017) Identification and characterization of natural sources of perchlorate. Final Report of Phase II. SERDP Project ER-1435

Hatzinger PB, Böhlke JK, Jackson WA, Gu BH, Mroczkowski SJ, Struchio NC (2022) Isotopic discrimination of natural and anthropogenic perchlorate sources in groundwater in a semi-arid region of northeastern Oregon (USA). Appl Geochem 139:105232

Hayden TS (2022) Assessing the volatile inventory and history of the moon using lunar meteorites. PhD Thesis. Open University, Milton Keyes, UK

Heckel B, Rodríguez-Fernández D, Torrentó C, Meyer A, Palau J, Domènech C, Rosell M, Soler A, Hunkeler D, Elsner M (2017) Compound-specific chlorine isotope analysis of tetrachloromethane and trichloromethane by gas chromatographyisotope ratio mass spectrometry vs gas chromatography-quadrupolemass spectrometry: method development and evaluation of precision and trueness. Anal Chem 89:3411–3420

Henkemans E, Frape SK, Ruskeeniemi T, Anderson NJ, Hobbs M (2018) A landscape-isotopic approach to the geochemical characterization of lakes in the Kangerlussuaq region, West Greenland. Arct Antarct Alp Res 50:e1420863

Herzog W, Dörnenburg E (1958) Die Verwendung van Methylchlorid zur massenspektrometrichen Isotopenanalyse von Chlor. Z Naturforschg 13a:51–52

Herzog W, Klemm A (1958) Die Temperaturabhängigkeit des Isotopie-Effekts bei der elektrolytischen Wanderungen der Chlorionen in geschmolzenem Thallium(I)-chlorid. Z Naturforschg 13a:7–16

Hoering TC, Parker PL (1961) The geochemistry of the stable isotopes of chlorine. Geochim Cosmochim Acta 23:186–199

Holmstrand H, Andersson P, Gustafsson Ö (2004) Chlorine isotope analysis of submicromole organochlorine samples by sealed tube combustion and thermal ionization mass spectrometry. Anal Chem 76:2336–2342

Holmstrand H, Gadomski D, Mandalakis M, Tysklind M, Irvine R, Andersson P, Gustafsson Ö (2006) Origin of PCDDs in ball clay assessed with compound-specific chlorine isotope analysis and radiocarbon dating. Environ Sci Technol 40:3730–3735

Holmstrand H, Mandalakis M, Zencak Z, Andersson P, Gustafsson O (2007) First compound-specific chlorine-isotope analysis of environmentally-bioaccumulated organochlorines indicates a degradation-relatable kinetic isotope effect for DDT. Chemosphere 69:1533–1539

Holmstrand H, Unger M, Carrizo D, Andersson P, Gustafsson Ö (2010a) Compound-specific bromine isotope analysis of brominated diphenyl ethers using GC-ICP-multicollector-MS. Rapid Commun Mass Spectrom 24:2135–2142

Holmstrand H, Zencak Z, Mandalakis M, Andersson P, Gustafsson Ö (2010b) Chlorine isotope evidence for the anthropogenic origin of tris-(4-chlorophenyl) methane. Appl Geochem 25:1301–1306

Holt BD, Sturchio NC, Abrajano TA, Heraty LJ (1997) Conversion of chlorinated volatile organic compounds to carbon dioxide and methyl chloride for isotopic analysis of carbon and chlorine. Anal Chem 69:2727–2733

Holt BD, Heraty LJ, Sturchio NC (2001) Extraction of chlorinated aliphatic hydrocarbons from groundwater at micromolar concentrations for isotopic analysis of chlorine. Environ Pollut 113:263–269

Horst A (2013) Stable bromine isotopic composition of methyl bromide. Method development and applications. PhD thesis. Stockholm University

Horst A, Renpenning J, Richnow HH, Gehre M (2017) Compound specific stable chlorine isotopic analysis of volatile aliphatic compounds using gas chromatography hyphenated with multiple collector inductively coupled plasma mass spectrometry. Anal Chem 89:9131–9138

Horst A, Bonifacie M, Bardoux G, Richnow HH (2019) Isotopic characterization (^{2}H, ^{13}C, ^{37}Cl, ^{81}Br) of abiotic degradation of methyl bromide and methyl chloride in water and implications for future studies. Environ Sci Technol 53:8813–8822

Hu S, Lin YT, Zhang JC, Hao JL, Xing WF, Zhang T, Yang W, Changela H (2019) Ancient geologic events on Mars revealed by zircons and apatites from the Martian regolith breccia NWA 7034. Meteorit Planet Sci 54:850–879

Hunkeler D, Van Breukelen BM, Elsner M (2009) Modelling chlorine isotope trends during sequential transformation of chlorinated ethenes. Environ Sci Technol 43:6750–6756

Hunkeler D, Aravena R, Shouakar-Stash O, Weisbrod N, Nasser A, Netzer L, Ronen D (2011) Carbon and chlorine isotope ratios of chlorinated ethenes migrating through a thick unsaturated zone of a sandy aquifer. Environ Sci Technol 45:8247–8253

Ivdra N, Ficher A, Herrero-Martin S, Giunta T, Bonifacie M, Richnow HH (2017) Carbon, hydrogen and chlorine stable isotope fingerprinting for forensic investigations of Hexachlorocyclohexanes. Environ Sci Technol 51:446–454

Jendrzejewski N, Eggenkamp HGM, Coleman ML (1997) Sequential determination of chlorine and carbon isotopic composition in single microliter samples of chlorinated solvent. Anal Chem 69:4259–4266

Jendrzejewski N, Eggenkamp HGM, Coleman ML (2001) Characterisation of chlorinated hydrocarbons from chlorine and carbon isotopic compositions: scope and application to environmental problems. Appl Geochem 16:1021–1031

John T, Layne GD, Haase KM, Barnes JD (2010) Chlorine isotope evidence for crustal recycling into Earth's mantle. Earth Planet Sci Lett 298:175–182

Kannath S, Adamczyk P, Wu LP, Richnow HH, Dybala-Defratyka A (2019) Can alkaline hydrolysis of γ-HCH serve as a model reaction to study its aerobic enzymatic Dehydrochlorination by LinA? Int J Mol Sci 20:5955

Kaown D, Shouakar-Stash O, Yang J, Hyun YJ, Lee KK (2014) Identification of multiple sources of groundwater contamination by dual isotopes. Ground Water 52(6):875–885. https://doi.org/10.1111/gwat.12130

Kaufmann RS (1984) Chlorine in groundwater: stable isotope distribution. Ph.D. thesis. University of Arizona, Tucson, Ariz

Kaufmann RS, Long A, Bentley H, Davis S (1984) Natural chlorine isotope variations. Nature 309:338–340

Kaufmann RS, Frape S, Fritz P, Bentley H (1987) Chlorine stable isotope composition of Canadian shield brines. In: Fritz P, Frape SK (eds) Saline water and gases in crystalline rocks. Geol Assoc Canada, Special Paper 33, pp 89–93

Kaufmann RS, Frape SK, McNutt R, Eastoe CJ (1993) Chlorine stable isotope distribution of Michigan basin formation waters. Appl Geochem 8:403–407

Klemm A, Lundén A (1955) Isotopenanreicherung beim Chlor durch electrolytische Überfürung in geschmolzenem Bleichlorid. Z Naturforschg 10a:282–284

Kohler G, Wassenaar LI (2009) The stable isotopic composition (^{37}Cl/^{35}Cl) of dissolved chloride in rainwater. Appl Geochem 25:91–96

Kohli P, Richnow HH, Lal R (2017) Uptake and transformation of Hexachlorocyclohexane isomers (HCHs) in tree growth rings at a contaminated field site. Indian J Microbiol 57:11–22

Kusebauch C, John T, Whitehouse MJ, Engvik AK (2015) Apatite as probe for the halogen composition of metamorphic fluids (Bamble Sector, SE Norway). Contrib Min Petrol 170:34

Lavastre V, Jendrzejewski N, Agrinier P, Javoy M, Evrard M (2005) Chlorine transfer out of a very low permeability clay sequence (Paris Basin, France): ^{35}Cl and ^{37}Cl evidence. Geochim Cosmochim Acta 69:4949–4961

Layne GD, Godon A, Webster JD, Bach W (2004) Secondary ion mass spectrometry for the determination of $\delta^{37}Cl$. Part I. Ion microprobe analysis of gasses and fluids. Chem Geol 207:277–289

Layne GD, Kent AJR, Bach W (2009) $\delta^{37}Cl$ systematics of a backarc spreading system: the Lau Basin. Geology 37:427–430

Le Gal La Salle C, Matray JM, Bensenouci F, Michelot JL, Dauzères A, Wittebroodt C, Frape S, Shouakar Stash O, Rebaix R, Lancelot J (2013) Modelling Cl-concentration and $\delta^{37}Cl$ profiles in porewater across a 250m-thick indurated clayrock at the tournemire URL (France). Proc Earth Planet Sci 7:471–474

Lenoir L, Blaise T, Chourio-Camacho D, Richard A, Tarentola A, Agrinier P, Rigaudier T, Monvoisin G, Bardoux G, Brigaud B, Barbarand J (2024) The origin of fluorite-barite mineralization at the interface between the Paris Basin and its Variscan basement: insights from fluid inclusion chemistry and isotopic (O, H, Cl) composition. Miner Deposita 59:397–417

Li XQ, Zhou AG, Liu YD, Ma T, Liu CF, Liu L, Yang J (2012) Stable isotope geochemistry of dissolved chloride in relation to hydrogeology of the strongly exploited quaternary aquifers, North China Plain. Appl Geochem 27:2031–2041

Lian S, Nikolausz M, Nijenhuis I, Leite FA, Richnow HH (2018) Biotransformation and inhibition effects of hexachlorocyclohexanes during biogas production from contaminated biomass characterized by isotope fractionation concepts. Bioresour Technol 250:683–690

Lihl C, Heckel B, Grybkowska A, Dybala-Defratyka A, Ponsin V, Torrető C, Hunkeler D, Elsner M (2020) Compound-specific chlorine isotope fractionation in biodegradation of atrazine. Environ Sci Process Impacts 22:792–801

Liotta M, Rizzo AL, Barnesw JD, D'Auria L, Mertelli M, Bobrowski N, Wittmer J (2017) Chlorine isotope composition of volcanic rocks and gases at Stromboli volcano (Aeolian Islands, Italy): inferences on magmatic degassing prior to 2014 eruption. J Volcanol Geotherm Res 336:168–178

Liu WG, Xiao YK, Wang QZ, Qi HP, Wang YH, Zhou YM, Shirodkar PV (1997) Chlorine isotopic geochemistry of salt lakes in the Qaidam Basin, China. Chem Geol 136:271–279

Liu YD, Zhou AG, Gan YQ, Liu CF, Yu TT, Li XQ (2013) An online method to determine chlorine stable isotope composition by continuous flow isotope ratio mass spectrometry (CF-IRMS) coupled with a Gasbench II. J South Cent Univ 20:193–198

Liu X, Wu LP, Kümmel S, Merbach I, Lal R, Richnow HH (2019) Compound-specific isotope analysis and enantiomer fractionation to characterize the transformation of Hexachlorocyclohexane isomers in a soil–wheat pot system. Environ Sci Technol 53:8715–8724

Liu X, Wang Y, Wei HZ (2020a) Advances in stable chlorine isotope geochemistry. Earth Sci Front 27:29–41. (in Chinese with English abstract)

Liu YQ, Kümmel S, Yao J, Nijenhuis I, Richnow HH (2020b) Dual C-Cl isotope analysis for characterizing the anaerobic transformation of α, β, γ, and δ-hexachlorocyclohexane in contaminated aquifers. Water Res 184:116128

Liu X, Wu LP, Kümmel S, Merbach I, Lol R, Richnow HH (2020c) Compound-specific isotope analysis and enantiomer fractionation to characterize the transformation of Hexachlorocyclohexane isomers in a soil-wheat pot system. Environ Sci Technol 54:8690–8698

Liu X, Wei HZ, Li YC, Williams-Jones AE, Lu JJ, Jiang SY, Dong G, Ma J, Eastoe CJ (2021a) Chlorine isotope mantle heterogeneity: constraints from theoretical first-principles calculations. Chem Geol 572:120193

Liu X, Wei HZ, Williams-Jones AE, Ma J, Lu JJ, Jiang SY, Li YC, Dong G (2021b) Chlorine isotope fractionation during serpentinization and hydrothermal mineralization: a density functional theory study. Chem Geol 581:120406

Liu X, Bonhomme J, Merbach I, Kümmel S, Richnow HH (2021c) Uptake of α-HCH by wheat from the gas phase and translocation to soil analyzed by a stable carbon isotope labeling experiment. Chemosphere 264:128489

Liu X, Wu LP, Kümmel S, Richnow HH (2021d) Characterizing the biotransformation of hexachlorocyclohexanes in wheat using compound-specific stable isotope analysis and enantiomer fraction analysis. J Hazard Mater 406:124301

Liu X, Wu LP, Kümmel S, Richnow HH (2022a) Stable isotope fractionation associated with the synthesis of hexachlorocyclohexane isomers for characterizing sources. Chemosphere 296:133938

Liu Y, Li QL, Tang GQ, Li XH (2022b) Simultaneous determination of δ18O and δ37Cl isotopic compositions and the concentrations of F, Cl, and S in apatite using SIMS. Crystals 12:383

Liu X, Kümmel S, Trapp S, Richnow HH (2023) Uptake and transformation of Hexachlorocyclohexane isomers (HCHs) in tree growth rings at a contaminated field site. Environ Sci Technol 57:8776–8784

Long A, Eastoe CJ, Kaufmann RS, Martin JG, Wirt L, Finley JB (1993) High-precision measurement of chlorine stable isotope ratios. Geochim Cosmochim Acta 57:2907–2912

Loomis JL, Coleman ML, Eggenkamp HGM (1997) Effect of ion exchange on solute transport in shallow groundwaters: an experimental study using chlorine isotopes. EUG IX. Terra Nova 9 Abst. Supp. 1:659

Louvat P, Bonifacie M, Giunta T, Michel A, Coleman M (2016) Determination of bromine stable isotope ratios from saline solutions by 'wet plasma' MC-ICPMS including a comparison between high and low-resolution modes, and three introduction systems. Anal Chem 88:3891–3898

Lundén A, Herzog W (1956) Isotopenanreicherung bei Chlor durch electrolytische Überführung in geschmolzenem Zinkchlorid. Z Naturforschg 11a:520

Lundén A, Lodding A (1960) Isotopenanreicherung bei Brom durch electrolytische Überführung in geschmolzenem Zinkbromid. Z Naturforschg 15a:320–322
Luo CG, Xiao YK, Ma HZ, Ma YQ, Zhang YL, He MY (2012) Stable isotope fractionation of chlorine during evaporation of brine from a saline lake. Chin Sci Bull 57:1833–1843
Luo CG, Xiao YK, Wen HJ, Ma HZ, Ma YQ, Zhang YL, Zhang YX, He MY (2014) Stable isotope fractionation of chlorine during the precipitation of single chloride minerals. Appl Geochem 47:141–149
Luo CG, Xiao YK, Wen HJ, Ma HZ, Ma YQ, Zhang YL, He MY (2015) Reply to the comment on the paper "stable isotope fractionation of chlorine during the precipitation of single chloride minerals". Appl Geochem 54:117–118
Luo CG, Wen HJ, Xiao YK, Ma HZ, Fa QH, Ma YK, Zhang YL, Yang XQ, He MY (2016) Chlorine isotopes in sediments of the Qarhan Playa of China and their paleoclimatic significance. Geochemistry 76:149–156
Ma YQ, Peng ZK, Chen YJ, Yang J, Xiao YK, Zhang YL (2016a) High precise determination of bromine isotopic ratios by positive thermal ionization mass spectrometry using static multicollection based on Cs_2Br^+ ions. Chin J Anal Chem 44:186–191
Ma YQ, Peng ZK, Han FQ, Liu WG, Yang J, Xiao YK, Zhang YL (2016b) Extension of mass range and application in the determination of bromine isotope by TIMS. J Chin Mass Spectrom Soc 37:465–470. (in Chinese with English abstract)
Magenheim AJ, Spivack AJ, Volpe C, Ransom B (1994) Precise determination of stable chlorine isotopic ratios in low-concentration natural samples. Geochim Cosmochim Acta 58:3117–3121
Magenheim AJ, Spivack AJ, Michael PJ, Gieskes JM (1995) Chlorine stable isotope composition of the oceanic crust: implications for Earth's distribution of chlorine. Earth Planet Sci Lett 131:427–432
Mandalakis M, Holmstrand H, Andersson P, Gustafsson Ö (2008) Compound-specific chlorine isotope analysis of polychlorinated biphenyls isolated from Aroclor and Clophen technical mixtures. Chemosphere 71:299–305
Manzini M, Bouvier AS, Barnes JD, Bonifacie M, Rose-Koga EF, Ulmer P, Métrich N, Bardoux G, Williams J, Layne GD, Straub S, Baumgartner L, John T (2017) SIMS chlorine isotope analyses in melt inclusions from arc settings. Chem Geol 449:112–122
Markl G, Musashi M, Bucher K (1997) Chlorine stable isotope composition of granulites from Lofoten, Norway: implications for the Cl isotope composition and for the source of Cl enrichment in the lower crust. Earth Planet Sci Lett 150:95–102
Martin JG (1993) Stable chlorine isotopes in the Milk River aquifer, Alberta, Canada: implications for the geochemical evolution of the aquifer system. MSc thesis. Univ Arizona, Tucson, Ariz
Musashi M, Markl G, Kreulen R (1998) Stable chlorine-isotope analysis of rock samples: new aspects of chlorine extraction. Anal Chim Acta 362:261–269
Musashi M, Oi T, Eggenkamp HGM (2004) Experimental determination of chlorine isotope separation factor by anion-exchange chromatography. Anal Chim Acta 508:37–40
Musashi M, Oi T, Eggenkamp HGM, Matsuo M (2008) Chlorine isotope fractionation associated with volcanic activity at the Kusatsu-Bandaiko hot spring in Japan. Isot Environ Health Stud 44:305–313
Nijenhuis I, Renpenning J, Kümmel S, Richnow HH, Gehre M (2015) Recent advances in multi-element compound-specific stable isotope analysis of organohalides: achievements, challenges and prospects for assessing environmental sources and transformation. Trends environ. Anal Chem 71:901–911
Owen HR, Schaeffer OA (1955) The isotope abundances of chlorine from various sources. J Am Chem Soc 77:898–899
Palau J, Jamin P, Badin A, Vanhecke N, Haerens B, Brouyère S, Hunkeler D (2016) Use of dual carbon-chlorine isotope analysis to assess the degradation pathways of 1,1,1-trichloroethane in groundwater. Water Res 92:235–243
Palau J, Yu R, Hatijah Mortan S, Shouakar-Stash O, Rosell M, Freedman DL, Sbarbati C, Fiorenza S, Aravena R, Marco-Urrea E, Elsner M, Soler A, Hunkeler D (2017) Distinct dual C-Cl isotope fractionation patterns during anaerobic biodegradation of 1,2-dichloroethane: potential to characterize microbial degradation in the field. Environ Sci Technol 51:2685–2694
Palau J, Trueba-Santiso A, Ye R, Mortan SH, Shouakar-Stash O, Freedman DL, Wasmund K, Hunkeler D, Marco-Urrea E, Rosell M (2023) Dual C-Br isotope fractionation indicates distinct reductive dehalogenation mechanisms of 1,2-Dibromoethane in Dehalococcoides- and Dehalogenimonas-containing cultures. Environ Sci Technol 57:1949–1958
Pekora M, Paneth A, Paneth P (2022) Unprecedently large $^{37}Cl/^{35}Cl$ equilibrium isotopic fractionation on nanoconfinement of chloride anion. Sci Rep 12:1768
Poghosyan A, Struchio NC, Morrison CG, Beloso AD, Guan YB, Eiler JM, Jackson WA, Hatzinger PB (2014) Perchlorate in the Great Lakes: isotopic composition and origin. Environ Sci Technol 48:11146–11153
Poghosyan A, Morel-Espinosa M, Valentin-Blasini L, Blount BC, Ferreccio C, Steinmaus CM, Struchio NC (2016) Chlorine isotopic composition of perchlorate in human urine as a means of distinguishing among exposure sources. J Expo Sci Environ Epidemiol 26:324–328
Ponsin V, Buschek TE, Hunkeler D (2018) How to apply compound-specific isotope analysis to complex environmental samples. Chimia 72:69
Ponsin V, Torrentó C, Lihl C, Elsner M, Hunkeler D (2019) Compound-specific chlorine isotope analysis

of the herbicides atrazine, acetochlor, and metolachlor. Anal Chem 91:14290–14298
Potts NJ, Barnes JJ, Tartèse R, Franchi IA, Anand M (2018) Chlorine isotopic compositions of apatite in Apollo 14 rocks: evidence for widespread vapor-phase metasomatism on the lunar nearside ~4 billion years ago. Geochim Cosmochim Acta 230:46–59
Ransom B, Spivack AJ, Kastner M (1995) Stable Cl isotopes in subduction-zone porewaters: implications for fluid rock reactions and the cycling of chlorine. Geology 23:715–718
Ranta E, Halldórsson SA, Barnes JD, Jónasson SA (2021) Chlorine isotope ratios record magmatic brine assimilation during rhyolite genesis. Geochem Perspect Lett 16:35–39
Rebeix R, Le Gal La Salle C, Jean-Baptiste P, Lavastre V, Fourré E, Bensenouci F, Matray JM, Landrein P, Shouakar-Stash O, Frape SK, Michelot JL, Lancelot J (2014) Chlorine transport processes through a 2000 m aquifer/aquitard system. Mar Pet Geol 53:102–116
Reich M, Barnes JD, Breecker DO, Barra F, Milojevic C, Drew D (2019) Chlorine isotope fractionation recorded in atacamite during supergene copper oxidation. Chem Geol 525:168–176
Reitz A, Pape T, Haeckel M, Schmidt M, Berner U, Scholz F, Liebetrau V, Aloisi G, Weise SM, Wallmann K (2011) Sources of fluids and gases expelled at cold seeps offshore Georgia, eastern Black Sea. Geochim Cosmochim Acta 75:3250–3268
Richard A, Banks DA, Mercadier J, Boiron MC, Cuney M, Cathelineau M (2011) An evaporated seawater origin for the ore-forming brines in unconformity-related uranium deposits (Athabasce Basin, Canada): Cl/Br and δ^{37}Cl analysis of fluid inclusions. Geochim Cosmochim Acta 75:2792–2810
Richet P, Bottinga Y, Javoy M (1977) A review of hydrogen, carbon, nitrogen, oxygen, Sulphur, and chlorine stable isotope fractionation among gaseous molecules. Annu Rev Earth Planet Sci 5:65–110
Rizzo AL, Caracausi A, Liotta M, Paonita A, Barnes JD, Corsaro RA, Martelli M (2013) Chlorine isotopic composition of volcanic gases and rocks at Mount Etna (Italy) and inferences on the local mantle source. Earth Planet Sci Lett 371:134–142
Rodríguez A, Eggenkamp HGM, Martínez-Cruz M, van Bergen MJ (2016) Chlorine isotope and Cl–Br fractionation in fluids of Poás volcano (Costa Rica): insight into an active volcanic–hydrothermal system. J Volcanol Geotherm Res 325:70–85
Rodríguez A, van Bergen MJ, Eggenkamp HGM (2018) Experimental evaporation of hyperacid brines: effects on chemical composition and chlorine isotope fractionation. Geochim Cosmochim Acta 222:467–484
Rodríguez-Fernández D, Torrentó C, Palau J, Marchesi M, Soler A, Hunkeler D, Domènech C, Rosell M (2018) Unravelling long-term source removal effects and chlorinated methanes natural attenuation processes by C and Cl stable isotopic patterns at a complex field site. Sci Total Environ 645:286–296
Rosenbaum JM, Cliff RA, Coleman ML (2000) Chlorine stable isotopes: a comparison of dual inlet and thermal ionization mass spectrometric measurements. Anal Chem 72:2261–2264
Saeed W, Shouakar-Stash O, Wood WW, Frape S (2021) Isotopically enriched Geogenic δ^{81}Br and δ^{37}Cl: primary evidence for the ascending brine model. Groundwater 59(5):671–676. https://doi.org/10.1111/gwat.13096
Sarafian AR, John T, Roszjar J, Whitehouse MJ (2017) Chlorine and hydrogen degassing in Vesta's magma ocean. Earth Planet Sci Lett 459:311–319
Schauble EA, Rossman GR, Taylor HP (2003) Theoretical estimates of equilibrium chlorine-isotope fractionations. Geochim Cosmochim Acta 67:3267–3281
Segee-Wright G, Barnes JD, Lassiter JC (2024) Chlorine isotope evidence for Farallon-derived metasomatism of the North American lithospheric mantle. Geochim Cosmochim Acta 365:70–84
Selverstone J, Sharp ZD (2011) Chlorine isotope evidence for multicomponent mantle metasomatism in the Ivrea Zone. Earth Planet Sci Lett 310:429–440
Selverstone J, Sharp ZD (2015) Chlorine isotope behavior during prograde metamorphism of sedimentary rocks. Earth Planet Sci Lett 417:120–131
Sharp ZD, Barnes JD (2004) Water-soluble chlorides in massive seafloor serpentinites: a source of chloride in subduction zones. Earth Planet Sci Lett 226:243–254
Sharp ZD, Barnes JD, Brearly AJ, Chaussidon M, Fisher TP, Kamenetsky VS (2007) Chlorine isotope homogeneity of the mantle, crust and carbonaceous chondrites. Nature 446:1062–1065
Sharp ZD, Barnes JD, Fischer TP, Halick M (2010a) An experimental determination of chlorine isotope fractionation in acid systems and applications to volcanic fumaroles. Geochim Cosmochim Acta 74:264–273
Sharp ZD, Shearer CK, McKeagan KD, Barnes JD, Wang YQ (2010b) The chlorine isotope composition of the Moon and implications for an anhydrous mantle. Science 329:1050–1053
Sharp ZD, Shearer CK, FM MC, Agee CB (2011) The Cl isotope composition of the Moon and Mars. In: 21th Annual V.M. Goldschmidt Conference, vol 75. Mineral Mag, Prague, Czech Republic, p 1848
Sharp ZD, Mercer JA, Jones RH, Brearley AJ, Selverstone J, Bekker A, Stachel T (2013) The chlorine isotope composition of chondrites and earth. Geochim Cosmochim Acta 107:189–204
Sharp Z, Williams J, Shearer C, Agee C, McKeegan K (2016) The chlorine isotope composition of Martian meteorites 2. Implications for the early solar system and the formation of Mars. Meteorit Planet Sci 51:2111–2126
Shearer CK, Messenger S, Sharp ZD, Burger PV, Nguyen AN, McCubbin FM (2018) Distinct chlorine isotopic reservoirs on Mars. Implications for character, extent and relative timing of crustal interactions with mantle-derived magmas, evolution of the martian atmosphere, and the building blocks of an early Mars. Geochim Cosmochim Acta 234:24–36

Shirodkar PV, Xiao YK, Hai L (2003) Boron and chlorine isotopic signatures of seawater in the Central Indian ridge. Current Sci 85:313–320

Shirodkar PV, Xiao YK, Sarkar A, Dalal SG, Chivas AR (2006) Influence of air-sea fluxes on chlorine isotopic composition of ocean water: Implications for constancy in δ^{37}Cl – A statistical inference. Environ Intern 32:235–239

Shouakar-Stash O (2008) Evaluation of stable chlorine and bromine isotopes in sedimentary formation fluids. PhD Thesis. University of Waterloo

Shouakar-Stash O, Drimmie RJ, Frape SK (2003) Stable hydrogen, carbon and chlorine isotope measurements of selected chlorinated organic solvents. J Contam Hydrol 60:211–228

Shouakar-Stash O, Drimmie RJ, Frape SK (2005a) Determination of inorganic chlorine stable isotopes by continuous flow isotope ratio mass spectrometry. Rapid Commun Mass Spectrom 19:121–127

Shouakar-Stash O, Frape SK, Drimmie RJ (2005b) Determination of bromine stable isotopes using continuous-flow isotope ratio mass spectrometry. Anal Chem 77:4027–4033

Shouakar-Stash O, Frape SK, Aravena R, Gargini B, Pasini M, Drimmie RJ (2009) Analysis of compound-specific chlorine stable isotopes of vinyl chloride by continuous flow-isotope ratio mass spectrometry (CF-IRMS). Environ Forensic 10:299–306

Stefánsson A, Barnes JD (2016) Chlorine isotope geochemistry of Icelandic thermal fluids: implications for geothermal system behavior at divergent plate boundaries. Earth Planet Sci Lett 449:69–78

Stephant A, Anand M, Zhao XC, Chan QHS, Bonifacie M, Franchi IA (2019) The chlorine isotopic composition of the Moon: insights from melt inclusions. Earth Planet Sci Lett 523:115715

Stephant A, Anand M, Carli C, Zhao X, Davidson J, Cuppone T, Pratesi G, Franchi IS (2024) Chondritic chlorine isotope composition of acapulcoites and lodranites. Geochem Perspect Lett 29:14–19

Stotler RL, Frape SK, Shouakar-Stash O (2010) An isotopic survey of δ^{81}Br and δ^{37}Cl of dissolved halides in the Canadian and Fennoscandian shields. Chem Geol 274:38–55

Stotler RL, Kirk MF, Newell KD, Goldstein RH, Frape SK, Gwynne R (2021) Stable bromine isotopic composition of coal bed methane (CBM) produced water, the occurrence of enriched ^{81}Br, and implications for fluid flow in the midcontinent, USA. Fortschr Mineral 11:358

Strydom J, Sterpenich J, Grgic D, Richard A, Eggenkamp HGM, Agrinier P, Louvat P, Mosser-ruck R, Gaire P, Gaucher E (2022) Experimental study of chemical evolution and isotope fractionation of Cl and Br in pore water expelled during strong clay compaction. Appl Geochem 140:105274

Sturchio NC, Böhlke JK, Beloso Jr AD, Streger SH, Heraty LJ, Hatzinger PB (2007) Oxygen and chlorine isotopic fractionation during perchlorate biodegradation: laboratory results and implications for forensics and natural attenuation studies. Environ Sci Technol 41:2796–2802

Sturchio NC, Hoaglund III JR, Marroquin RJ, Beloso Jr AD, Heraty LJ, Bortz SE, Patterson TL (2012) Isotopic mapping of groundwater perchlorate plumes. Ground Water 50:94–102

Sturchio NC, Beloso Jr A, Heraty LJ, Wheatcraft S, Schumer R (2014) Isotopic tracing of perchlorate sources in groundwater from Pomona, California. Appl Geochem 43:80–87

Tan HB, Ma HZ, Xiao YK, Wei HZ, Zhang XY, Ma WD (2005) Characteristics of chlorine isotope distribution and analysis on sylvinite deposit formation based on ancient salt rock in western Tarim basin. Sci China (ser. D) 48:1913–1920

Tan HB, Ma HZ, Wei HZ, Xu JX, Li TW (2006) Chlorine, sulfur and oxygen isotopic constraints on ancient evaporite deposit in the Western Tarim Basin, China. Geochem J 40:569–577

Tan HB, Ma HZ, Zhang XY, Xu JX, Xiao YK (2009) Fractionation of chlorine isotope in salt mineral sequences and application: research on sedimentary stage of ancient salt rock deposit in Tarim Basin and western Qaidam Basin. Acta Petrol Sinica 25:955–962. (in Chinese with English abstract)

Tang CM, Peng XZ (2017) Theoretical evaluation of intermolecular and intramolecular isotope fractionations in fragmentation. https://doi.org/10.48550/arXiv.1704.05607

Tartèse R, Anand M, Franchi IA (2019) H and Cl isotope characteristics of indigenous and late hydrothermal fluids on the differentiated asteroidal parent body of Grave Nunataks 06128. Geochem Cosmochem Acta 266:529–543

Taylor JW, Grimsrud EP (1969) Chlorine isotopic ratios by negative ion mass spectrometry. Anal Chem 41:805–810

Torrentó C, Palau J, Rodríguez-Fernández D, Heckel B, Meyer A, Doménech C, Rosell M, Soler A, Elsner M, Hunkeler D (2017) Carbon and chlorine isotope fractionation patterns associated with different engineered chloroform transformation reactions. Environ Sci Technol 51:6174–6184

Torrentó C, Ponsin V, Lihl C, Hofstetter TB, Baran N, Elsner M, Hunkeler D (2021) Triple-element compound-specific stable isotope analysis (3D-CSIA): added value of Cl isotope ratios to assess herbicide degradation. Environ Sci Technol 55:13891–13901

Turnquist CR, Taylor JW, Grimsrud EP, Williams RC (1973) Temperature dependence of chlorine kinetic isotope effects for aliphatic chlorides. J Am Chem Soc 95:4133–4138

Urey HC (1947) The thermodynamic properties of isotopic substances. J Chem Soc 1947:562–581

Van Acker MRMD, Shahar A, Young ED, Coleman ML (2006) GC/multiple collector-ICPMS method for chlorine stable isotope analysis of chlorinated aliphatic hydrocarbons. Anal Chem 78:4663–4667

Van Warmerdam EM, Frape SK, Aravena R, Drimmie RJ, Flatt H, Cherry JA (1995) Stable chlorine and carbon isotope measurements of selected chlorinated organic solvents. Appl Geochem 10:547–552

Volker A, Van der Molen WH (1991) The influence of groundwater currents on diffusion processes in a lake bottom: an old report reviewed. J Hydrol 126:159–169

Volpe C, Spivack AJ (1994) Stable chlorine isotopic composition of marine aerosol-particles in the Western Atlantic Ocean. Geophys Res Lett 21:1161–1164

Volpe C, Wahlen M, Pszenny AAP, Spivack AJ (1998) Chlorine isotopic composition of marine aerosols: implications for the release of reactive chlorine and HCl cycling rates. Geophys Res Lett 25:3831–3834

Wanner P, Hunkeler D (2015) Carbon and chlorine isotopologue fractionation of chlorinated hydrocarbons during diffusion in water and low permeability sediments. Geochim Cosmochim Acta 157:198–212

Wanner P, Hunkeler D (2019) Molecular dynamic simulations of carbon and chlorine Isotopologue fractionation of Chlorohydrocarbons during diffusion in liquid water. Environ Sci Technol Lett 6:681–685

Wassenaar LI, Koehler G (2004) On-line technique for the determination of the $\delta^{37}Cl$ of inorganic and total organic Cl in environmental samples. Anal Chem 76:6384–6388

Wei W, Kastner M, Spivack A (2008a) Chlorine stable isotopes and halogen concentrations in convergent margins with implications for the Cl isotopes cycle in the ocean. Earth Planet Sci Lett 266:90–104

Wei W, Kastner M, Spivack A (2008b) Reply to comment on: "chlorine stable isotopes and halogen concentrations in convergent margins with implications for the Cl isotopes cycle in the ocean". Earth Planet Sci Lett 274:535

Wei HZ, Jiang SY, Hemming NG, Yang JH, Xiao YK, Yang T, Yan X, Yan Y (2011) Emission mechanism of polyatomic ions Cs_2Cl^+ and $Cs_2BO_2^+$ in thermal ionization mass spectrometry with various carbon materials. J Phys Chem A 115:14570–14580

Wei HZ, Jiang SY, Xiao YK, Wang J, Lu H, Wu B, Wu HP, Li Q, Luo CG (2012) Precise determination of the absolute isotopic abundance ratio and the atomic weight of chlorine in three international reference materials by the positive thermal ionization mass spectrometer-Cs_2Cl^+-graphite method. Anal Chem 84:10350–10358

Wei HZ, Jiang SY, Hemming NG, Yang JH, Yang T, Wu HP, Yang TL, Yan X, Pu W (2014a) An improved procedure for separation/purification of boron from complex matrices and high-precision measurement of boron isotopes by positive thermal ionization and multicollector inductively coupled plasma mass spectrometry. Talanta 123:51–160

Wei HZ, Jiang SY, Xiao YK, Hemming NG (2014b) Boron isotopic fractionation and trace element incorporation in various species of modern corals in Sanya Bay, South China Sea. J Earth Sci 25:431–444

Wei HZ, Jiang SY, Tan HB, Zhang WJ, Li BK, Yang TL (2014c) Boron isotope geochemistry of salt sediments from the Dongtai salt lake in qaidam basin: boron budget and sources. Chem Geol 380:74–83

Wei HZ, Lei F, Jiang SY, Lu HY, Xiao YK, Zhang HZ, Sun XF (2015a) Implication of boron isotope geochemistry for the pedogenic environments in loess and paleosol sequences of Central China. Quat Res (United States) 83:243–255

Wei HZ, Jiang SY, Zhu ZY, Yang T, Yang JH, Yan X, Wu HP, Yang TL (2015b) Improvements on high-precision measurement of bromine isotope ratios by multicollector inductively coupled plasma mass spectrometry. Talanta 143:302–306

Willey JF, Taylor JW (1978) Capacitive integration to produce high precision isotope ratio measurements on methyl chloride and bromide. Anal Chem 50:1930–1933

Willey JF, Taylor JW (1980) Temperature-dependence of bromine kinetic isotope effects for reactions of normal-butyl and *tert*-butyl bromides. J Am Chem Soc 102:2387–2391

Williams JT, Shearer CK, Sharp ZD, Burger PV, McCubbin FM, Santos AR, Agee CB, McKeegan KD (2016) The chlorine isotopic composition of Martian meteorites 1: chlorine isotope composition of Martian mantle and crustal reservoirs and their interactions. Meteoroit Planet Sci 51:2092–2110

Wirt L (1988) The origin of chloride in groundwater in the Stripa granite, Sweden. MSc thesis. Univ Arizona, Tucson, Ariz

Woulé Ebongué V, Jendrzejewski N, Walgenwitz F, Pineau F, Javoy M (2005) Chlorine isotope residual salt analysis: a new tool to investigate formation waters from core analyses. AAPG Bull 89:1005–1018

Wu LP, Moses S, Liu YQ, Renpenning J, Richnow HH (2019) A concept for studying the transformation reaction of hexachlorocyclohexanes in food webs using multi-element compound-specific isotope analysis. Anal Chim Acta 1064:56–64

Wudarska A, Słaby E, Wiedenbeck M, Birski Ł, Wirth R, Götze J, Lepland A, Kusebauch C, Kocjan I (2020) Chlorine isotope composition of apatite from the >3.7 Ga Isua Supracrustal Belt, SW Greenland. Minerals 10:27

Wudarska A, Słaby E, Wiedenbeck M, Barnes JD, Bonifacie M, Sturchio NC, Bardoux G, Couffignal F, Glodny J, Heraty L, John T, Kusebauch C, Mayanna S, Wilke FDH, Deput E (2021) Inter-laboratory characterisation of apatite reference materials for chlorine isotope analysis. Geostand Geoanal Res 45:121–142

Xiao YK, Zhang CG (1992) High precision isotopic measurement of chlorine by thermal ionization mass spectrometry of the Cs_2Cl^+ ion. Int J Mass Spectrom Ion Processes 116:183–192

Xiao YK, Beary ES, Fassett JD (1988) An improved method for the high-precision isotopic measurement of boron by thermal ionization mass-spectrometry. Int J Mass Spectrom Ion Processes 85:203–213

Xiao YK, Jin L, Qi HP (1991) Investigation of thermal ion emission characteristics of graphite. Int J Mass Spectrom Ion Processes 107:205–213

Xiao YK, Liu WG, Qi HP, Zhang CG (1993) A new method for the high precision isotopic measurement of bromine by thermal ionization mass spectrometry. Int J Mass Spectrom Ion Processes 123:117–123

Xiao YK, Liu WG, Zhou YM, Sun DP (1997) Isotope compositions of chlorine in brine and saline minerals. Chin Sci Bull 42:406–409

Xiao YK, Liu WG, Zhou YM, Wang YH, Shirodkar PV (2000) Variations in isotopic compositions of chlorine in evaporation controlled salt lake brines of Qaidam Basin, China. Chin J Oceanol Limn 18:169–177

Xu JH (2020) Modelling of liquid water and ionic solutions by first-principles simulations. PhD Thesis. Temple University, Philadelphia, PA, USA

Yan X, Jiang SY, Wei HZ, Yan Y, Wu HP, Pu W (2012) Extraction and determination of boron isotopic composition in tourmalines. Fenxi Huaxue / Chin J Anal Chem 40:1654–1660

Yankelson I (2019) Microbial and abiotic degradation assessment of 1-bromo-2-chloroethane using multi-elemental C-Br-Cl isotope analysis: laboratory and field study. PhD Thesis. Hebrew University, Jerusalem, Israel

Yu HT, Ma T, Du Y, Chen LZ (2019) Genesis of formation water in the northern sedimentary basin of South China Sea: clues from hydrochemistry and stable isotopes (D, ^{18}O, ^{37}Cl and ^{81}Br). J Geochem Explor 196:57–65

Yu HD, Zou H, Wu JH, Bagas L, Gong DX, Cheng HF, Huang CC, Zhu H (2023) Prospects for Eocene potash deposition in the east Tethyan domain: new evidence from fluid inclusions and chlorine isotope of halite crystals in the Kuqa Basin, NW China. Ore Geol Rev 154:105313

Zakon Y, Halicz L, Gelman F (2013) Bromine and carbon isotope effects during photolysis of brominated phenols. Environ Sci Technol 47:14147–14153

Zakon Y, Halicz L, Gelman F (2014) Isotope analysis of sulfur, bromine, and chlorine in individual anionic species by ion chromatography/multicollector-ICPMS. Anal Chem 86:6495–6500

Zakon Y, Halicz L, Lev O, Gelman F (2016) Compound-specific bromine isotope ratio analysis using gas chromatography/quadrupole mass spectrometry. Rapid Commun Mass Spectrom 30:1951–1956

Zhang MM, Wei HZ, Lu JJ, Williams-Jones AE, Eastoe CJ, Li WQ, Xia ZG, Hohi SV, Ma J, Jiang SY, Wang JL (2024) Estimates of chlorine isotope fractionation factors using density functional theory: applications to ore-forming systems. Geochim Cosmochim Acta 367:1–15

Zhao Y, Wei HZ, Liu X, Wang YJ, Jiang SY, Eastoe CJ, Peryt TM (2021) Isotope evidence for multiple sources of B and Cl in middle Miocene (Badenian) evaporites, Carpathian Mountains. Appl Geochem 124:104819

Zimmerman J, Halloran LJS, Hunkeler D (2020) Tracking chlorinated contaminants in the subsurface using compound-specific chlorine isotope analysis: a review of principles, current challenges and applications. Chemosphere 244:125476

The manufacturer's authorised representative in the EU is Springer Nature Customer Service Centre GmbH, Europaplatz 3, 69115 Heidelberg, Germany. If you have any concerns regarding our products, please contact ProductSafety@springernature.com

Printed and bound by CPI Group (UK) Ltd, Croydon, CR0 4YY
01/10/2026
02225914-0020